Sustaining the Earth
An Integrated Approach
FOURTH EDITION

G. TYLER MILLER, JR.

President, Earth Education and Research
Adjunct Professor of Human Ecology
St. Andrews Presbyterian College

Brooks/Cole Publishing Company
I(T)P® *An International Thomson Publishing Company*

Pacific Grove • Albany • Belmont • Boston • Cincinnati • Johannesburg • London • Madrid •
Melbourne • Mexico City • New York • Scottsdale • Singapore • Tokyo • Toronto

Publisher: *Jack Carey*
Assistant Editor: *Kristin Milotich*
Editorial Assistant: *Susan Lussier*
Production Coordinator: *Tessa McGlasson Avila*
Production Management: *Electronic Publishing Services Inc., NYC*
Marketing: *Tami Cueny*
Cover Design: *Vernon T. Boes*

Cover Photo: *Schafer & Hill/Peter Arnold, Inc.*
Interior Illustration: *Electronic Publishing Services Inc., NYC; Precision Graphics; Sarah Woodward; Darwin and Vally Hennings;Tasa Graphic Arts, Inc.; Alexander Teshin Associates; John and Judith Waller; Raychel Ciemma; and Victor Royer*
Typesetting: *Electronic Publishing Services Inc., NYC*
Printing and Binding: *R.R. Donnelley—Crawfordsville*

Two trees have been planted in a tropical rain forest for every tree used to make this book, courtesy of G. Tyler Miller, Jr., and Brooks/Cole Publishing Company. The author also sees that 50 trees are planted to compensate for the paper he uses and that several hectares of tropical rain forest are protected.

For more information, contact:

BROOKS/COLE PUBLISHING COMPANY
511 Forest Lodge Road
Pacific Grove, CA 93950
USA

International Thomson Publishing Europe
Berkshire House 168-173
High Holborn
London WC1V 7AA
England

Thomas Nelson Australia
102 Dodds Street
South Melbourne, 3205
Victoria, Australia

Nelson Canada
1120 Birchmount Road
Scarborough, Ontario
Canada M1K 5G4

International Thomson Editores
Seneca 53
Col. Polanco
11560 México, D. F., México

International Thomson Publishing GmbH
Königswinterer Strasse 418
53227 Bonn
Germany

International Thomson Publishing Asia
60 Albert Street
#15-01 Albert Complex
Singapore 189969

International Thomson Publishing Japan
Hirakawacho Kyowa Building, 3F
2-2-1 Hirakawacho
Chiyoda-ku, Tokyo 102
Japan

Printed in the United States of America.

10 9 8 7 6 5 4 3 2 1

Library of Congress Cataloging-in-Publication Data

Miller, G. Tyler (George Tyler).
    Sustaining the Earth : an integrated approach / G. Tyler Miller,
Jr. — 4th ed.
        p. cm.
    Includes bibliographical references and index.
    ISBN 0-534-56286-8 (pbk. : alk. paper)
    1. Environmental sciences.   2. Environmental protection.
I. Title
GE105.M55   1999
363.7—dc21   99-21140
                                                   CIP

This book is printed on acid-free recycled paper with the highest available content of post-consumer waste.

FOR INSTRUCTORS AND STUDENTS

How I Became Involved In 1966 I heard a scientist give a lecture on the problems of overpopulation and environmental abuse. Afterward I went to him and said, "If even a fraction of what you have said is true, I will feel ethically obligated to give up my research on the corrosion of metals and devote the rest of my life to research and education on environmental problems and solutions. Frankly, I don't want to believe a word you have said, and I'm going into the literature to try to prove that your statements are either untrue or grossly distorted."

After 6 months of study I was convinced of the seriousness of these problems. Since then, I have been studying, teaching, and writing about them. This book summarizes what I have learned in almost three decades of trying to understand environmental principles, problems, connections, and solutions.

Concepts, Problems, Connections, and Solutions This book treats environmental science as an *interdisciplinary* study, combining ideas and information from natural sciences (such as biology, chemistry, and geology) and social sciences (such as economics, politics, and ethics) to present a general idea of how nature works and how things are interconnected. It is a study of *connections in nature.*

In this book, I use scientific laws, principles, models, and concepts to help us understand environmental and resource problems and possible solutions—and how these concepts, problems, and solutions are connected. I have introduced only the concepts and principles necessary for understanding the material in the book, and I have tried to present them simply but accurately.

My aim is to provide a readable and accurate introduction to environmental science without the use of mathematics or complex scientific information. To help ensure that the material is accurate and up-to-date, I have consulted more than 10,000 research sources in the professional literature. In writing this book, I have also benefited from the more than 200 experts and teachers who have provided detailed reviews of the editions of this and my other three books in this field.

After Chapters 1 and 2 have been covered, the rest of the book can be used in almost any order. In addition, many sections within chapters can be moved around or omitted to accommodate courses with different lengths and emphases. The book's 223 illustrations are designed to present complex ideas in understandable ways and to relate learning to the real world. To save space and improve clarity, I have converted important ideas to detailed drawings instead of using photographs.

In addition, 40 case studies (some in the basic text and others in special boxes) provide an in-depth look at specific environmental problems and their possible solutions. Other special boxes include *Pro/Con boxes* that present both sides of controversial environmental issues; *Connections boxes* that show connections in nature and among environmental concepts, problems, and solutions; *Solutions boxes* that summarize a variety of solutions (some of them controversial) to environmental problems proposed by various analysts; *Spotlight boxes* that highlight and give insights into key environmental problems and concepts; and *Individuals Matter boxes* that describe what individuals have done to help solve environmental problems. To encourage critical thinking and integrate it throughout the book, all boxes (except Individuals Matter) end with critical thinking questions.

To reduce student costs this book is printed in black-and-white and has a soft cover. It is also printed on acid-free recycled paper with the maximum content of postconsumer waste currently available at an affordable cost.

Instructors wanting books covering this material with a different emphasis, organization, and length can use one of my three other books written for various types of environmental science courses: *Living in the Environment,* 11th edition (815 pages, Wadsworth, 2000); *Environmental Science,* 7th edition (566 pages, Wadsworth, 1999, a shorter version of *Living in the Environment*); and *Environment: Problems and Solutions* (150 pages, Wadsworth, 1994).

An Integrated Approach This is the first environmental science textbook designed to fully integrate environmental concepts, problems, connections, and solutions. This book is integrated in several ways:

- An overview of the problems of population, pollution, and resource use and how they are connected is provided in Chapter 1.

- Key scientific concepts are discussed and integrated in Chapter 2, and then used as needed throughout the book.

- Major topics are discussed and integrated in each of the remaining chapters, except for the keystone concept of biodiversity, which is discussed in three chapters: Chapter 5 (biodiversity in ecosystems), Chapter 6 (biodiversity of wild species), and Chapter 7 (biodiversity in soils and food-producing systems). Problems and solutions to human population growth are integrated and discussed in Chapter 3, energy in Chapter 4, human health and risks in Chapter 8, air (including atmosphere, weather, climate, global warming, ozone depletion, and indoor and outdoor pollution) in Chapter 9, water (including use of water resources and water pollution) in Chapter 10, wastes and resource conservation (including solid waste, hazardous waste, recycling, reuse, waste reduction, and pollution prevention) in Chapter 11, and the social science aspects of environmental problems and solutions (economics, politics, and worldviews) in Chapter 12.

- In addition to vertical integration within chapters, horizontal integration is achieved by using several major themes as connecting threads woven throughout the book: *biodiversity and earth capital, sustainability pollution prevention and waste reduction, population, exponential growth, energy and energy efficiency, economics and environment, politics and environmental laws, and individual action.*

- Throughout the book material is connected by cross-references.

Major Changes in the Fourth Edition   Major changes include the following:

- Updated and revised material throughout the book.

- Addition of critical thinking questions to material in all boxes (except Individuals Matter boxes)

- The addition or expansion of many topics, including, survival ability of cockroaches, a solar village in Colombia (Gaviotas), economic values of forests and nature's ecological services, population problems from snow geese, environmental and health effects of *Pfiesteria* microbes, fuel cells in cars, attempts to reduce protection of public lands in the United States,

sustainability of industrial forestry, changes in management of U.S. national forests, future water shortages in China, more sustainable management of fisheries, the Kyoto, Japan international conference on global warming, and running out of water in Las Vegas, Nevada.

- A greatly expanded and improved interactive World Wide Web site that can be used as a source of further information and ideas (see description on p. v).

- *Biolink*, an instructor presentation tool (see description on p. v).

- *InfoTrac College Edition*, a fully searchable online university library (see description on p. v.

- *Online Regional Articles for Environmental Science* (see description on p. v).

- *Thomson World Class Software* (see description on p. v).

- *CNN Today Videos for Environmental Science* (see description on p. v).

- Booklet on *Study Skills for Science Students* (see description on p. v).

Welcome to Uncertainty, Controversy, and Challenge   There are no easy solutions to the environmental problems and challenges we face. We will never have complete agreement about what we should do because science advances through continuous controversy and careful scrutiny of its results until there is general consensus about their validity. What is important is not what the experts disagree on (the frontiers of knowledge that are still being developed, tested, and argued about), but what they generally agree on—the *scientific consensus*—on concepts, problems, and possible solutions.

Despite considerable research, we still know little about how nature works at a time when we are altering nature at an accelerating pace. This uncertainty, as well as the complexity and importance of these issues to current and future generations of humans and other species, makes many of these issues highly controversial. Intense controversy also arises because environmental science is a dynamic blend of natural and social sciences that sometimes questions the ways we view and act in the world around us. This can often be a threatening process.

Rosy optimism and gloom-and-doom pessimism are traps; both usually lead to denial, indifference, and inaction. I have tried to avoid these two extremes and give a realistic yet hopeful view of the future. This book is filled with stories of people who have acted to help sustain the earth's life-support systems for us and for all life, and whose actions inspire us to do better. It's an exciting time to be alive as we enter into a new, more cooperative relationship with the planet that is our only home.

Interact with and Help Me Improve This Book

I urge you to interact with this book as a way to make learning more interesting and effective. When I read books and articles, I mark key sentences and paragraphs with a highlighter or pen. I put an asterisk in the margin next to something I think is important and double asterisks next to something I think is especially important. I write comments in the margins, such as *Beautiful*, *Confusing*, *Bull*, or *Wrong*. I fold down the top corner of pages with highlighted passages and the top and bottom corners of especially important pages. This way, I can flip through a book and quickly review the key passages. I urge you to interact in such ways with this book.

Let me know how you think this book can be improved; if you find any errors, bias, or confusing explanations please send them to Jack Carey, Biology Publisher, Brooks/Cole Publishing Company, 10 Davis Drive, Belmont, CA 94002. He will forward them to me. Most errors can be corrected in subsequent printings of this edition, rather than waiting for a new edition.

Study Aids   When a new term is introduced and defined, it is printed in boldface type. A glossary of all key terms is located at the end of the book.

Factual recall questions (with answers) are listed at the bottom of most pages. You might cover the answer (on the right-hand page) with a piece of paper and then try to answer the question on the left-hand page. (These questions are not necessarily related to the chapter in which they are found.)

Each chapter ends with a set of questions designed to encourage students to think critically and to apply what they have learned in their lives. Some ask students to take sides on controversial issues and to back up their conclusions and beliefs.

Readers who become especially interested in a particular topic can consult the list of further readings for each chapter, given in the back of the book. For the most complete list of such readings, consult my largest book, *Living in the Environment*. The appendix contains a list of important publications and a list of some key environmental organizations.

Supplements   The following supplements are available:

■ *Biolink*, an instructor presentation tool that allows the quick, easy assembly of media files into a multimedia presentation. Through an easy-to-use interface, users can assemble, edit, publish, and present custom lectures built from an extensive multimedia database. It includes all illustrations from the book and art from other Brooks/Cole biology books and CD-ROMs. Biolink also has a browser with an easy drop-and-drag feature that allows file export into such presentation tools as PowerPoint. Upon its creation, a file or lecture created with Biolink can be posted to the Web, where students can access it for reference or study.

■ The *Online Regional Articles for Environmental Science* is a collection of online articles from *InfoTrac College Edition* organized by region: west, southwest, rocky mountains, midwest, east, and south. These regional articles allow students to learn about local environmental issues in the region of the country in which they live. Students can access the *Online Regional Articles* on the Biology Resource Center site or directly at http://www.upcloser.com.

■ *InfoTrac College Edition* is a fully searchable online university library available free with each copy of this book. It gives students access to full-length articles from over 700 scholarly and popular journals, updated daily and dating back as much as 4 years. An online *Student Guide to InfoTrac College Edition* is located on the Brooks/Cole Biology Resource Center Web site. It has critical thinking questions and a set of electronic readings for each chapter to invite deeper examination of the issues.

■ The *Brooks/Cole Biology Resource Center* is arranged by chapter. Every month it has new BioUpdates on relevant applications and hyperlinks. It also has an average of 40 practice quiz questions per chapter, descriptions of degrees and careers in biology and environmental science, a student feedback site, clip art, ideas for teaching on the Web, and a forum in which instructors can share ideas on teaching courses. It also includes flashcards for all glossary terms, critical thinking exercises, newsgroups, a variety of search engines, and Internet exercises for each chapter. An event of the quarter will include an ongoing experiment in which students and instructors can participate. The address for the Brooks/Cole Biology Resource Center is http://www.brookscole.com/biology.

■ *Thomson World Class Course* software enables instructors to create their own Web sites. Instructors can post course information, office hours, related Internet links, downloaded materials, lesson information, assignments, and sample tests or quizzes. More information is available at http://www.worldclasslearning.com

■ *CNN Today Videos for Environmental Science* are short clips of current news footage that make great lecture launchers. A new tape is offered every year.

- *Study Skills for Science Students*, by Dan Chiras, is an 86-page booklet that explains how to develop good study habits, sharpen memory and learning, prepare for tests, and produce term papers.

- *Internet Booklet*, by Daniel J. Kurland, and Jane Heinze-Fry. An introduction to the Internet and World Wide Web plus selected sites to visit and learning exercises.

- *Critical Thinking and the Environment: A Beginners' Guide*, by Jane Heinze-Fry and G. Tyler Miller, Jr. An introduction to different critical thinking approaches, with questions by chapter using these approaches.

- *Green Lives, Green Campuses*, written by Jane Heinze-Fry. This hands-on workbook contains projects to help students evaluate the environmental impact of their own lives and guide them in making an environmental audit of their campuses.

- *Laboratory Manual*, by C. Lee Rockett (Bowling Green State University) and Kenneth J. Van Dellen (Macomb Community College).

- A set of 100 color acetates and more than 600 black-and-white transparency masters for making overhead transparencies or slides of line art (including concept maps for each chapter), available to adopters.

- *Watersheds: Classic Cases in Environmental Ethics*, 2d ed., by Lisa H. Newton and Catherine K. Dillingham (Wadsworth, 1997). Nine balanced case studies that amplify material in this book.

- *Environmental Ethics*, by Joseph R. Des Jardins (Wadsworth, 1993). A very useful survey of environmental ethics. Brief case studies and many specific examples are included.

- *Radical Environmentalism*, by Peter C. List (Wadsworth, 1993). A series of readings on environmental politics and philosophy.

- *A Beginner's Guide to Scientific Method*, by Stephen S. Carey (Wadsworth, 1994). A concise, hands-on introduction that helps students develop critical thinking skills essential to understanding the scientific process.

- *The Game of Science*, 5th edition, by Garvin McCain and Erwin M. Segal (Brooks/Cole, 1988). An accurate, lively, and up-to-date view of what science is, who scientists are, and how they approach science.

Acknowledgments   I wish to thank the many students and teachers who responded so favorably to the three previous editions of *Sustaining the Earth: An Integrated Approach*, the eleven editions of *Living in the Environment*, the seven editions of *Environmental Science*, and *Environment: Problems and Solutions*—and who corrected errors and offered many helpful suggestions for improvement. I am also deeply indebted to the reviewers, who pointed out errors and suggested many important improvements in this book. Any errors and deficiencies left are mine.

The members of the talented production team, listed on the copyright page, have made vital contributions as well. I especially appreciate the competence and cheerfulness of production editor Rob Anglin and the helpful inputs by copyeditor Carol Anne Peschke. My thanks also go to Brooks/Cole's dedicated sales staff, and to Jane Heinze-Fry for being such a delight to work with and for her outstanding work on *Green Lives, Green Campuses, Environmental Articles, Critical Thinking and the Environment: A Beginners Guide* and *Internet Booklet*. Thanks also go to C. Lee Rockett and Kenneth J. Van Dellen for developing the *Laboratory Manual* to accompany this book. I also wish to thank the people who have translated this book into five different languages for use throughout much of the world.

My deepest thanks go to Jack Carey, Biology Publisher at Brooks/Cole, for his encouragement, help, 30 years of friendship, and superb reviewing system. It helps immensely to work with the best and most experienced editor in college textbook publishing. I dedicate this book to the earth that sustains us all.

G. Tyler Miller, Jr.

CONTENTS

1 ENVIRONMENTAL PROBLEMS AND THEIR CAUSES

Alone in space, alone in its life-supporting systems, powered by inconceivable energies, mediating them to us through the most delicate adjustments, wayward, unlikely, unpredictable, but nourishing, enlivening, and enriching in the largest degree—is this not a precious home for all of us? Is it not worth our love?

BARBARA WARD AND RENÉ DUBOS

1-1 LIVING SUSTAINABLY

What Is Exponential Growth? Once there were two kings from Babylon who enjoyed playing chess, with the winner claiming a prize from the loser. After one match, the winning king asked the loser to pay him by placing one grain of wheat on the first square of the chessboard, two on the second, four on the third, and so on. The number of grains was to double each time until all 64 squares were filled.

The losing king, thinking he was getting off easy, agreed with delight. It was the biggest mistake he ever made. He bankrupted his kingdom and still could not produce the 2^{63} grains of wheat he had promised. In fact, it's probably more than all the wheat that has ever been harvested!

This is an example of **exponential growth**, in which a quantity increases by a fixed percentage of the whole in a given time. As the losing king learned, exponential growth is deceptive. It starts off slowly, but after only a few doublings it grows to enormous numbers because each doubling is more than the total of all earlier growth.

Any quantity growing by a fixed percentage, even as small as 0.001% or 0.1%, is undergoing exponential growth. If this growth continues, the quantity will experience extraordinary growth as its base of growth doubles again and again. If plotted on a graph, continuing exponential growth eventually yields a graph shaped somewhat like the letter *J* (Figure 1-1).

Here is another example. Fold a piece of paper in half to double its thickness. If you could do this 42 times, the stack would reach from the earth to the moon, 386,400 kilometers (240,000 miles) away. If you could double it 50 times, the folded paper would almost reach the sun, 149 million kilometers (93 million miles) away!

The environmental problems we face—population growth, wasteful use of resources, destruction and degradation of wildlife habitats, extinction of plants and animals, poverty, and pollution—are interconnected and are growing exponentially. For example, world population has more than doubled in only 47 years, from 2.5 billion in 1950 to 5.9 billion in 1998. Unless death rates rise sharply, it may reach 8 billion by 2025, 10–11 billion by 2050, and 14 billion by 2100 (Figure 1-1). Global economic output, much of it environmentally damaging, has increased almost sixfold since 1950.

Each year more forests, grasslands, and wetlands disappear and some deserts grow larger. According to a recent study by Conservation International, *human activities have modified or disturbed 73% of the earth's land area* (if we exclude uninhabitable areas of rock, ice, desert, and steep mountain terrain). Vital topsoil is washed or blown away from farmland, cleared forests, and construction sites, clogging streams, lakes, and reservoirs with sediment. Many grasslands have been overgrazed and fisheries overharvested to the point of collapse. In a growing number of places underground water is pumped from wells faster than it can be replenished. An increasing number of countries are squabbling over access to shared but limited water supplies. Oceans, streams, and the atmosphere are used as trash cans for a variety of wastes, many of them toxic. We drive an estimated two to eight wildlife species to extinction every hour, mostly because of loss of their habitats. Within the next 40–50 years, the earth's climate may become warm enough to disrupt agricultural productivity, alter water distribution, drive countless species to extinction, and cause economic chaos because of the release of heat-trapping gases into the lower atmosphere from burning fossil fuels and cutting down forests. Extracting and burning fossil fuels (oil, coal, and natural gas) also pollutes the air and water and disrupts the land. Other chemicals we add to the air drift into the upper atmosphere and deplete a gas (ozone), which filters out much of the sun's harmful ultraviolet radiation. Toxic wastes from factories and mines poison the air, water, and soil. Agricultural pesticides contaminate some of our drinking water and food.

There is also some exciting good news. Mainly because of improved sanitation and medical advances, average human life expectancy has doubled and global

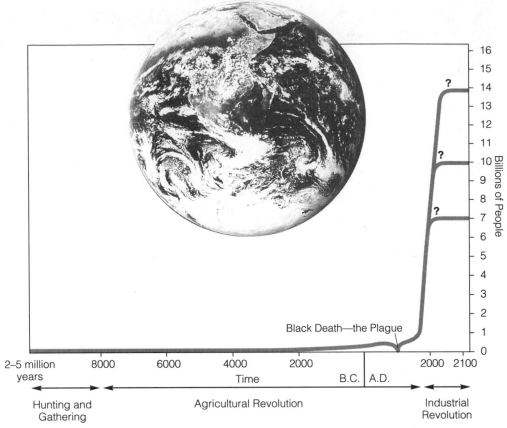

Figure 1-1 The J-shaped curve of past exponential world population growth, with projections beyond 2100. Notice that exponential growth starts off slowly, but as time passes the curve becomes increasingly steep. World population has more than doubled in only 48 years, from 2.5 billion in 1950 to 5.9 billion in 1998. Unless death rates rise sharply, it may reach 8 billion by 2025, 10–11 billion by 2050, and 14 billion by 2100. (This figure is not to scale.) (Data from World Bank and United Nations; photo courtesy of NASA)

Black Death—the Plague

2–5 million years 8000 6000 4000 2000 B.C. A.D. 2000 2100

Time

Hunting and Gathering

Agricultural Revolution

Industrial Revolution

Billions of People

infant mortality has dropped by almost two-thirds during this century. Since the 1960s global food production has outpaced population growth, thanks mostly to new high-yield forms of agriculture.

Because of improved mining technology, there have been significant increases in proven deposits of virtually all of the earth's fossil fuel and mineral resources since 1950. Since 1970 air and water pollution levels in most industrialized countries have dropped because of new pollution control laws and technologies. Recently, industrial nations have developed international treaties to eventually phase out production of chemicals that deplete ozone in the upper atmosphere.

What Are Solar Capital, Earth Capital, and Sustainability? Our existence, lifestyles, and economies depend completely on the sun and the earth, a blue and white island in the black void of space. We can think of energy from the sun as **solar capital**, and we can think of the planet's air, water, soil, wildlife, minerals, and natural purification, recycling, and pest control processes as **earth capital** (Figure 1-2). The term **environment** is often used to describe these life-support systems; in effect, it's another term for describing solar capital and earth capital.

The concept of earth capital means that we and all organisms are interdependent and interconnected parts of nature and are completely dependent on nature. Our survival and health, our economies, and the survival and health of all living things depend on the earth and its natural systems (Figure 1-2). The air you breathe, the water you drink, the food you eat, and all of your possessions are derived from solar energy and the earth's air, water, soil, plants and animals, minerals, energy resources (such as oil and coal), and life-sustaining processes.

A **sustainable society** manages its economy and population size without exceeding all or part of the planet's ability to absorb environmental insults, replenish its resources, and sustain human and other forms of life over a specified period, usually hundreds to thousands of years. During this period, it satisfies the needs of its people without degrading or depleting earth capital and thereby jeopardizing the prospects of current and future generations.

Living sustainably means living off of income and not depleting the capital that supplies the income. Imagine that you inherit $1 million. If you invest this capital at 10% interest, you will have a sustainable annual income of $100,000; that is, you can spend up to $100,000 a year without touching your capital.

Q: How many people are there in the world?

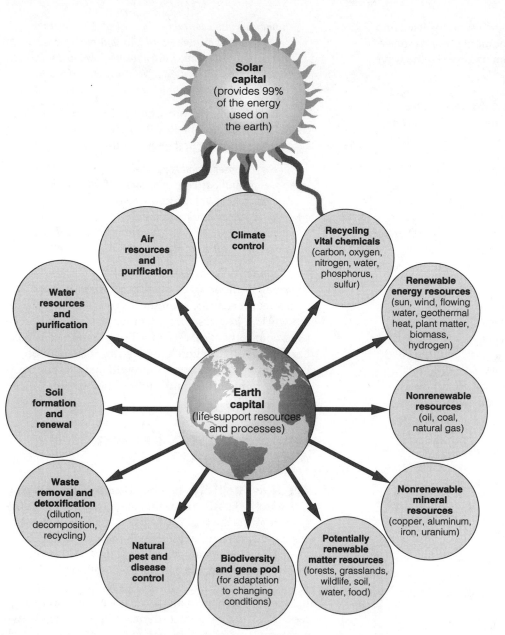

Figure 1-2 Solar and earth capital consist of the life-support resources and processes provided by the sun and the planet for use by us and other species. These two forms of capital support and sustain all life and all economies on the earth.

Suppose you develop a taste for diamonds or a yacht, or all of your relatives move in with you. If you spend $200,000 a year, your $1 million will be gone early in the 7th year; even if you spend just $110,000 a year, you will be bankrupt early in the 18th year. The lesson here is a very old one: *Don't eat the goose that lays the golden egg.* Deplete your capital, and you move from a sustainable to an unsustainable lifestyle.

The same lesson applies to earth capital, the natural ecological services that support all life on the earth (Figure 1-2). With the help of solar energy, natural processes developed over billions of years can indefinitely renew the topsoil, water, air, forests, grasslands, and wildlife on which we and other forms of life de-

pend, as long as we don't use these potentially renewable resources faster than they are replenished.

Some of the earth's natural processes also provide flood prevention, build and renew soil, slow soil erosion, and keep the populations of at least 95% of the species we consider pests under control—all at no cost. Such ecological services are key components of earth capital (Figure 1-2).

The *bad news* is the environmental problems we face and their root causes, as outlined in this chapter. The *good news* is how much has been done and that it's not too late to replace our earth-degrading actions with earth-sustaining ones, as discussed throughout this book. The key is acquiring *earth wisdom*—learning as much as we can about how the earth sustains itself and adapts to ever-changing environmental conditions—and integrating such lessons from nature into the ways we think and act.

Is Our Present Course Sustainable? The Scientific Consensus Environmentalists and many leading scientists believe that we are depleting and degrading the earth's natural capital at an accelerating rate as our population (Figure 1-1) and demands on the earth's resources and natural processes increase exponentially (Figure 1-3).

On November 18, 1992, some 1,680 of the world's senior scientists from 70 countries, including 102 of the 196 living scientists who are Nobel laureates, signed and sent an urgent warning to government leaders of all nations. According to this warning,

The environment is suffering critical stress. . . . Our massive tampering with the world's interdependent web of life—coupled with the environmental damage inflicted by deforestation, species loss, and climate change—could trigger widespread adverse effects, including unpredictable collapses of critical biological systems whose interactions and dynamics we only imperfectly understand. Uncertainty over the extent of these effects cannot excuse complacency or delay in facing the threats. . . . No more than one or a few decades remain before the chance to avert the threats we now confront will be lost and the prospects for

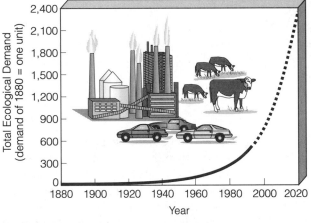

Figure 1-3 *J*-shaped curve of exponential growth in the total ecological demand on the earth's resources from agriculture, mining, and industry between 1880 and 1998. Projections to 2020 assume that resource use will continue to increase at the current rate of 5.5% per year. At that rate, our total ecological demand on the earth's resources doubles every 13 years. If global economic output grew by only 3% a year, resource consumption would still double every 23 years. (Data from United Nations, World Resources Institute, and Carrol Wilson, *Man's Impact on the Global Environment*, Cambridge, Mass.: MIT Press, 1970)

humanity immeasurably diminished. . . . Whether industrialized or not, we all have but one lifeboat. No nation can escape injury when global biological systems are damaged. . . . We must recognize the earth's limited capacity to provide for us.

Also in 1992, the prestigious U.S. National Academy of Sciences and the Royal Society of London issued a joint report—their first ever—which began,

If current predictions of population growth prove accurate and patterns of human activity on the planet remain unchanged, science and technology may not be able to prevent either irreversible degradation of the environment or continued poverty for much of the world.

These two major warnings represent not the views of a small number of scientists but the consensus of the mainstream scientific community, consisting of most of the world's key researchers on environmental problems. Some other analysts, mostly economists, disagree. They contend that there are no limits to human population growth and economic growth that can't be overcome by human ingenuity and technology.

1-2 POPULATION GROWTH AND THE WEALTH GAP

How Rapidly Is the Human Population Growing? Fossil and anthropological evidence suggests that the current form of our species, *Homo sapiens sapiens*, has walked the earth for only about 60,000 years (some recent evidence suggests 90,000 years), an instant in the planet's estimated 4.6-billion-year existence. Until about 12,000 years ago, we were mostly hunter–gatherers who moved as needed to find enough food for survival. Since then, there have been two major cultural shifts: the *agricultural revolution*, which began 10,000–12,000 years ago, and the *industrial revolution*, which began about 275 years ago.

These cultural revolutions have given us much more energy (Figure 1-4) and new technologies with which to alter and control more of the planet to meet our basic needs and increasing desires. By expanding food supplies, lengthening life spans, and raising living standards for many people, each cultural shift contributed to the expansion of the human population. More people fed, bred, and spread (Figure 1-5). However, the results include skyrocketing resource use, pollution, and accelerating environmental degradation.

The increasing size of the human population is one example of exponential growth. As the population base grows, the number of people on the earth soars. If such exponential growth continues (even at a low percentage

Q: How many people are added to the world's population each day?

of annual growth), eventually the population growth curve rounds a bend and heads almost straight up, creating a *J*-shaped curve (Figure 1-1).

It took at least 60,000 years to reach a billion people, 130 years to add the second billion, 30 years for the third, 15 years for the fourth, and only 12 years for the fifth billion. At current growth rates, the sixth billion will be added by the end of 1999, the seventh by 2012, and the eighth by 2025. Between 1900 and 1999 the human population has grown from 1 billion to 6 billion.

The relentless ticking of this population clock means that in 1998 the world's population of 5.9 billion grew by 84 million people (5.9 billion × 0.0143 = 84 million), an average increase of 230,000 people a day, 9,600 an hour. At this 1.43% annual rate of exponential growth, it takes about 5 days to add the number of Americans killed in all U.S. wars, 4 months to add as many people as live in Los Angeles, 2 years to add the 167 million people killed in all wars fought during the past 200 years, 3 years to add 270 million people (the population of the United States in 1998), and only 15 years to add 1.24 billion people (the population of China, the world's most populous country, in 1998).

What Is Economic Growth? Virtually all countries seek **economic growth**: an increase in their capacity to provide goods and services for people's final use. Such growth is normally achieved by increasing the flow or **throughput** of matter and energy resources used to produce goods and services through an economy. This is accomplished by means of population growth (more consumers and producers), or more consumption per person, or both.

Economic growth is usually measured by an increase in a country's **gross national product (GNP)**: the market value in current dollars of all goods and services produced within and outside of a country by the country's businesses for final use during a year. **Gross domestic product (GDP)** is the market value in current dollars of all goods and services produced *within* a country for final use during a year. To show

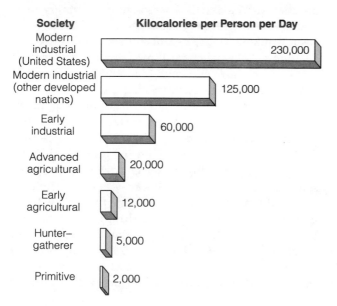

Society	Kilocalories per Person per Day
Modern industrial (United States)	230,000
Modern industrial (other developed nations)	125,000
Early industrial	60,000
Advanced agricultural	20,000
Early agricultural	12,000
Hunter–gatherer	5,000
Primitive	2,000

Figure 1-4 Average direct and indirect per capita daily energy use at various stages of human cultural development. A *calorie* is the amount of energy needed to raise the temperature of 1 gram of water 1°C (1.8°F); a *kilocalorie* is 1,000 calories. Food calories or calories expended during exercise are kilocalories, sometimes called *Calories* (with a capital C).

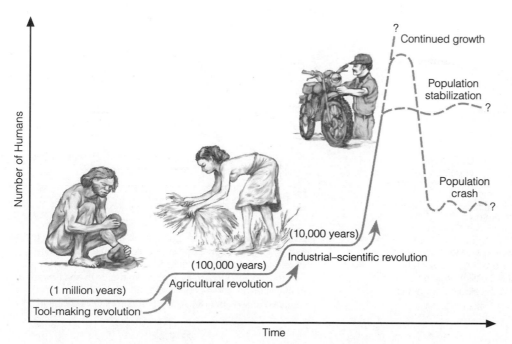

Figure 1-5 Expansion of the earth's carrying capacity for humans. Technological innovation has led to major cultural changes, and we have displaced and depleted numerous species that compete with us for—and provide us with—resources. Dashed lines represent three alternative futures: **(1)** uninhibited human population growth, **(2)** population stabilization, and **(3)** growth followed by a crash and stabilization at a much lower level.

A: An average of about 236,000

one person's slice of the economic pie, economists often calculate the **per capita GNP**: the GNP divided by the total population.

The United Nations broadly classifies the world's countries as economically developed or developing. The **developed countries** are highly industrialized. Most (except the countries of the former Soviet Union) have high average per capita GNPs (above $4,000). These countries, with 1.2 billion people (20% of the world's population in 1998), command about 85% of the world's wealth and income and use about 88% of its natural resources. They generate about 75% of its pollution and wastes (including about 90% of the world's estimated hazardous waste). Three developed countries—the United States, Japan, and Germany—together account for more than half of the world's economic output.

All other nations are classified as **developing countries**, with low to moderate industrialization and per capita GNPs. Most are in Africa, Asia, and Latin America. Their 4.7 billion people (80% of the world's population in 1998) have only about 15% of the wealth and income and use only about 12% of the world's natural resources.

More than 95% of the projected increase in world population is expected to take place in developing countries (Figure 1-6), *where 1 million people are added every 4 days.* By 2010, the combined population of Asia and Africa is projected to be 5.3 billion—almost as many as now live on the entire planet. The primary reason for such rapid population growth in developing countries (1.7% compared to 0.1% in developed countries in 1998) is the *large percentage of people who are under age 15* (35%, compared to 20%

in developed countries in 1998) and who will be moving into their prime reproductive years over the next several decades.

What Is the Wealth Gap? Since 1960, and especially since 1980, the gap between the per capita GNP of the rich, middle income, and poor has widened (Figure 1-7). Today, one person in five lives in luxury, the next three get by, and the fifth struggles to survive on less than $1 a day. One person in six is hungry, malnourished, or severely undernourished and lacks clean drinking water, decent housing, and adequate health care. One of every three people lacks enough fuel to keep warm and cook food and more than half of humanity lacks sanitary toilets.

Daily life for the estimated 1.3 billion desperately poor people in developing countries is a harsh struggle for survival. Parents, some with nine or more children, are struggling to live on the equivalent of $1 a day or less. Having many children makes good sense to most poor parents because their children are a form of economic security, helping them grow food, gather fuel (mostly wood and dung), haul drinking water, tend livestock, or beg in the streets. The desperately poor tend to have many offspring because many of their children die at an early age. The two or three who live to adulthood will help their parents

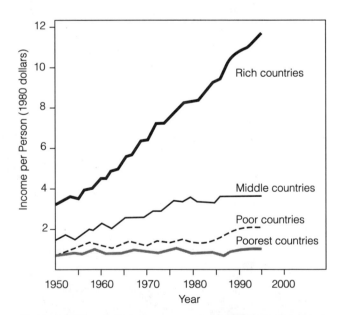

Figure 1-7 The wealth gap: changes in the distribution of global per capita GNP in high-income, middle-income, low-income, and very-low-income countries, 1950–96. Instead of trickling down, most of the income from economic growth has flowed up, with the situation worsening since 1980. More than 1 billion people survive on less than $1 a day. (Data from United Nations)

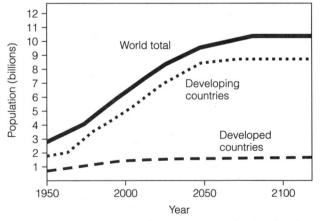

Figure 1-6 Past and projected population size for developed countries, developing countries, and the world, 1950–2120. Over 95% of the projected addition of 3.6 billion people between 1990 and 2030 is projected to occur in developing countries. (Data from United Nations)

Q: What is the projected population of the world in 2050?

survive in old age (their 50s or 60s). Another 1.7 billion poor people struggle to survive on a cash income of about $3 per day.

However, when many poor families have several children, the result is often far more people than local resources can support. To survive now, even though they know this may lead to disaster in the long run, they may deplete and degrade local forests, soil, grasslands, wildlife, and water supplies.

The poor often have little choice but to live in areas with the highest levels of air and water pollution and with the greatest risk of natural disasters such as floods, earthquakes, hurricanes, and volcanic eruptions. They are also the ones who must take jobs (if they can find them) that often subject them to unhealthy and unsafe working conditions at very low pay.

Each year, at least 10 million of the desperately poor, or an average of 27,400 people per day (half of them children under age 5), die of malnutrition (lack of protein and other nutrients needed for good health) or related diseases and contaminated drinking water. *This premature dying of human beings is equivalent to 69 jumbo jet planes, each carrying 400 passengers, crashing every day with no survivors.* A 1997 study by Johns Hopkins University researchers put this annual death toll from poverty at about 18 million a year, or an average of 49,300 premature deaths per day.

1-3 RESOURCES

What Is a Resource? Ecological Versus Economic Resources An **ecological resource** is anything required by an organism for normal maintenance, growth, and reproduction. Examples include habitat, food, water, and shelter. On our short human time scale, we classify material resources as renewable, potentially renewable, or nonrenewable (Figure 1-8).

Some resources, such as solar energy, fresh air, wind, fresh surface water, fertile soil, and wild edible plants, are directly available for use by us and other organisms. Other resources, such as petroleum (oil), iron, groundwater (water found underground), and modern crops, aren't directly available. They become useful to us only with some effort and technological ingenuity. Petroleum, for example, was a mysterious fluid until we learned how to find, extract, and refine it into gasoline, heating oil, and other products that could be sold at affordable prices.

What Are Renewable Resources? Solar energy is called a **renewable** or **perpetual resource** because on a human time scale this solar capital (Figure 1-2) is essentially inexhaustible. It is expected to last at least 6 billion years as the sun completes its life cycle.

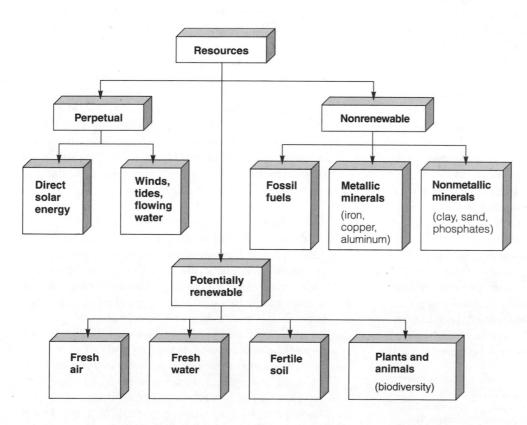

Figure 1-8 Major types of material resources. This scheme isn't fixed; potentially renewable resources can become nonrenewable resources if used for a prolonged period at a faster rate than they are renewed by natural processes.

A: 10–11 billion (about twice the current population)

A **potentially renewable resource*** can be replenished fairly rapidly (hours to several decades) through natural processes. Examples of such resources are forest trees, grassland grasses, wild animals, fresh lake and stream water, groundwater, fresh air, and fertile soil.

One important potentially renewable resource for us and other species is **biological diversity**, or **biodiversity**, which consists of the different life forms (species) that can best survive the variety of conditions currently found on the earth. The earth's vast inventory of life forms and biological communities are a vital part of the earth capital that supports life on the planet (Figure 1-2).

However, potentially renewable resources can be depleted. The highest rate at which a potentially renewable resource can be used *indefinitely* without reducing its available supply is called its **sustainable yield**. If a resource's natural replacement rate is exceeded, the available supply begins to shrink, a process known as **environmental degradation**. Several types of environmental degradation can change potentially renewable resources into nonrenewable or unusable resources (Figure 1-9).

🌐 **Connections: Renewable Resources and the Tragedy of the Commons** One cause of environmental degradation is the overuse of **common-property resources**, which are owned by no one (or jointly by everyone in a country or area) but are available to all users free of charge. Most are potentially renewable. Examples include clean air, the open ocean and its fish, migratory birds, publicly owned lands (such as national forests, national parks, and wildlife refuges), gases of the lower atmosphere, and space.

In 1968, biologist Garrett Hardin called the degradation of common-property resources the **tragedy of the commons**. It happens because each user reasons, "If I don't use this resource, someone else will. The little bit I use or pollute is not enough to matter." With only a few users, this logic works. However, the cumulative effect of many people trying to exploit a common-property resource eventually exhausts or ruins it. Then no one can benefit from it, and therein lies the tragedy.

One solution is to use common-property resources at rates below their sustainable yields or overload limits by reducing population, regulating access, or both. Unfortunately, it is difficult to determine the sustainable yield of forest, grassland, or an animal population, partly because yields vary with weather, climate, and unpredictable biological factors, and because tracking such data is expensive.

---

*Most sources use the term *renewable resource*. I have added the word *potentially* to emphasize that these resources can be depleted if we use them faster than natural processes renew them.

These uncertainties mean that *it is best to use a potentially renewable resource at a rate well below its estimated sustainable yield*. This is a *prevention or precautionary approach* designed to reduce the risk of environmental degradation. This approach is rarely used because it requires hard-to-enforce regulations that restrict resource use and thus conflict with the drive for short-term profit or pleasure.

Another approach is to convert common-property resources to private ownership. The reasoning behind this is that owners of land or some other resource have a strong incentive to see that their investment is protected. However, this approach is not practical for global common resources, such as the atmosphere, the open ocean, and migratory birds that cannot be divided up and converted to private property. Experience has also shown that private ownership can lead to short-term exploitation instead of long-term sustainability.

Some believe that privatization is a better way to protect nonrenewable and potentially renewable resources found on publicly owned lands than relying on command-and-control government regulations and bureaucracies. Most environmentalists disagree. They point out that widespread pollution and environmental degradation have resulted from the removal of nonrenewable mineral resources and unsustainable use of potentially renewable resources on privately owned lands.

Many environmentalists agree that the command-and-control approach to use of resources on public lands (such as national parks, wildlife refuges, and wilderness areas) has some serious problems. They and some free-market economists are seeking *users-pay* solutions to replace the current *taxpayers-pay* approach to use of such publicly owned resources. This would involve marketplace incentives coupled with regulations that require users to pay a fair price for all resources extracted from public lands and to be responsible for preventing or cleaning up any environmental damage caused by resource extraction or use.

🌐 **What Are Nonrenewable Resources?**
Resources that exist in a fixed quantity in the earth's crust and thus theoretically can be completely used up are called **nonrenewable**, or **exhaustible**, **resources**. On a time scale of millions to billions of years, such resources can be renewed by geological processes. However, on the much shorter human time scale of hundreds to thousands of years, these resources can be depleted much faster than they are formed.

These exhaustible resources include *energy resources* (coal, oil, natural gas, and uranium, which cannot be recycled), *metallic mineral resources* (iron, copper, and aluminum, which can be recycled), and *nonmetallic mineral resources* (salt, clay, sand, and phosphates, which

Q: Where does everything that supports your life come from?

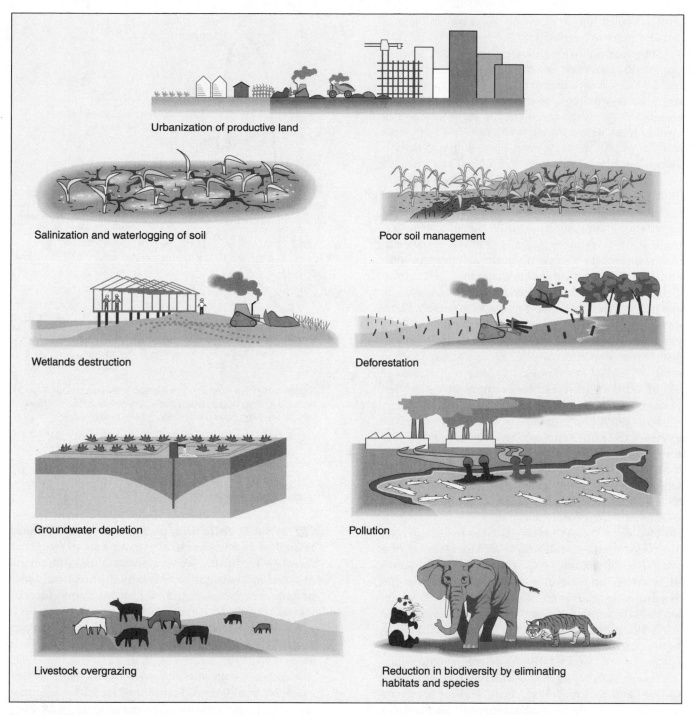

Figure 1-9 Major types of environmental degradation that can convert potentially renewable resources into nonrenewable resources.

are usually difficult or too costly to recycle). A **mineral** is any hard, usually crystalline material that is formed naturally. Soil and most rocks consist of two or more minerals. We know how to find and extract more than 100 nonrenewable minerals from the earth's crust. We convert these raw materials into many everyday items and then we discard, reuse, or recycle them.

In practice, we never completely exhaust a nonrenewable mineral resource. However, such a resource becomes *economically depleted* when the costs of exploiting what is left exceed its economic value. At that point, we have five choices: recycle or reuse existing supplies (except for nonrenewable energy resources, which cannot be recycled or reused), waste less, use less, try to develop a substitute, or do without and wait millions of years for more to be produced.

Some nonrenewable material resources, such as copper and aluminum, can be recycled or reused to extend supplies. **Recycling** involves collecting and reprocessing a resource into new products. For example, glass bottles can be crushed and melted to make new bottles or other glass items. **Reuse** involves using a resource over and over in the same form. For example, glass bottles can be collected, washed, and refilled many times.

Recycling nonrenewable metallic resources requires much less energy, water, and other resources and produces much less pollution and environmental degradation than exploiting virgin metallic resources. Reuse of such resources requires even less energy and other resources than recycling, and it results in less pollution and environmental degradation.

Nonrenewable *energy* resources, such as coal, oil, and natural gas, can't be recycled or reused. Once burned, the useful energy in these fossil fuels is gone, leaving behind waste heat and polluting exhaust gases. Most of the per capita economic growth shown in Figure 1-7 has been fueled by cheap nonrenewable oil, which is expected to be economically depleted within 40–80 years.

Most published estimates of the supply of a given nonrenewable resource refer to **reserves**: known deposits from which a usable mineral can be extracted profitably at current prices. Reserves can be increased when new deposits are found or when price increases make it profitable to extract identified deposits that were previously considered too expensive to exploit.

Depletion time is the time it takes to use up a certain proportion—usually 80%—of the reserves of a mineral at a given rate of use (Figure 1-10). The shortest depletion time assumes no recycling or reuse and no increase in reserves (curve A, Figure 1-10). A longer depletion time assumes that recycling will stretch existing reserves and that better mining technology, higher prices, and new discoveries will increase reserves (curve B, Figure 1-10). An even longer depletion time assumes that new discoveries will further expand reserves and that recycling, reuse, and reduced consumption will extend supplies (curve C, Figure 1-10). Finding a substitute for a resource dictates a new set of depletion curves for the new resource.

Some environmentalists and resource experts believe that the greatest danger may not be the exhaustion of nonrenewable resources but the damage that their extraction, processing, and conversion to products do to the environment in the form of energy use, land disturbance, soil erosion, water pollution, and air pollution (Figure 1-11).

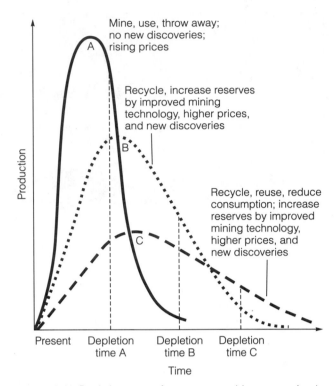

Figure 1-10 Depletion curves for a nonrenewable resource (such as aluminum or copper) using three sets of assumptions. Dashed vertical lines represent times when 80% depletion occurs.

1-4 POLLUTION

What Is Pollution, and Where Does It Come From? Any addition to air, water, soil, or food that threatens the health, survival, or activities of humans or other living organisms is called **pollution**. Most pollutants are solid, liquid, or gaseous by-products or wastes produced when a resource is extracted, processed, made into products, or used. Pollution can also take the form of unwanted energy emissions, such as excessive heat, noise, or radiation.

Pollutants can enter the environment naturally (for example, from volcanic eruptions) or through human (anthropogenic) activities (for example, from burning coal). Most pollution from human activities occurs in or near urban and industrial areas, where pollutants are concentrated. Some pollutants contaminate the areas where they are produced; others are carried by winds or flowing water to other areas. Pollution does not respect local, state, or national boundaries.

Some pollutants come from single, identifiable sources, such as the smokestack of a power plant, the drainpipe of a meat-packing plant, or the exhaust pipe

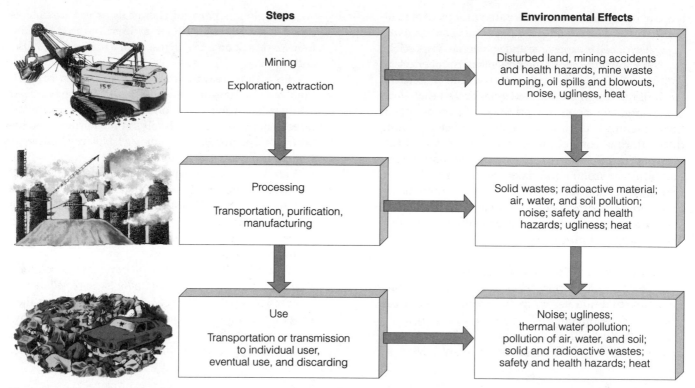

Steps	Environmental Effects
Mining Exploration, extraction	Disturbed land, mining accidents and health hazards, mine waste dumping, oil spills and blowouts, noise, ugliness, heat
Processing Transportation, purification, manufacturing	Solid wastes; radioactive material; air, water, and soil pollution; noise; safety and health hazards; ugliness; heat
Use Transportation or transmission to individual user, eventual use, and discarding	Noise; ugliness; thermal water pollution; pollution of air, water, and soil; solid and radioactive wastes; safety and health hazards; heat

Figure 1-11 Some harmful environmental effects of resource extraction, processing, and use. The energy used to carry out each step causes additional pollution and environmental degradation. Harm could be minimized by requiring mining companies to include the full costs of the pollution and environmental degradation in the prices of their products. Many of these *external* costs are now passed on to society in the form of poorer health, increased health and insurance costs, and increased taxes to deal with pollution and environmental degradation.

of an automobile. These are called **point sources**. Other pollutants come from dispersed (and often difficult to identify) **nonpoint sources**. Examples are the runoff of fertilizers and pesticides (from farmlands, golf courses, and suburban lawns and gardens) into streams and lakes and pesticides sprayed into the air or blown by the wind into the atmosphere. It is much easier and cheaper to identify and control pollution from point sources than from widely dispersed nonpoint sources.

What Types of Harm Are Caused by Pollutants? Unwanted effects of pollutants include disruption of life-support systems for humans and other species, damage to wildlife, damage to human health, damage to property, and nuisances such as noise and unpleasant smells, tastes, and sights.

Three factors determine how severe the harmful effects of a pollutant are. One is its *chemical nature*: how active and harmful it is to living organisms. Another is its **concentration**: the amount per unit of volume or weight of air, water, soil, or body weight. One way to lower the concentration of a pollutant is to dilute it in a large

volume of air or water. Until we started overwhelming the air and waterways with pollutants, dilution was *the* solution to pollution. Now it is only a partial solution.

The third factor is a pollutant's *persistence*: how long it stays in the air, water, soil, or body. **Degradable**, or **nonpersistent, pollutants** are broken down completely or reduced to acceptable levels by natural physical, chemical, and biological processes. Complex chemical pollutants broken down (metabolized) into simpler chemicals by living organisms (usually by specialized bacteria) are called **biodegradable pollutants**. Human sewage in a river, for example, is biodegraded fairly quickly by bacteria if the sewage is not added faster than it can be broken down.

Many of the substances we introduce into the environment take decades or longer to degrade. Examples of these **slowly degradable**, or **persistent, pollutants** include the insecticide DDT and most plastics.

Nondegradable pollutants cannot be broken down by natural processes. Examples include the toxic elements lead and mercury. The best ways to deal with nondegradable pollutants (and slowly degradable pollutants) are to avoid releasing them into the environment or to re-cycle or reuse them. Removing them from contaminated air, water, or soil is expensive and sometimes impossible.

We know little about the possible harmful effects of 90% of the 72,000 synthetic chemicals now in commercial use and the roughly 1,000 new ones added each year. Our knowledge about the effects of the other 10% of these chemicals is limited, mostly because it is quite difficult, time-consuming, and expensive to get this information. Even if we determine the main health and other environmental risks associated with a par-ticular chemical, we know little about its possible in-teractions with other chemicals or about the effects of such interactions on human health, other organisms, and life-support processes.

Solutions: What Can We Do About Pollution?
There are two basic approaches to dealing with pollu-tion: prevent it from reaching the environment or clean it up if it does (Figure 1-12). **Pollution prevention**, or **input pollution control**, is a *throughput solution*. It slows or eliminates the production of pollutants, often by switching to less harmful chemicals or processes. Pol-lution can be prevented (or at least reduced) by the four Rs of resource use: *refuse (don't use), reduce, reuse,* and *recycle*.

Pollution cleanup, or **output pollution control**, involves cleaning up pollutants after they have been produced. However, environmentalists have identified three major problems with relying primarily on pollu-tion cleanup. First, *it is often only a temporary bandage as long as population and consumption levels continue to grow without corresponding improvements in pollution control technology*. For example, adding catalytic converters to cars has reduced air pollution, but increases in the number of cars and in the total distance each travels (increased throughput) have reduced the effectiveness of this cleanup approach.

Second, *pollution cleanup often removes a pollutant from one part of the environment only to cause pollution in another*. We can collect garbage, but the garbage is then either burned (perhaps causing air pollution and leav-ing a toxic ash that must be put somewhere); dumped into streams, lakes, and oceans (perhaps causing water

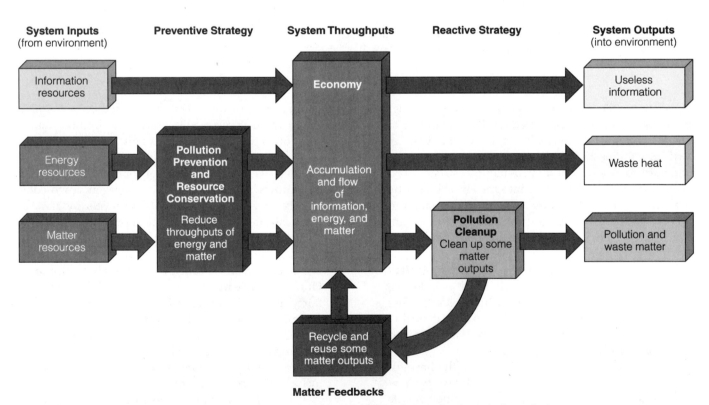

Figure 1-12 Inputs, throughputs, and outputs of a system (an economy) with two strategies for reducing pollution. Pollution prevention, or input pollution control, is based on reducing or eliminating pollution and conserving re-sources by reducing the throughputs (flows) of matter and energy resources through an economy. Throughput can also be reduced by recycling or reusing some or most of the output of pollution or waste matter. Pollution cleanup, or output pollution control, involves trying to reduce pollutants to acceptable levels after they have been produced. Both approaches are needed, but environmentalists and a growing number of economists believe that much greater emphasis should be placed on pollution prevention and resource conservation.

 Vitousek, Peter M. 1997. "Human Domination of Earth's Ecosystems." *Science*, vol. 275, no. 5325, 494(6).

pollution); or buried (perhaps causing soil and ground-water pollution). Third, *once pollutants have entered and become dispersed in the air and water (and in some cases, the soil) at harmful levels, it usually costs too much to reduce them to acceptable concentrations.*

Both pollution prevention and pollution cleanup are needed, but environmentalists and some economists urge us to emphasize prevention because it works better and is cheaper than cleanup. For widely dispersed and difficult-to-identify nonpoint pollution, hazardous wastes, and slowly degradable and nondegradable pollutants, pollution prevention is the most effective (perhaps the only) approach. As Benjamin Franklin reminded us long ago, "An ounce of prevention is worth a pound of cure."

An increasing number of businesses have found that *pollution prevention pays*. So far, however, about 99% of environmental spending in the United States (and in most other developed countries) is devoted to pollution cleanup and only 1% to pollution preven-

tion, a situation that environmental scientists and some economists believe must be reversed as soon as possible.

Both pollution prevention and pollution cleanup can be encouraged by the *carrot approach* of using incentives such as subsidies and tax write-offs or by the *stick approach* of regulations and taxes. Most analysts believe that a mix of both approaches is probably best because excessive regulation and too much taxation can incite resistance and cause a political backlash. Achieving the right balance is difficult.

1-5 ENVIRONMENTAL AND RESOURCE PROBLEMS: CAUSES AND CONNECTIONS

What Are Key Environmental Problems and Their Root Causes? We face a number of interconnected environmental and resource problems

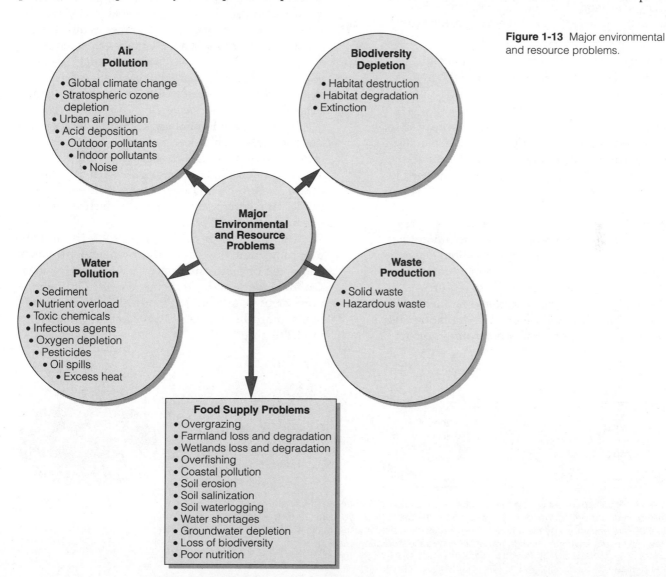

Figure 1-13 Major environmental and resource problems.

Hint: Enter the search terms *environment, effect of man* using the Subject Guide.

(Figure 1-13). The first step in dealing with these problems is to identify their underlying causes. According to environmentalists, these include the following:

- Rapid population growth

- Rapid and wasteful use of resources with too little emphasis on pollution prevention and waste reduction

- Simplification and degradation of parts of the earth's life-support systems

- Poverty, which can drive poor people to use potentially renewable resources unsustainably for short-term survival and often exposes the poor to health risks and other environmental risks

- Failure of economic and political systems to encourage earth-sustaining forms of economic development and discourage earth-degrading forms of economic growth

- Failure of economic and political systems to have market prices include the overall environmental cost of an economic good or service

- Our urge to dominate and manage nature for our use with far too little knowledge about how nature works

How Are Environmental Problems and Their Root Causes Connected? Once we have identified environmental problems and their root causes, the next step is to understand how they are connected to one another. The three-factor model in Figure 1-14 is a good starting point.

According to this simple model, a given area's total environmental degradation and pollution (that is, the environmental impact of population) depends on three factors: the number of people (population size, P), the average number of units of resources each person uses (per capita consumption or affluence, A), and the amount of environmental degradation and pollution produced for each unit of resource used (the environmental destructiveness of the technologies used to provide and consume resources, T). This model, developed in the early 1970s by biologist Paul Ehrlich and physicist John Holdren, can be summarized as

$$\text{Population} \times \text{Affluence} \times \text{Technology} = \text{Environmental impact}$$

$$\text{or} \quad P \quad \times \quad A \quad \times \quad T \quad = \quad I$$

Figure 1-15 shows how the three factors depicted in Figure 1-14 can interact in developing countries and developed countries. In developing countries, population size and the resulting degradation of potentially renewable resources (as the poor struggle to stay alive) tend to be the key factors in total environmental impact.

In developed countries, high rates of per capita resource use (and the resulting high levels of pollution and environmental degradation per person) are believed to be the key factors determining overall environmental impact. For example, it is estimated that the average U.S. citizen consumes 35 times as much as the average citizen of India and 100 times as much as the average person in the world's poorest countries. *Thus, poor parents in a developing country would need 70–200 children to have the same lifetime environmental impact as 2 children in a typical U.S. family.*

Some forms of technology, such as polluting factories and motor vehicles and energy-wasting devices, increase environmental impact by raising the T factor in the equation. Other technologies, such as pollution control, solar cells, and more energy-efficient devices, can lower environmental impact by decreasing the T factor in the equation.

The three-factor model in Figure 1-15 can help us understand how key environmental problems and some of their causes are connected. However, the interconnected problems we face involve a number of poorly understood interactions among many more factors than those in the three-factor model (Figure 1-16).

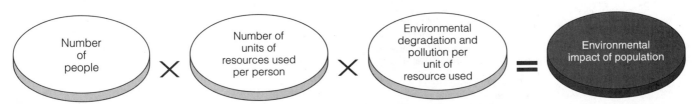

Figure 1-14 Simplified model of how three factors—population, affluence, and technology—affect the environmental impact of population. According to this model, the damage we do to the earth is equal to the number of people there are, multiplied by the amount of resources each person uses, multiplied by the amount of pollution, resource waste, and environmental degradation involved in extracting, making, and using each resource unit.

Meisner, Mark. 1997. "Green on the Screen." *Alternatives Journal*, vol. 23, no. 1, 9(2).

Developing countries

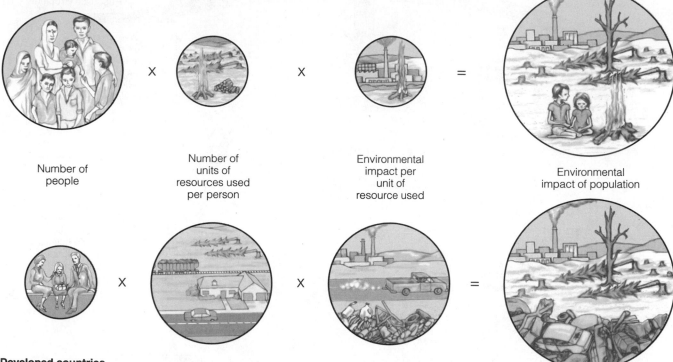

Number of people		Number of units of resources used per person		Environmental impact per unit of resource used		Environmental impact of population
	X		X		=	

Developed countries

Figure 1-15 Environmental impact of developing countries (top) and developed countries (bottom) based on the relative importance of the factors in the model shown in Figure 1-14. Circle size shows the relative importance of each factor. The size of the *T* factor can be reduced by improved technology for controlling and preventing pollution, resource waste, and environmental degradation.

1-6 SOLUTIONS: WORKING WITH THE EARTH

This chapter has presented an overview of the serious problems most environmentalists and many of the world's most prominent scientists believe we face and their root causes. It has also summarized key arguments about how serious environmental problems are. The rest of this book presents a more detailed analysis of these problems, the controversies they have created, and solutions proposed by various scientists, environmentalists, and other analysts.

Here are some guidelines various analysts have suggested for working with the earth:

- Leave the earth as good as or better than we found it.
- Take no more than we need.
- Try not to harm life, air, water, or soil.
- Sustain biodiversity.
- Help maintain the earth's capacity for self-repair.
- Don't use potentially renewable resources (soil, water, forests, grasslands, and wildlife) faster than they are replenished.

- Don't waste resources.
- Don't release pollutants into the environment faster than the earth's natural processes can dilute or assimilate them.
- Emphasize pollution prevention and waste reduction.
- Slow the rate of population growth.
- Reduce poverty.

Specific things you can do to work with the earth by trying to implement such guidelines are listed in Appendix 4.

The key to dealing with our environmental problems and challenges lies in recognizing that *individuals matter*. Anthropologist Margaret Mead has summarized our potential for change: "Never doubt that a small group of thoughtful, committed citizens can change the world. Indeed, it is the only thing that ever has."

What's the use of a house if you don't have a decent planet to put it on?

HENRY DAVID THOREAU

Hint: Enter the search terms *environment, information* using Key Words.

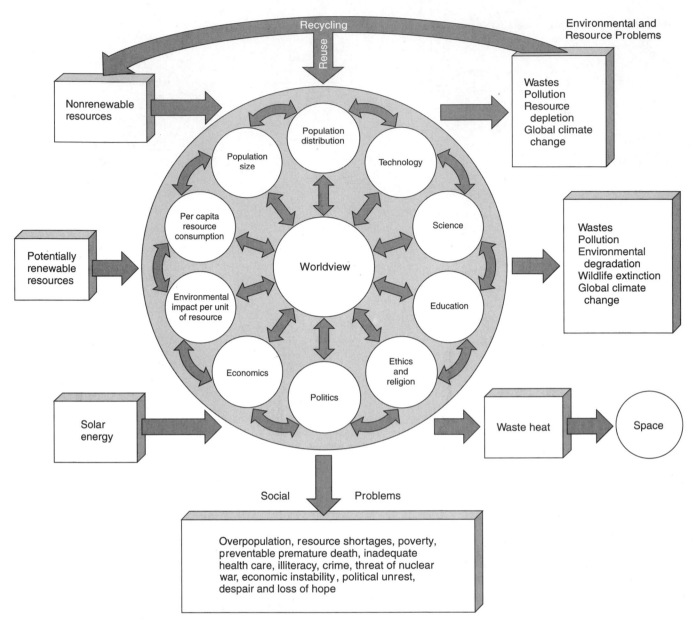

Figure 1-16 Environmental, resource, and social problems are caused by a complex, poorly understood mix of interacting factors, as illustrated by this simplified model.

CRITICAL THINKING

1. Is the world overpopulated? Explain. Is the United States overpopulated? Explain.

2. (a) Do you believe that the society you live in is on an unsustainable path? Explain. What about the world as a whole? **(b)** Do you believe that it is possible for the society you live in to become a sustainable society within the next 50 years? Explain.

3. Do you favor instituting policies designed to reduce population growth and stabilize **(a)** the size of the world's population as soon as possible and **(b)** the size of the U.S. population as soon as possible? Explain. If you agree that population stabilization is desirable, what three major policies do you believe should be implemented to accomplish this goal?

4. Explain why you agree or disagree with the following propositions:
  a. High levels of resource use by the United States and other developed countries are more beneficial than harmful.

Q: How many of the world's people attempt to survive on an annual income of about $1 per day?

b. The economic growth from high levels of resource use in developed countries provides money for more financial aid to developing countries for reducing pollution, environmental degradation, and poverty.

c. Stabilizing population is not desirable because without more consumers, economic growth would stop.

5. Explain why you agree or disagree with the following proposition: The world will never run out of renewable resources and most currently used nonrenewable resources because technological innovations will produce substitutes, reduce resource waste, or allow use of lower grades of scarce nonrenewable resources.

6. Would you support greatly increasing the amount of land designated as wilderness to protect it from economic development, even if the land contained valuable minerals, oil, natural gas, timber, or other resources? Explain.

7. Do you believe that your current lifestyle is sustainable? If the answer is yes, explain why and include the impact of the world's other 5.9 billion people on your ability to sustain your current lifestyle. If your answer is no, explain why and indicate the five most important things you could do now to make your lifestyle more sustainable. Which of these things do you actually plan to do?

*8. Make a list of the resources you truly need. Then make another list of the resources you use each day only because you want them. Finally, make a third list of resources you want and hope to use in the future. Compare your lists with those compiled by other members of your class, and relate the overall result to the tragedy of the commons.

*9. Write two-page scenarios describing what your life and that of any children you choose to have might be like 50 years from now if **(a)** we continue on our present path, or **(b)** we shift to sustainable societies throughout most of the world.

---

*Questions preceded by asterisks are laboratory exercises or individual or class projects.

2 SCIENCE, MATTER, ENERGY, AND ECOLOGY: CONNECTIONS IN NATURE

When we try to pick out anything by itself, we find it hitched to everything else in the universe.

JOHN MUIR

2-1 SCIENCE AND ENVIRONMENTAL SCIENCE

What Is Science and What Do Scientists Do? **Science** is based on the assumption that there is discoverable order in nature. It is an attempt to discover that order and use that knowledge to make predictions about what will happen in nature. As Albert Einstein once said, "The whole of science is nothing more than a refinement of everyday thinking." Figure 2-1 summarizes the more systematic version of the everyday thinking process used by scientists.

The first thing scientists must do is ask a question or identify a problem to be investigated. Then scientists working on this problem collect **scientific data**, or facts, by making observations and taking measurements. Such facts must be verified or confirmed by repeated observations and measurements, ideally by several different investigators.

The primary goal of science is not facts themselves, but a new idea, principle, or model that connects and explains certain facts and leads to useful predictions about what should happen in nature. Scientists working on a particular problem try to come up with a variety of possible explanations, or **scientific hypotheses**, of what they (or other scientists) observe in nature.

To be accepted, a scientific hypothesis not only must explain scientific data and phenomena but also should make predictions that can be used to test the validity of the hypothesis. Once a scientific hypothesis is invented, *experiments* are conducted (and repeated to be sure they are reproducible) to test the deductions or predictions. Experiments can eliminate (disprove) various hypotheses, but they can never prove that any hypothesis is the best (most useful) or the only explanation.

One method scientists use to test a hypothesis is to develop a **model**, an approximate representation or simulation of a system being studied. There are many types of models: mental, conceptual, graphic, physical, and mathematical.

If many experiments by different scientists support a particular hypothesis, it becomes a **scientific theory**: an idea, principle, or model that usually ties together and explains many facts that previously appeared to be unrelated and is supported by a large body of evidence. To scientists, theories are not to be taken lightly. They are ideas or principles stated with a high degree of certainty because they are supported by a great deal of evidence and are considered the greatest achievements in science.

Nonscientists often use the word *theory* incorrectly when they mean to refer to a *scientific hypothesis*, a tentative explanation that needs further evaluation. The statement, "Oh, that's just a theory," made in everyday conversation, implies a lack of knowledge and careful testing—the opposite of the scientific meaning of the word.

Figure 2-1 What scientists do: a summary of the scientific process, a form of critical thinking. Facts (data) are gathered and verified by repeated experiments; data are analyzed to see whether there is a consistent pattern of behavior that can be summarized as a scientific law; hypotheses are proposed to explain the data; deductions or predictions are made and tested to evaluate each hypothesis. A hypothesis that is supported by a great deal of evidence and is widely accepted by the scientific community becomes a scientific theory.

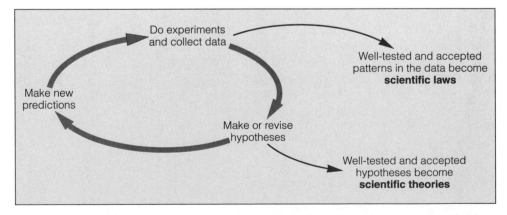

Do experiments and collect data

Well-tested and accepted patterns in the data become **scientific laws**

Make new predictions

Make or revise hypotheses

Well-tested and accepted hypotheses become **scientific theories**

Soule, Michael E. and Daniel Press. 1998. "What Is Environmental Studies?" *BioScience*, vol. 48, no. 5, 397(9).

Another important end result of science is a **scientific law**: a description of what we find happening in nature over and over in the same way, without known exception. For example, after making thousands of observations and measurements over many decades, scientists discovered what is called the *second law of energy* or *thermodynamics*. One simple way of stating this law is that heat always flows spontaneously from hot to cold—something you learned the first time you touched a hot object.

We often hear about *the* scientific method. In reality, there are many scientific methods: ways scientists gather data and formulate and test scientific hypotheses, models, theories, and laws (Figure 2-1).

New discoveries happen in many ways. Some follow this sequence: data → law → hypothesis → theory. At other times, scientists simply follow a hunch, bias, or belief and then do experiments to test their idea or hypothesis. According to physicist Albert Einstein, "There is no completely logical way to a new scientific idea." Intuition, imagination, and creativity are as important in science as they are in poetry, art, music, and other great adventures of the human spirit that awaken us to the wonder, mystery, and beauty of life, the earth, and the universe.

Most processes or parts of nature that scientists seek to understand are influenced by a number of *variables* or *factors*. One way scientists test a hypothesis about the effects of a particular variable is to conduct a *controlled experiment*. This is done by setting up two groups: an *experimental group*, in which the chosen variable is changed in a known way, and a *control group*, in which the chosen variable is not changed. The experiment is designed so that all components of each group are as identical as possible and experience the same conditions, except for the single factor being varied in the experimental group. If the experiment is designed properly, any difference between the two groups should result from a variable that was changed in the experimental group.

A basic problem is that many of nature's components and processes, especially those investigated by environmental scientists, involve a huge number of variables interacting in often poorly understood ways. In such cases, it is very difficult or impossible to carry out meaningful controlled experiments.

How Valid Are the Results of Science? Scientists can disprove things, and they can establish that a particular model, theory, or law has a very high degree of validity and is extremely useful in examining how nature works and predicting what will happen in nature. However, like scholars in virtually any field, scientists cannot prove that their ideas are *absolutely* true.

Although it may be extremely low, there's always some uncertainty involved in any scientific model, theory, or law. The goal of the rigorous scientific process is to reduce the degree of uncertainty as much as possible. However, the more complex the system being studied, the greater the degree of uncertainty or unpredictability about its behavior.

The standard for evaluating a scientific model, theory, or law is not absolute truth or proof. Instead, the validity of such ideas is based on how *useful* they are in helping us understand how nature works and making predictions about what will happen in nature.

How Does Frontier Science Differ from Consensus Science? News reports often focus on new scientific "breakthroughs" and on disputes among scientists over the validity of preliminary (untested) data, hypotheses, and models (which are by definition tentative). This aspect of science, controversial because it has not been widely tested and accepted, is called **frontier science**.

By contrast, **consensus science** consists of data, theories, and laws that are widely accepted by scientists considered experts in the field involved. This aspect of science is very reliable but is rarely considered newsworthy. One way to find out what scientists generally agree on is to seek out reports by scientific bodies such as the U.S. National Academy of Sciences and the British Royal Society that attempt to summarize consensus among experts in key areas of science. Sometimes a new discovery or hypothesis goes against consensus science and causes a change in the consensus view. However, because such occasions are extremely rare, the consensus view on a particular widely accepted scientific model or theory represents the best available information.

What Is Environmental Science and What Are Its Limitations? **Environmental science** is the study of how we and other species interact with one another and with the nonliving environment (matter and energy). It is a *physical and social science* that integrates knowledge from a wide range of disciplines including physics, chemistry, biology (especially ecology), geology, meteorology, geography, resource technology and engineering, resource conservation and management, demography (the study of population dynamics), economics, politics, sociology, psychology, and ethics. In other words, it is a study of how the parts of nature and human societies operate and interact—a study of *connections* and *interactions*.

There is controversy over some of the knowledge provided by environmental science, for much of it falls into the realm of frontier science. One problem involves *arguments over the validity of data*. There is no way to measure accurately how many metric tons of soil are eroded worldwide, how many hectares of tropical forest are cut, how many species become extinct, or how many metric tons of certain pollutants are emitted into the atmosphere or aquatic systems each year.

We may legitimately argue over the numbers, but the point environmental scientists want to make is that the trends in these phenomena are significant enough to be evaluated and addressed. Such environmental data should not be dismissed because they are "only estimates" (which are all we can ever have). However, this does not relieve investigators from the responsibility of getting the best estimates possible and pointing out that they *are* estimates.

Another limitation is that *most environmental problems involve so many variables and such complex interactions that we don't have enough data or sufficiently sophisticated models to aid in understanding them very well.* Reputable scientists in a field may state contradictory opinions about the meaning of the data from experiments (especially frontier scientific results) and the validity of various hypotheses.

Because environmental problems won't go away, at some point we must evaluate the available (but always inadequate) information and make political and economic decisions. Without sufficient evidence, such decisions are often based primarily on individual and societal values, which is why environmental aspects of differing worldviews are at the heart of most environmental controversies, as discussed in Section 12-7. People with different environmental worldviews and values can take the same information, come to completely different conclusions, and still be logically consistent.

 ## 2-2 MATTER AND ENERGY

What Are Nature's Building Blocks? Matter is anything that has mass (the amount of material in an object) and takes up space. Scientists classify matter as existing in various levels of organization (Figure 2-2).

Matter includes the solids, liquids, and gases around us and within us. Matter is found in two *chemical forms*: **elements** (the distinctive building blocks of matter that make up every material substance) and **compounds** (two or more different elements held together in fixed proportions by attractive forces called *chemical bonds*). Various elements, compounds, or both can be found together in **mixtures**.

All matter is built from the 112 known chemical elements (92 of them occurring naturally and 20 of them made in the laboratory from existing elements). To simplify things, chemists represent each element by a one- or two-letter symbol: hydrogen (H), carbon (C), oxygen (O), nitrogen (N), phosphorus (P), sulfur (S), chlorine (Cl), fluorine (F), bromine (Br), sodium (Na), calcium (Ca), lead (Pb), mercury (Hg), and uranium (U), to mention but a few.

If we had a supermicroscope capable of looking at individual elements and compounds, we could see that they are made up of three types of building blocks: **atoms** (the smallest unit of matter that is unique to a particular element), **ions** (electrically charged atoms or combinations of atoms), and **molecules** (combinations of two or more atoms of the same or different elements held together by chemical bonds). Because ions and molecules are formed from atoms, *atoms are the ultimate building blocks for all matter.*

If we increased the magnification of our supermicroscope, we would find that each different type of atom contains a certain number of *subatomic particles*. The main building blocks of an atom are positively charged **protons** (p), uncharged **neutrons** (n), and negatively charged **electrons** (e). Each atom consists of an extremely small center, or **nucleus**, containing protons and neutrons, and one or more electrons in rapid motion somewhere around the nucleus.

Each atom has an equal number of positively charged protons (inside its nucleus) and negatively charged electrons (outside its nucleus). Because these electrical charges cancel one another, *the atom as a whole has no net electrical charge.*

Each element has its own specific **atomic number**, equal to the number of protons in the nucleus of each of its atoms. The simplest element, hydrogen (H), has only 1 proton in its nucleus, so its atomic number is 1. Carbon (C), with 6 protons, has an atomic number of 6, whereas uranium (U), a much larger atom, has 92 protons and an atomic number of 92.

Because electrons have so little mass compared with the mass of a proton or a neutron, *most of an atom's mass is concentrated in its nucleus.* We describe the mass of an atom in terms of its **mass number**: the total number of neutrons and protons in its nucleus.

Although all atoms of an element have the same number of protons in their nuclei, they may have different numbers of uncharged neutrons in their nuclei, and thus may have different mass numbers. Various forms of an element having the same atomic number but a different mass number are called **isotopes** of that element. Isotopes are identified by attaching their mass numbers to the name or symbol of the element. For example, hydrogen has three isotopes: hydrogen-1 (H-1), hydrogen-2 (H-2, common name deuterium), and hydrogen-3 (H-3, common name tritium). A natural sample of an element contains a mixture of its isotopes in a fixed proportion or percentage abundance by weight (Figure 2-3).

Most matter exists as compounds. Chemists use a shorthand *chemical formula* to show the number of atoms (or ions) of each type in a compound. The formula contains the symbols for each of the elements present and uses subscripts to represent the number of atoms or ions of each element in the compound's basic structural unit. Each molecule of water, for example, consists of two hydrogen atoms chemically bonded to

 Kestenbaum, David. 1998. "Gentle Force of Entropy Bridges Disciplines." *Science*, vol. 279, no. 5358, 1849(1).

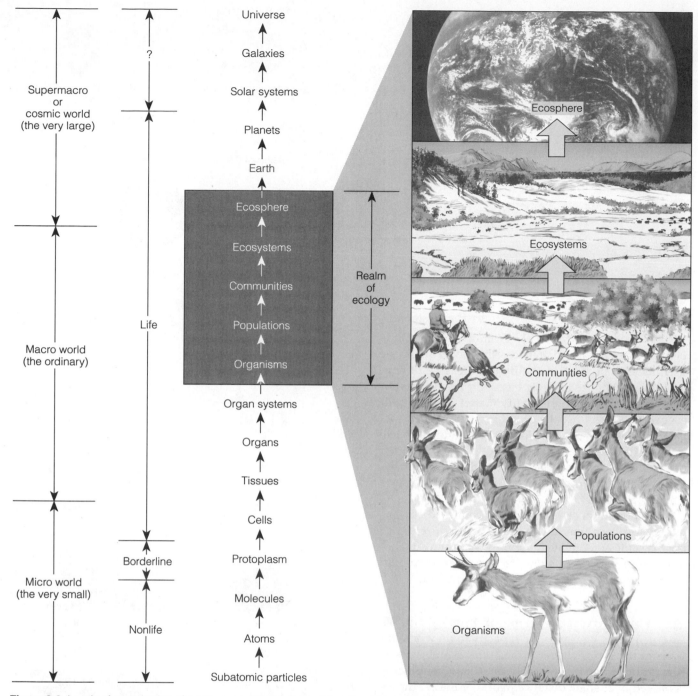

Figure 2-2 Levels of organization of matter according to size and function. This is one way scientists classify patterns of matter found in nature. Note that ecology focuses on five levels of this hierarchical model.

an oxygen atom, yielding H_2O (read as "H-two-O") molecules. Other examples you will encounter in this book are oxygen (O_2), ozone (O_3), nitrogen (N_2), nitrous oxide (N_2O), nitric oxide (NO), hydrogen sulfide (H_2S), carbon monoxide (CO), carbon dioxide (CO_2), nitrogen dioxide (NO_2), sulfur dioxide (SO_2), ammonia (NH_3), sulfuric acid (H_2SO_4), nitric acid (HNO_3), methane (CH_4), and glucose ($C_6H_{12}O_6$).

Matter is also found in three *physical states*: solid, liquid, and gas. Water, for example, exists as ice, liquid water, and water vapor, depending on its temperature and pressure. The three physical states of matter differ in the relative spacing and orderliness of its atoms, ions, or molecules, with solids having the most compact and orderly arrangement and gases the least compact and orderly arrangement.

Hint: Enter the search term *energy* using the Subject Guide and then select the hotkey to *force and energy*.

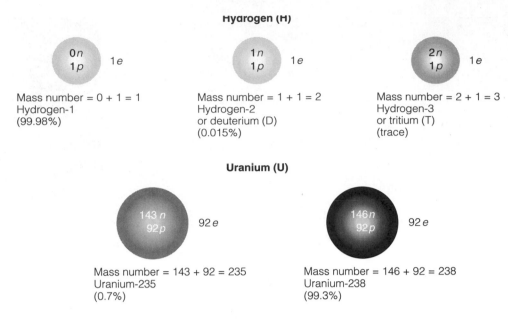

Figure 2-3 Isotopes of hydrogen and uranium. All isotopes of hydrogen have an atomic number of 1 because each has one proton in its nucleus; similarly, all uranium isotopes have an atomic number of 92. However, each isotope of these elements has a different mass number because its nucleus contains a different number of neutrons. Figures in parentheses indicate the percentage abundance by weight of each isotope in a natural sample of the element.

Hydrogen (H)

$0n$ $1p$ $1e$

Mass number = 0 + 1 = 1
Hydrogen-1
(99.98%)

$1n$ $1p$ $1e$

Mass number = 1 + 1 = 2
Hydrogen-2
or deuterium (D)
(0.015%)

$2n$ $1p$ $1e$

Mass number = 2 + 1 = 3
Hydrogen-3
or tritium (T)
(trace)

Uranium (U)

$143n$ $92p$ $92e$

Mass number = 143 + 92 = 235
Uranium-235
(0.7%)

$146n$ $92p$ $92e$

Mass number = 146 + 92 = 238
Uranium-238
(99.3%)

What Is Matter Quality? From a human standpoint, we can classify matter according to its quality or usefulness to us. **Matter quality** is a measure of how useful a matter resource is, based on its availability and concentration. **High-quality matter** is organized, concentrated, and usually found near the earth's surface, and has great potential for use as a matter resource; **low-quality matter** is disorganized, dilute, and often deep underground or dispersed in the ocean or the atmosphere, and usually has little potential for use as a matter resource (Figure 2-4).

An aluminum can is a more concentrated, higher-quality form of aluminum than aluminum ore containing the same amount of aluminum. That's why it takes less energy, water, and money to recycle an aluminum can than to make a new can from aluminum ore.

What Different Forms of Energy Do We Encounter? Energy is the capacity to do work and transfer heat. Work is performed when an object—be it a grain of sand, this book, or a giant boulder—is moved over some distance. Work, or matter movement, also is needed to boil water (to change it into the more dispersed and faster-moving water molecules in steam) or to burn natural gas to heat a house or cook food. Energy is also the heat that flows automatically from a hot object to a cold object when they come in contact.

Energy comes in many forms: light; heat; electricity; chemical energy stored in the chemical bonds in coal, sugar, and other materials; the mechanical energy of moving matter such as flowing water, wind (air masses), and joggers; and nuclear energy emitted from the nuclei of certain isotopes.

High Quality | Low Quality

Solid | Gas

Salt | Solution of salt in water

Coal | Coal-fired power-plant emissions

Gasoline | Automobile emissions

Aluminum can | Aluminum ore

Figure 2-4 Examples of differences in matter quality. High-quality matter (left-hand column) is fairly easy to extract and is concentrated; low-quality matter (right-hand column) is more difficult to extract and is more dispersed than high-quality matter.

Q: What percentage of the world's population lives in the United States?

Scientists classify energy as either kinetic or potential. **Kinetic energy** is the energy that matter has because of its mass and its speed or velocity. It is energy in action or motion. Wind (a moving mass of air), flowing streams, falling rocks, heat flowing from a body at a high temperature to one at a lower temperature, electricity (flowing electrons), moving cars—all have kinetic energy. Radio waves, TV waves, microwaves, infrared radiation, visible light, ultraviolet radiation, X rays, gamma rays, and cosmic rays are all forms of kinetic energy known as **electromagnetic radiation** (Figure 2-5).

Potential energy is stored energy that is potentially available for use. A rock held in your hand, an unlit stick of dynamite, still water behind a dam, the gasoline in a car tank, and the nuclear energy stored in the nuclei of atoms all have potential energy because of their position or the position of their parts. Potential energy can be changed to kinetic energy. When you drop a rock, its potential energy changes into kinetic energy. When you burn gasoline in a car engine, the potential energy stored in the chemical bonds of its molecules changes into heat, light, and mechanical (kinetic) energy that propels the car.

What Is Energy Quality? From a human standpoint, the measure of an energy source's ability to do useful work is called its **energy quality** (Figure 2-6). **High-quality energy** is organized or concentrated and can perform much useful work. Examples are electricity, coal, gasoline, concentrated sunlight, nuclei of uranium-235 used as fuel in nuclear power plants, and heat concentrated in small amounts of matter so that its temperature is high.

By contrast, **low-quality energy** is disorganized or dispersed and has little ability to do useful work. An example is heat dispersed in the moving molecules of a large amount of matter (such as the atmosphere or a large body of water) so that its temperature is low. Thus, even though the total amount of heat stored in the Atlantic Ocean is greater than the amount of high-quality chemical energy stored in all the oil deposits of Saudi Arabia, the ocean's heat is so widely dispersed that it can't be used to move things or to heat things to high temperatures.

We use energy to accomplish certain tasks, each requiring a certain minimum energy quality. It makes sense to match the quality of an energy source with the quality of energy needed to perform a particular task (Figure 2-6) because doing so saves energy and usually money.

What Is the Difference Between a Physical and a Chemical Change? A **physical change** involves no change in chemical composition. Cutting a piece of aluminum foil into small pieces is one example. Changing a substance from one physical state to another is a second example. For example, when solid water (ice) is melted or liquid water is boiled, none of the H_2O molecules involved are altered; instead, the molecules are organized in different spatial (physical) patterns.

In a **chemical change** or **chemical reaction**, on the other hand, the chemical compositions of the elements or compounds are altered. For example, when coal burns completely, the solid carbon (C) it contains combines with oxygen gas (O_2) to form the gaseous compound carbon dioxide (CO_2). We can represent this chemical reaction in shorthand form as $C + O_2 \longrightarrow CO_2 + energy$.

Energy is given off in this reaction, making coal a useful fuel. The reaction also shows how the complete burning of coal (or any of the carbon-containing compounds in wood, natural gas, oil, and gasoline) gives off carbon dioxide gas to the atmosphere.

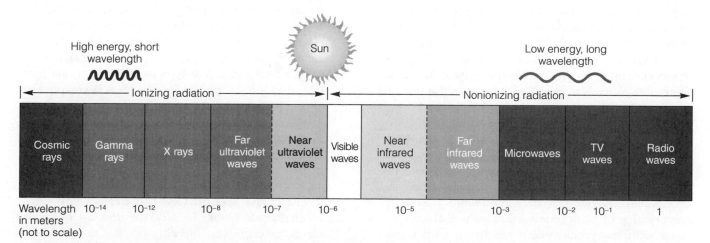

Figure 2-5 The electromagnetic spectrum: the range of electromagnetic waves, which differ in wavelength (distance between successive peaks or troughs) and energy content.

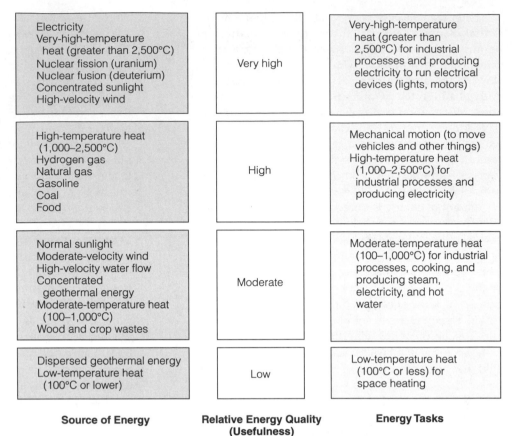

Figure 2-6 Categories of the quality (usefulness for performing various energy tasks) of different sources of energy. *High-quality energy* is concentrated and has great ability to perform useful work; *low-quality energy* is dispersed and has little ability to do useful work. To avoid unnecessary energy waste, it is best to match the quality of an energy source with the quality of energy needed to perform a task.

Source of Energy	Relative Energy Quality (Usefulness)	Energy Tasks
Electricity Very-high-temperature heat (greater than 2,500°C) Nuclear fission (uranium) Nuclear fusion (deuterium) Concentrated sunlight High-velocity wind	Very high	Very-high-temperature heat (greater than 2,500°C) for industrial processes and producing electricity to run electrical devices (lights, motors)
High-temperature heat (1,000–2,500°C) Hydrogen gas Natural gas Gasoline Coal Food	High	Mechanical motion (to move vehicles and other things) High-temperature heat (1,000–2,500°C) for industrial processes and producing electricity
Normal sunlight Moderate-velocity wind High-velocity water flow Concentrated geothermal energy Moderate-temperature heat (100–1,000°C) Wood and crop wastes	Moderate	Moderate-temperature heat (100–1,000°C) for industrial processes, cooking, and producing steam, electricity, and hot water
Dispersed geothermal energy Low-temperature heat (100°C or lower)	Low	Low-temperature heat (100°C or less) for space heating

What Is The Law of Conservation of Matter? Why There Is No "Away" People commonly talk about consuming or using up material resources, but the truth is that *we don't consume matter—we only use some of the earth's resources for a while.* We take materials from the earth, carry them to another part of the globe, and process them into products that are used and then discarded, burned, buried, reused, or recycled.

In so doing, *we may change various elements and compounds from one physical or chemical form to another, but in no physical and chemical change can we create or destroy any of the atoms involved.* All we can do is rearrange them into different spatial patterns (physical changes) or different combinations (chemical changes). The italicized statement in this paragraph, based on many thousands of measurements, is known as the **law of conservation of matter**.

The law of conservation of matter means that there really is no "away" in "to throw away." *Everything we think we have thrown away is still here with us in one form or another.* Even though we can make the environment cleaner and convert some potentially harmful chemicals into less harmful physical or chemical forms, the law of conservation of matter means that we will always be faced with the problem of what to do with some quantity of wastes. By placing much greater emphasis on pollution prevention, waste reduction, and more efficient use of resources, we can greatly reduce the amount of wastes we add to the environment.

What Is the First Law of Energy? You Can't Get Something for Nothing Scientists have observed energy being changed from one form to another in millions of physical and chemical changes, but they have never been able to detect the creation or destruction of any energy (except in nuclear changes). The results of their experiments have been summarized in the **law of conservation of energy**, also known as the **first law of energy** or the **first law of thermodynamics**: *In all physical and chemical changes, energy is neither created nor destroyed, but it may be converted from one form to another.*

This scientific law tells us that when one form of energy is converted to another form in any physical or chemical change *energy input always equals energy output.* No matter how hard we try or how clever we are, we can't get more energy out of a system than we put in; in other words, *we can't get something for nothing in terms of energy quantity.*

What Is the Second Law of Energy? You Can't Even Break Even Because the first law of energy states that energy can be neither created nor destroyed, it's tempting to think that there will always be enough energy, yet if we fill a car's tank with gasoline

Q: What percentage of the world's mineral resources and nonrenewable energy is used by the United States?

and drive around, or use a flashlight battery until it is dead, something has been lost. If it isn't energy, what is it? The answer is *energy quality* (Figure 2-6), the amount of energy available that can perform useful work.

Countless experiments have shown that when energy is changed from one form to another, a decrease in energy quality always occurs. The results of these experiments have been summarized in what is called the **second law of energy** or the **second law of thermodynamics**: *When energy is changed from one form to another, some of the useful energy is always degraded to lower-quality, more dispersed, less useful energy.* This degraded energy usually takes the form of heat given off at a low temperature to the surroundings (environment). It is dispersed by the random motion of air or water molecules and becomes even more disorderly and less useful. Another way to state the second law of energy is that *heat always flows spontaneously from hot (high-quality energy) to cold (low-quality energy).*

Basically, this law says that in any energy conversion, we always end up with *less* usable energy than we started with. So not only can we not get something for nothing in terms of energy quantity, *we can't even break even in terms of energy quality because energy always goes from a more useful to a less useful form.* The more energy we use, the more low-grade energy (heat) we add to the environment. No one has ever found a violation of this fundamental scientific law.

Consider three examples of the second energy law in action. First, when a car is driven, only about 10% of the high-quality chemical energy available in its gasoline fuel is converted into mechanical energy (to propel the vehicle) and electrical energy (to run its electrical systems); the remaining 90% is degraded to low-quality heat that is released into the environment and eventually lost into space. Second, when electrical energy flows through filament wires in an incandescent light bulb, it is changed into about 5% useful light and 95% low-quality heat that flows into the environment; this so-called *light bulb* is really a *heat bulb*. Third, in living systems, solar energy is converted into chemical energy (photosynthesis and food) and then into mechanical energy (moving, thinking, and living); high-quality energy is degraded during this change of forms (Figure 2-7).

The second law of energy also means that *we can never recycle or reuse high-quality energy to perform useful work.* Once the concentrated energy in a serving of food, a liter of gasoline, a lump of coal, or a chunk of uranium is released, it is degraded to low-quality heat that is dispersed into the environment. We can heat air or water at a low temperature and upgrade it to high-quality energy, but the second law of energy tells us that it will take more high-quality energy to do this than we get in return.

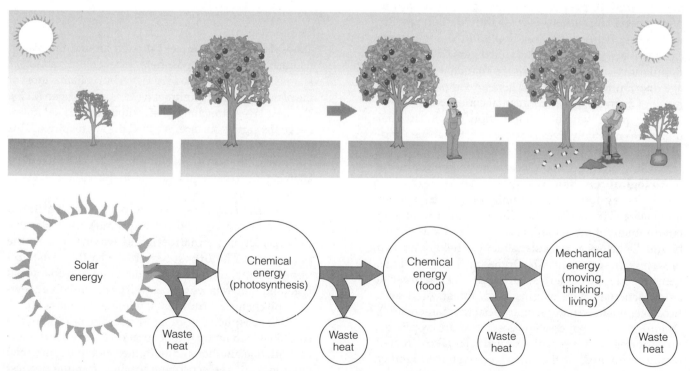

Figure 2-7 The second energy law in action in living systems. Each time energy is changed from one form to another, some of the initial input of high-quality energy is degraded, usually to low-quality heat that disperses into the environment.

2-3 EARTH'S LIFE-SUPPORT SYSTEMS: FROM ORGANISMS TO THE ECOSPHERE

What Is Ecology? Ecology (from the Greek words *oikos*, "house" or "place to live," and *logos*, "study of") is the study of how organisms interact with one another and with their nonliving environment (including such factors as sunlight, temperature, moisture, and vital nutrients). Ecology deals mainly with interactions among organisms, populations, communities, ecosystems, and the ecosphere (Figure 2-2).

An **organism** is any form of life. Organisms can be classified into species, groups of organisms that resemble one another in appearance, behavior, chemistry, and genetic endowment. Organisms that reproduce sexually are classified in the same species if, under natural conditions, they can actually or potentially breed with one another and produce live, fertile offspring.

We don't know how many species exist on the earth. Estimates range from 5 million to 100 million, most of them insects and microorganisms. So far biologists have identified and named only about 1.8 million species. Biologists know a fair amount about roughly one-third of the known species, but detailed roles and interactions of only a few.

A **population** consists of all members of the same species occupying a given area at the same time. Examples are all sunfish in a pond, all white oak trees in a forest, and all people in a country. In most natural populations, individuals vary slightly in their genetic makeup, which is why they don't all look or behave exactly alike—a phenomenon called **genetic diversity**. Populations are dynamic groups that change in size, age distribution, density, and genetic composition as a result of changes in environmental conditions.

The place where a population (or individual organism) normally lives is known as its **habitat**. Populations of all the different species occupying and interacting in a particular place make up a **community**, or **biological community**.

An **ecosystem** is a community of different species interacting with one another and with their nonliving environment of matter and energy. An ecosystem may be small, such as a particular stream or field or a patch of woods, desert, or marsh. Or the units may be large, generalized types of terrestrial (land) ecosystems such as a particular type of grassland, forest, or desert. Ecosystems can be natural or artificial (human-created). Examples of human-created ecosystems are cropfields, farm ponds, and reservoirs or artificial lakes created behind dams. All of the earth's ecosystems together make up what we call the **biosphere**, or **ecosphere**.

What Are the Major Parts of Earth's Life-Support Systems? We can think of the earth as being made up of several layers or concentric spheres (Figure 2-8). The

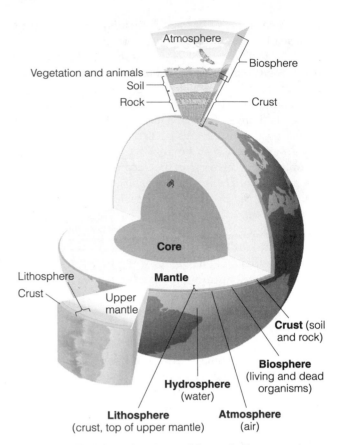

Figure 2-8 The general structure of the earth. The atmosphere consists of several layers, including the troposphere (innermost layer) and the stratosphere (second layer).

atmosphere is a thin envelope of air around the planet. Its inner layer, the **troposphere**, extends only about 17 kilometers (11 miles) above sea level but contains most of the planet's air, mostly nitrogen (78%) and oxygen (21%). The next layer, stretching 17–48 kilometers (11–30 miles) above the earth's surface, is called the **stratosphere**. This layer's lower portion contains enough ozone (O_3) to filter out most of the sun's harmful ultraviolet radiation, thus allowing life to exist on land and in the surface layers of bodies of water.

The **hydrosphere** consists of the earth's liquid water (both surface and underground), ice (polar ice, icebergs, ice in permafrost), and water vapor in the atmosphere. The **lithosphere** is the earth's crust and upper mantle; the crust contains nonrenewable fossil fuels and minerals we use as well as potentially renewable soil chemicals (nutrients) required for plant life.

The **ecosphere**, or **biosphere**, is the portion of the earth in which living (biotic) organisms exist and interact with one another and with their nonliving (abiotic) environment. The ecosphere reaches from the deepest ocean floor 20 kilometers (12 miles) below sea level to the tops of the highest mountains. If the earth were an apple, the ecosphere would be no thicker than the apple's skin. *The goal of ecology is to understand the*

Rykiel, Ed. 1997. "Ecosystem Science for the Twenty-First Century." *BioScience*, vol. 47 no. 10, 705(3).

interactions in this thin, life-supporting global skin of air, water, soil, and organisms.

What Sustains Life on Earth? Life on the earth depends on three interconnected factors (Figure 2-9):

■ The *one-way flow of high-quality (usable) energy* from the sun, first through materials and living things in their feeding interactions, then into the environment as low-quality energy (mostly heat dispersed into air or water molecules at a low temperature), and eventually back into space as infrared radiation (Figure 2-10)

■ The *cycling of types of matter or nutrients* needed for survival by living organisms through parts of the ecosphere

■ *Gravity*, which allows the planet to hold onto its atmosphere and causes the downward movement of chemicals in the matter cycles

🌐 Why Is Biodiversity Such an Important Ecosystem Service? As environmental conditions have changed over billions of years, many species have become extinct and new ones have formed. The result of these changes is **biological diversity**, or **biodiversity**: the forms of life that can best survive the variety of conditions currently found on the earth. Biodiversity includes **genetic diversity** (variability in the genetic makeup among individuals in a single species, **species diversity** (the variety of species in different habitats on the earth), and **ecological diversity** (the variety of biological communities that interact with one another and with their nonliving environments).

Another term for biodiversity is **wildness**: the existence of wild gene pools, species, and ecosystems that are completely or mostly undisturbed by human activities. In the words of Henry David Thoreau, "In wildness is the preservation of the world."

We are utterly dependent on this mostly unknown biocapital. This rich variety of genes, species, and ecosystems gives us food, wood, fibers, energy, raw materials, industrial chemicals, and medicines, and it pours hundreds of billions of dollars yearly into the global economy.

The earth's life-forms and ecosystems also provide recycling, purification, and natural pest control (Figure 1-2). Every species here today contains genetic information that represents thousands to millions of years of adaptation to the earth's changing environmental conditions

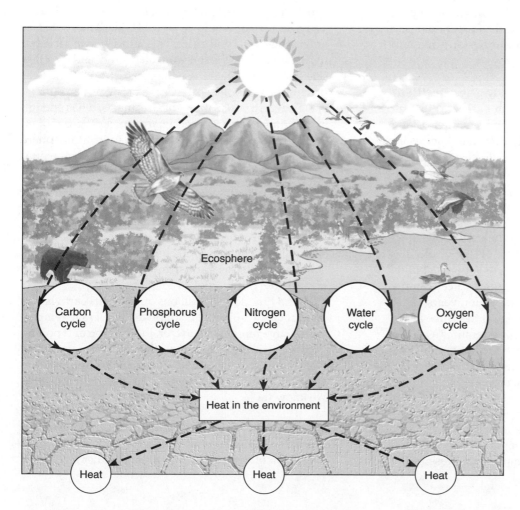

Figure 2-9 Life on the earth depends on the *one-way flow of energy* (dashed lines) from the sun through the ecosphere, the *cycling of crucial elements* (solid lines around circles), and *gravity*, which keeps atmospheric gases from escaping into space and draws chemicals downward in the matter cycles. This simplified conceptual model depicts only a few of the many cycling elements.

Ecosphere

Carbon cycle

Phosphorus cycle

Nitrogen cycle

Water cycle

Oxygen cycle

Heat in the environment

Heat   Heat   Heat

Hint: Enter the search terms *ecosystem, science* using Key Words.

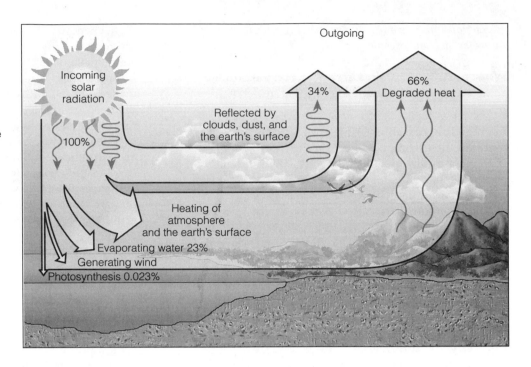

Figure 2-10 The flow of energy to and from the earth. The ultimate source of energy in most ecosystems is sunlight. Solar energy also powers the cycling of matter and drives the climate and weather systems that distribute heat and fresh water over earth's surface. How fast this heat flows through the atmosphere and back into space is affected by tropospheric heat-trapping (greenhouse) gases such as water vapor, carbon dioxide, methane, nitrous oxide, and ozone. Without this atmospheric thermal blanket, known as the *natural greenhouse effect*, the earth would be nearly as cold as Mars, and life as we know it could not exist.

and is the raw material for future adaptations. Biodiversity is nature's insurance policy against disasters.

Some people also include *human cultural diversity* as part of the earth's biodiversity. The variety of human cultures represents numerous social and technological solutions that have enabled us to survive and adapt to and work with the earth.

Connections: Earth, the Just-Right, Resilient Planet How did life on the earth evolve to its present system of diverse species living in an interlocking network of matter cycles, energy flow, and species interactions? We don't know the full answer to this question, but a widely accepted body of evidence indicates that through chemical and physical changes taking place over billions of years, conditions for life as we know it to exist on the earth developed. This body of evidence suggests that life on the earth developed in two phases: *chemical evolution* of the molecules needed for the earth's primitive cells (about 1 billion years) and *biological evolution* of primitive cells into a variety of organisms (about 3.7–3.8 billion years).

Like Goldilocks tasting porridge at the Three Bears' house, life on the earth as we know it requires a certain temperature range: Venus is much too hot and Mars is much too cold, but the earth is *just right*. (Otherwise, you wouldn't be reading these words.)

Life as we know it depends on liquid water. Again, temperature is crucial; life on the earth requires average temperatures between the freezing and boiling points of water, between 0°C and 100°C (32°F and 212°F) at the earth's range of atmospheric pressures.

The earth's orbit around the sun is the right distance from the sun to provide these conditions. If the earth were much closer, it would be too hot—like Venus—for water vapor to condense to form rain. If it were much farther away, its surface would be so cold—like Mars—that its water would exist only as ice. The earth also spins; if it didn't, the side facing the sun would be too hot and the other side too cold for water-based life to exist. So far, the temperature has been, like Baby Bear's porridge, just right.

The earth is also the right size; that is, it has enough gravitational mass to keep its iron–nickel core molten and to keep the gaseous molecules in its atmosphere from flying off into space. (A much smaller earth would be unable to hold onto an atmosphere consisting of such light molecules as N_2, O_2, CO_2, and H_2O.) The slow transfer of its internal heat (geothermal energy) to the surface also helps keep the planet at the right temperature for life. And thanks to the development of photosynthesizing bacteria over 2 billion years ago, an ozone sunscreen protects us and many other forms of life from an overdose of ultraviolet radiation.

On a time scale of millions of years, the earth is also enormously resilient and adaptive. Its average temperatures have remained between the freezing and boiling points of water even though the sun's energy output has increased by about 30% over the 3.6 billion years since life arose. In short, the earth is just right for life as we know it.

What Types of Organisms Are Found on Earth? On the basis of their cell structure, biologists classify all of the earth's organisms as either eukaryotic

Q: What is a scientific theory?

or prokaryotic. All organisms except bacteria are **eukaryotic**: Their cells are surrounded by a membrane and have a *nucleus* (a membrane-bounded structure - containing genetic material in the form of DNA) and several other internal parts. Bacterial cells are **prokaryotic**: They are surrounded by a membrane but have no distinct nucleus or other internal parts enclosed by membranes. Although most familiar organisms are eukaryotic, they could not exist without hordes of microscopic prokaryotic organisms (bacteria).

Scientists group organisms into various categories based on their common characteristics, a process called *taxonomic classification*. The largest category is the *kingdom*, which includes all organisms that have one or perhaps several common features. The earth's organisms are commonly classified into five kingdoms: monera, protists, fungi, plants, and animals (Figure 2-11). Most bacteria, fungi, and protists are **microorganisms**: organisms that are so small that they can be seen only by using a microscope.

Monera (**bacteria** and **cyanobacteria**) are single-celled, microscopic prokaryotic organisms. A few bacteria can cause diseases in humans. However, your body is inhabited by billions of beneficial bacteria that help keep you healthy by aiding in food digestion and by crowding out disease-causing bacteria.

Protists (protista) are mostly single-celled eukaryotic organisms such as diatoms, dinoflagellates, amoebas, golden brown and yellow-green algae, and protozoans. Some protists cause human diseases such as malaria, sleeping sickness, and Chagas's disease. **Fungi** are mostly many-celled (sometimes microscopic) eukaryotic organisms such as mushrooms, molds, mildews, and yeasts.

Plants (plantae) are mostly many-celled eukaryotic organisms such as red, brown, and green algae and mosses, ferns, conifers, and flowering plants (whose flowers produce seeds that perpetuate the species). Some plants such as corn and marigolds are **annuals**, which complete their life cycles in one growing season; others are **perennials**, which can live for more than two years, such as roses, grapes, elms, and magnolias.

Animals (animalia) are also many-celled, eukaryotic organisms. Most, called **invertebrates**, have no backbones. They include sponges, jellyfish, worms, arthropods (insects, shrimp, and spiders), mollusks (snails, clams, and octopuses), and echinoderms (sea urchins and sea stars). Insects play roles that are vital to our existence (Connections, below). **Vertebrates** (animals with backbones and a brain protected by skull bones) include fishes (sharks and tuna), amphibians (frogs and salamanders), reptiles (crocodiles and

Have You Thanked the Insects Today?

CONNECTIONS

Insects have a bad reputation. We classify many insect species as *pests* because they compete with us for food, spread human diseases (such as malaria), and invade our lawns, gardens, and houses. Some people have "bugitis," fear all insects, and think that the only good bug is a dead bug. However, this view fails to recognize the vital roles insects play in helping sustain life on earth.

A large proportion of the earth's plant species depend on insects to pollinate their flowers. Plants also benefit when insects help loosen the soil around their roots and decompose dead tissue into nutrients the plants need. In turn, we and other land-dwelling animals depend on plants for food, either by eating them or by consuming animals that eat them.

Insects that eat other insects help control the populations of at least half the species of insects we call pests. This free pest control service is an important part of the earth capital that helps sustain us.

Suppose all insects disappeared today. Within a year most of the earth's amphibians, reptiles, birds, and mammals would become extinct because of the disappearance of so much plant life. The earth would be covered with rotting vegetation and animal carcasses being decomposed by unimaginably huge hordes of bacteria and fungi. The land, largely devoid of animal life, would be covered by mats of wind-pollinated vegetation and intermittent clumps of small trees and bushes.

Fortunately, this is not a realistic scenario because insects, which have been around for at least 400 million years, are phenomenally successful

forms of life. Today they are by far the planet's most diverse, abundant, and successful group of animals.

Some people have wondered whether insects will take over the world if the human race extinguishes itself. This is the wrong question. Insects are already in charge. The approximately billion billion (10^{18}) insects alive at any given time have already dominated much of the earth's land surface for millions of years, and they're likely to be here for millions of years more. Insects can thrive without newcomers such as us, but we and most other land organisms would quickly perish without them.

Critical Thinking

Are you afraid of most insects? If so, try to trace where this fear might have originated.

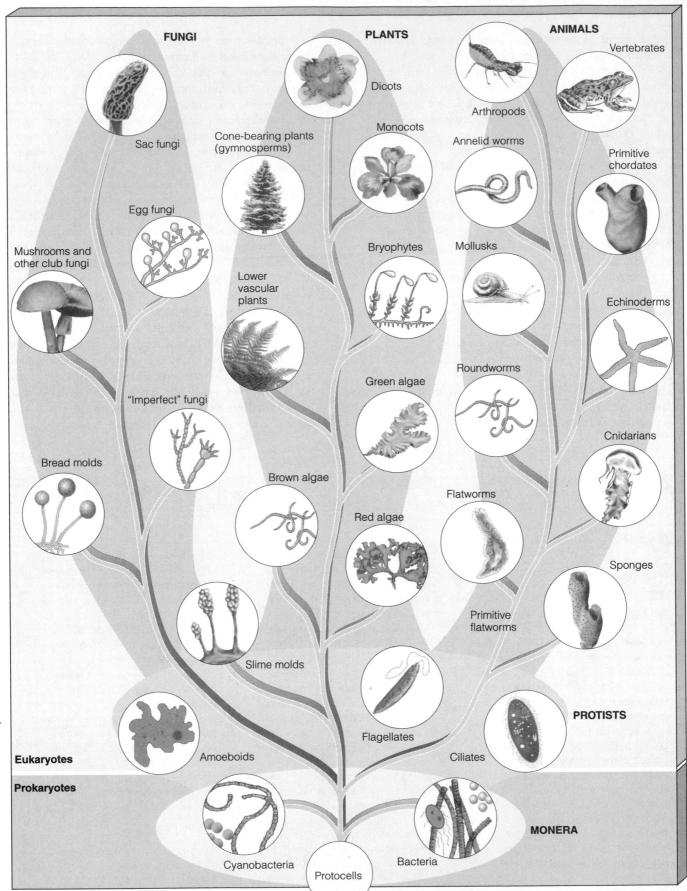

FUNGI   PLANTS   ANIMALS

Sac fungi

Dicots

Vertebrates

Cone-bearing plants
(gymnosperms)

Monocots

Arthropods

Primitive
chordates

Egg fungi

Annelid worms

Mushrooms and
other club fungi

Bryophytes

Mollusks

Lower
vascular
plants

Echinoderms

"Imperfect" fungi

Roundworms

Green algae

Cnidarians

Bread molds

Brown algae

Flatworms

Red algae

Sponges

Primitive
flatworms

Slime molds

Flagellates

PROTISTS

Eukaryotes

Amoeboids

Ciliates

Prokaryotes

MONERA

Cyanobacteria

Protocells

Bacteria

Figure 2-11 Kingdoms of the living world, one of several schemes for classifying the earth's diverse species into major groups. Most biologists believe that protocells gave rise to single-celled prokaryotes and that these in turn evolved into the more complex protists, fungi, plants, and animals that make up the earth's stunning biodiversity.

Q: Can scientists prove anything absolutely?

snakes), birds (eagles and robins), and mammals (bats, elephants, whales, and humans).

What Are the Major Components of Ecosystems?
The ecosphere and its ecosystems can be separated into two parts: **(1) abiotic**, or nonliving, components (water, air, nutrients, and solar energy), and **(2) biotic**, or living, components (plants, animals, and microorganisms, sometimes referred to as *biota*). Figures 2-12 and 2-13 are greatly simplified diagrams of some of the biotic and abiotic components in a freshwater aquatic ecosystem and in a terrestrial ecosystem.

What Are the Major Nonliving Components of Ecosystems?
The nonliving, or abiotic, components of an ecosystem are the physical and chemical factors that influence living organisms. Some important physical factors affecting land (terrestrial) ecosystems are sunlight, temperature, precipitation, wind, latitude (distance from the equator), altitude (distance above sea level), frequency of fire, and nature of the soil. For aquatic ecosystems, major physical factors include water currents, concentrations of dissolved nutrients (such as nitrogen and phosphorus), and the amount of suspended solid material.

Different species thrive under different physical conditions. Some need bright sunlight, others thrive better in shade. Some require a hot environment, others a cool or cold one. Some do best under wet conditions, others under dry conditions.

Each population in an ecosystem has a **range of tolerance** to variations in its physical and chemical environment (Figure 2-14). Individuals within a population may also have slightly different tolerance ranges for temperature or other factors because of small differences in genetic makeup, health, and age. Thus, although a trout population may do best within a narrow band of temperatures (*optimum level or range*), a few individuals can survive above and below that band. As Figure 2-14 shows, tolerance has its limits, beyond which none of the trout can survive.

These observations are summarized in the **law of tolerance**: *The existence, abundance, and distribution of a species in an ecosystem are determined by whether the levels of one or more physical or chemical factors fall within the range tolerated by that species.* In other words, there are minimum and maximum limits for physical conditions (such as temperature) and concentrations of chemical substances, called **tolerance limits**, beyond which no members of a particular species can survive.

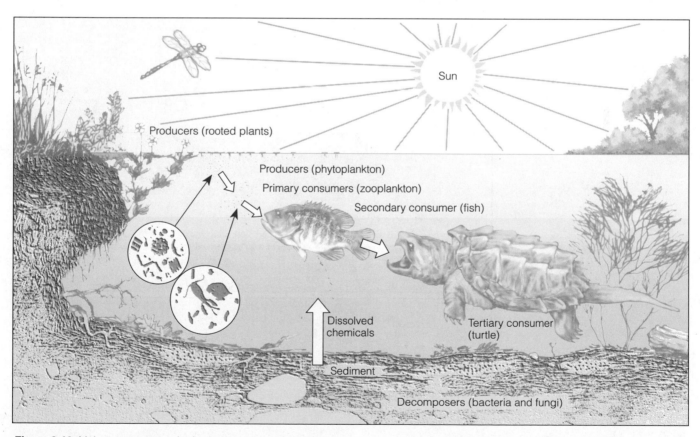

Figure 2-12 Major components of a freshwater aquatic ecosystem.

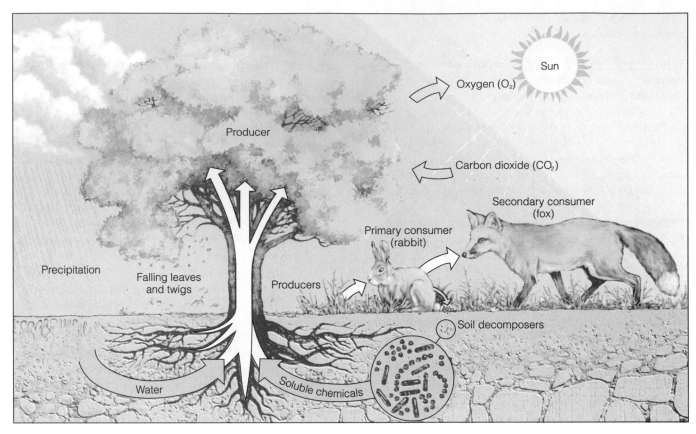

Figure 2-13 Major components of a terrestrial ecosystem.

Figure 2-14 Range of tolerance for a population of organisms to an abiotic environmental factor—in this case, temperature.

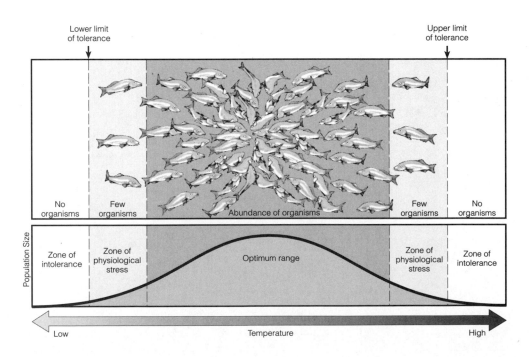

Q: What is the law of conservation of matter?

A species may have a wide range of tolerance to some factors and a narrow range of tolerance to others. Most organisms are least tolerant during juvenile or reproductive stages of their life cycles. Highly tolerant species can live in a variety of habitats with widely different conditions.

Some species can adjust their tolerance to physical or chemical factors if change is gradual, just as you can tolerate a hotter bath by adding hot water slowly. This adjustment to slowly changing new conditions, or **acclimation,** is a useful adaptation. However, acclimation has limits, and as change occurs a species comes closer to its absolute limit. Then, suddenly, without warning, the next small change triggers a **threshold effect,** a harmful or even fatal reaction as the tolerance limit is exceeded. It is like the proverbial straw that breaks an already heavily loaded camel's back.

The threshold effect explains why many environmental problems seem to arise suddenly. For example, when spruce trees suddenly begin dying in large numbers, the cause may be decades of exposure to numerous air pollutants, which can make the trees more vulnerable to drought, cold weather, diseases, and insects. Prevention of pollution and environmental degradation is the best way to keep thresholds from being exceeded.

Although the organisms in a population are affected by a variety of environmental factors, one factor, known as the **limiting factor,** often turns out to be more important than others in regulating population growth. This ecological principle, related to the law of tolerance, is called the **limiting factor principle**: *Too much or too little of any abiotic factor can limit or prevent growth of a population, even if all other factors are at or near the optimum range of tolerance.*

On land, precipitation often is the limiting factor. Lack of water in a desert limits the growth of plants. Soil nutrients can also act as a limiting factor on land. Suppose a farmer plants corn in phosphorus-poor soil. Even if water, nitrogen, potassium, and other nutrients are at optimum levels, the corn will stop growing when it uses up the available phosphorus. Here, the amount of phosphorus in the soil limits corn growth.

Too much of an abiotic factor can also be limiting. For example, plants can be killed by too much water or too much fertilizer, a common mistake of many beginning gardeners.

Important limiting factors for aquatic ecosystems include temperature, sunlight, **dissolved oxygen content** (the amount of oxygen gas dissolved in a given volume of water at a particular temperature and pressure), and availability of nutrients. Another limiting factor in aquatic ecosystems is **salinity** (the amounts of various salts dissolved in a given volume of water).

What Are the Major Living Components of Ecosystems? Living organisms in ecosystems are usually classified as either *producers* or *consumers,* based on how they get food. **Producers,** sometimes called **autotrophs** (self-feeders), make their own food from compounds obtained from their environment. On land, most producers are green plants. In freshwater and marine ecosystems, algae and plants are the major producers near shorelines; in open water the dominant producers floating and drifting are *phytoplankton,* most of them microscopic. Only producers make their own food; all other organisms are consumers, which depend directly or indirectly on food provided by producers.

Most producers use sunlight to make complex compounds (such as glucose, $C_6H_{12}O_6$) by **photosynthesis**. Although a sequence of hundreds of chemical changes takes place during photosynthesis, the overall reaction can be summarized as follows:

$$\text{carbon dioxide + water + } \textbf{solar energy} \longrightarrow \text{glucose + oxygen}$$

A few producers, mostly specialized bacteria, can convert simple compounds from their environment into more complex nutrient compounds without sunlight, a process called **chemosynthesis**.

All other organisms in an ecosystem are **consumers** or **heterotrophs** ("other-feeders"), which get their energy and nutrients by feeding on other organisms or their remains. There are several classes of consumers, depending on their primary source of food. **Herbivores** (plant eaters) are called **primary consumers** because they feed directly on producers. **Carnivores** (meat eaters) feed on other consumers; those called **secondary consumers** feed only on primary consumers (herbivores). Most secondary consumers are animals, but a few (such as the Venus flytrap plant) trap and digest insects. **Tertiary (higher-level) consumers** feed only on other carnivores. **Omnivores** are consumers that eat both plants and animals; examples are pigs, rats, foxes, bears, cockroaches, and humans.

Other consumers, called **scavengers,** feed on dead organisms that were either killed by other organisms or died naturally. Vultures, flies, crows, hyenas, and some species of sharks, beetles, and ants are examples of scavengers. **Detritivores** (detritus feeders and decomposers) live off **detritus** (pronounced di-TRI-tus), parts of dead organisms and cast-off fragments and wastes of living organisms (Figure 2-15). **Detritus feeders,** such as crabs, carpenter ants, termites, earthworms, and wood beetles, extract nutrients from partly decomposed organic matter in leaf litter, plant debris, and animal dung.

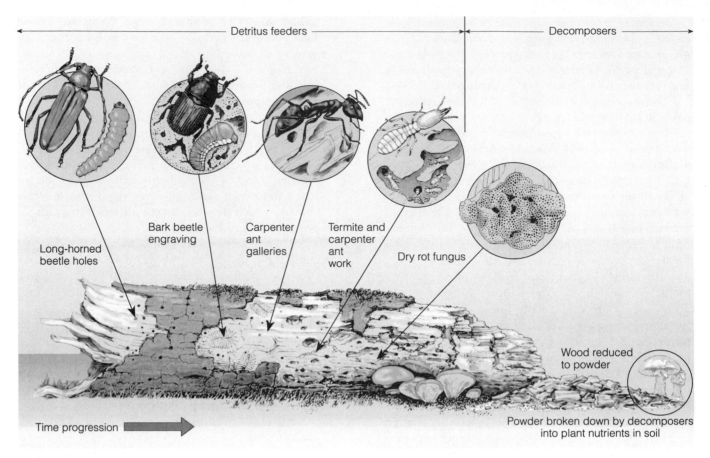

Detritus feeders ——————————————————————— Decomposers ——————

Long-horned
beetle holes

Bark beetle
engraving

Carpenter
ant
galleries

Termite and
carpenter
ant
work

Dry rot fungus

Wood reduced
to powder

Time progression

Powder broken down by decomposers
into plant nutrients in soil

Figure 2-15 Some detritivores, called *detritus feeders*, directly consume tiny fragments of this log. Other detritivores, called *decomposers* (mostly fungi and bacteria), digest complex organic chemicals in fragments of the log into simpler inorganic nutrients. If these nutrients are not washed away or otherwise removed from the system, they can be used again by producers near this location.

Figure 2-16 The main structural components (energy, chemicals, and organisms) of an ecosystem are linked by matter recycling and the one way flow of high-quality energy from the sun, through organisms, and then into the environment as low-quality heat. Each type of organism in an ecosystem plays a unique role in the processes of energy flow and matter cycling.

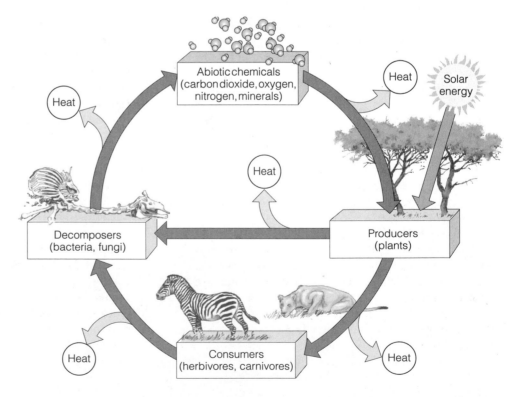

Abiotic chemicals
(carbon dioxide, oxygen,
nitrogen, minerals)

Heat

Solar
energy

Heat

Heat

Decomposers
(bacteria, fungi)

Producers
(plants)

Heat

Consumers
(herbivores, carnivores)

Heat

Q: What is the first law of energy or thermodynamics?

Decomposers, mostly certain types of bacteria and fungi, are consumers that recycle organic matter in ecosystems by breaking down dead organic material (detritus) to get nutrients and releasing the resulting simpler compounds into the soil and water, where they can be taken up as nutrients by producers. In turn, decomposers are important food sources for worms and insects living in the soil and water. When we say that something is **biodegradable**, we mean that it can be broken down by decomposers.

Both producers and consumers use the chemical energy stored in glucose and other compounds to drive their life processes. In most cells, this energy is released by the process of **aerobic respiration**, which uses oxygen to convert organic nutrients back into carbon dioxide and water. The net effect of the hundreds of steps in this complex process is represented by the following reaction:

$$\text{glucose} + \text{oxygen} \longrightarrow \text{carbon dioxide} + \text{water} + \textbf{energy}$$

Although the detailed steps differ, the net chemical change for aerobic respiration is the opposite of that for photosynthesis.

The survival of any individual organism depends on the *flow of matter and energy* through its body. However, an ecosystem as a whole survives primarily through a combination of *matter recycling* (rather than one-way flow) and *one-way energy flow* (Figure 2-16). Decomposers complete the cycle of matter by breaking down detritus into inorganic nutrients that are usable by producers. Without decomposers, the entire world would soon be knee-deep in plant litter, dead animal bodies, animal wastes, and garbage. Most life as we know it would no longer exist.

2-4 CONNECTIONS: ROLES OF SPECIES IN ECOSYSTEMS

What Role Does a Species Play in an Ecosystem? The **ecological niche**, or simply **niche** (pronounced nitch), of a species is its way of life or functional role—everything it does to survive and reproduce—in an ecosystem. This includes **(1)** the range of tolerance for various physical and chemical conditions, such as temperature or water availability (Figure 2-14); **(2)** the types of resources it uses, such as food or nutrients; **(3)** how it interacts with other living and nonliving components of the ecosystems in which it is found, such as what it eats or is eaten by; and **(4)** the role it plays in the flow of energy and cycling of matter in an ecosystem.

Is It Better to Be a Specialist or a Generalist Species? Species can be broadly classified as specialists or generalists, according to their niches. **General-**

Cockroaches: Nature's Ultimate Survivors

Cockroaches, the bugs many people love to hate, have been around for about 350 million years and are one of the great success stories of evolution. The major reason they are so successful is that they are *generalists*, able to eat almost anything (including algae, dead insects, salts in tennis shoes, electrical cords, glue, paper, soap, and—when times are bad—other, weaker cockroaches). The 4,000 known cockroach species can live and breed almost anywhere except polar regions.

Some species can go for months without food, last a month without water, and withstand massive doses of radiation. One species can survive being frozen for 48 hours. The antennae of most cockroach species (which can detect minute movements of air), the vibration sensors in their knee joints, and their lightning-fast response times (faster than you can blink) allow them to evade predators and a human foot in hot pursuit. Some even have wings.

High reproductive rates also aid the survival of cockroaches. In only a year, a single Asian cockroach (especially prevalent in Florida) and its young can add about 10 million new cockroaches to the world. Their high reproductive rate also helps them quickly develop genetic resistance to almost any poison we throw at them. Most cockroaches also sample food before it enters their mouths and learn to shun foul-tasting poisons.

Only about 25 species of cockroach live in homes, but such species can carry viruses and bacteria that cause such diseases as hepatitis, polio, typhoid fever, plague, and salmonella. They can cause people to have allergic reactions ranging from watery eyes to severe wheezing. Indeed, about 60% of the 12 million Americans suffering from asthma are allergic to dead or live cockroaches.

Critical Thinking

How do you feel about cockroaches? If you could, would you exterminate them? What might be some ecological consequences of doing this?

ist species have broad niches: They can live in many different places, eat a variety of foods, and tolerate a wide range of environmental conditions. Flies, cockroaches (Spotlight, above), mice, rats, white-tailed deer, black bears, raccoons, coyotes, channel catfish, and humans are all generalist species.

Specialist species have narrow niches: They may be able to live in only one type of habitat, tolerate only a narrow range of climatic and other environmental conditions, or use only one or a few types of food, which makes them more prone to becoming endangered when environmental conditions change. Examples of specialists are *tiger salamanders*, which can breed only in fishless ponds so their larvae won't be eaten; *red-cockaded woodpeckers*, which carve nest-holes almost exclusively in old (at least 75 years) longleaf pines; *spotted owls*, which require old-growth forests in the Pacific Northwest for food and shelter; and the endangered giant panda (about 800 left in the wild in China), which gets 99% of its food from bamboo.

In a tropical rain forest, an incredibly diverse array of species survives by occupying specialized ecological niches in various distinct layers of vegetation exposed to different levels of light (Figure 2-17). The widespread clearing and degradation of such forests is dooming millions of such specialized species to extinction.

Is it better then to be a generalist than a specialist? It depends. When environmental conditions are fairly constant, as in a tropical rain forest, specialists have an advantage because they have fewer competitors. When environments are changing rapidly, however, the adaptable generalist is usually better off than the specialist.

How Can We Classify the Roles Various Species Play in Ecosystems? When examining ecosystems, ecologists often apply particular labels—such as *native*, *nonnative*, *indicator*, or *keystone*—to various species to clarify some of the ecological roles they play. Any given species may function as more than one of these four types in a particular ecosystem.

Species that normally live and thrive in a particular ecosystem are known as **native species**. Others that migrate into an ecosystem or are deliberately or accidentally introduced into an ecosystem by humans are called **nonnative species**, **exotic species**, or **alien species**. Some of these introduced species (such as crop species and game for sport hunting) are beneficial to humans, but some thrive and crowd out many native species.

Species that serve as early warnings that a community or an ecosystem is being damaged are called **indicator species**. Birds are excellent biological indicators because they are found almost everywhere and respond quickly to environmental change. Research indicates that a major factor in the current decline of migratory, insect-eating songbirds in North America is loss or fragmentation of habitat. The tropical forests of Latin America and the Caribbean that are the birds' winter habitats are rapidly disappearing. Their summer habitats in North America are also disappearing or are being fragmented into patches that make the birds more vulnerable to attack by predators and parasites.

Some ecologists call species whose roles in an ecosystem are much more important than their abundance suggests **keystone species**, although this desig-

Figure 2-17 Stratification of specialized plant and animal niches in various layers of a tropical rain forest. The presence of these specialized niches enables species to avoid or minimize competition for resources and results in the coexistence of a great variety of species (biodiversity).

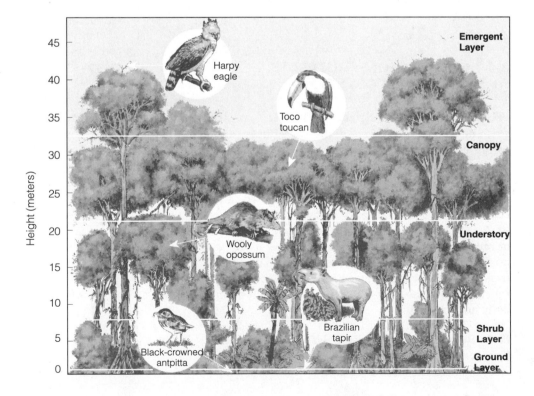

Q: What is the second law of energy or thermodynamics?

nation is controversial.* In tropical forests, various species of bees, bats, ants, and hummingbirds play keystone roles by pollinating flowering plants or dispersing seed. Some keystone species—including the wolf, leopard, lion, giant anteater, and great white shark (Case Study, p. 38)—are top predators that exert a stabilizing effect on their ecosystems by feeding on and regulating the populations of certain species.

The loss of a keystone species can lead to population crashes and extinctions of other species that depend on it for certain services, a ripple or domino effect that spreads throughout an ecosystem. According to biologist Edward O. Wilson, "The loss of a keystone species is like a drill accidentally striking a power line. It causes lights to go out all over."

Connections: Why Should We Care About the American Alligator? The American alligator, North America's largest reptile, has no natural predators except humans. Hunters once killed large numbers of these animals for their exotic meat and their supple belly skin, used to make shoes, belts, and pocketbooks.

Other people considered alligators to be useless, dangerous vermin and hunted them for sport or out of hatred. Between 1950 and 1960, hunters wiped out 90% of the alligators in Louisiana, and by the 1960s the alligator population in the Florida Everglades was also near extinction.

People who say "So what?" are overlooking the alligator's important ecological roles in subtropical wetland ecosystems. Alligators dig deep depressions, or gator holes, that collect fresh water during dry spells, serve as refuges for aquatic life, and supply fresh water and food for many animals. Large alligator nesting mounds provide nesting and feeding sites for species of herons and egrets. Alligators also eat large numbers of predatory gar fish and thus help maintain populations of game fish such as bass and bream.

As alligators move from gator holes to nesting mounds, they help keep areas of open water free of invading vegetation. Without these ecosystem services, freshwater ponds and coastal wetlands found in the alligator's habitat would be filled in by shrubs and trees, and dozens of species would disappear. Some ecologists classify the North American alligator as a *keystone species* because of its important ecological roles in helping maintain the structure, function, and sustainability of its natural ecosystems.

---

*All species play some role in their ecosystems and thus are important. Whereas some scientists consider all species equally important, others consider certain species to be more important than others, at least in helping maintain the ecosystems they are a part of.

In 1967 the U.S. government placed the American alligator on the endangered species list. Protected from hunters, the alligator population made a strong comeback in many areas by 1975—too strong, according both to those who find alligators in their backyards and swimming pools and to duck hunters, whose retriever dogs are sometimes eaten by alligators.

In 1977, the U.S. Fish and Wildlife Service reclassified the American alligator from an *endangered* to a *threatened* species in Florida, Louisiana, and Texas, where 90% of the animals live. In 1987 this reclassification was extended to seven other states.

Alligators now number perhaps 3 million, most in Florida and Louisiana. It is generally illegal to kill members of a threatened species, but limited kills by licensed hunters are allowed in some areas of Florida, Louisiana, and South Carolina to control the population. The comeback of the American alligator is an important success story in wildlife conservation.

The increased demand for alligator meat and hides has created a booming business in alligator farms, especially in Florida. By controlling diet and other conditions, alligator farm operators have quadrupled the species's reproductive rate, doubled its growth rate, and reduced mortality from 35% to 1%. Such success reduces the need for illegal hunting of wild alligators.

How Do Species Interact? When different species in an ecosystem have activities or resource requirements in common, they may interact with one another. The principal types of species interactions are *interspecific competition, predation, parasitism, mutualism,* and *commensalism.*

As long as commonly used resources are abundant, different species can share them, allowing each species to come closer to occupying its **fundamental niche**: the full potential range of physical, chemical, and biological conditions and resources it could theoretically use if there were no competition from other species.

However, most species face competition from other species for one or more limited resources (such as food, sunlight, water, soil nutrients, space, nest sites, and good places to hide). Because of such **interspecific competition** (competition among species), parts of the fundamental niches of different species overlap. As a result, a species usually occupies only part of its fundamental niche in a particular community or ecosystem—what ecologists call its **realized niche**. By analogy, you may be capable of being president of a particular company (your *fundamental professional niche*), but competition from others may mean that you may become only a vice president (your *realized professional niche*).

Over a long time scale species that compete for the same resources may evolve adaptations that reduce or avoid competition or overlap of their fundamental niches. One way this happens is through **resource partitioning,**

The world's 350 shark species range in size from the dwarf dog shark, about the size of a large goldfish, to the whale shark, the world's largest fish at 18 meters (60 feet) long. Various shark species, feeding at the top of food webs, cull injured and sick animals from the ocean and thus play an important ecological role. Without such shark species the oceans would be overcrowded with dead and dying fish.

Many of us, influenced by movies and popular novels, think of sharks as people-eating monsters. But the two largest species—the whale shark and the basking shark—sustain their enormous bulk by filtering out and swallowing huge quantities of *plankton* (small free-floating sea creatures).

Every year, members of a few species of shark—mostly great white, bull, tiger, gray reef, lemon, and blue—injure about 100 people worldwide and kill between 5 and 10. Most attacks are by great white sharks, which feed on sea lions and other marine mammals and sometimes mistake divers in wet suits and swimmers on surfboards for their usual prey. A typical ocean-goer is 150 times more likely to be killed by lightning than by a shark.

For every shark that injures a person, we kill 500,000 to 1 million sharks, for a total of 50 to 100 million sharks each year. Sharks are killed mostly for their fins, widely used in Asia as a soup ingredient and as a pharmaceutical cure-all.

Sharks are also killed for their livers, meat (especially mako and thresher), and jaws (especially great whites), or just because we fear them. Some sharks (especially blue, mako, and oceanic whitetip) die when they are trapped in nets deployed to catch swordfish, tuna, shrimp, and other commercially important species.

Sharks also help save human lives. In addition to providing people with food, they are helping us learn how to fight cancer (which sharks almost never get), bacteria, and viruses. Their highly effective immune system is being studied because it allows wounds to heal without becoming infected.

Sharks have several natural traits that make them prone to population declines from overfishing. Unlike most other fish, they have only a few offspring (between 2 and 10) once every year or two. Depending on the species, sharks require 10 to 15 years (and in some cases, 24 years) to reach sexual maturity and begin reproducing. Sharks also have long gestation (pregnancy) periods—as much as 24 months for some species.

With more than 400 million years of evolution behind them, sharks have had a long time to get things right. Preserving this evolutionary genetic wisdom begins with the recognition that sharks don't need us, but we and other species need them.

Critical Thinking

Do you fear sharks? Why? How do you feel about killing sharks in large numbers? Why?

---

the dividing up of scarce resources so that species with similar requirements use them at different times, in different ways, or in different places (Figure 2-17). In effect, they evolve traits that allow them to share the wealth, with each competing species occupying a realized niche that makes up only part of its fundamental niche.

For example, where lions and leopards live in the same area, lions take mostly larger animals as prey; leopards take smaller ones. Hawks and owls feed on similar prey, but hawks hunt during the day; owls hunt at night. Some species of birds, such as warblers and tanagers, avoid competition by hunting for insects in different parts of the same coniferous trees.

Research has shown that two species with identical fundamental niches cannot coexist indefinitely in an ecosystem in which there is not enough of a particular resource to meet the needs of both species. This finding is called the **competitive exclusion principle**. As a result, one of the competing species must either migrate to another area (if possible), shift its feeding habits or behavior, suffer a sharp population decline, or become extinct in that area.

The most obvious form of species interaction is **predation**, in which members of a *predator* species feed on all or parts of members of a *prey* species, but they do not live on or in the prey. Together, the two kinds of organisms, such as lions (the predator) and zebras (the prey), are said to have a **predator–prey relationship,** as depicted in Figures 2-12 and 2-13.

Some people tend to view predators with contempt. When a hawk tries to capture and feed on a rabbit, some tend to root for the rabbit. Yet the hawk (like all predators) is merely trying to get enough food to feed itself and its young; in the process, it is merely doing what it is genetically programmed to do. When people hunt or fish or pick and eat vegetation, they act as predators. When we buy meat or vegetables from a grocery store we are indirect predators who have someone else kill, skin, and cut up cattle, chickens, or other sources of meat for us and pick and process vegetables and other plants for us.

Another interaction that in some ways resembles predation is **parasitism**, an interaction in which a member of one species (the *parasite*) obtains its nourishment by living on, in, or near a member of another species (its *host*) over an extended time. Unlike a conventional predator, a parasite **(1)** is usually smaller than its host (prey); **(2)** remains closely associated with, draws nourishment from, and may gradually weaken its host over time; and **(3)** rarely kills its host.

Tapeworms, disease-causing microorganisms (pathogens), and other parasites live *inside* their hosts. Other parasites, such as ticks, fleas, mosquitoes, mistletoe plants, and fungi (that cause diseases such as athlete's foot), attach themselves to the *outside* of their hosts. Some parasites move from one host to another, as fleas and ticks do; others, such as tapeworms, spend their adult lives with a single host.

From the host's point of view parasites are harmful, but parasites play important ecological roles. Collectively, the incredibly complex matrix of parasitic relationships in an ecosystem acts somewhat like a glue that helps hold the species in an ecosystem together. Parasites living within their hosts also help dampen drastic swings in population sizes and parasites promote biodiversity by helping prevent some organisms from becoming too plentiful.

Mutualism is a type of species interaction in which both participating species generally benefit. The honeybee and certain flowers have a mutualistic relationship: The honeybee feeds on a flower's nectar and in the process picks up pollen, pollinating female flowers when it feeds on them. Other important mutualistic relationships exist between animals and the vast armies of bacteria in their stomachs or intestines that break down (digest) their food. The bacteria gain a safe home with a steady food supply; the animal gains access to a large source of energy. In addition to helping digest your food, bacteria in your intestines synthesize vitamin K and the B-complex vitamins, which you can't make.

Research indicates that mutualism is more common when resources become scarce. In other words, when the going gets tough, the tough often survive by evolving mutually beneficial relationships with other species. It is tempting to think of mutualism as an example of cooperation between species, but it actually involves each species benefiting by exploiting the other.

In another type of species interaction, called **commensalism**, one species benefits while the other is neither helped nor harmed to any significant degree.

An example is the intimate relationship in tropical waters between various species of clownfish and sea anemones, marine animals with stinging tentacles that paralyze most fish that touch them. The clownfish are not harmed by sea anemones. They gain protection by living among the deadly tentacles, and they feed on the detritus left from the meals of the anemone. The sea anemones seem to neither benefit nor suffer harm from this relationship.

On land there are commensalistic relationships between various trees and *epiphytes* (such as various types of orchids and bromeliads) that attach themselves to the trunks or branches of large trees. These so-called air plants benefit by living in an elevated spot that gives them better access to sunlight. Their position in the tree allows them to get most of their water from the humid air and from rain collecting in their usually cupped leaves. They also absorb nutrient salts falling from the tree's upper leaves and limbs and from the dust in rainwater. Epiphytes benefit by obtaining water and nutrients from air or bark surfaces without penetrating or harming the tree.

2-5 CONNECTIONS: ENERGY FLOW IN ECOSYSTEMS

How Does Energy Flow Through Ecosystems? Food Chains and Food Webs All organisms, whether dead or alive, are potential sources of food for other organisms. A caterpillar eats a leaf, a robin eats the caterpillar, and a hawk eats the robin. When leaf, caterpillar, robin, and hawk have all died, they in turn are consumed by decomposers. As a result, *there is little waste in natural ecosystems.*

The sequence of organisms, each of which is a source of food for the next, is called a **food chain**. It determines how energy and nutrients move from one organism to another through the ecosystem (Figure 2-18). Ecologists assign each of the organisms in an ecosystem to a *feeding level*, or **trophic level** (from the Greek word *trophos*, "nourishment"), depending on whether it is a producer or a consumer and on what it eats or decomposes. Producers belong to the first trophic level, primary consumers to the second trophic level, secondary consumers to the third, and so on. Detritivores process detritus from all trophic levels.

Real ecosystems are more complex than this. Most consumers feed on more than one type of organism, and most organisms are eaten by more than one type of consumer. Because most species participate

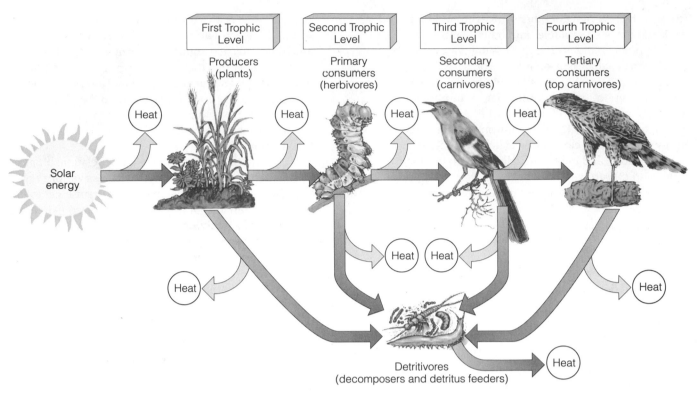

Figure 2-18 A food chain. The arrows show how chemical energy in food flows through various *trophic levels* or energy transfers; most of the energy is degraded to heat in accordance with the second law of energy. Food chains rarely have more than four trophic levels.

in several different food chains, the organisms in most ecosystems form a complex network of inter-connected food chains called a **food web** (Figure 2-19). Trophic levels can be assigned in food webs just as in food chains. Each trophic level in a food chain or web contains a certain amount of **biomass**, the combined dry weight of all organic matter contained in its organisms.

In a food chain or web, energy stored in biomass is transferred from one trophic level to another, with some usable energy degraded and lost to the environment as low-quality heat in each transfer. Thus, only a small portion of what is eaten and digested is actually converted into an organism's bodily material or biomass, and the amount of usable energy available to each successive trophic level declines.

The percentage of usable energy transferred from one trophic level to the next varies from 5% to 20% (that is, a loss of 80–95%), depending on the types of species and the ecosystem involved. The percentage of energy transferred from one trophic level to another is called **ecological efficiency**. Assuming 10% ecological efficiency (90% loss) at each trophic transfer, if green plants in an area manage to capture 10,000 units of energy

from the sun, then only about 1,000 units of energy are available to support herbivores and only about 100 units to support carnivores.

The more trophic levels or steps in a food chain or web, the greater the cumulative loss of usable energy as energy flows through the various trophic levels. The **pyramid of energy flow** in Figure 2-20 illustrates this energy loss for a simple food chain, assuming a 90% energy loss with each transfer. Figure 2-21 shows the actual pyramid of energy flow during one year for an aquatic ecosystem in Silver Springs, Florida.

Energy flow pyramids explain why the earth can support more people if they eat at lower trophic levels by consuming grains, vegetables, and fruits directly (for example, grain ⟶ human) rather than passing such crops through another trophic level and eating grain-eaters (grain ⟶ steer ⟶ human).

How Rapidly Do Producers in Different Ecosystems Produce Biomass? The *rate* at which an ecosystem's producers convert solar energy into chemical energy as biomass is the ecosystem's **gross primary productivity (GPP)**. However, to stay alive, grow, and reproduce, an ecosystem's producers must

Q: What are the three components of biodiversity?

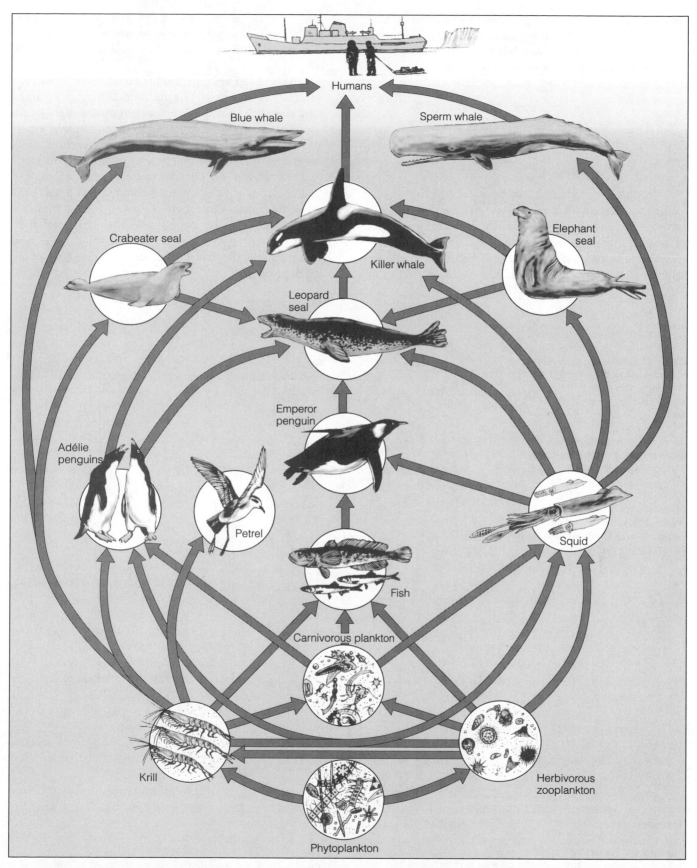

Figure 2-19 Greatly simplified food web in the Antarctic. Many more participants in the web, including an array of decomposer organisms, are not depicted here.

use some of the total biomass they produce for their own respiration. Only what is left, called **net primary productivity (NPP)**, is available for use as food by other organisms (consumers) in an ecosystem:

Net primary productivity	=	Rate at which producers store chemical energy as biomass (produced by photosynthesis)	−	Rate at which producers use chemical energy stored as biomass (through aerobic respiration)

Net primary productivity is the *rate* at which energy for use by consumers is stored in new biomass (cells, leaves, roots, and stems). It is usually measured in units of the energy or biomass available to consumers in a specified area over a given time. It is measured in kilocalories per square meter per year (kcal/m²/yr) or grams of biomass created per square meter per year (g/m²/yr). *The earth's total net primary productivity is the upper limit determining the planet's carrying capacity for all species.*

Various ecosystems and life zones differ in their net primary productivity (Figure 2-22).Estuaries, swamps and marshes, and tropical rain forests are highly productive; open ocean, tundra (arctic and alpine grasslands), and desert are the least productive. It is tempting to conclude that to feed our hungry millions we should harvest plants in estuaries, swamps, and marshes, or clear tropical forests and plant crops. However, the grasses in estuaries, swamps, and marshes cannot be eaten by people and they are vital food sources (and spawning areas) for fish, shrimp, and other aquatic life that provide us and other consumers with protein.

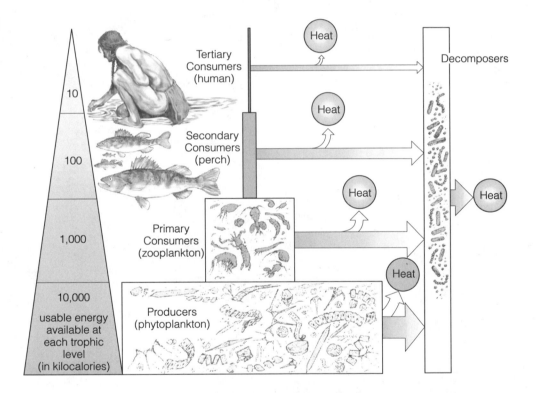

Figure 2-20 Generalized pyramid of energy flow, showing the decrease in usable energy available at each succeeding trophic level in a food chain or web. This conceptual model assumes a 10% ecological efficiency (90% loss in usable energy to the environment, in the form of low-quality heat) with each transfer from one trophic level to another. In nature, ecological efficiency varies from 5% to 20%. Because of the degradation of energy quality required by the second law of energy, these models always have a pyramidal shape.

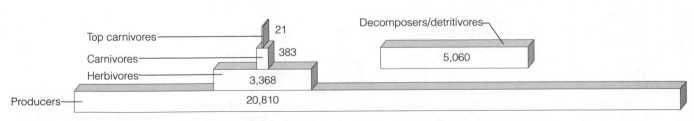

Figure 2-21 Annual pyramid of energy flow (in kilocalories per square meter per year) for an aquatic ecosystem in Silver Springs, Florida. (Used by permission from Cecie Starr and Ralph Taggart, *Biology: The Unity and Diversity of Life*, 6th ed., Belmont, Calif.: Wadsworth, 1992)

Wagener, S. M., M. W. Oswood, and J. P. Schimel. "Rivers and Soils." 1998. *BioScience*, vol. 48, no. 2, 104(5).

In tropical forests, most nutrients are stored in the vegetation rather than in the soil. When the trees are removed, the nutrient-poor soils are rapidly depleted of their nutrients by frequent rains and growing crops. Crops can be grown only for a short time without massive and expensive applications of commercial fertilizers. This explains why many ecologists urge us to protect, not clear, large areas of tropical forest to supply food.

Agricultural land is a highly modified ecosystem in which we try to increase the net primary productivity and biomass of selected crop plants by adding water (irrigation) and nutrients (fertilizers, usually containing nitrogen and phosphorus because they are most often the nutrients limiting crop growth). Despite such inputs, the net primary productivity of agricultural land is not particularly high compared to that of other ecosystems (Figure 2-22).

Ecologists have estimated that humans now use, waste, or destroy about 27% of the earth's total potential net primary productivity and 40% of the net primary productivity of the planet's terrestrial ecosystems. This is the main reason why we are crowding out or eliminating the habitats and food supplies for a growing number of other species. If current estimates of our use of the earth's annual net primary productivity are reasonably correct, what will happen to us and to other species if the human population doubles over the next 40–50 years and per capita consumption rises sharply?

2-6 CONNECTIONS: MATTER CYCLING IN ECOSYSTEMS

How Do Nutrient Cycles Sustain Life? Any atom, ion, or molecule an organism needs to live, grow, or reproduce is called a **nutrient**. Some elements (such as carbon, oxygen, hydrogen, nitrogen, phosphorus, sulfur, potassium, calcium, magnesium, and iron) are needed in fairly large amounts, whereas others (such as sodium, zinc, copper, chlorine, and iodine) are needed in small or even trace amounts.

These nutrient atoms, ions, and molecules are continuously cycled from the nonliving environment (air, water, soil, and rock) to living organisms (biota) and then back again in what are called **nutrient cycles**, or **biogeochemical cycles** (literally, life–earth–chemical cycles). These cycles, driven directly or indirectly by incoming solar energy and gravity, include the carbon, oxygen, nitrogen, phosphorus, and hydrologic (water) cycles (Figure 2-9).

The earth's chemical cycles also connect past, present, and future forms of life. Some of the carbon atoms in your skin may once have been part of a leaf, a dinosaur's skin, or a layer of limestone rock. Some of the oxygen molecules you just inhaled may have been inhaled by your grandmother, Plato, or a hunter–gatherer who lived 25,000 years ago.

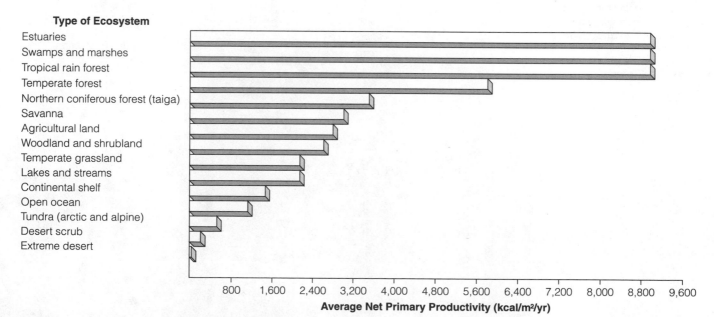

Figure 2-22 Estimated annual average net primary productivity per unit of area in major life zones and ecosystems, expressed as kilocalories of energy produced per square meter per year (kcal/m²/yr). (Data from R. H. Whittaker, *Communities and Ecosystems*, 2d ed., New York: Macmillan, 1975)

Hint: Enter the search term *stream ecology* using the Subject Guide.

How Is Carbon Cycled in the Ecosphere? The **carbon cycle** is based on carbon dioxide gas, which makes up only 0.036% of the volume of the troposphere and is also dissolved in water. As a heat-trapping gas, carbon dioxide is a key component of nature's thermostat. If the carbon cycle removes too much CO_2 from the atmosphere, the earth will cool; if the cycle generates too much, the earth will get warmer. Thus, even slight changes in the carbon cycle can affect climate and ultimately the types of life that can exist on various parts of the planet.

Terrestrial producers remove CO_2 from the atmosphere and aquatic producers remove it from the water. They then use photosynthesis to convert CO_2 into complex carbohydrates such as glucose ($C_6H_{12}O_6$). The cells in oxygen-consuming producers, consumers, and decomposers then carry out aerobic respiration, which breaks down glucose and other complex organic compounds and converts the carbon back to CO_2 in the atmosphere or water, for reuse by producers. This linkage between photosynthesis in producers and aerobic respiration in producers, consumers, and decomposers

circulates carbon in the ecosphere and is a major part of the global carbon cycle (Figure 2-23). Oxygen and hydrogen, the other elements in carbohydrates, cycle almost in step with carbon.

Since 1800 and especially since 1950, as world population and resource use have soared, humans have disturbed the carbon cycle in two ways that add more carbon dioxide to the atmosphere than oceans and plants have been able to remove: **(1)** Forest and brush removal has left less vegetation to absorb CO_2 through photosynthesis and **(2)** burning fossil fuels and wood produces CO_2 that flows into the atmosphere.

Computer models of the earth's climate systems suggest that increased concentration of CO_2 (and of other heat-trapping gases we're adding to the atmosphere) could enhance the planet's natural greenhouse effect. This could in turn alter climate patterns for hundreds to thousands of years as the carbon cycle adjusts to these rapid inputs. It could also disrupt global food production and wildlife habitats and possibly raise the average sea level, as discussed in more detail in Chapter 9.

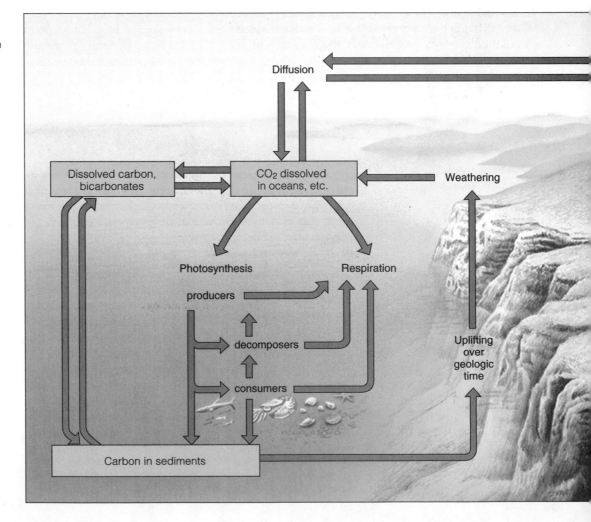

Figure 2-23 Simplified conceptual model of the global carbon cycle. The left portion shows the movement of carbon through marine ecosystems and the right portion shows its movement through terrestrial ecosystems. Carbon reservoirs are shown as boxes. (Modified by permission from Cecie Starr and Ralph Taggart, *Biology: The Unity and Diversity of Life*, 7th ed., Belmont, Calif.: Wadsworth, 1995)

Q: What are the three most productive types of ecosystems?

How Is Nitrogen Cycled in the Ecosphere? Bacteria in Action Although chemically unreactive nitrogen gas (N_2) makes up 78% of the volume of the troposphere, it cannot be absorbed and used directly as a nutrient by multicellular plants or animals. Fortunately, lightning and certain bacteria convert nitrogen gas into compounds that can enter food webs as part of the **nitrogen cycle** (Figure 2-24).

In the first step in the nitrogen cycle, called *nitrogen fixation*, specialized bacteria convert gaseous nitrogen (N_2) to ammonia (NH_3) that can be used by plants. This is done mostly by cyanobacteria in soil and water and by *Rhizobium* bacteria living in small nodules (swellings) on the root systems of a wide variety of plant species, including soybeans, alfalfa, and clover. In a two-step process called *nitrification*, most of the ammonia in soil is converted by specialized aerobic bacteria to nitrite ions (NO_2^-), which are toxic to plants, and then to nitrate ions (NO_3^-), which are easily taken up by plants as a nutrient.

After nitrogen has served its purpose in living organisms, vast armies of specialized decomposer bacteria convert the complex nitrogen-rich compounds, wastes, cast-off particles, and dead bodies of organisms into simpler nitrogen-containing compounds. This process is known as *ammonification*. In a process called *denitrification*, other specialized bacteria then convert these forms of nitrogen back into nitrogen gas, which is released to the atmosphere to begin the cycle again.

Humans intervene in the nitrogen cycle in several ways. *First*, we emit large quantities of nitric oxide (NO) into the atmosphere when we burn any fuel. (Most of this NO is produced when nitrogen and oxygen molecules in the air combine at high temperatures.) In the atmosphere, this nitric oxide combines with oxygen to form nitrogen dioxide (NO_2) gas, which can react with water vapor to form nitric acid (HNO_3). Droplets of HNO_3 dissolved in rain or snow are components of *acid deposition*, which is commonly called *acid rain*. Nitric acid, along with other air pollutants, can damage and weaken trees, upset aquatic ecosystems, corrode metals, and damage marble, stone, and other types of building materials.

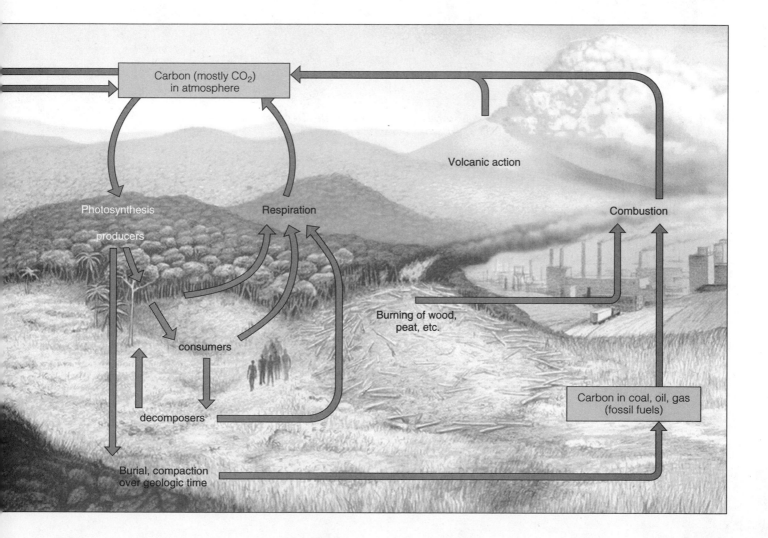

A: Estuaries, swamps and marshes, and tropical rain forests

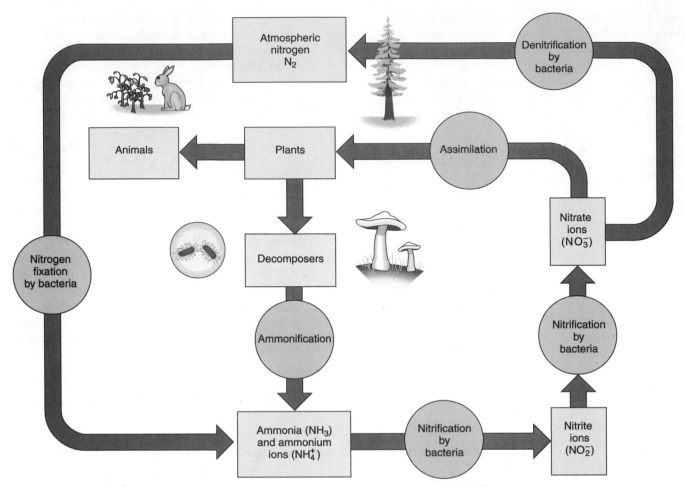

Figure 2-24 Simplified conceptual model of the nitrogen cycle in a terrestrial ecosystem. Nitrogen reservoirs are shown as boxes, and processes changing one form of nitrogen to another are shown in circles.

Second, human activities emit heat-trapping nitrous oxide gas (N_2O) into the atmosphere through the action of anaerobic bacteria on livestock wastes and commercial inorganic fertilizers applied to the soil. When N_2O reaches the stratosphere it contributes to depletion of the earth's ozone shield that filters out harmful ultraviolet radiation from the sun.

Third, we remove nitrogen from the earth's crust when we mine nitrogen-containing mineral deposits (mostly nitrates) for fertilizers, deplete nitrogen from topsoil by harvesting nitrogen-rich crops, and leach water-soluble nitrate ions from soil through irrigation.

Fourth, we remove nitrogen from topsoil when we burn grasslands and clear forests before planting crops. At the same time, we emit nitrogen oxides into the atmosphere.

Fifth, we add excess nitrogen compounds to aquatic ecosystems in agricultural runoff and discharge of municipal sewage. This excess of plant nutrients stimulates rapid growth of photosynthesizing algae and other aquatic plants. The subsequent breakdown of dead algae by aerobic decomposers can deplete the water of dissolved oxygen and can disrupt aquatic ecosystems by killing some types of fish and other oxygen-requiring (aerobic) organisms.

How Is Phosphorus Cycled in the Ecosphere?
Phosphorus circulates through water, Earth's crust, and living organisms in the **phosphorus cycle** (Figure 2-25). In this cycle, phosphorus moves slowly from phosphate deposits on land and in shallow ocean sediments to living organisms, and then much more slowly back to the land and ocean. Bacteria are less important here than in the nitrogen cycle. Unlike carbon and nitrogen, very little phosphorus circulates in the atmosphere because at the earth's normal temperatures and pressures, phosphorus and its compounds are not gases.

Because most soils contain little phosphate, it is often the limiting factor for plant growth on land unless phosphorus (as phosphate salts mined from the earth) is applied to the soil as a fertilizer. Phosphorus

Q: What percentage of the world's net primary productivity on land is used, wasted, or destroyed by the human population?

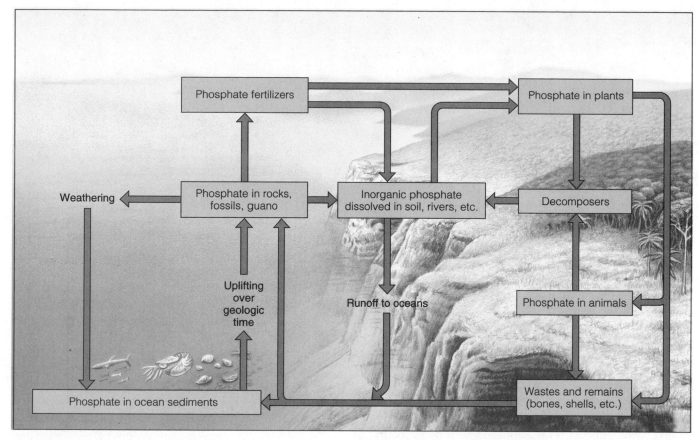

Figure 2-25 Simplified conceptual model of the phosphorus cycle. Phosphorus reservoirs are shown as boxes; the processes shown change one form of phosphorus to another. (Used by permission from Cecie Starr and Ralph Taggart, *Biology: The Unity and Diversity of Life,* 7th ed., Belmont, Calif.: Wadsworth, 1995)

also limits the growth of producer populations in many freshwater streams and lakes because phosphate salts are only slightly soluble in water.

Humans intervene in the phosphorus cycle in three main ways. *First*, we mine large quantities of phosphate rock for use in commercial inorganic fertilizers and detergents.

Second, we are sharply reducing the available phosphate and primary productivity of tropical forests by cutting them. In such ecosystems, hardly any phosphorus is found in the soil. Rather, it's in the ecosystem's plant and animal life, which is rapidly recycled from dead plants and animals by hordes of decomposers. When such forests are cut and burned, most remaining phosphorus and other soil nutrients are readily washed away by heavy rains and the soil becomes unproductive.

Third, we add excess phosphate to aquatic ecosystems in runoff of animal wastes from livestock feedlots, runoff of commercial phosphate fertilizers from cropland, and discharge of municipal sewage. Too much of this nutrient causes explosive growth of cyanobacteria, algae, and aquatic plants, disrupting life in aquatic ecosystems.

🔷 How Is Water Cycled in the Ecosphere?

The **hydrologic cycle**, or **water cycle**, which collects, purifies, and distributes the earth's fixed supply of water, is shown in simplified form in Figure 2-26. The main processes in this water recycling and purifying cycle are *evaporation* (conversion of water into water vapor), *transpiration* (evaporation from leaves of water extracted from soil by roots and transported throughout the plant), *condensation* (conversion of water vapor into droplets of liquid water), *precipitation* (rain, sleet, hail, and snow), *infiltration* (movement of water into soil), *percolation* (downward flow of water through soil and permeable rock formations to groundwater storage areas called aquifers), and *runoff* (downslope surface movement back to the sea where it can be evaporated to cycle).

The water cycle is powered by energy from the sun and by gravity. Incoming solar energy evaporates water

A: About 40% (27% including terrestrial and aquatic productivity)

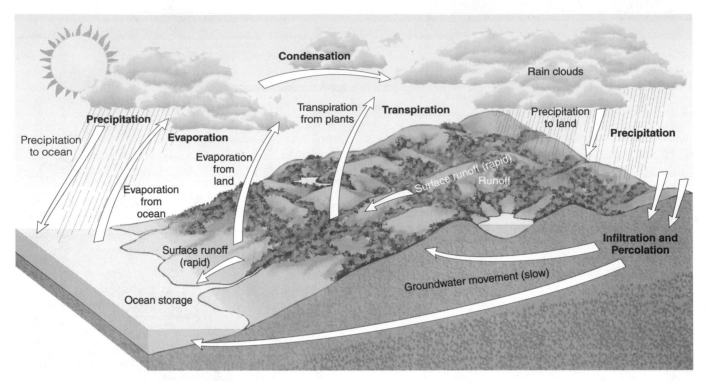

Figure 2-26 Simplified diagram of the hydrologic cycle.

from oceans, streams, lakes, soil, and vegetation. About 84% of water vapor in the atmosphere comes from the oceans, which cover about 71% of the earth's surface; the rest comes from land.

Besides replenishing streams and lakes, surface runoff also causes soil erosion, which moves soil and weathered rock fragments from one place to another. Water is thus the primary sculptor of the earth's landscape. Also, rainwater that is acidic, either naturally or from pollution, reacts chemically with atoms and ions on the earth's surface, releasing water-soluble metallic salts that are carried by rivers to the sea. This is one of the reasons the ocean is so salty. Because water dissolves many nutrient compounds, it is a major medium for transporting nutrients within and among ecosystems.

Humans intervene in the water cycle in three main ways: *First*, we withdraw large quantities of fresh water from streams, lakes, and underground sources. In heavily populated or heavily irrigated areas, withdrawals have led to groundwater depletion or intrusion of ocean salt water into underground water supplies.

Second, we clear vegetation from land for agriculture, mining, road and building construction, and other activities. This increases runoff and reduces infiltration that recharges groundwater supplies; it also increases the risk of flooding and accelerates soil erosion and landslides.

Third, we modify water quality, in particular by adding nutrients (such as phosphates) and other pollutants and by changing ecological processes that naturally purify water.

How Are the Earth's Three Types of Rock Recycled? The Rock Cycle Rock is any material that makes up a large, natural, continuous part of the earth's crust. Geologic processes constantly redistribute the chemical elements within and at the earth's surface. Based on the way it forms, rock is placed in three broad classes: igneous, sedimentary, and metamorphic.

Igneous rock can form below or on the earth's surface when molten rock material (magma) wells up from the earth's upper mantle or deep crust, cools, and hardens into rock. Examples are granite (formed underground) and lava rock (formed above ground when molten lava cools and hardens). Although often covered by sedimentary rocks or soil, igneous rocks form the bulk of the earth's crust. They also are the main source of many nonfuel mineral resources.

Sedimentary rock forms from sediment in several ways. Most such rocks are formed when preexisting rocks are weathered and eroded into small pieces, transported from their sources, and deposited in a body of surface water. As these deposited layers become buried and compacted, the resulting pressure causes their parti-

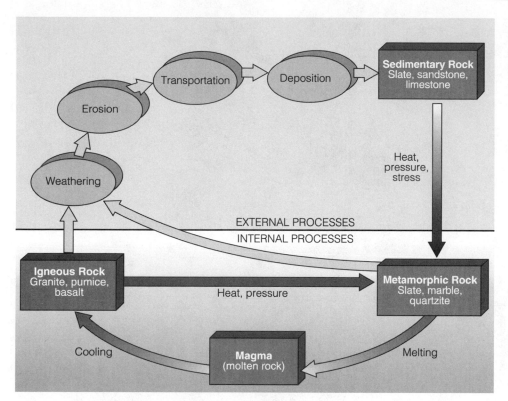

Figure 2-27 The rock cycle, the slowest of the earth's cyclic processes. The earth's materials are recycled over millions of years by three processes: *melting, erosion,* and *metamorphism,* which produce *igneous, sedimentary,* and *metamorphic* rocks. Rock of any of the three classes can be converted to rock of either of the other two classes (or can even be recycled within its own class).

cles to bond together to form sedimentary rocks such as sandstone and shale. Some sedimentary rocks, such as dolomite and limestone, are formed from the compacted shells, skeletons, and other remains of dead organisms. Two types of coal—lignite and bituminous coal—are sedimentary rocks derived from plant remains.

Metamorphic rock is produced when a preexisting rock is subjected to high temperatures (which may cause it to melt partially), high pressures, chemically active fluids, or a combination of these agents. Examples are anthracite (a form of coal), slate, and marble.

Rocks are constantly exposed to various physical and chemical conditions that can change them over time. The interaction of processes that change rocks from one type to another is called the **rock cycle** (Figure 2-27). Recycling material over millions of years, the slowest of the earth's cyclic processes is responsible for concentrating the planet's nonrenewable mineral resources on which humans depend.

2-7 LIFE ON LAND AND IN WATER ENVIRONMENTS

What Are the Major Types of Land Vegetation? Biologists have classified the terrestrial (land) portion of the ecosphere into **biomes,** large regions (such as forests, deserts, and grasslands) characterized by a distinct climate and specific life-forms—especially vegetation adapted to it. Each biome consists of many ecosystems whose communities have adapted to differences in climate, soil, and other environmental factors throughout the biome.

Why is one area of the earth's land surface a desert, another a grassland, and another a forest? Why are there different types of deserts, grasslands, and forests? The general answer to these questions is differences in **climate:** the average long-term weather of an area. It is a region's general pattern of atmospheric or weather conditions, including seasonal variations and weather extremes (such as hurricanes or prolonged drought or rain) averaged over a long period (at least 30 years). Such differences result primarily from differences in average temperature and precipitation. Figure 2-28 shows the global distribution of biomes. Figure 2-29 shows major biomes in the United States as one moves through different climates along the 39th parallel.

For plants, *precipitation is generally the limiting factor that determines whether a land area is desert, grassland, or forest.* A **desert** is an area where evaporation exceeds precipitation. Precipitation is typically less than 25 centimeters (10 inches) a year and is often scattered unevenly throughout the year. Deserts have sparse, widely spaced, mostly low vegetation.

Grasslands are regions with enough average annual precipitation to allow grass (and in some areas, a few trees) to prosper, but with precipitation

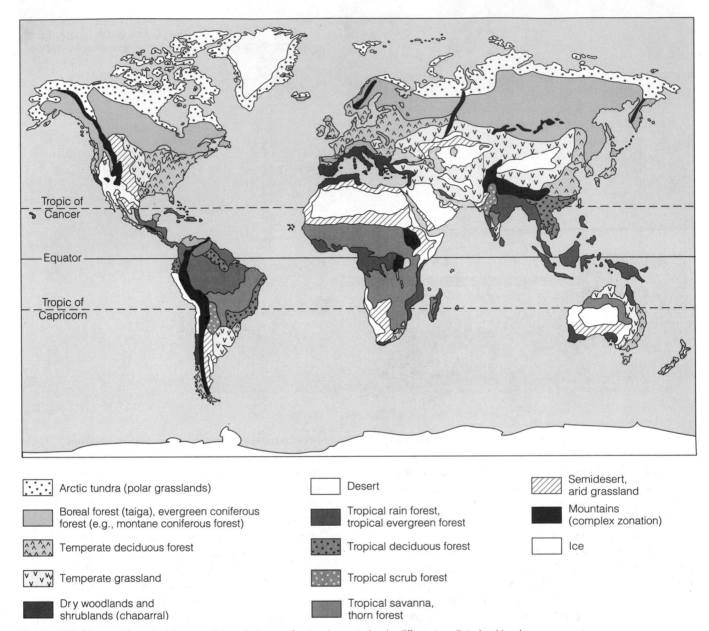

Arctic tundra (polar grasslands)

Boreal forest (taiga), evergreen coniferous forest (e.g., montane coniferous forest)

Temperate deciduous forest

Temperate grassland

Dry woodlands and shrublands (chaparral)

Desert

Tropical rain forest, tropical evergreen forest

Tropical deciduous forest

Tropical scrub forest

Tropical savanna, thorn forest

Semidesert, arid grassland

Mountains (complex zonation)

Ice

Figure 2-28 The earth's major biomes—the main types of natural vegetation in different undisturbed land areas—result primarily from differences in climate. Each biome contains many ecosystems whose communities have adapted to differences in climate, soil, and other environmental factors. In reality, people have removed or altered much of this natural vegetation for farming, livestock grazing, harvesting lumber and fuelwood, mining, and constructing villages and cities, thereby altering the biomes.

so erratic that drought and fire prevent large stands of trees from growing. Grasses (many of them perennials) in these biomes are renewable resources, if not overgrazed, because grass plants grow out from the bottom; thus, their stems can grow again after being nibbled off by grazing animals. Undisturbed areas with moderate to high average annual precipitation tend to be covered with **forest**, which contains various species of trees and smaller forms of vegetation.

Taken together, average annual precipitation and temperature (along with soil type) are the most important factors in producing tropical, temperate, or polar deserts, grasslands, and forests (Figure 2-30). Climate and vegetation both vary with latitude (distance from the equator) and altitude (elevation above sea level). If you travel from the equator toward either pole, you will generally encounter ever colder climates and zones of vegetation adapted to those climates. Similarly, as elevation above sea level increases, climate becomes

Q: What factor is most important in determining whether a land area is a desert, grassland, or forest?

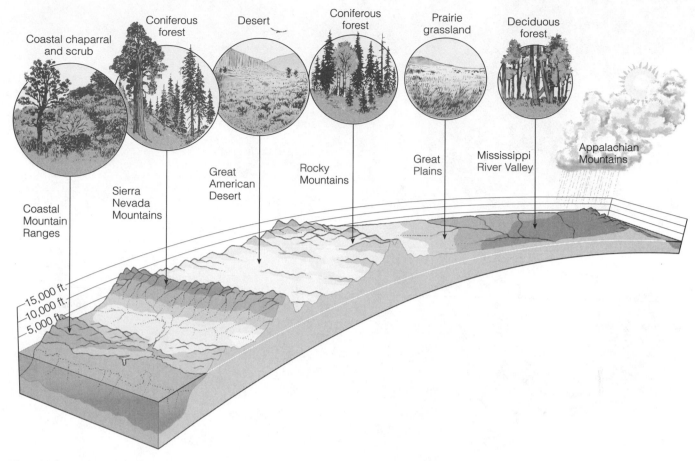

Figure 2-29 Major biomes found along the 39th parallel across the United States. The differences reflect changes in climate, mainly differences in average annual precipitation and temperature (not shown).

colder. Thus, if you climb a tall mountain from its base to its summit, you can observe changes in plant life similar to those you would encounter in traveling from equator to poles (Figure 2-31). Figure 2-32 (p. 54) shows some major components and interactions in a temperate desert biome.

What Are the Two Major Types of Aquatic Life Zones? The aquatic equivalents of biomes are called *aquatic life zones*. The major types of organisms found in aquatic environments are determined by the water's *salinity* (the amounts of various salts such as sodium chloride (NaCl) dissolved in a given volume of water). As a result, aquatic life zones are divided into two major types: *saltwater* or *marine* (such as estuaries, coastlines, coral reefs, coastal marshes, mangrove swamps, and the deep ocean) and *freshwater* (such as mostly nonflowing lakes and ponds, flowing streams, and inland wetlands).

Most aquatic life zones can be divided into surface, middle, and bottom layers. Important factors determining the types and numbers of organisms found in these layers are *temperature, access to sunlight for photosynthesis, dissolved oxygen content,* and *availability of nutrients* such as carbon (as dissolved CO_2 gas), nitrogen (as nitrate), and phosphorus (mostly as phosphate) for producers.

What Are the Major Saltwater Life Zones? A more accurate name for earth would be *Ocean* because saltwater oceans cover about 71% of its surface (Figure 2-33, p. 55). Because solar heat is distributed through ocean currents and because ocean water evaporates as part of the global hydrologic cycle, oceans play a major role in regulating the earth's climate. They also participate in other important nutrient cycles.

By serving as a gigantic reservoir for carbon dioxide, oceans help regulate the temperature of the troposphere. Oceans provide habitats for about 250,000 known species of marine plants and animals, which are food for many other organisms (including humans). In addition, many human-produced wastes that flow into or are dumped into the ocean are dispersed by currents (thus often diluted to less harmful levels).

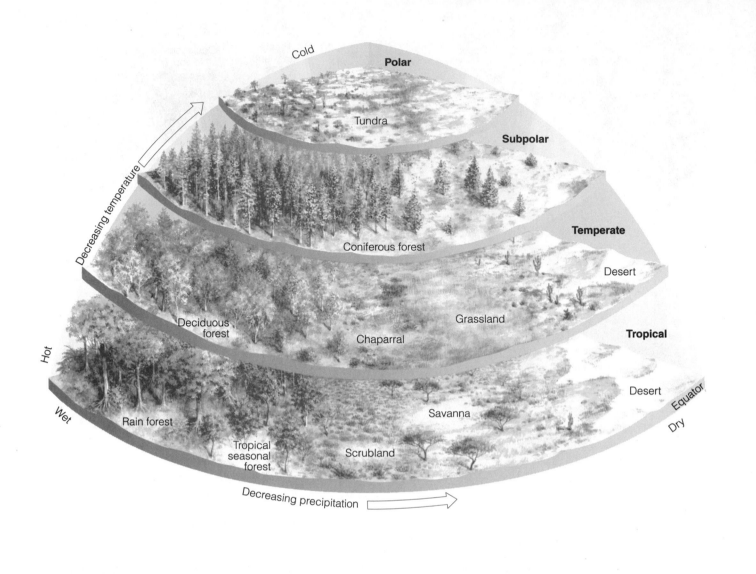

Cold

Polar

Tundra

Subpolar

Decreasing temperature

Coniferous forest

Temperate

Desert

Deciduous forest

Grassland

Chaparral

Tropical

Hot

Desert

Equator

Wet

Rain forest

Savanna

Dry

Tropical seasonal forest

Scrubland

Decreasing precipitation

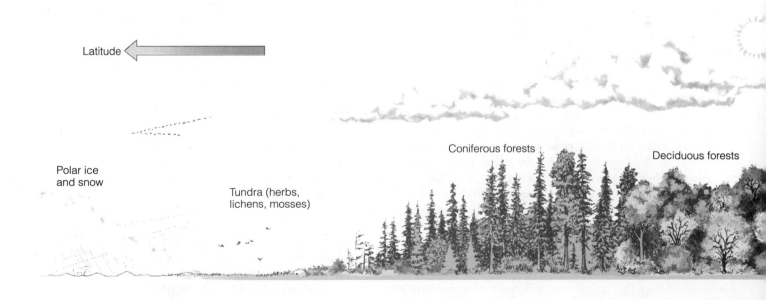

Latitude

Coniferous forests

Deciduous forests

Polar ice and snow

Tundra (herbs, lichens, mosses)

Q: What percentage of the earth's surface is covered by oceans?

Oceans have two major life zones: the **coastal zone** and the **open sea** (Figure 2-34). Although it makes up less than 10% of the ocean's area, the coastal zone contains 90% of all marine species and is the site of most of the large commercial marine fisheries. Most ecosystems found in the coastal zone have a very high primary productivity and net primary productivity per unit of area because of the zone's ample supplies of sunlight and plant nutrients (deposited from land and stirred up by wind and ocean currents).

Several types of highly productive ecosystems are found in the coastal zone. **Coral reefs** are often found in the shallow coastal zones of warm tropical and subtropical oceans. These beautiful natural wonders are among the world's oldest and most diverse and productive ecosystems and are homes for one-fourth of all marine species. These reefs are formed by massive colonies of tiny animals called *polyps* that secrete a protective crust of limestone (calcium carbonate)

around their soft bodies. By forming limestone shells, coral polyps take up carbon dioxide as part of the carbon cycle. The reefs also act as natural barriers that help protect 15% of the world's coastlines from battering waves and storms. These ecosystems grow slowly and are easily disrupted. Despite their ecological importance, coral reefs are disappearing and being degraded at an alarming rate by human activities.

One highly productive area in the coastal zone is an **estuary**, a partially enclosed area of coastal water where seawater mixes with fresh water and nutrients from rivers, streams, and runoff from land. According to one estimate, just 0.4 hectare (1 acre) of tidal estuary provides an estimated $75,000 worth of free waste treatment and has a value of about $83,000 when recreation and fish for food are included. By comparison, 0.4 hectare (1 acre) of prime farmland in Kansas has a top value of about $1,200 and an annual production value of about $600.

Areas of coastal land that are covered all or part of the year with salt water are called **coastal wetlands**. They are breeding grounds and habitats for a variety of waterfowl and other wildlife; they also serve as popular areas for recreational activities such as boating, fishing, and hunting. They also help maintain the quality of coastal waters by diluting, filtering, and settling out sediments, excess nutrients, and pollutants. In addition, coastal wetlands protect lives and property during floods by slowing the flow of water, and during storms they buffer shores against damage and erosion.

Figure 2-30 Average precipitation and average temperature, acting together as limiting factors over a period of 30 or more years, determine the type of desert, grassland, or forest biome in a particular area. Although the actual situation is much more complex, this simplified diagram explains how climate determines the types and amounts of natural vegetation found in an area left undisturbed by human activities. (Used by permission of Macmillan Publishing Company, from Derek Elsom, *The Earth*, New York: Macmillan, 1992. Copyright © 1992 by Marshall Editions Developments Limited)

Figure 2-31 Generalized effects of latitude and altitude on climate and biomes. Parallel changes in vegetation type occur when we travel from the equator to the poles or from lowlands to mountaintops.

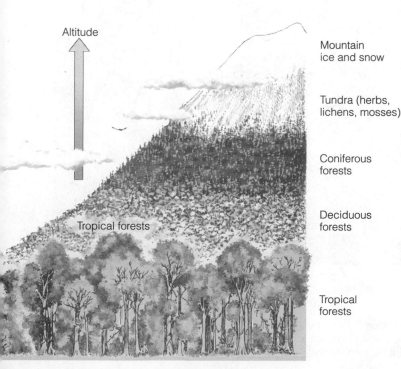

Altitude

Mountain ice and snow

Tundra (herbs, lichens, mosses)

Coniferous forests

Deciduous forests

Tropical forests

Tropical forests

| Producer to primary consumer | Primary to secondary consumer | Secondary to higher-level consumer | All producers and comsumers to decomposers |

Figure 2-32 Some components and interactions in a temperate desert biome. When these organisms die their organic matter is broken down by decomposers into minerals used by plants. Transfers of matter and energy among producers, primary consumers (herbivores), and secondary (or higher-level) consumers (carnivores) are indicated by arrows (see key).

Q: What percentage of the earth's coastal and inland wetlands have been destroyed or polluted?

In temperate areas, these wetlands usually consist of a mix of bays, lagoons, salt flats, mud flats, and salt marshes, in which grasses are the dominant vegetation.

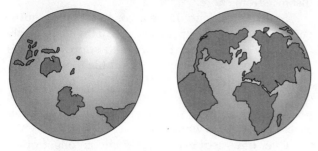

Figure 2-33 The ocean planet. The salty oceans cover about 71% of the earth's surface. About 97% of the earth's water is in the interconnected oceans, which cover 90% of the planet's mostly ocean hemisphere (left) and 50% of its land–ocean hemisphere (right). The average depth of the world's oceans is 3.8 kilometers (2.4 miles).

These highly productive ecosystems serve as nurseries and habitats for shrimp and many other aquatic animals. Since 1900, the world has lost approximately half of its coastal wetlands, primarily through coastal development.

Along warm tropical coasts where there is too much silt for coral reefs to grow we find highly productive **mangrove swamps**, dominated by about 55 species of salt-tolerant trees or shrubs known as mangroves. These swamps help protect the coastline from erosion and reduce damage from typhoons and hurricanes. They also trap sediment washed off the land and provide breeding, nursery, and feeding grounds for some 2,000 species of fish, invertebrates, and plants.

Along some coasts (such as most of North America's Atlantic and Gulf coasts) are **barrier islands**: long, thin, low offshore islands of sediment that generally run parallel to the shore. These islands help protect the mainland, estuaries, lagoons, and coastal wetlands by

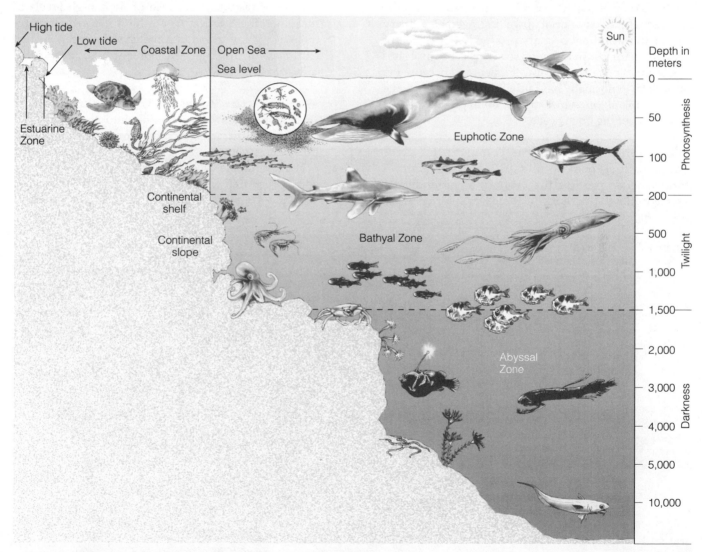

Figure 2-34 Major life zones in an ocean. (Not drawn to scale. Actual depths of zones may vary.)

dispersing the energy of approaching storm waves. Their low-lying beaches are constantly shifting: Gentle waves build them up, and storms flatten and erode them. Sooner or later many of the structures humans build on low-lying barrier islands, such as Atlantic City, Miami Beach, and Ocean City, Maryland, are damaged or destroyed by flooding, severe beach erosion, or wind from major storms (including hurricanes).

Despite their ecological importance all of these coastal ecosystems are under severe stress from human activities. Coastal zones are among our most densely populated and most intensely used, and polluted, ecosystems. Currently, nearly two-thirds of the world's population live along coasts or within 160 kilometers (100 miles) of a coast. By 2025, it's estimated that 75%, or 6.2 billion people, will reside on or near coastal areas.

The **open sea** is divided into three vertical zones—euphotic, bathyal, and abyssal—based primarily on the penetration of sunlight (Figure 2-34). This vast volume contains only about 10% of all marine species. Except at an occasional equatorial upwelling, average primary productivity and net primary productivity per unit of area are quite low in the open sea. This is because sunlight cannot penetrate the lower layers and because the surface layer normally has fairly low levels of nutrients for phytoplankton, which are the main photosynthetic producers of the open ocean.

What Zones Are Found in Freshwater Lakes?
Large natural bodies of standing fresh water—formed when precipitation, runoff, or groundwater seepage fills depressions in the earth's surface—are called **lakes**. Lakes normally consist of distinct zones (Figure 2-35), providing habitats and niches for different species, that are defined by their depth and distance from shore.

Ecologists classify lakes according to their nutrient content and their primary productivity. A newly formed lake generally has a small supply of plant nutrients and is called an **oligotrophic** (poorly nourished) **lake** (Figure 2-36, bottom). This type of lake is often deep, with steep banks. Because of its low net primary productivity, such a lake usually has crystal-clear blue or green water.

Over time, sediment washes into an oligotrophic lake, and plants grow and decompose to form bottom sediments. A lake with a large or excessive supply of nutrients (mostly nitrates and phosphates) needed by producers is called a **eutrophic** (well-nourished) **lake** (Figure 2-36, top). Such lakes are typically shallow, and their water is generally a murky brown or green with very poor visibility. Because of their high levels of nutrients, these lakes have a high net primary productivity. Many lakes fall somewhere between the two extremes of nutrient enrichment and are called **mesotrophic lakes**.

What Are the Zones in Freshwater Streams?
Precipitation that doesn't sink into the ground or evaporate is **surface water**. It becomes **runoff** when it flows into streams and eventually to the ocean as part of the hydrologic cycle (Figure 2-26). This entire land area, which delivers water, sediment, and dissolved substances via small streams to a larger stream or

Figure 2-35 The distinct zones of life in a temperate-zone lake.

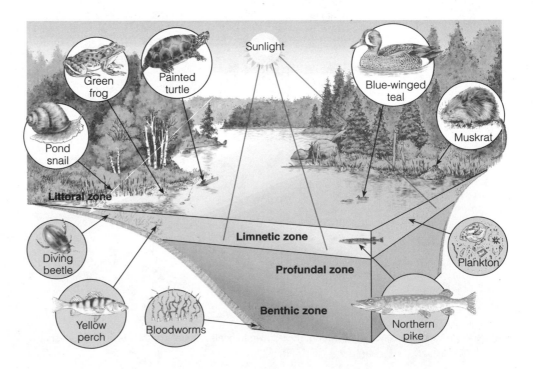

Q: How many species are there on the earth?

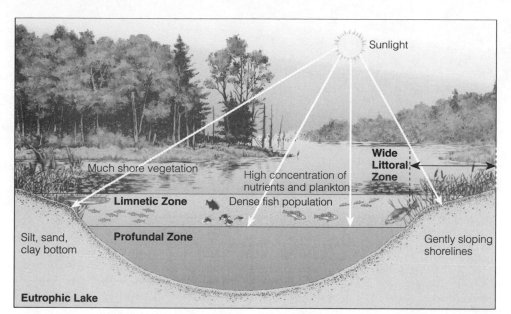

Figure 2-36 A eutrophic, or nutrient-rich, lake (top) and an oligotrophic, or nutrient-poor, lake (bottom). Mesotrophic lakes fall between these two extremes of nutrient enrichment.

Sunlight

Much shore vegetation

Wide Littoral Zone

High concentration of nutrients and plankton

Limnetic Zone

Dense fish population

Silt, sand, clay bottom

Profundal Zone

Gently sloping shorelines

Eutrophic Lake

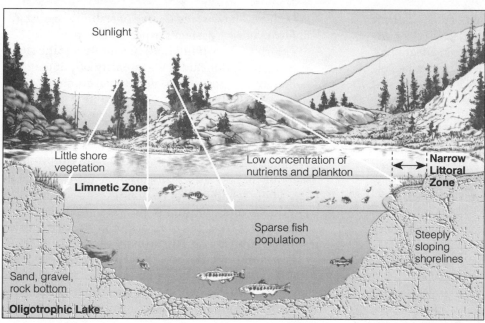

Sunlight

Little shore vegetation

Low concentration of nutrients and plankton

Narrow Littoral Zone

Limnetic Zone

Sparse fish population

Steeply sloping shorelines

Sand, gravel, rock bottom

Oligotrophic Lake

river and ultimately to the sea, is called a **watershed** or **drainage basin**.

The downward flow of water from mountain highlands to the sea takes place in three zones (Figure 2-37). Because of different environmental conditions in each zone, a *river system* is actually a series of different ecosystems with different average depths, flow rates, dissolved oxygen levels, and temperatures.

As streams flow downhill, they become powerful shapers of land. Over millions of years, the friction of moving water levels mountains and cuts deep canyons; the rock and soil the water removes are deposited as sediment in low-lying areas.

Why Are Freshwater Inland Wetlands Important? Lands covered with fresh water all or part of the time (excluding lakes, reservoirs, and streams) and located away from coastal areas are called **inland wetlands**. They include bogs, marshes, prairie potholes (depressions carved out by glaciers), swamps (dominated by trees and shrubs), mud flats, floodplains, wet meadows, and the wet arctic tundra in summer.

Some wetlands are covered with water year-round; others, such as prairie potholes, floodplain wetlands, and bottomland hardwood swamps are *seasonal wetlands*, usually underwater or soggy for only a short time each year. Some stay dry for years before filling with water again.

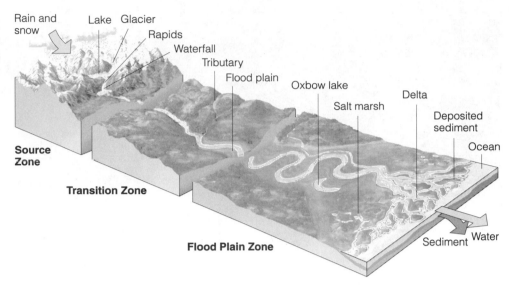

Figure 2-37 The three zones in the downhill flow of water: mountain (headwater) streams; wider, lower-elevation streams; and rivers, which empty into the ocean.

Inland wetlands provide food and habitats for fish, migratory waterfowl, and other wildlife, and they improve water quality by filtering, diluting, and degrading toxic wastes, excess nutrients, sediments, and other pollutants. The Audubon Society conservatively estimates that inland wetlands in the United States provide water-quality protection worth at least $1.6 billion per year.

Floodplain wetlands near rivers reduce flooding and erosion by absorbing stormwater and releasing it slowly and by absorbing overflows from streams and lakes. According to Audubon Society estimates, if the remaining wetlands in the United States were destroyed, additional flood-control costs would be $7.7–31 billion per year. Inland wetlands also help replenish groundwater supplies, a primary water source for over 50% of the U.S. population. In addition, they play significant roles in the global carbon, nitrogen, sulfur, and water cycles.

Despite the ecological importance of year-round and seasonal inland wetlands, many are drained, dredged, filled in, or covered over. Each year some 475 square kilometers (183 square miles) of inland wetland in the United States are lost, about 80% to agriculture and the rest to mining, forestry, oil and gas extraction, highways, and urban development. Other countries have suffered similar losses.

2-8 POPULATION DYNAMICS, EVOLUTION, AND BIODIVERSITY

What Is the Major Driving Force of Nature? Nature is characterized by constant change caused by natural and human-related forces and the adjustments to various environmental stresses. However, despite changes in environmental conditions all living systems—from single-celled organisms to the ecosphere—have some degree of stability or sustainability over each system's expected life span.

Such sustainability does not imply a static situation. Instead, it includes the capacity of populations, communities, ecosystems, and human economic, political, and social systems to adapt to new conditions. *There is no balance of nature—only constant changes in response to changes in environmental conditions.*

Populations, communities, and ecosystems are so complex and variable that ecologists have little understanding of how they maintain some degree of stability while continually responding to changes in environmental conditions. However, scientists have learned that the signs of ill health in stressed ecosystems include (1) a drop in primary productivity, (2) increased nutrient losses, (3) decline or extinction of indicator species, (4) larger populations of insect pests or disease organisms, (5) a decline in species diversity, and (6) the presence of contaminants.

What Limits Population Growth? Four variables—births, deaths, immigration, and emigration—govern changes in population size. A population gains individuals by birth and immigration and loses them by death and emigration:

Population change = (Births + Immigration) − (Deaths + Emigration)

Variables in turn depend on changes in resource availability or other environmental changes.

Populations vary in their capacity for growth, also known as the **biotic potential** of the population. The **intrinsic rate of increase (r)** is the rate at which a population could grow if it had unlimited resources. Generally, individuals in populations with a high biotic potential *reproduce early in life*, *have short generation times* (the time between successive generations), *can reproduce*

Thorson, Bruce. 1998. "Boom and bust." *Canadian Geographic*, vol. 118, no. 2, 68(7).

many times (have a long reproductive life), and *produce many offspring each time they reproduce.*

No population can grow exponentially indefinitely. In the real world, a rapidly growing population reaches some size limit imposed by a shortage of one or more limiting factors, such as light, water, space, or nutrients. *There are always limits to population growth in nature.*

Environmental resistance consists of all the factors acting jointly to limit the growth of a population. The population size of a species in a given place and time is determined by the interplay between its biotic potential and environmental resistance. Together biotic potential and environmental resistance determine the **carrying capacity (K)**, the number of individuals of a given species that can be sustained indefinitely in a given space (area or volume).

Any population growing exponentially starts out slowly and then goes through a rapid, mostly unrestricted exponential growth phase. If plotted, this sequence yields a *J*-shaped exponential growth curve (Figure 1-1). However, because of environmental resistance its growth tends to level off once the carrying capacity of the area is reached. (In most cases, the size of such a population fluctuates slightly above and below the carrying capacity.) An idealized plot of this type of growth yields a sigmoid or *S-shaped curve* (Figure 2-38).

The populations of some species don't make such a smooth transition from a *J*-shaped curve to an *S*-shaped curve. Instead, they temporarily use up their resource base (for example, by eating more plants or animals than can be replenished). Thus, the population temporarily *overshoots* or exceeds the carrying capacity of its habitat. This overshoot occurs because of a *reproductive time lag*, the period required for the birth rate to fall and the death rate to rise in response to resource overconsumption. Unless the excess individuals switch to new resources or move to an area with more favorable conditions, the population will suffer a *dieback* or *crash* (Figure 2-39), falling to a lower level that typically fluctuates around the area's new (lowered) carrying capacity.

Humans are not exempt from overshoot and dieback. Ireland, for example, experienced a population crash after a fungus destroyed the potato crop in 1845. About 1 million people died and 3 million people emigrated to other countries. Technological, social, and other cultural changes have extended earth's carrying capacity for the human species (Figure 1-5). We have increased food production and used large amounts of energy and matter resources to make normally uninhabitable areas of the earth habitable. However, there is growing concern about how long we will be able to keep doing this on a planet with a finite size and resources but an exponentially growing population and per capita resource use.

Carrying capacity is not a simple, fixed quantity, but rather a variable determined by many factors. Examples include competition within and among species, immigration and emigration, natural and human-caused

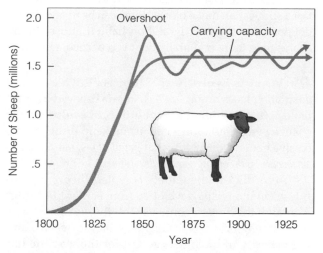

Figure 2-38 *S*-shaped population change curve of sheep population on the island of Tasmania between 1800 and 1925. After sheep were introduced in 1800 their population grew exponentially because of ample food. By 1855, however, they overshot the land's carrying capacity. Their numbers then stabilized and oscillated around a carrying capacity of about 1.6 million sheep. Sheep are large and long-lived and reproduce slowly. As a result, their population size does not vary much from year to year once they reach the carrying capacity of their habitat (assuming that other environmental factors do not lower the carrying capacity).

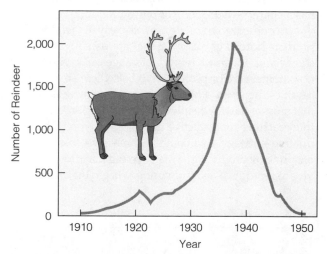

Figure 2-39 Exponential growth, overshoot, and population crash of reindeer introduced to a small island off the southwest coast of Alaska. In 1910, 26 reindeer (24 of them female) were introduced onto an island of the Aleutian chain off the southwest coast of Alaska. When the reindeer were first introduced, lichens, mosses, and other food sources were plentiful. By 1935 the herd's population had soared to 2,000, overshooting the island's carrying capacity. So many reindeer eating vegetation faster than it could be replenished caused overgrazing and resulted in a population crash, with the herd plummeting to only 8 reindeer by 1950.

Hint: Enter the search term *lemmings* using the Subject Guide.

catastrophic events, and seasonal fluctuations in the supply of food, water, hiding places, and nesting sites.

What Kinds of Population Change Curves Do We Actually Find in Nature? In nature we find that over time species have three generalized types of *population cycles*: stable, irruptive, and cyclic (Figure 2-40). A species whose population size fluctuates only slightly above and below its carrying capacity has a fairly stable population size (Figure 2-38). Such stability is characteristic of many species found in undisturbed tropical rain forests, where there is little variation in average temperature and rainfall.

Some species, such as the raccoon, normally have a fairly stable population that may occasionally explode, or *irrupt*, to a high peak, and then crash to a more stable lower level or in some cases to a very low level (Figure 2-39). The population explosion is caused by some factor that temporarily increases carrying capacity for the population, such as more food or fewer predators.

Other species undergo sharp increases in their numbers followed by seemingly periodic crashes. Predators are sometimes blamed, but the actual causes of such boom–bust cycles are poorly understood.

What Reproductive Strategies Do Species Use to Survive? Reproductive individuals in populations of all species are engaged in a struggle for genetic immortality by trying to have as many members as possible in the next generation carrying their genes. Each species has a characteristic mode of reproduction.

At one extreme are species that early in their life cycle produce hordes of offspring that are usually small and short-lived, and reach reproductive age early with little or no parental care or protection to help them survive. The result of this *many-small-and-unprotected-young strategy* is that most of the tiny, helpless offspring die or are eaten before reaching reproductive age. Species with this reproductive strategy overcome the massive loss of their offspring by producing so many young that a few will survive to reproduce many offspring to begin the cycle again.

Algae, bacteria, rodents, annual plants (such as dandelions), many bony fish, and most insects are examples. Such species tend to be *opportunists*, reproducing and dispersing rapidly when conditions are favorable or when a new habitat or niche becomes available. Changing or unfavorable environmental conditions can cause such populations to crash. Hence, such species tend to go through boom–bust cycles.

At the other extreme are species that tend to produce late and have few offspring with long generation times. Typically these offspring develop inside their mothers (where they are safe). They are fairly large and mature slowly. They are cared for and protected by one or both parents until they reach reproductive age. This *few-but-large-young reproductive strategy* results in a few big and strong individuals that can compete for resources and reproduce a few young to begin the cycle again.

Such species tend to maintain their population size near their habitat's carrying capacity. Their populations typically follow an *S*-shaped growth curve (Figures 2-38). Examples are most large mammals (such as elephants, whales, and humans), birds of prey, and large and long-lived plants (such as the saguaro cactus, redwood trees, and most tropical rain forest trees). The reproductive strategies of most species fall somewhere between these two extremes.

The reproductive strategy of a species may give it a temporary advantage, but *the availability of suitable habitat for individuals of a population in a particular area is what determines its ultimate population size.* Regardless of how fast a species can make babies, there can be no more dandelions than there is dandelion habitat and no more zebras than there is zebra habitat in a particular area.

✵ What Is Evolution? The major driving force of adaptation to environmental change is believed by most biologists to be **biological evolution,** or **evolution:** the change in a population's genetic makeup through successive generations. Note that *populations, not individuals, evolve by becoming genetically different.*

According to the widely accepted **theory of evolution,** all life-forms developed from earlier life-forms. Although this theory conflicts with the creation stories of most religions, it is the explanation the overwhelming majority of biologists accept for the way life has changed over the past 3.7–3.8 billion years (Figure 2-41) and for why life is so diverse today.

Evolutionary change occurs as a result of the interplay of genetic variation in a population and changes in environmental conditions. The first step in evolution is the development of *genetic variability* in a population. Genetic information in *chromosomes* is contained in various sequences of chemical units (called *nucleotides*) in DNA molecules. **Genes** found in chromosomes are

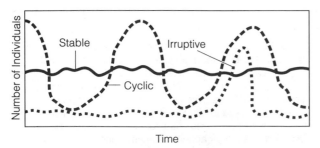

Figure 2-40 General types of population change curves found in nature.

Q: What is biological evolution?

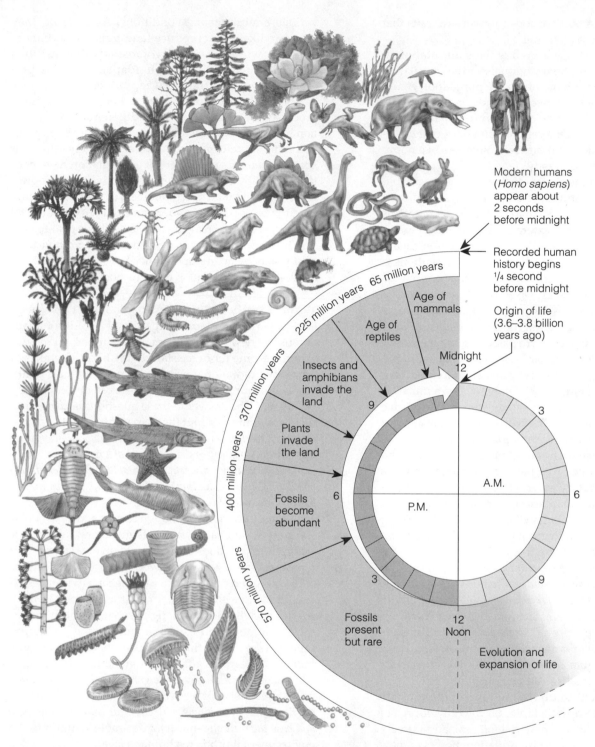

Figure 2-41 Greatly simplified overview of the biological evolution of life on the earth, which was preceded by about 0.5–1 billion years of chemical evolution. The early span of biological evolution on the earth, between 3.7 billion and 570 million years ago, was dominated by microorganisms (mostly bacteria and, later, protists) that lived in water. Plants and animals evolved first in the seas and moved onto land about 400 million years ago. Humans arrived on the scene only a very short time ago. If we compress the earth's roughly 3.7-billion-year history of biological evolution to a 24-hour time scale, the first human species (*Homo habilis*) appeared about 47 seconds before midnight and our species (*Homo sapiens sapiens*) appeared about 1.4 seconds before midnight. Agriculture began only 0.25 second before midnight, and the industrial revolution has been around for only 0.007 second. (Adapted from George Gaylord Simpson and William S. Beck, *Life: An Introduction to Biology,* 2d ed., New York: Harcourt Brace Jovanovich, 1965)

A: The change in the genetic makeup of a population through successive generations

segments of DNA that are coded for certain traits that can be passed on to offspring.

A population's **gene pool** is the sum total of all genes possessed by the individuals of the population of a species. Although members of a population generally have the same number and kinds of genes, a particular gene may have two or more different molecular forms, called **alleles**. Different combinations of these alleles are inherited so that different members of a population have genetic diversity.

The source of the alleles leading to genetic variability is **mutations**: random changes in the structure or number of DNA molecules in a cell. One way mutations occur is by exposure to external agents such as radioactivity, X rays, and natural and human-made chemicals (called *mutagens*). Another source of mutations is random mistakes that are sometimes made in coded genetic instructions when DNA molecules are copied (each time a cell divides and whenever an organism reproduces). Mutations can occur in any cells, but only those in reproductive cells are passed on to offspring.

Some mutations are harmless, but many are harmful, altering traits in such a way that an individual cannot survive (lethal mutations). Every so often a mutation is beneficial. The result is new genetic traits that give their bearer and its offspring better chances for survival and reproduction, either under existing environmental conditions or when such conditions change. Any genetically controlled trait that helps an organism survive and reproduce under a given set of environmental conditions is called an **adaptation**, or **adaptive trait**.

Structural adaptations include *coloration* (allowing more individuals to hide from predators or to sneak up on prey), *mimicry* (looking like a poisonous or dangerous species), *protective cover* (shell, thick skin, bark, or thorns), and *gripping mechanisms* (hands with opposable thumbs). *Physiological adaptations* include the ability to hibernate during cold weather and to give off chemicals that poison or repel prey. *Behavioral adaptations* include the ability to fly to a warmer climate during winter, niche specialization (Figure 2-17), and species interactions such as parasitism, mutualism, and commensalism.

It is important to understand that mutations are **(1)** random and unpredictable, **(2)** the only source of totally new genetic raw material (alleles), and **(3)** very rare events. Once created by mutation, however, new alleles can be shuffled together or recombined *randomly* to create new combinations of genes in populations of sexually reproducing species.

What Role Does Natural Selection Play in Evolution? Because of the random shuffling or recombination of various alleles, certain individuals in a population may by chance have one or more beneficial adaptations that help them to survive under various environmental conditions. As a result, they are more likely to reproduce (and thus produce more offspring with the same favorable adaptations) than are individuals without such adaptations. This process is known as **differential reproduction**.

Natural selection occurs when the combined processes of adaptation and differential reproduction result in a particular beneficial gene or set of genes becoming more common in succeeding generations. It occurs when some members of a population have heritable traits that enable them to survive and produce more offspring than other members of the population.

It is important to understand that environmental conditions do not create favorable heritable characteristics. Instead, natural selection favors some individuals over others by acting upon inherited genetic variations (alleles) already present in the gene pool of a population.

What Limits Adaptation? Shouldn't evolution lead to perfectly adapted organisms? Shouldn't adaptations to new environmental conditions allow our skin to become more resistant to the harmful effects of ultraviolet radiation, our lungs to cope with air pollutants, and our livers to become better at detoxifying pollutants? The answer to these questions is *no* because there are limits to adaptations in nature. First, *a change in environmental conditions can lead to adaptation only for traits already present in the gene pool of a population.*

Second, *even if a beneficial heritable trait is present in a population, that population's ability to adapt can be limited by its reproductive capacity.* If members of a population can't reproduce quickly enough to adapt to a particular environmental change, all of its members can die. For example, populations of genetically diverse species that reproduce quickly—such as weeds, mosquitoes, rats, or bacteria—often can adapt to a change in environmental conditions in a short time. In contrast, populations of species such as elephants, tigers, sharks, and humans, which cannot produce large numbers of offspring rapidly, take a long time (typically thousands or even millions of years) to adapt through natural selection.

Finally, *even if a favorable genetic trait is present in a population, most of its members would have to die or become sterile so that individuals with the trait could predominate and pass the trait on*—hardly a desirable solution to the environmental problems humans face.

How Do New Species Evolve? Under certain circumstances natural selection can lead to an entirely new species. In this process, called **speciation**, two species arise from one in response to changes in environmental conditions.

The most common mechanism of speciation (especially among animals) takes place in two phases: geographic isolation and reproductive isolation.

Q: Does evolution involve tooth-and-claw competition and survival of the strongest?

Geographic isolation occurs when two populations of a species or two groups of the same population become physically separated for fairly long periods into areas with different environmental conditions. For example, part of a population may migrate in search of food and then begin living in another area with different environmental conditions (Figure 2-42). Populations may also become separated by a physical barrier (such as a mountain range, stream, lake, or road) by a change such as a volcanic eruption or earthquake, or when a few individuals are carried to a new area by wind or water.

The second phase of speciation is **reproductive isolation**. It occurs as mutation and natural selection operate independently in two geographically isolated populations and change the allele frequencies in different ways—a process called *divergence*. If divergence continues long enough, members of the geographically and reproductively isolated populations may become so different in genetic makeup that they can't interbreed—or if they do, they can't produce live, fertile offspring. Then one species has become two, and *speciation* has occurred through *divergent evolution*.

What Are Three Common Misconceptions About Evolution? One misconception about evolution arises from the interpretation of the expression "survival of the fittest" (which biologists almost never use), sometimes used to describe how natural selection works. This has often been misinterpreted as "survival of the strongest." To biologists, however, *fitness* is a measure of reproductive success, so that the fittest individuals are those that leave the most descendants. Instead of "tooth and claw" competition, natural selection favors populations of species that *avoid* direct competition by producing offspring that can occupy niches different from those of other species.

Some people have misinterpreted the theory of evolution's assertion that all species share a common ancestry to mean that "humans evolved from apes." The theory of evolution makes no such claim. Instead, it states that apes and humans are descended from a common ancestor. In other words, at some time in the distant past a particular population of organisms had descendants, with some of them evolving into apes and others evolving into the human species.

A third misconception is that evolution involves some grand plan of nature in which species become progressively more perfect. This ignores the fact that the mutations and other processes that drive microevolution occur as a result of random, unpredictable events. From a scientific standpoint, there is no plan or goal of perfection in the evolutionary process.

How Do Species Become Extinct? After evolution, the second process affecting the number and types of species on the earth is **extinction**. When environmental conditions change, a species may either evolve (become better adapted) or cease to exist (become extinct). Extinction is the ultimate fate of all species, just as death is for all individual organisms. Biologists estimate that more than 99.9% of all the species that have ever existed are now extinct.

Some species inevitably disappear at some low rate, called **background extinction**, as local conditions change. In contrast, **mass extinction** is an abrupt rise in extinction rates above the background level. It is a catastrophic, widespread (often global) event in which large groups of existing species (perhaps 25–70%) are wiped out. Most mass extinctions are believed to result from global climate changes that kill many species and leave behind those able to adapt to the new conditions. Fossil and geological evidence indicates

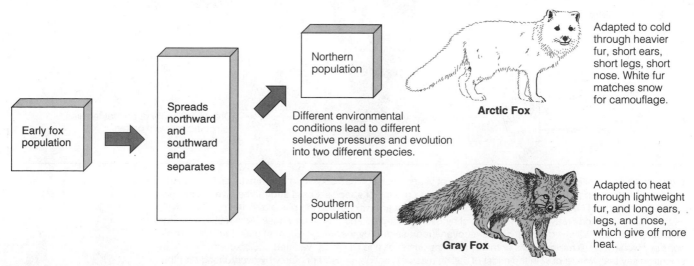

Figure 2-42 How geographic isolation can lead to reproductive isolation, divergence, and speciation.

that the earth's species have experienced five great mass extinctions (20 to 60 million years apart) during the past 500 million years (Figure 2-43).

A crisis for one species is an opportunity for another. The fact that millions of species exist today means that speciation, on average, has kept ahead of extinction. Evidence shows that the earth's mass extinctions have been followed by periods of recovery called **adaptive radiations**, in which numerous new species have evolved over several million years to fill new or vacated ecological niches (Figure 2-43).

How Do Speciation and Extinction Affect Biodiversity? Speciation minus extinction equals *biodiversity*, the planet's genetic raw material for future evolution in response to changing environmental conditions. In this long-term give-and-take between extinction and speciation, mass extinctions temporarily reduce biodi-

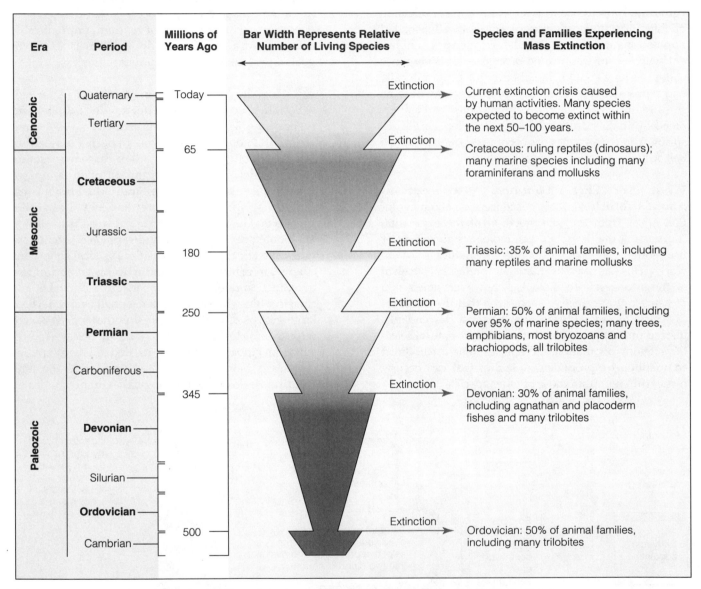

Figure 2-43 Over millions to hundreds of millions of years, evolution has consisted of dramatic exits (extinctions) and grand entrances (speciation and radiations) of large groups of species. Fossil and radioactive dating evidence indicate that five major mass extinctions (indicated by arrows) have taken place over the past 500 million years. Mass extinctions leave large numbers of niches unoccupied and create new ones. As a result, each mass extinction has been followed by periods of recovery (represented by the wedge shapes) called *adaptive radiations* in which (over 10 million years or more) new species evolve to fill new or vacated ecological niches. Many scientists say that we are now in the midst of a sixth mass extinction, caused primarily by overhunting and the increasing elimination, degradation, and fragmentation of wildlife habitats as a result of human activities.

Q: What percentage of all species that have ever lived on earth have become extinct?

versity. However, they also create evolutionary opportunities for surviving species to undergo adaptive radiations to fill unoccupied and new niches (Figure 2-43).

Although extinction is a natural process, humans have become a major force in the premature extinction of species. As population and resource consumption increase over the next 50 years and we take over more and more of the earth's surface and net primary productivity, we may cause the extinction of up to a quarter of the earth's species, each the product of millions to billions of years of evolution. If this happens it will constitute a sixth mass extinction, caused by us.

Many scientists fear that our immense and rapidly growing abilities to exploit nature could backfire. By reducing and degrading the earth's life-support systems (earth or ecological capital), we could make our own species more vulnerable to extinction, or at least to a massive dieback. What do you think?

2-9 ECOLOGICAL SUCCESSION

How Do Communities and Ecosystems Respond to Change? One characteristic of all communities and ecosystems is that their structures, especially their vegetation, are constantly changing in response to changing environmental conditions. The gradual and fairly predictable change in species composition of a given area is called **ecological succession**. Succession involves a complex set of species interactions over time. During succession some species colonize and their populations become more numerous, while populations of other species decline and even disappear.

Ecologists recognize two types of ecological succession: primary and secondary, depending on the conditions present at the beginning of the process. **Primary succession** involves the gradual establishment of biotic communities in an area that was not occupied by life before. In contrast, **secondary succession**, the more common type of succession, involves the *reestablishment* of a biotic community in an area where a different biotic community was previously present.

Primary succession begins with an essentially lifeless area, where there is no soil in a terrestrial ecosystem or no bottom sediment in an aquatic ecosystem (Figure 2-44). Examples of such areas include the rock or mud exposed by a retreating glacier or a mudslide, newly cooled lava, an abandoned highway or parking lot, or a newly created shallow pond or reservoir.

Before a community of plants (producers), consumers, and decomposers can become established on land, there must be *soil*: a complex mixture of rock particles, decaying organic matter, air, water, and living organisms. Depending mostly on the climate, it takes natural processes several hundred to several thousand years to produce fertile soil. Soil formation begins when a rock or other lifeless area is colonized by a few hardy **pioneer species** (microbes, mosses, and lichens). They are usually species with the ability to establish large populations quickly in a new area, species that biologist Edward O. Wilson calls "nature's sprinters."

The more common type of succession is secondary succession. This begins in an area where the natural community of organisms has been disturbed, removed, or destroyed, but the soil or bottom sediment remains. Candidates for secondary succession include abandoned farmlands, burned or cut forests, heavily polluted streams, and land that has been dammed or flooded. Because some soil or sediment is present, new vegetation can usually sprout within a few weeks.

In the central (Piedmont) region of North Carolina, European settlers cleared the mature native oak and hickory forests and replanted the land with crops. Some of the land was subsequently abandoned because of erosion and loss of soil nutrients. Figure 2-45 shows how such abandoned farmland, still covered with a thick layer of soil, has undergone secondary succession.

Because primary and secondary succession involve changes in the makeup of biological communities, it is not surprising that the various stages of succession have different patterns of species diversity, trophic structure, niches, nutrient cycling, and energy flow and efficiency (Table 2-1).

At any time during primary or secondary succession, disturbances such as natural or human-caused fires or deforestation can convert a particular stage of succession to an earlier stage. Such disturbances create new conditions that encourage some species and discourage or eliminate others.

How Predictable Is Succession? It is tempting to conclude that ecological succession is an orderly sequence in which each stage leads predictably to the next, more stable stage. According to this classic view, succession proceeds until an area is occupied by a predictable type of *climax community* dominated by a few long-lived plant species (which Edward O. Wilson calls "nature's long-distance runners").

However, research has shown that the sequences of species and community types that appear during primary or secondary succession can be highly variable and unpredictable. We cannot predict the course of a given succession or view it as some preordained progress toward an ideally adapted climax community. Rather, succession reflects the ongoing struggle by different species for enough light, nutrients, food, and space to survive and to give them a reproductive advantage over other species.

Figure 2-44 Primary succession over several hundred years of plant communities on bare rock exposed by a retreating glacier on Isle Royal in northern Lake Superior.

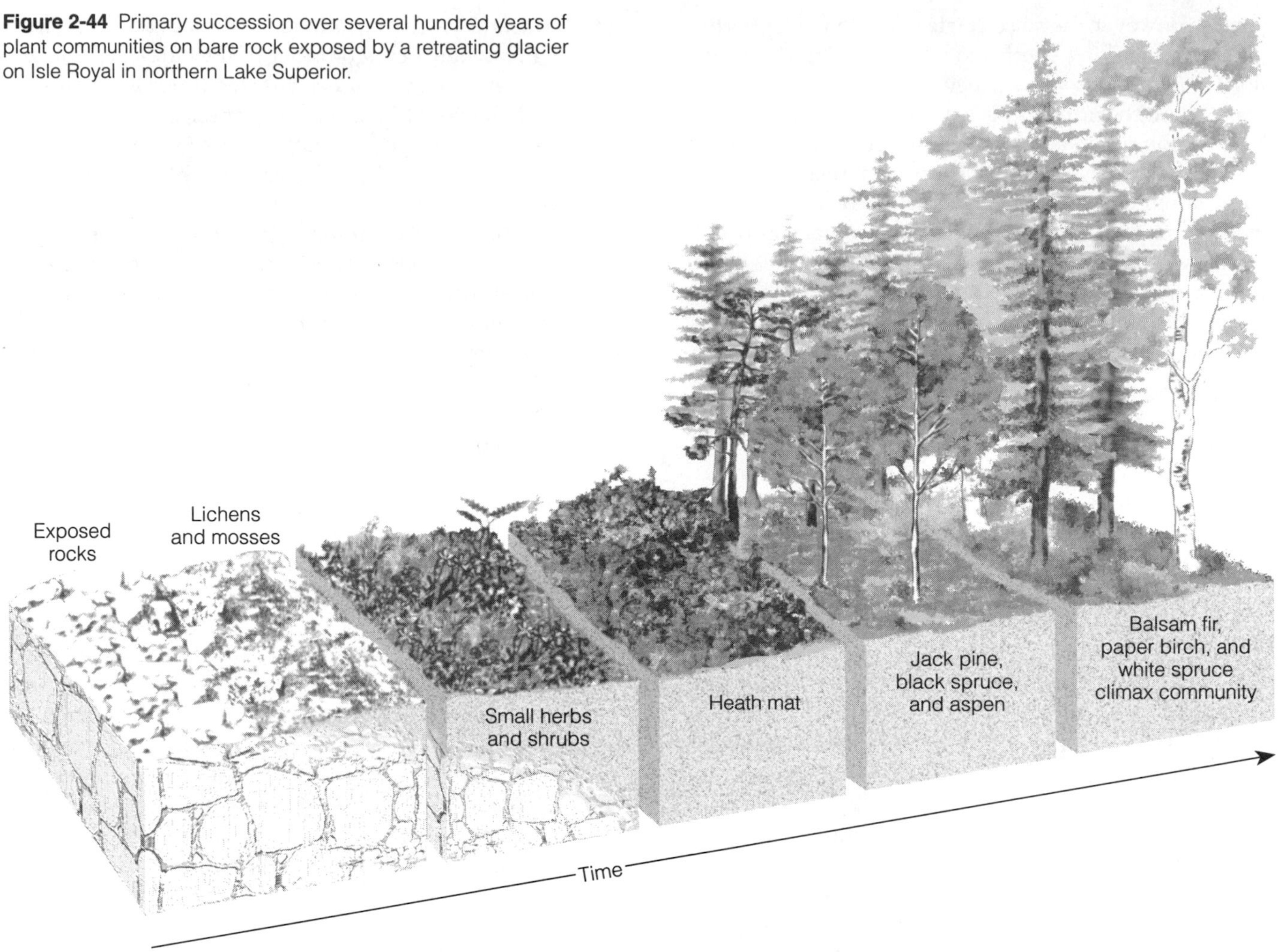

Exposed rocks

Lichens and mosses

Small herbs and shrubs

Heath mat

Jack pine, black spruce, and aspen

Balsam fir, paper birch, and white spruce climax community

Time

2-10 HUMAN IMPACT ON ECOSYSTEMS: LEARNING FROM NATURE

How Have Humans Modified Natural Ecosystems? To survive and support growing numbers of people, we have greatly increased the number and area of the earth's natural systems that we have modified, cultivated, built on, or degraded. We have used technology to severely alter much of the rest of nature in several ways by

■ *Fragmenting and degrading habitat.* Our landscape is rapidly being transformed from one with vast expanses of continuous forest, grassland, or other natural ecosystems into a patchwork of remnant habitat fragments surrounded by land developed for farmland, urban areas, and other human uses. The remaining habitat, even when it is continuous, is often degraded and not suitable for many native species.

■ *Simplifying natural ecosystems.* Humans eliminate some wildlife habitats by plowing grasslands, clearing forests, and filling in wetlands, often replacing their thousands of interrelated plant and animal species with one crop or one kind of tree—called *monocultures*—or with buildings, highways, and parking lots. Then we spend a lot of time, energy, and money trying to protect such monocultures from invasion by opportunist species of plants (weeds), pests (mostly insects), and pathogens (fungi, viruses, or bacteria that harm the plants we want to grow).

■ *Strengthening some populations of pest species and disease-causing bacteria by speeding up natural selection and causing genetic resistance.* When fast-breeding insect species begin to undergo natural selection and develop genetic resistance to pesticides, the salespeople urge farmers to use larger doses or switch to a new pesticide. This increases natural selection of

Q: What is ecological succession?

Figure 2-45 Secondary ecological succession of plant communities on an abandoned farm field in North Carolina. It took about 150–200 years after the farmland was abandoned for the area to be covered with a mature oak and hickory forest. A new disturbance such as deforestation or fire would create conditions favoring early successional species. In the absence of new disturbances, secondary succession would again occur over time, although not necessarily in the same sequence or patterns shown here.

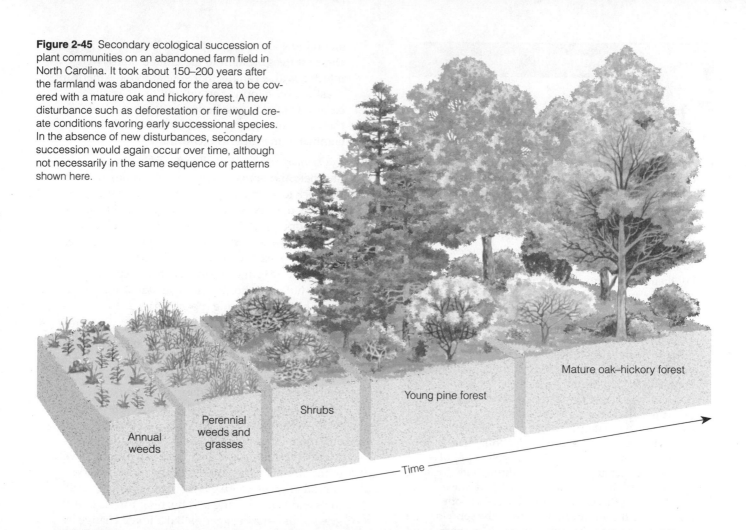

Annual weeds

Perennial weeds and grasses

Shrubs

Young pine forest

Mature oak–hickory forest

Time

Table 2-1 Ecosystem Characteristics at Immature and Mature Stages of Ecological Succession		
Characteristic	**Immature Ecosystem**	**Mature Ecosystem**
Ecosystem Structure		
Plant size	Small	Large
Species diversity	Low	High
Trophic structure	Mostly producers, few decomposers	Mixture of producers, consumers, and decomposers
Ecological niches	Few, mostly generalized	Many, mostly specialized
Community organization (number of interconnecting links)	Low	High
Ecosystem Function		
Food chains and webs	Simple, mostly plant → herbivore with few decomposers	Complex, dominated by decomposers
Efficiency of nutrient recycling	Low	High
Efficiency of energy use	Low	High

A: The gradual change in species composition of a given area

Ecological Surprises

Malaria once infected 9 out of 10 people in North Borneo, now known as Brunei. In 1955 the World Health Organization (WHO) began spraying the island with dieldrin (a DDT relative) to kill malaria-carrying mosquitoes. The program was so successful that the dreaded disease was virtually eliminated.

Other, unexpected things began to happen, however. The dieldrin also killed other insects, including flies and cockroaches living in houses. At first the islanders applauded this turn of events, but then small lizards that also lived in the houses died after gorging themselves on dieldrin-contaminated insects. Next, cats began dying after feeding on the lizards. Then, in the absence of cats, rats flourished and overran the villages. When the people became threatened by sylvatic plague carried by rat fleas, WHO parachuted healthy cats onto the island to help control the rats.

Then the villagers' roofs began to fall in. The dieldrin had killed wasps and other insects that fed on a type of caterpillar that either avoided or was not affected by the insecticide. With most of its predators eliminated, the caterpillar population exploded, munching its way through its favorite food: the leaves used in thatched roofs.

Ultimately, this episode ended happily: Both malaria and the unexpected effects of the spraying program were brought under control. Nevertheless, the chain of unforeseen events emphasizes the unpredictability of interfering with an ecosystem.

Critical Thinking

Do you believe that the beneficial effects of spraying pesticides in North Borneo outweighed the resulting unexpected and harmful effects? Explain.

---

the pests until eventually the chemicals become ineffective.

- *Eliminating some predators.* Ranchers, who don't want bison or prairie dogs competing with their sheep for grass, want to eradicate those species. They also want to eliminate wolves, coyotes, eagles, and other predators that occasionally kill sheep. Big game hunters also push for elimination of predators that prey on game species.

- *Deliberately or accidentally introducing new species,* some beneficial and some harmful to us

and other species. In the late 1800s several Chinese chestnut trees brought to the United States were infected with a fungus that spread to the American chestnut, once found throughout much of the eastern United States. Between 1910 and 1940, this accidentally introduced fungus virtually eliminated the American chestnut.

- *Overharvesting potentially renewable resources.* Ranchers and nomadic herders sometimes allow livestock to overgraze grasslands until erosion converts these ecosystems to less productive semideserts or deserts. Farmers sometimes deplete the soil of nutrients by excessive crop growing. Species of fish are overharvested. Wildlife species with economically valuable parts (such as elephant tusks, rhinoceros horns, and tiger skins) are endangered by illegal hunting (poaching).

- *Interfering with the normal chemical cycling and energy flows (throughputs) in ecosystems* (Figure 2-9). Soil nutrients can be easily eroded from monoculture crop fields, tree farms, construction sites, and other simplified ecosystems and can overload and disrupt other ecosystems such as lakes and coastal ecosystems. Chlorofluorocarbons (CFCs) released into the environment can increase the flow of ultraviolet energy reaching the earth, by reducing ozone levels in the stratosphere. Emissions of carbon dioxide and other greenhouse gases—from burning fossil fuels and from clearing and burning forests and grasslands—may trigger global climate change by disrupting energy flow through the atmosphere.

To survive we must exploit and modify parts of nature. However, we are beginning to understand that any human intrusion into nature has multiple effects, most of them unpredictable (Connections, left).

Solutions: What Can We Learn from Nature About Living Sustainably? Biologists have formulated several important principles that can help guide us in our search for more sustainable lifestyles:

- *We are part of, not apart from, the earth's dynamic web of life.*

- *Our lives, lifestyles, and economies are totally dependent on the sun and the earth.*

- *We can never do merely one thing*—what biologist Garrett Hardin calls the **first law of human ecology.**

- *Everything is connected to everything else; we are all in it together.* We are connected to all living organisms through the long evolutionary history contained in our DNA. We are connected to the earth through our interactions with air, water, soil, and other living

---

Q: At what rate was the world's population growing in 1997?

organisms making up the ever-changing web of life. The destiny of all species is a shared one. The primary goal of ecology is to discover which connections in nature are the strongest, most important, and most vulnerable to disruption.

We need not—and indeed cannot—stop growing food or building cities. Indeed, concentrating people in cities and increasing food supplies by raising the yields per area of cropland are both ways to help protect much of the earth's biodiversity from being destroyed or degraded, as long as the harmful environmental side effects of such activities are kept under control. The challenge is to maintain a balance between simplified, human-altered ecosystems and the neighboring, more complex natural ecosystems on which we and other forms of life depend, and to slow down the rates at which we are altering nature for our purposes.

If we simplify and degrade too much of the planet to meet our needs and wants, what's at risk is not the earth but our own species. According to biodiversity expert E. O. Wilson, "If this planet were under surveillance by biologists from another world, I think they would look at us and say, 'Here is a species in the mid-stages of self-destruction.'" The evolutionary lesson to be learned from nature is that no species can get "too big for its britches," at least not for long.

According to environmentalist David Brower, we need to focus on "global CPR—that's conservation, preservation, and restoration." This means **(1)** building societies based on conservation, not waste, **(2)** preserving what we can't replace, and **(3)** working with nature to help restore what we have degraded or destroyed.

Solutions: How Can We Develop Sustainable Societies? As a result of the law of conservation of matter and the second law of energy, individual resource use automatically adds some waste heat and waste matter to the environment. Most of today's advanced industrialized countries are **high-waste** (or high-throughput) **societies** that attempt to sustain ever-increasing economic growth by increasing the *throughput* of matter and energy resources in their economic systems (Figure 2-46). These resources flow through the economies of such societies to planetary *sinks* (air, water, soil, organisms), where pollutants and wastes end up and can accumulate to harmful levels.

However, there is an important lesson from nature to be learned from the scientific laws of matter and energy and the ecological concepts and prin-

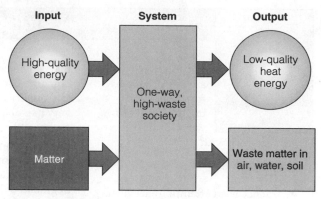

Figure 2-46 The eventually unsustainable high-waste or high-throughput societies of most developed countries are based on maximizing the rates of energy and matter flow. This process is rapidly converting the world's high-quality matter and energy resources into waste, pollution, and low-quality heat.

ciples discussed in this chapter (Spotlight, p. 70). They tell us that if more and more people continue to use and waste more and more energy and matter resources at an increasing rate, eventually the capacity of the environment to dilute and degrade waste matter and absorb waste heat will be exceeded. Thus, *at some point high-waste or high-throughput societies become unsustainable.*

A stopgap solution to this problem is to convert an unsustainable high-throughput society to a **matter-recycling society**. The goal of such a conversion is to allow economic growth to continue without depleting matter resources or producing excessive pollution and environmental degradation.

Even though recycling matter saves energy, the two laws of energy tell us that *recycling matter resources always requires expenditure of high-quality energy (which cannot be recycled) and adds waste heat to the environment.* For the long run, a matter-recycling society based on continuing population growth and per capita resource consumption must have an inexhaustible supply of affordable high-quality energy, and its environment must have an infinite capacity to absorb and disperse waste heat and to dilute and degrade waste matter.

There is also a limit to the number of times some materials, such as paper fiber, can be recycled before they become unusable. Changing to a matter-recycling society is an important way to buy some time, but it does not allow more and more people to use more and more resources indefinitely, even if all of them were somehow perfectly recycled.

The scientific laws governing matter and energy changes and the ecological concepts discussed in this

The following are key principles for understanding how the earth works and how we can work with it.

Science

- Science is an attempt to discover order in nature and then to use that knowledge to describe, explain, and predict what happens in nature.

- Scientific theories and laws are well tested and widely accepted principles with a high degree of certainty.

Matter

- Matter cannot be created or destroyed; it can only be changed from one form to another. Everything we think we have thrown away is still with us in one form or another; there is no "away" (*law of conservation of matter*).

- Organized and concentrated matter is high-quality matter that can usually be converted into useful resources at an affordable cost; disorganized and dispersed matter is low-quality matter that often costs too much to convert to a useful resource (*principle of matter quality*).

Energy

- Energy cannot be created or destroyed; it can only be changed from one form to another. We can't get energy for nothing; in terms of energy quantity, it takes energy to get energy (*first law of energy or thermodynamics* or *law of conservation of energy*).

- Organized or concentrated energy is high-quality energy that can be used to do things; disorganized or diluted energy is low-quality energy that is not very useful (*principle of energy quality*).

- In any conversion of energy from one form to another, high-quality, useful energy is always degraded to lower-quality, less useful energy that can't be recycled to give high-quality energy; we can't break even in terms of energy quality (*second law of energy or thermodynamics*).

- Ideally, high-quality energy should not be used to do something that can be done with lower-quality energy; we don't need to use a chain saw to cut butter (*principle of energy efficiency*).

Life

- Life on earth depends on the one-way flow of high-quality energy from the sun, through earth's life-support systems, and eventually back into space as low-quality heat; gravity; and the recycling of vital chemicals by a combination of biological, geological, and chemical processes (*principle of energy flow, gravity, and matter recycling*).

- Each species and each individual organism can tolerate only a certain range of environmental conditions (*range-of-tolerance principle*).

- Too much or too little of a physical or chemical factor can limit or prevent the growth of a population in a particular place (*limiting factor principle*).

- Every species has a specific role to play in nature (*ecological-niche principle*).

- Species interact through competition for resources, predation, parasitism, mutualism, and commensalism (*principle of species interactions*).

- When possible species reduce or avoid competition with one another by dividing up scarce resources so that species with similar requirements use them at different times, in different ways, or in different places (*principle of resource partitioning*).

- The size, growth rate, age structure, density, and distribution of a species's population are controlled by its interactions with other species and with its non-living environment (*principle of population dynamics*).

- No population can keep growing indefinitely (*carrying capacity principle*).

- Average precipitation and temperature are the major factors determining whether a particular land area supports a desert, grassland, or forest (*climate–biome principle*).

- Individuals of a population of a species vary in their genetic makeup as a result of random mutations or inheritable changes in molecules making up their genes (*principle of genetic variability*).

- Individuals of a population of a species that possess genetically controlled characteristics enhancing their ability to survive under existing environmental conditions have a greater chance of surviving and producing more offspring than do those lacking such traits (*principle of adaptation and natural selection*).

- As environmental conditions change, the number and types

Q: What two countries have the world's largest populations?

of species present in a particular area change and, if not disturbed, can often form more complex communities (*principle of ecological succession*).

■ All species eventually become extinct by disappearing or by evolving into one or more new species in response to environmental changes brought about by natural processes or by human action (*principle of evolution*).

■ Over billions of years, changes in environmental conditions have led to development of a variety of species (species diversity), genetic variety within species (genetic diversity), and a variety of natural systems (ecosystem diversity) through a mixture of extinction and formation of new species (*biodiversity principle*).

■ The earth's atmosphere, hydrosphere, lithosphere, and forms of life are continually changing in response to changes in solar input, heat flows from the earth's interior, movements of the earth's crust, other natural changes, and changes brought about by humans and other living organisms (*principle of adaptability*).

■ The earth's life-support systems can withstand much stress and abuse, but there are limits to how much can be tolerated (*principle of limits*).

Humans and Environment

■ Our survival, life quality, and economies are totally dependent on the sun and the earth; the earth can get along without us, but we can't get along without the earth (*principle of earth capital*).

■ We should try to understand and work with the rest of nature to sustain the ecological integrity, biodiversity, and adaptability of earth's life-support systems for us and other species (*sustainability principle*).

■ We can learn a lot about how nature works, but nature is so incredibly complex and dynamic that such knowledge will always be limited (*principle of complexity*).

■ In nature, we can never do just one thing; everything we do creates effects that are often unpredictable (*first law of human ecology*).

■ Everything is connected to and intermingled with everything else; we are all in this together; we need to understand these connections and discover which connections are most important for sustaining life on earth (*principle of interdependence or connectedness*).

■ Most resources are limited and should not be wasted; there is not always more (*principle of resource conservation*).

■ Living off renewable solar energy and renewable matter resources is a sustainable human lifestyle; using renewable matter resources faster than they are replenished and living off nonrenewable matter and energy resources degrade and deplete earth capital and ultimately constitute an unsustainable lifestyle (*principle of sustainable living*).

■ Renewable resources should be used no faster than they are replenished by natural processes (*principle of sustainable use*).

■ Increases in population, resource use, or both can eventually overwhelm attempts to control pollution and manage wastes (*environmental impact principle*).

■ Pollution and wastes should not be put into the environment faster than the environment can degrade and recycle them or render them harmless (*principle of optimum pollution*).

■ The best and cheapest way to reduce pollution and waste is to not produce so much (*principle of pollution prevention and waste reduction*).

■ The best way to protect species and individual organisms is to protect the ecosystems in which they live and to help restore those we have degraded (*principle of ecosystem protection and restoration*).

■ We should change earth-degrading and earth-depleting manufacturing processes, products, and businesses into earth-sustaining ones by using economic incentives and penalties (*principle of economic–ecological sustainability*).

■ The market price of a product should include all estimated present and future costs of any pollution, environmental degradation, or other harmful effects connected with it that are passed on to society, the environment, and future generations (*principle of full-cost pricing*).

■ Anticipating and preventing problems is cheaper and more effective than reacting to and trying to rectify them; an ounce of prevention is worth a pound of cure (*precautionary principle*).

Critical Thinking

1. List any of the above principles you disagree with and explain why you disagree with them.

2. Can you add any principles to this list?

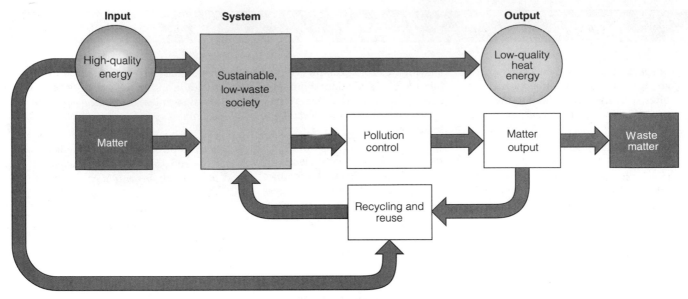

Figure 2-47 A sustainable low-waste or low-throughput society, based on energy flow and matter recycling, works with nature to reduce throughput. This is done by **(1)** reusing and recycling most nonrenewable matter resources, **(2)** using potentially renewable resources no faster than they are replenished, **(3)** using matter and energy resources efficiently, **(4)** reducing unnecessary consumption, **(5)** emphasizing pollution prevention and waste reduction, and **(6)** controlling population growth.

chapter suggest that the best long-term solution to our environmental and resource problems is to shift from a society based on maximizing matter and energy flow (throughput) to a sustainable **low-waste** (or low-throughput) **society**. The major features of such a society are summarized in Figure 2-47.

According to many scientists, the scientific laws and principles discussed in this chapter reveal that we all depend on one another and on the rest of nature for our survival and that we must learn how to work with, not against, the rest of nature.

If we love our children, we must love the earth with tender care and pass it on, diverse and beautiful, so that on a warm spring day 10,000 years hence they can feel peace in a sea of grass, can watch a bee visit a flower, can hear a sandpiper call in the sky, and can find joy in being alive.

HUGH H. ILTIS

CRITICAL THINKING

1. Respond to the following statements:
   a. It has never been absolutely scientifically proven that anyone has ever died from smoking cigarettes.
   b. The greenhouse theory—that certain gases (such as water vapor and carbon dioxide) trap heat in the atmosphere and thus influence the average temperature and climate of the earth—is not a reliable idea because it is only a scientific theory.

2. See whether you can find an advertisement or an article describing or using some aspect of science in which **(a)** the concept of scientific proof is misused, **(b)** the term *theory* is used when it should have been *hypothesis*, and **(c)** a consensus scientific finding is dismissed or downplayed because it is "only a theory."

3. If there is no "away," why isn't the world filled with waste matter?

4. Use the second energy law to explain why a barrel of oil can be used as a fuel only once.

5. **(a)** A bumper sticker asks, "Have you thanked a green plant today?" Give two reasons for appreciating a green plant. **(b)** Trace the sources of the materials that make up the bumper sticker and then decide whether the sticker itself is a sound application of the slogan.

6. Explain how decomposers help keep you alive.

7. Using the second law of energy, explain why there is such a sharp decrease in usable energy as energy flows through a food chain or web. Doesn't an energy loss at each step violate the first law of energy? Explain.

8. Using the second law of energy, explain why many poor people in developing countries live mostly on a vegetarian diet.

Q: How many people are added to the world's population each year?

9. Why are coastal and inland wetlands and coral reefs such important ecosystems? Why have so many of these vital ecosystems been destroyed by human activities?

10. Someone tells you not to worry about air pollution because through natural selection the human species will develop lungs that can detoxify pollutants. How would you reply?

11. Explain why a simplified ecosystem such as a cornfield is usually much more vulnerable to harm from insects and plant diseases than a more complex, natural ecosystem such as a grassland.

12. (a) Use the law of conservation of matter to explain why a matter-recycling society will sooner or later be nec-essary. (b) Use the first and second laws of energy to explain why, in the long run, we will need a low-waste or low-throughput society, not just a matter-recycling society.

13. (a) Imagine that you have the power to violate the law of conservation of energy (the first energy law) for one day. What are the three most important things you would do with this power? (b) Repeat this process, imagining that you have the power to violate the second law of energy for one day.

14. Imagine that after you die you are somehow given the choice to live again as any organism you choose except a human being. What organism would you choose to be, and why?

3 THE HUMAN POPULATION: SIZE AND DISTRIBUTION

We shouldn't delude ourselves: The population explosion will come to an end before very long. The only remaining question is whether it will be halted through the humane method of birth control, or by nature wiping out our surplus.

PAUL H. EHRLICH

3-1 FACTORS AFFECTING HUMAN POPULATION SIZE

How Is Population Size Affected by Birth Rates and Death Rates? Populations grow or decline through the interplay of three factors: births, deaths, and migration. **Population change** is calculated by subtracting the number of people leaving a population (through death and emigration) from the number entering it (through birth and immigration) during a specified period of time (usually a year):

$$\text{Population change} = \begin{pmatrix} \text{Births} \\ + \\ \text{Immigration} \end{pmatrix} - \begin{pmatrix} \text{Deaths} \\ + \\ \text{Emigration} \end{pmatrix}$$

When births plus immigration exceed deaths plus emigration, population increases; when the reverse is true, population declines. When these factors balance out, population size remains stable, a condition known as **zero population growth (ZPG)**.

Instead of using the total numbers of births and deaths per year, demographers use two statistics: the **birth rate**, or **crude birth rate** (the number of live births per 1,000 people in a population in a given year), and the **death rate**, or **crude death rate** (the number of deaths per 1,000 people in a population in a given year). Figure 3-1 shows the crude birth and death rates for various groupings of countries in 1998.

Birth rates and death rates are coming down worldwide, but death rates have fallen more sharply than birth rates. As a result, there are more births than deaths; every time your heart beats three more babies are added to the world's population. At this rate, each day we share the earth and its resources with about 236,000 more people than the day before. In 1998 there were about 84 million more mouths to feed, and 12 years from now there will be 1 billion more people.

The rate of the world's annual population change (excluding migration) is usually expressed as a percentage:

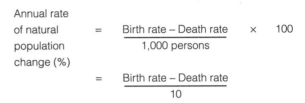

$$\begin{aligned}
\text{Annual rate of natural population change (\%)} &= \frac{\text{Birth rate} - \text{Death rate}}{1,000 \text{ persons}} \times 100 \\
&= \frac{\text{Birth rate} - \text{Death rate}}{10}
\end{aligned}$$

The rate of the world's annual population growth (natural increase) dropped 35% between 1963 and 1998, from 2.2% to 1.43%. This is good news, but during the same period the population base rose by about 85%,

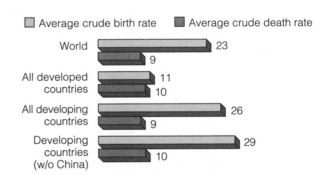

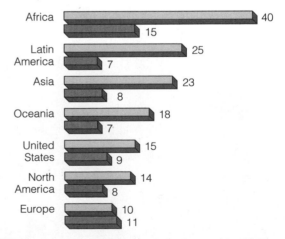

Figure 3-1 Average crude birth and death rates for various groupings of countries in 1998. (Data from Population Reference Bureau)

Ehrlich, Paul R., and Anne H. Ehrlich. 1997. "The Population Explosion." *Environmental Law*, vol. 27, no. 4, 1187(22).

from 3.2 billion to 5.93 billion. This 35% drop in the rate of population increase is roughly analogous to learning that the truck heading straight at you has slowed from 100 kilometers per hour to 65 while its weight increased by 85%. In other words, the problem of exponential population growth has not disappeared; it's just occurring at a slower rate.

An annual natural increase rate of 1–3% may seem small, but the current annual natural increase rate of 1.43% is equal to adding another Los Angeles every 3 weeks, another New York City every month, a Germany every year, and a United States every 3 years. Despite the drop in the rate of population growth, the larger base of population means that the 84 million people added in 1998 was much higher than the 69 million added in 1963 when the world's population growth rate reached its peak.

In numbers of people, China (with 1.24 billion in 1998, about one of every five people in the world) and India (with 989 million) dwarf all other countries. Together they make up 38% of the world's population. The United States, with 270 million people in 1998 had the world's third largest population but only 4.6% of the world's people in 1997. Figure 3-2 gives projected population growth in various regions between 1998 and 2025; more than 95% of this growth is projected to take place in developing countries, where hunger and poverty have become a way of life for almost a billion people.

🌐 How Have Global Fertility Rates Changed?

Two types of fertility rates affect a country's population size and growth rate. The first type, **replacement-level fertility**, is the number of children a couple must bear to replace themselves. It is slightly higher than two children per couple (2.1 in developed countries and as high as 2.5 in some developing countries), mostly because some female children die before reaching their reproductive years.

Lowering fertility rates to replacement level does not mean an immediate halt in population growth (zero population growth); there are so many future parents already alive that if each had an average of 2.1 children and their children also had 2.1 children, the population would continue to grow for 50 years or more (assuming that death rates don't rise). About 3 billion women—equal to the world's population in 1960—will enter their childbearing years in the next few years, thus causing population growth to continue for decades.

The second type of fertility rate, and the most useful measure of fertility for projecting future population change, is the **total fertility rate (TFR)**: an estimate of the average number of children a woman will have during her childbearing years under current age-specific birth rates. In 1998, the worldwide average TFR was 2.9 children per woman. It was 1.6 in developed countries (down from 2.5 in 1950) and 3.3 in developing countries (down from 6.5 in 1950). This drop in the average number of children born to women in developing countries is an impressive decline, but this level of fertility is still far above the replacement level.

Population experts expect TFRs in developed countries to remain around 1.6 and those in developing countries to drop to around 2.3 by 2025; these rates are the basis of the population projections in Figure 3-2. That is good news, but it will still lead to a projected world population of around 8 billion by 2025, with more than 90% of this growth taking place in developing countries (Figure 1-6).

Case Study: How Have Fertility Rates Changed in the United States? The population of the United States has grown from 76 million in 1900 to 270 million in 1998—a 2.6-fold increase—even though the country's TFR has oscillated wildly (Figure 3-3). In 1957, the peak of the post–World War II baby boom, the TFR reached 3.7 children per woman. Since then it

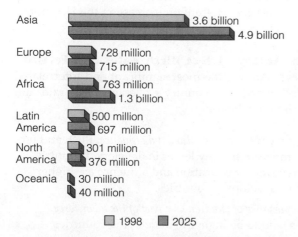

Figure 3-2 Population projections by region, 1998–2025. (Data from United Nations and Population Reference Bureau)

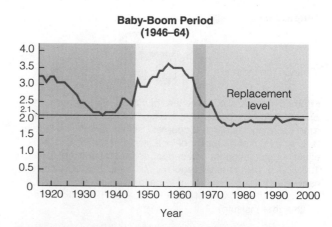

Figure 3-3 Total fertility rate for the United States between 1917 and 1998. (Data from Population Reference Bureau and U.S. Census Bureau)

Hint: Enter the search term *zero population growth* using the Subject Guide.

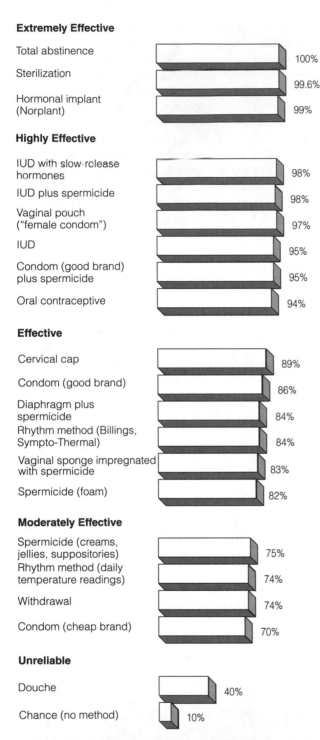

Extremely Effective

Total abstinence — 100%

Sterilization — 99.6%

Hormonal implant (Norplant) — 99%

Highly Effective

IUD with slow-release hormones — 98%

IUD plus spermicide — 98%

Vaginal pouch ("female condom") — 97%

IUD — 95%

Condom (good brand) plus spermicide — 95%

Oral contraceptive — 94%

Effective

Cervical cap — 89%

Condom (good brand) — 86%

Diaphragm plus spermicide — 84%

Rhythm method (Billings, Sympto-Thermal) — 84%

Vaginal sponge impregnated with spermicide — 83%

Spermicide (foam) — 82%

Moderately Effective

Spermicide (creams, jellies, suppositories) — 75%

Rhythm method (daily temperature readings) — 74%

Withdrawal — 74%

Condom (cheap brand) — 70%

Unreliable

Douche — 40%

Chance (no method) — 10%

Figure 3-4 Typical effectiveness of birth control methods in the United States. Percentages are based on the number of undesired pregnancies per 100 couples using a specific method as their sole form of birth control for a year. For example, a 94% effectiveness rating for oral contraceptives means that for every 100 women using the pill regularly for 1 year, 6 will get pregnant. Effectiveness rates tend to be lower in developing countries, primarily because of lack of education. (Data from Alan Guttmacher Institute)

has generally declined, remaining at or below replacement level since 1972.

The drop in the TFR has led to a decline in the rate of population growth in the United States. However, the country's population is still growing faster than that of most developed countries and is not even close to zero population growth. Including immigration, the U.S. population of 270 million grew by 1.17% in 1998—more than double the mean rate of the world's industrialized nations. This growth added about 3.1 million people: 1.8 million more births than deaths (accounting for about 60% of the growth), 935,000 legal immigrants and refugees, and an estimated 400,000 illegal immigrants. This is equivalent to adding another California every 10 years.

According to U.S. Bureau of Census projections, the U.S. population will increase from 270 million to 383 million by the year 2050—a 42% increase, with no stabilization on the horizon. This is a moderate projection. A less conservative estimate projects a population of 507 million by 2050, almost double the population in 1998. Because of a high per capita rate of resource use, each addition to the U.S. population has an enormous environmental impact (Figure 1-15).

The main reasons for this projected growth are **(1)** the large number of baby-boom women who are still in their childbearing years (even though the TFR has remained at or below replacement level for 25 years, there has been a large increase in the number of potential mothers), **(2)** an increase in the number of unmarried mothers (including teenagers), **(3)** a continuation of higher fertility rates for women in some racial and ethnic groups than for Caucasian women, **(4)** high levels of legal and illegal immigration (which accounts for about 43% of current U.S. population growth), and **(5)** inadequate family planning services (especially for the poor, who often can't afford such services). If immigration continues at current levels, new immigrants and their descendants would account for 80 million, or 72% of the projected 111-million increase in the U.S. population between 1999 and 2050.

What Factors Affect Birth Rates and Fertility Rates? Among the most significant and interrelated factors affecting a country's average birth rate and TFR are the following:

- *Average level of education and affluence.* Birth and fertility rates are usually lower in developed countries, where levels of education and affluence are higher than in developing countries.

- *Importance of children as a part of the labor force.* Rates tend to be higher in developing countries (especially in rural areas, where children begin working at an early age).

Q: How much did global average life expectancy increase between 1900 and 1998

- *Urbanization.* People living in urban areas usually have better access to family planning services and tend to have fewer children than those living in rural areas, where children are needed to perform essential tasks.

- *Cost of raising and educating children.* Rates tend to be lower in developed countries, where raising children is much more costly because children don't enter the labor force until their late teens or early 20s.

- *Educational and employment opportunities for women.* Rates tend to be low when women have access to education and paid employment outside the home. TFRs tend to decline as the female literacy rate increases.

- *Infant mortality rate.* In areas with low infant mortality rates, people tend to have smaller families because fewer children die at an early age.

- *Average age at marriage* (or, more precisely, the average age at which women have their first child). Women normally have fewer children when their average age at marriage is 25 or older.

- *Availability of private and public pension systems.* Pensions eliminate the need of parents to have many children to help support them in old age.

- *Availability of legal abortions.* There are an estimated 30 million legal abortions and 11–22 million illegal abortions worldwide each year.

- *Availability of reliable methods of birth control* (Figure 3-4).

- *Religious beliefs, traditions, and cultural norms.* In some countries, these factors favor large families and strongly oppose abortion and some forms of birth control.

What Factors Affect Death Rates? The rapid growth of the world's population over the past 100 years is not the result of a rise in the crude birth rate; rather, it has been caused largely by a decline in crude death rates (especially in developing countries; Figure 3-5). More people started living longer (and fewer infants died) because of increased food supplies and distribution, better nutrition, improvements in medical and public health technology (such as immunizations and antibiotics), improvements in sanitation and personal hygiene, and safer water supplies (which have curtailed the spread of many infectious diseases).

Two useful indicators of overall health in a country or region are **life expectancy** (the average number of years a newborn infant can expect to live) and the **infant mortality rate** (the number of babies out of every 1,000 born each year who die within a year of birth; Figure 3-6). In most cases, a low life expectancy

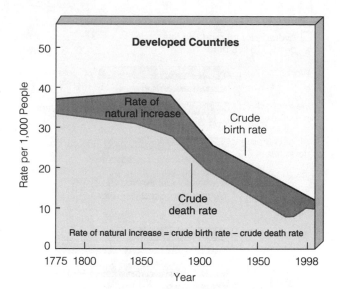

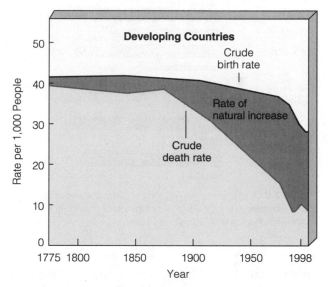

Figure 3-5 Changes in crude birth and death rates for developed and developing countries between 1775 and 1998, and projected rates (dashed lines) to 2000. (Data from Population Reference Bureau and United Nations)

in an area is the result of high infant mortality. Life expectancy has increased since 1965 to an average of 75 years in developed countries and 63 years in developing countries in 1998. Globally, life expectancy at birth increased from 48 in 1955 to 66 in 1998 and is projected to reach 73 by 2025. But in the world's 41 poorest countries, mainly in Asia and Africa, life expectancy is only about 50 years.

Because it reflects the general level of nutrition and health care, infant mortality is probably the single most important measure of a society's quality of life. A high infant mortality rate usually indicates insufficient food (undernutrition), poor nutrition (malnutrition), and a high incidence of infectious disease (usually from

A: From 30 to 66

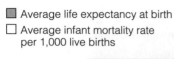

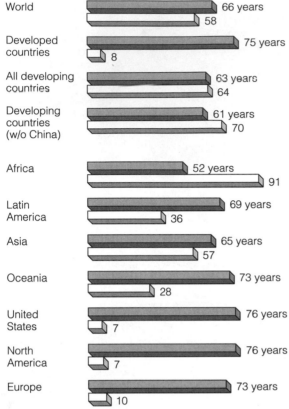

Figure 3-6 Average life expectancy at birth and average infant mortality rates for various groupings of countries in 1998. (Data from Population Reference Bureau)

contaminated drinking water). Between 1965 and 1998, the world's infant mortality rate dropped from 20 per 1,000 live births to 8 in developed countries, and from 118 to 64 in developing countries. This is an impressive achievement, but it still means that at least 9 million infants die of preventable causes during their first year of life—an average of 24,700 unnecessary infant deaths per day.

Although the U.S. infant mortality rate of 7.0 per 1,000 in 1998 was low by world standards, 32 other countries had lower rates in 1998. Three factors that keep the U.S. infant mortality rate higher than it could be are inadequate health care (for poor women during pregnancy and for their babies after birth), drug addiction among pregnant women, and the high birth rate among teenage women. The United States has the highest teenage pregnancy rate of any industrialized country. Babies born to teenagers are more likely to have low birth weights, the most important factor in infant deaths.

What Are Age Structure Diagrams? As mentioned earlier, even if the replacement-level fertility rate of 2.1 were magically achieved globally tomorrow, the world's population would keep growing for at least another 50 years, stabilizing at about 8.6 billion (assuming no increase in death rates). The reason for this is the **age structure** of a population, or the proportion of the population (or of each sex) at each age level.

Demographers typically construct a population age structure diagram by plotting the percentages or numbers of males and females in the total population in each of three age categories: *prereproductive* (ages 0–14), *reproductive* (ages 15–44), and *postreproductive* (ages 45 and up). Figure 3-7 presents generalized age structure diagrams for countries with rapid, slow, zero, and negative population growth rates.

Any country with many people below age 15 (represented by a wide base in Figure 3-7, left) has a powerful built-in *momentum* to increase its population size unless death rates rise sharply. The number of births rises even if women have only one or two children because of the large number of women who will soon be moving into their reproductive years.

In 1998, half the world's 3.0 billion women were in the reproductive age group, and 32% of the people on the planet were under 15 years old, poised to move into their prime reproductive years. In developing countries the number is even higher: 35% compared with 19% in developed countries. This powerful force for continued population growth, mostly in developing countries, will be slowed only by an effective program to reduce birth rates, or by a catastrophic rise in death rates.

How Can Age Structure Diagrams Be Used to Make Population and Economic Projections? The 78-million-person increase that occurred in the U.S. population between 1946 and 1964, known as the *baby boom* (Figure 3-8), will continue to move up through the country's age structure as the members of this group grow older (Figure 3-8). Baby boomers now make up nearly half of all adult Americans. As a result, they dominate the population's demand for goods and services and play an increasingly important role in deciding who gets elected and what laws are passed. Baby boomers who created the youth market in their teens and 20s are now creating the 50-something market.

The economic burden of helping support so many retired baby boomers will fall on the *baby-bust generation*: people born since 1965 (when TFRs fell sharply and have remained below 2.1 since 1970; Figure 3-3).

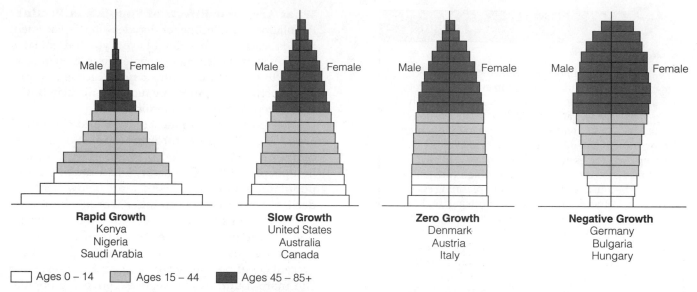

Figure 3-7 Generalized population age structure diagrams for countries with rapid, slow, zero, and negative population growth rates. (Data from Population Reference Bureau)

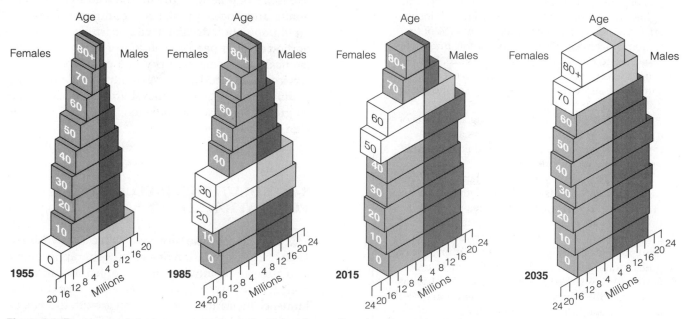

Figure 3-8 Tracking the baby-boom generation in the United States. (Data from Population Reference Bureau and U.S. Census Bureau)

Retired baby boomers may use their political clout to force the smaller number of people in the baby-bust generation to pay higher income, health-care, and Social Security taxes. This could lead to much resentment and conflicts between the two generations.

In other respects the baby-bust generation should have an easier time than the baby-boom generation. Fewer people will be competing for educational op-portunities, jobs, and services, and labor shortages may drive up their wages, at least for jobs requiring education or technical training beyond high school. On the other hand, members of the baby-bust group may find it difficult to get job promotions as they reach middle age because most upper-level positions will be occupied by members of the much larger baby-boom group. Many baby boomers may delay retirement because of

Hint: Enter the search term *demographic transition* using the Subject Guide.

The Graying of Japan

In only 7 years, between 1949 and 1956, Japan cut its birth, total fertility, and population growth rates in half. The main reason was widespread access to family planning implemented by the post–World War II U.S. occupation forces and the Japanese government.

Since 1956 these rates have declined further, mostly because of access to family planning services and other factors: cramped housing, high land prices, late marriage ages, and high costs of education. In 1949 Japan's total fertility rate was 4.5; in 1998 it was 1.4, one of the world's lowest. If this trend continues and immigration doesn't rise, Japan's population should begin decreasing around 2006 and could shrink to 65–96 million by 2090, depending on its fertility rate and immigration rate (which is currently negligible).

As Japan approaches zero population growth, it is beginning to face some of the problems of an aging population. Japan's universal health insurance and pension systems used about 42% of the national income in 1997. This economic burden is projected to rise to 60% or higher in 2020. Japanese economists worry that the steep taxes needed to fund these services could discourage economic growth.

Since 1980 Japan has been feeling the effects of a declining workforce. This is one reason it has invested heavily in automation and encouraged women to work outside the home.

The population of Japan is 99% Japanese. Fearing a breakdown in its social cohesiveness, the government has been unwilling to increase immigration to provide more workers. Despite this official policy, the country is becoming increasingly dependent on illegal immigrants to keep its economic engines running. How Japan deals with these problems will be watched closely by other countries as they make the transition to zero population growth and, eventually, to population decline.

Critical Thinking

What do you believe are the three most important things Japan should do about the problems associated with population decline?

---

improved health and the need to accumulate adequate retirement funds.

From these few projections we can see that any booms or busts in the age structure of a population create social and economic changes that ripple through a society for decades.

What Are Some Effects of Population Decline? Populations can decline for decades after replacement level is reached if they don't have a youth-dominated age structure. When more people are in their postreproductive years than in their reproductive and prereproductive years, there are more deaths than births, even at replacement-level fertility.

The populations of most of the world's countries are projected to grow throughout most of the 21st century. By 1998, however, 37 countries with 684 million people—12% of humanity—had roughly stable populations (annual growth rates below 0.3%) or declining populations. In other words, about one-eighth of humanity has achieved a stable population. As the projected age structure of the world's population changes between 1998 and 2150, and the percentage of people age 65 or older increases, more and more countries will begin experiencing population declines.

If population decline is gradual, its negative effects can usually be managed. But rapid population decline, like rapid population growth, can lead to severe economic and social problems. Countries undergoing rapid population decline have a sharp rise in the proportion of older people, who consume a large share of medical care, Social Security, and other costly public services (Case Study, left). A country with a declining population can also face labor shortages unless it relies on greatly increased automation, immigration of foreign workers, or both.

3-3 SOLUTIONS: INFLUENCING POPULATION SIZE

How Is Population Size Affected by Migration? The population of a given geographic area is also affected by movement of people into (immigration) and out of (emigration) that area. Most countries influence their rates of population growth to some extent by restricting immigration. Only a few countries—chiefly Canada, Australia, and the United States (Case Study, right)—allow large annual increases in population from immigration.

Only about 1% of the annual population growth in developing countries is absorbed by developed countries through international migration. Thus, population change for most countries is determined mainly by the difference between their birth rates and death rates.

Migration within countries, especially from rural to urban areas, plays an important role in the population dynamics of cities, towns, and rural areas, as discussed in Section 3-5.

Q: What is the size of the U.S. population?

CASE STUDY

Between 1820 and 1996, the United States admitted almost twice as many immigrants and refugees as all other countries combined. However, the number of legal immigrants has varied during different periods because of changes in immigration laws and rates of economic growth (Figure 3-9).

Between 1820 and 1960, most legal immigrants to the United States came from Europe; since then, most have come from Asia and Latin America. If current trends continue, by 2050 almost half of the U.S. population will be Spanish speaking (up from 10% in 1993).

In 1995 the U.S. Commission on Immigration Reform recommended reducing the number of legal immigrants and refugees to about 700,000 per year for a transition period and then to 550,000 a year. Some demographers and environmentalists go further and call for lowering the annual ceiling for legal immigrants and refugees into the United States to 300,000–450,000, or for limiting legal immigration to about 20% of annual population growth.

Most of these analysts also support efforts to sharply reduce illegal immigration, although some are concerned that a crackdown on illegal immigrants can also lead to

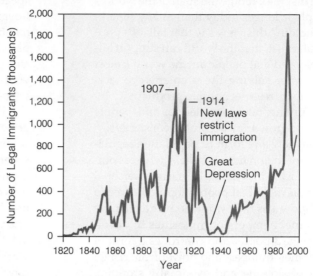

Figure 3-9 Legal immigration to the United States, 1820–1997. The large increase in immigration since 1989 resulted mostly from the Immigration Reform and Control Act of 1986, which granted legal status to illegal immigrants who could show that they had been living in the country for several years. In 1997, the United States received about 935,000 legal immigrants and refugees and 400,000 illegal immigrants, together accounting for 40% of the country's population growth. The Immigration and Naturalization Service estimates that there are about 1–3 million illegal immigrants in the United States. (Data from U.S. Immigration and Naturalization Service)

discrimination against legal immigrants. Proponents argue that such policies would allow the United States to stabilize its population sooner and help reduce the country's enormous environmental impact. Others oppose reducing current levels of legal immigration, arguing that it would diminish the historic role of the United States

as a place of opportunity for the world's poor and oppressed.

Critical Thinking

Should the United States reduce its current level of legal immigration and tighten up on illegal immigration? Explain.

What Are the Pros and Cons of Reducing Birth Rates? Because raising the death rate is not desirable, lowering the birth rate is the focus of most efforts to slow population growth. Today about 93% of the world's population (and 91% of the people in developing countries) live in countries with fertility reduction programs. The funding for and effectiveness of these programs vary widely; few governments spend more than 1% of the national budget on them.

The unprecedented projected doubling of the human population from 5 to 10 billion or more between 1985 and 2050 raises some important questions. Can we

provide enough food, energy, water, sanitation, education, health care, and housing for twice as many people? Can we reduce already serious poverty so that people can get enough food and other basic necessities without being forced to use potentially renewable resources unsustainably to survive? Can the world provide an adequate standard of living for twice as many people without causing massive environmental damage?

There is intense controversy over these questions, whether the earth is overpopulated, and what measures, if any, should be taken to slow population growth. To some the planet is already overpopulated, but others disagree. Some analysts—mostly economists—argue that we should encourage population growth.

Those who don't believe that the earth is overpopulated point out that the average life span of the world's 5.93 billion people is longer today than at any time in the past. Those holding this view say that talk of a population crash is alarmist, that the world can support billions more people, and that people are the world's most valuable resource for solving the problems we face. Large populations increase economic productivity by creating and applying new knowledge, and people stimulate economic growth by becoming consumers.

Some people view any form of population regulation as a violation of their religious beliefs, whereas others see it as an intrusion into their privacy and personal freedom. They believe that all people should be free to have as many children as they want. Some developing countries and some members of minorities in developed countries regard population control as a form of genocide to keep their numbers and power from rising.

Proponents of slowing and eventually stopping population growth point out that we fail to provide the basic necessities for one out of five people on the earth today. If we can't (or won't) do this now, how will we be able to do this for twice as many people within the next 49 years? They contend that if we don't sharply lower birth rates, we are deciding by default to raise death rates for humans and greatly increase environmental harm (Figure 3-10). In 1992, for example, the highly respected U.S. National Academy of Sciences and the Royal Society of London issued the following joint statement: "If current predictions of population growth and patterns of human activity on the planet remain unchanged, science and technology may not be able to prevent either irreversible degradation of the environment or continued poverty for much of the world."

Proponents of this view recognize that population growth is not the only cause of our environmental and resource problems. However, they argue that adding several hundred million more people in developed countries and several billion more in developing countries can only intensify many environmental and social problems. They call for drastic changes to prevent accelerating environmental decline (Figure 3-11).

Proponents of slowing and eventually halting population growth believe that the United States and other developed countries should establish an official goal of stabilizing their populations as soon as possible. Proponents also believe that developed countries have a better chance of influencing developing countries to reduce their population growth more rapidly if the more affluent countries officially recognize the need to

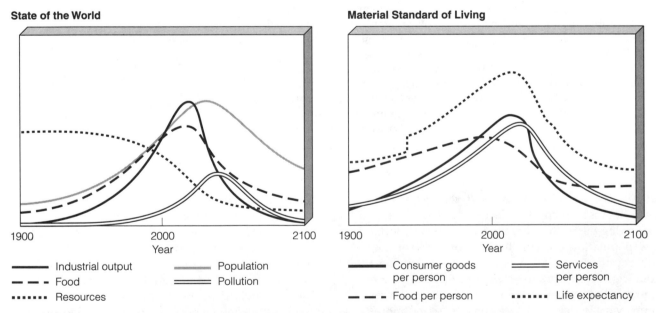

Figure 3-10 Plots of computer model projecting what might happen if the world's population and economy continue growing exponentially at 1990 levels, assuming no major policy changes or technological innovations. This scenario projects that the world has already overshot some of its limits and that if current trends continue unchanged, we face global economic and environmental collapse sometime in the next century. (Used by permission from Donella Meadows et al., *Beyond the Limits: Confronting Global Collapse, Envisioning a Sustainable Future*, White River Junction, Vt.: Chelsea Green, 1992)

Q: What is the average number of children per woman in the United States?

stabilize their own populations and reduce their unnecessary waste of the earth's resources.

These analysts believe that people should have the freedom to produce as many children as they want. However, such freedom would apply only if it did not reduce the quality of other people's lives now and in the future, either by impairing the earth's ability to sustain life or by causing social disruption. They point out that limiting the freedom of individuals to do anything they want in order to protect the freedom of other individuals is the basis of most laws in modern societies. What is your opinion on this issue?

How Can Economic Development Help Reduce Births? Demographers have examined the birth and death rates of western European countries that industrialized during the nineteenth century, and from these data they developed a hypothesis of population change known as the **demographic transition**: As countries become industrialized, first their death rates and then their birth rates decline.

According to this hypothesis, the transition takes place in four distinct stages (Figure 3-12). In the *preindustrial stage*, harsh living conditions lead to a high birth rate (to compensate for high infant mortality) and a high death rate. Thus, there is little population growth.

In the *transitional stage, industrialization* begins, food production rises, and health care improves. Death rates drop and birth rates remain high, so the population grows rapidly (typically 2.5–3% a year).

In the *industrial stage*, industrialization is widespread. The birth rate drops and eventually approaches the death rate. Reasons for this convergence of rates include better access to birth control, decline in the infant mortality rate, increased job opportunities for women, and the high costs of raising children who don't enter the workforce until after high school or college. Population growth continues, but at a slower and perhaps fluctuating rate, depending on economic conditions. Most developed countries are now in this third stage, and a few developing countries are entering this stage.

In the *postindustrial stage*, the birth rate declines even further, equaling the death rate and thus reaching zero population growth. Then the birth rate falls below the death rate and total population size slowly decreases. Emphasis shifts from unsustainable to sustainable forms of economic development. Some 37 countries (most of them in western Europe) containing about 12% of the world's population have entered this stage. To most population experts, the challenge is to help the remaining 88% of humanity reach this stage.

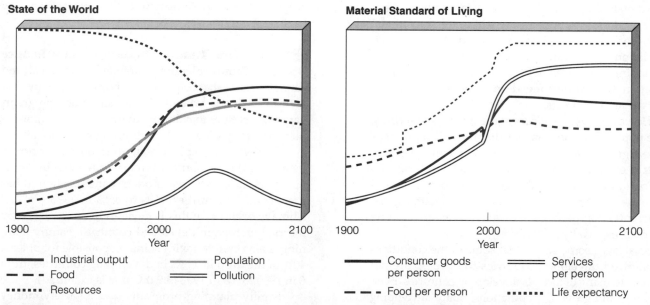

State of the World

1900   2000   2100
Year

——— Industrial output
– – – Food
······· Resources
——— Population
═══ Pollution

Material Standard of Living

1900   2000   2100
Year

——— Consumer goods per person
– – – Food per person
═══ Services per person
······· Life expectancy

Figure 3-11 Computer-generated scenario projecting how we can avoid overshoot and collapse and make a fairly smooth transition to a sustainable future. It assumes that **(1)** technology allows us to double supplies of non-renewable resources, double crop and timber yields, cut soil erosion in half, and double the efficiency of resource use within 20 years; **(2)** 100% effective birth control was made available to everyone by 1995; **(3)** no couple has more than two children, beginning in 1995; and **(4)** per capita industrial output is stabilized at 1990 levels. Another computer run projects that waiting until 2015 to implement these changes would lead to collapse and overshoot sometime around 2075, followed by a transition to sustainability by 2100. (Used by permission from Donella Meadows et al., *Beyond the Limits: Confronting Global Collapse, Envisioning a Sustainable Future*, White River Junction, Vt.: Chelsea Green, 1992)

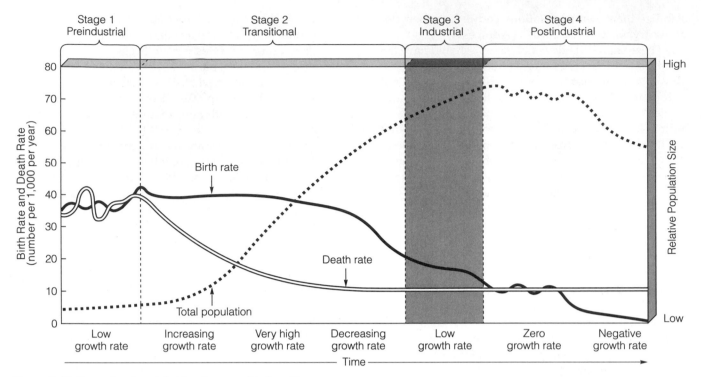

Stage 1
Preindustrial

Stage 2
Transitional

Stage 3
Industrial

Stage 4
Postindustrial

Birth Rate and Death Rate
(number per 1,000 per year)

Relative Population Size

High

80

70

60

50

40

Birth rate

30

20

Death rate

10

Total population

0

Low

Low
growth rate

Increasing
growth rate

Very high
growth rate

Decreasing
growth rate

Low
growth rate

Zero
growth rate

Negative
growth rate

— Time —

Figure 3-12 Generalized model of the demographic transition.

In most developing countries today, death rates have fallen much more than birth rates. In other words, these developing countries—mostly in Southeast Asia, Africa, and Latin America—are still in the transitional stage, halfway up the economic ladder, with high population growth rates. Some economists believe that developing countries will make the demographic transition over the next few decades without increased family-planning efforts.

However, despite encouraging declines in fertility, some population analysts fear that the still-rapid population growth in many developing countries will outstrip economic growth and overwhelm local life-support systems. This could cause many of these countries to be caught in a *demographic trap*, something that is currently happening in a number of developing countries, especially in Africa.

Analysts also point out that some of the conditions that allowed developed countries to develop are not available to many of today's developing countries. Even with large and growing populations, many developing countries do not have enough skilled workers to produce the high-tech products needed to compete in the global economy. Most low- and middle-income developing countries also lack the capital and resources needed for rapid economic development. Furthermore, the amount of economic assistance for developing countries, which are struggling under tremendous debts, has been de-

creasing since 1980. Indeed, since the mid-1980s, developing countries have paid developed countries $40–50 billion a year more (mostly in debt interest) than they have received from these countries.

How Can Family Planning Help Reduce Births? Family planning provides educational and clinical services that help couples choose how many children to have and when to have them. Such programs vary from culture to culture, but most provide information on birth spacing, birth control, and prenatal care.

Family planning has been an important factor in increasing the proportion of married women in developing countries who use modern contraception. It has gone from 10% in the 1960s to 49% in 1998 (36%, if China is excluded); this is close to the 61% usage by married women in developed countries. Family planning is also credited with being responsible for at least 40% of the drop in TFRs in developing countries, from 6 in 1960 to 3.3 in 1998 (3.9 if China is excluded).

Family-planning programs save a society money by reducing the need for children's social services. Proponents also argue that providing access to family planning throughout the world would bring about a sharp decline in the estimated 50 million legal and illegal abortions per year.

The effectiveness of family planning varies with program design and funding. It has been a significant

Q: How many people were added to the U.S. population in 1998?

factor in reducing birth and fertility rates in populous countries such as China (p. 86), Indonesia, Brazil, and Bangladesh. Family planning has also played a major role in reducing population growth in Japan, Thailand (p. 87), Mexico, South Korea, Taiwan, and several other countries with moderate to small populations. These successful programs, based on committed leadership, local implementation, and wide availability of contraceptives, demonstrate that population rates can decrease significantly within 15–30 years.

Despite such successes, family planning has had moderate to poor results in more populous developing countries such as India, Egypt, Pakistan, and Nigeria. Results have also been poor in 79 less populous developing countries, especially in Africa and Latin America, in which population growth rates are usually very high.

According to UN studies, an estimated 300 million women in developing countries want to limit the number and determine the spacing of their children, but they lack access to services. Extending family-planning services to these women and to those who will soon be entering their reproductive years could prevent an estimated 5.8 million births a year and over 5 million abortions a year. Other analysts call for expanding existing family-planning programs to include teenagers and sexually active unmarried women, who now are often excluded.

Some analysts urge that programs be broadened to educate males about the importance of having fewer children and taking more responsibility for raising them. Proponents also argue that much more research is needed on developing new, more effective, and acceptable methods of birth control for men. What do you think about this proposal?

Family planning could be provided in developing countries to all couples who want it for about $17 billion a year, the equivalent of less than a week's worth of worldwide military expenditures. This would help reduce world population by about 2.7 billion people, shrinking average family size from 3.4 to 2.1 children. However, in 1996 the U.S. Congress cut international family-planning assistance funds by 87% from 1995 levels and by 60% for 1997. According to the Alan Guttmacher Institute, the 1996 funding cuts increased abortions and led to 134,000 additional infant deaths and 8,000 deaths among women during pregnancy and childbirth.

How Can Economic Rewards and Penalties Be Used to Help Reduce Births? Some population experts argue that family planning, even coupled with economic development, cannot lower birth and fertility rates quickly enough to avoid a sharp rise in death rates in many developing countries. They point to studies showing that most couples in developing countries want three or four children, which is well above the replacement-level fertility required to bring about eventual population stabilization.

These analysts believe that we must go beyond family planning and use economic rewards and penalties to help slow population growth. About 20 countries offer small payments to people who agree to use contraceptives or to be sterilized; however, such payments are most likely to attract people who already have all the children they want. These countries also pay doctors and family-planning workers for each sterilization they perform and each IUD they insert.

Some countries, including China, penalize couples who have more than one or two children by raising their taxes, charging other fees, or eliminating income tax deductions for a couple's third child (as in Singapore, Hong Kong, Ghana, and Malaysia). Families who have more children than the prescribed limit may also lose health-care benefits, food allotments, and job options.

Economic rewards and penalties designed to lower birth rates work best if they encourage (rather than require) people to have fewer children, reinforce existing customs and trends toward smaller families, do not penalize people who produced large families before the programs were established, and increase a poor family's economic status. Once a country's population growth is out of control, however, it may be forced to use coercive methods to prevent mass starvation and hardship, as has been the case for China (Section 3-4).

How Can Empowering Women Help Reduce Births? Studies show that women tend to have fewer and healthier children and live longer when they have access to education and to paying jobs outside the home, and when they live in societies in which their individual rights are not suppressed.

Women, roughly half of the world's population, do almost all of the world's domestic work and child care and provide more health care with little or no pay than all the world's organized health services combined. They also do more than half the work associated with growing food, gathering fuelwood, and hauling water. As one Brazilian woman put it, "For poor women the only holiday is when you are asleep."

Women work two-thirds of all hours worked, but receive only one-tenth of the world's income and own a mere 0.01% of the world's property. In most developing countries, women don't have the legal right to own land or borrow money to increase agricultural productivity. Women also make up 70% of the world's poor and almost two-thirds of the more than 960 million adults who can neither read nor write.

Most analysts believe that women everywhere should have full legal rights and the opportunity to become educated and earn income outside the home. This would not only slow population growth but also promote human rights and freedom. However, empowering women by seeking gender equality will require some major social changes that will be difficult to achieve in male-dominated societies.

3-4 CASE STUDIES: SLOWING POPULATION GROWTH IN INDIA, CHINA, AND THAILAND

What Success Has India Had in Controlling Its Population Growth? The world's first national family-planning program began in India in 1952, when its population was nearly 400 million. In 1998, after 46 years of population control efforts, India was the world's second most populous country, with a population of 989 million—3.5 times as many people as the United States.

In 1952, India added 5 million people to its population; in 1998 it added 18 million—49,300 more mouths to feed each day. With 36% of its people under age 15, India's population is projected to reach 1.4 billion by 2025 and possibly 1.9 billion before leveling off early in the 22nd century.

India's people are among the poorest in the world, with an average per capita income of about $340 a year; for 30% of the population it is less than $100 a year, or 27¢ a day. Nearly half of India's labor force is unemployed or can find only occasional work. Although India is currently self-sufficient in food-grain production, about 40% of its population today suffers from malnutrition, mostly because of poverty. Life expectancy is only 59 years, and the infant mortality rate is 72 deaths per 1,000 live births.

Some analysts fear that India's already serious malnutrition and health problems will worsen as its population continues to grow rapidly. With 17% of the world's people, India has just 2.3% of the world's land resources and 1.7% of the world's forests. Some 40% of India's cropland is degraded as a result of soil erosion, waterlogging, salinization, overgrazing, and deforestation. About 70% of India's water is seriously polluted, and sanitation services are often inadequate.

Without its long-standing family-planning program, India's population and environmental problems would be growing even faster. Still, to its supporters the results of the program have been disappointing because of poor planning, bureaucratic inefficiency, the low status of women (despite constitutional guarantees of equality), extreme poverty, and a lack of administrative and financial support.

Even though the government has provided information about the advantages of small families for years, Indian women still have an average of 3.4 children because most couples believe they need many children to do work and care for them in old age. Many social and cultural norms favor large families, including the strong preference for male children; some couples keep having children until they produce one or more boys. These factors in part explain why even though 90% of Indian couples know of at least one modern birth control method, only 36% actually use one.

What Success Has China Had in Controlling Its Population Growth? Since 1970 China has made impressive efforts to feed its people and bring its population growth under control. Between 1972 and 1998 China achieved a remarkable drop in its crude birth rate, from 32 to 17 per 1,000 people, and its TFR dropped from 5.7 to 1.8 children per woman. Since 1985 its infant mortality rate has been almost one-half the rate in India. Life expectancy in China is 71 years, 12 years higher than in India. China's per capita income of $750 is almost twice that in India. Despite these achievements, with the world's largest population (1.24 billion) and a growth rate of 1.0%, China had about 12 million more mouths to feed in 1998. Its population is projected to reach 1.6 billion by 2025.

To achieve its sharp drop in fertility, China has established the most extensive, intrusive, and strict population control program in the world. Couples are strongly urged to postpone the age at which they marry and to have no more than one child. Married couples have ready access to free sterilization, contraceptives, and abortion. Paramedics and mobile units ensure access even in rural areas.

Couples who pledge to have no more than one child are given extra food, larger pensions, better housing, free medical care, and salary bonuses; their child will be given free school tuition and preferential treatment in employment when he or she enters the job market. Couples who break their pledge lose all the benefits. The result is that 81% of married women in China are using modern contraception, compared to 60% in developed countries and only 36% in other developing countries.

Government officials realized in the 1960s that the only alternative to strict population control was mass starvation. China is a dictatorship, and thus, unlike India, it has been able to impose a consistent population policy throughout society. Moreover, Chinese society is fairly homogeneous and has a widespread common written language, which aids in educating

Q: What percentage of the world's population is under age 15?

people about the need for family planning and in implementing policies for slowing population growth.

China's large and still growing population has an enormous environmental impact that could reduce its ability to produce enough food and threaten the health of many of its people. It is encouraging that between 1986 and 1996 the Chinese government almost doubled its expenditures on environmental protection. However, most of the nation's rivers, especially in urban areas, are seriously polluted, and air pollution in many of its cities is causing widespread health problems.

China has 21% of the world's population, but only 7% of its fresh water and cropland, 3% of its forests, and 2% of its oil. Soil erosion in China is serious and appears to be getting worse.

Most countries prefer to avoid the coercive elements of China's program. Coercion is not only incompatible with democratic values and notions of basic human rights, but ineffective in the long run because sooner or later people resist. However, other parts of this program could be used in many developing countries. Especially useful is the practice of localizing the program rather than asking the people to go to distant centers. Perhaps the best lesson for other countries is to act to curb population growth before they face the choice between mass starvation and coercive measures that severely restrict human freedom.

What Success Has Thailand Had in Controlling Its Population Growth? Can a country sharply reduce its population growth in only 15 years? Thailand did.

In 1971, Thailand adopted a national policy to control and reduce its population growth. When the program began the country's population was growing at a rate of 3.2% per year and the average Thai family had 6.4 children. Fifteen years later, the country's population growth rate had been cut in half to 1.6%. By 1998 the rate had fallen to 1.1%, and the average number of children per family was 2.0.

There are several reasons for this impressive feat: the creativity of the government-supported family-planning program, high literacy among women (90%), an increasing economic role for women, advances in women's rights, better health care for mothers and children, the openness of the Thai people to new ideas, the willingness of the government to encourage and financially support family planning and to work with the private, nonprofit Population and Community Development Association (PCDA), and support of family planning by the country's religious leaders (95% of Thais are Buddhist). Buddhist scripture teaches that "many children make you poor."

This remarkable transition was catalyzed by the charismatic leadership of Mechai Viravidaiya, a public relations genius and former government economist who launched the PCDA in 1974 to help make family planning a national goal. He established an imaginative, high-profile program to persuade Thai families to have fewer children. By 1998, 70% of the married women in Thailand were using some form of modern birth control—higher than the 61% usage in developed countries and the 49% usage in developing countries.

Mechai also set up an economic development program that has been a factor in doubling of the country's per capita income between 1971 and 1996. A German-financed revolving loan program was established to enable people participating in family-planning programs to install toilets and drinking water systems. Low-rate loans were also offered to farmers practicing family planning.

All is not completely rosy. Although Thailand has done well in slowing population growth and raising per capita income, it has been less successful in reducing pollution and improving public health, especially maternal health and control of AIDS and other sexually transmitted diseases. Its capital, Bangkok, remains one of the world's most polluted and congested cities. The typical motorist in Bangkok spends 44 days per year sitting in traffic, costing $2.3 billion per year in lost time.

How Could Population Growth Be Reduced to More Sustainable Levels? Many of the world's leading scientists and other analysts have concluded that we are exceeding the carrying capacity for humans in parts of the world and eventually for the entire world. If their analysis is correct, then reducing the current rate of population growth in both developed and developing countries and eventually stabilizing the world's population are essential goals for an ecologically and economically sustainable future.

In 1994 the United Nations held its third once-in-a-decade Conference on Population and Development in Cairo, Egypt. One of the conference's goals was to encourage action to stabilize the world's population at 7.8 billion by 2050, instead of the projected 9–12.5 billion.

The good news is that the experience of Japan, Thailand, South Korea, Taiwan, and China indicates that a country can achieve replacement-level fertility within 15–30 years. Such experience also suggests that the best way to slow population growth is a combination of investing in family planning, reducing poverty, and elevating the status of women.

Because countries differ in population growth rates, use and availability of resources, and social structure, the mix of these factors must be tailored to each country's situation. Furthermore, most analysts believe that government policy makers should devise policies that minimize the environmental impact of population growth in their effort to achieve sustainability.

3-5 POPULATION DISTRIBUTION: URBANIZATION AND URBAN PROBLEMS

How Fast Are Urban Areas Growing? An **urban area** is often defined as a town or city with a population of more than 2,500 people (although some countries set the minimum at 10,000–50,000 residents). A **rural area** is usually defined as an area with a population of less than 2,500 people.

Today about 44% of the world's population lives in urban areas, and by 2025 this figure is expected to increase to 61%. About 90% of this urban growth will occur in developing countries.

During the 1990s, more than 70% of the world's population increase is expected to occur in urban areas, adding about 63 million people a year to these already overburdened areas. This is equivalent to adding about four cities the size of New York each year. Every day roughly 150,000 people are added to the urban population of developing countries.

Because cities are the main centers for new jobs, higher income, education, innovation, culture, better health care, and trade, people are *pulled* to urban areas in search of jobs, a better life, and freedom from the constraints of village cultural life. They may also be *pushed* from rural areas into urban areas by factors such as poverty, lack of land, declining agricultural work, famine, and war.

Modern mechanized agriculture, for example, uses fewer farm laborers and allows large landowners to buy out subsistence farmers who cannot afford to modernize. Without jobs or land, these people are forced to move to cities. The urban poor fortunate enough to find employment usually must work long hours for low wages at jobs that may expose them to dust, hazardous chemicals, excessive noise, and dangerous machinery. Urban growth in developing countries is also fueled by government policies that distribute most income and social services to urban dwellers (especially in capital cities where a country's leaders live) at the expense of rural dwellers.

The number of large cities is mushrooming. In 1960 there were 111 cities with populations of more than 1 million; today there are 293, and some analysts expect this number to increase to at least 400 by 2025. Currently, 1 person out of every 10 lives in a city with a million or more inhabitants, and many live in the world's 15 *megacities*, with 10 million or more people (Case Study, right). The United Nations projects that by 2000 the world will have 21 megacities, 17 of them in developing countries.

As they grow outward, separate urban areas may merge to form a *megalopolis*. For example, the remaining open space between Boston and Washington,

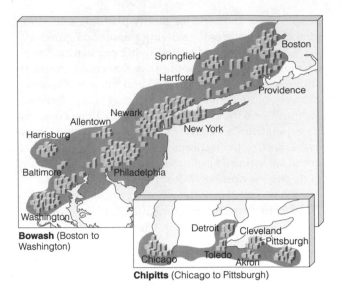

Figure 3-13 Two megalopolises: Bowash, consisting of urban sprawl and coalescence between Boston and Washington, D.C., and Chipitts, extending from Chicago to Pittsburgh.

D.C., is rapidly urbanizing and coalescing. This sprawling, 800-kilometer-long (500-mile-long) urban area, sometimes called *Bowash* (Figure 3-13), contains almost 60 million people, more than twice Canada's entire population.

Poverty is becoming increasingly urbanized as more poor people migrate from rural to urban areas, especially in Latin America. By the year 2000, half of the world's absolute poor will be living in urban areas. The United Nations estimates that at least 1 billion people—17% of the world's population—live either in the crowded *slums* of inner cities, or in vast, mostly illegal *squatter settlements* and *shantytowns*, where people move onto undeveloped land (usually without the owner's permission) and build shacks made of packing crates, plastic sheets, corrugated metal pipes, or whatever they can find.

Half of all urban children under age 15 in developing countries live in conditions of extreme poverty, and about one-fifth of them are street children with little or no family support. In Cairo, Egypt, children of kindergarten age can be found digging through clods of ox dung, looking for undigested kernels of corn to eat.

Many cities do not provide squatter settlements and shantytowns with adequate drinking water, sanitation facilities, electricity, food, health care, housing, schools, or jobs. Not only do these cities lack the needed money, but their officials fear that improving services will attract even more of the rural poor.

Despite joblessness, squalor, overcrowding, environmental hazards, and rampant disease, most squatter and slum residents are better off than the rural poor.

Q: What percentage of current annual U.S. population growth is due to legal and illegal immigration?

Mexico City

About 20 million people—about one of every six Mexicans—live in the country's capital, Mexico City, the world's fourth most populous city. Every day an additional 2,000 poverty-stricken rural peasants pour into the city, hoping to find a better life. This adds about 750,000 new people per year, equivalent to having to provide food, water, sanitation, jobs, and other services for a new city the size of Baltimore or San Francisco every year.

Mexico City suffers from severe air pollution, high unemployment (close to 50%), deafening noise, congestion, and a soaring crime rate. More than one-third of its residents live in crowded slums (called *barrios*) or squatter settlements, without running water or electricity. At least 8 million people have no sewer facilities. This means that huge amounts of human waste are deposited in gutters and vacant lots every day, attracting armies of rats and swarms of flies. When the winds pick up dried excrement, a *fecal snow* often falls on parts of the city, leading to widespread salmonella and hepatitis infections, especially among children.

Some 4 million motor vehicles and 30,000 factories spew pollutants into the atmosphere. Air pollution is intensified because the city lies in a basin surrounded by mountains, and frequent thermal inversions trap pollutants at ground level. Since 1982 the amount of contamination in the city's smog-choked air has more than tripled; breathing the air is said to be roughly equivalent to smoking two packs of cigarettes a day.

Writer Carlos Fuentes has nicknamed this megacity "Makesicko City." Because of the air pollution, many foreign companies and governments give imported workers additional hazard pay for working in Mexico City.

The Mexican government is industrializing other parts of the country in an attempt to slow migration to Mexico City. Cars have been banned from a 50-block central zone. Taxis built before 1985 have been taken off the streets, and trucks can run only on liquefied petroleum gas (LPG). The government began phasing in unleaded gasoline in 1991, but it will be years before millions of older vehicles, which burn leaded gas, are eliminated. The government has also planted 25 million trees to help clean the air and has purchased some land to provide green space for the city.

The city's air and water pollution cause an estimated 100,000 premature deaths a year. These problems, already at crisis levels, will become even worse if the city grows as projected to 25.8 million people by the end of this century.

Critical Thinking

If you were in charge of Mexico City, what would you do?

With better access to family-planning programs, they tend to have fewer children, who have better access to schools. Many squatter settlements provide a sense of community and a vital safety net of neighbors, friends, and relatives for the poor.

How Urbanized Is the United States? In 1800 only 5% of Americans lived in cities. Since then, four major internal population shifts have taken place in the United States. As a result of the first shift, *migration to large central cities*, about 75% of Americans live in 350 *metropolitan areas* (cities and towns with at least 50,000 people). Nearly half of the country's population lives in consolidated metropolitan areas containing 1 million or more residents (Figure 3-14).

In the second shift, more people began *migrating from large central cities to suburbs and smaller cities*. Since 1970 this type of migration has followed new jobs to such areas. Today about 41% of the country's urban dwellers live in central cities and 59% live in suburbs.

The third shift, which has been taking place for several decades, is *migration from the North and East to the South and West*. Since 1980 about 80% of the U.S. population increase has occurred in the South and West, particularly near the coasts. This shift is expected to continue.

In a recent fourth shift, people have been *migrating from urban areas back to rural areas*. According to the Census Bureau, between 1900 and 1997 rural counties had a net influx of about 2 million people, virtually all of it from domestic migration.

What Are the Major Urban Problems in the United States? Here is some great news. Since 1920, many of the worst urban environmental problems in the United States (and other developed countries) have been reduced significantly. Most people have better working and housing conditions and air and water quality have improved. Better sanitation, public water supplies, and medical care have slashed death rates and the prevalence of sickness from malnutrition

A: About 40%

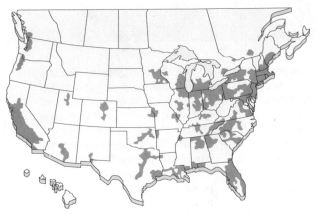

Figure 3-14 Major urban regions in the United States by the year 2000. Nearly half (48%) of Americans live in *consolidated metropolitan areas* with 1 million or more people. (Data from U.S. Census Bureau)

and transmittable diseases such as measles, diphtheria, typhoid fever, pneumonia, and tuberculosis. Furthermore, concentrating most of the population in urban areas has helped protect the country's biodiversity by reducing the destruction and degradation of wildlife habitat.

The biggest problems facing many cities in the United States (especially older ones) are deteriorating services, aging infrastructures (streets, schools, bridges, housing, sewers), budget crunches from lost tax revenues and rising costs as businesses and more affluent people move out, and rising poverty in many central city areas. As a result, violence, drug traffic and abuse, crime, decay, and blight have increased in parts of central cities. Unemployment rates in some inner-city areas are typically 50% or higher.

How Do Transportation Systems Affect Urban Growth and Development? If a city cannot spread outward, it must grow vertically—upward and downward (below ground)—so that it occupies a small land area with a high population density. Most people living in such compact cities walk, ride bicycles, or use energy-efficient mass transit. Residents often live in multistory apartment buildings; with few outside walls in many apartments, heating and cooling costs are reduced. Many European cities and urban areas such as Hong Kong and Tokyo are compact and tend to be more energy-efficient than the dispersed cities of the United States, Canada, and Australia, where ample land is often available for outward expansion.

A combination of cheap gasoline, plentiful land, and a network of highways produces dispersed, automobile-oriented cities with low population density,

often called *urban sprawl*. Most people living in such urban areas live in single-family houses with unshared walls that lose and gain heat rapidly unless they are well insulated and airtight. Urban sprawl also gobbles up unspoiled natural habitats, paves over fertile farmland, and promotes heavy dependence on the automobile.

What Are the Pros and Cons of Motor Vehicles? There are two main types of ground transportation: *individual* (such as cars, bicycles, motor scooters, and walking) and *mass* (mostly buses and rail systems). Despite having only 4.6% of the world's people, the United States has 35% of the world's cars and trucks. In the United States the car is used for 98% of all urban transportation and 86% of travel to work. Americans drive 3 billion kilometers (2 billion miles) each year—as far as the rest of the world combined. No wonder British author J. B. Priestley remarked, "In America, the cars have become the people."

Currently, only about 8% of the world's population own cars and only 10% can afford to. In developing countries as few as 1% of the people can afford a car; they travel mostly by foot, bicycle, or motor scooter. However, the number of cars in developing countries experiencing economic growth is increasing.

The automobile provides convenience and unprecedented mobility. To many people, cars are also symbols of power, sex, excitement, social status, and success. Moreover, much of the world's economy is built on producing motor vehicles and supplying roads, services, and repairs for them. In the United States, one of every four dollars spent and one of every six nonfarm jobs is connected to the automobile.

Despite their economic and personal benefits, motor vehicles have many destructive effects on people and the environment. Since 1885, when Karl Benz built the first automobile, almost 18 million people have been killed in motor vehicle accidents. According to the World Health Organization, this global death toll increases by an estimated 885,000 people per year (42,000 in the United States), and annually about 15 million people are injured or permanently disabled in motor vehicle accidents. *More Americans have been killed by cars than all the country's wars.*

Motor vehicles are also the largest source of air pollution, laying a haze of smog over the world's cities. In the United States they produce at least 50% of the air pollution, even though emission standards are as strict as any in the world. Two-thirds of the oil used in the United States and one-third of the world's total oil consumption are devoted to transportation.

By making long commutes and shopping trips possible, automobiles and highways have helped create urban sprawl and reduced use of more efficient forms

Q: What percentage of the world's people live in urban areas?

of transportation. Worldwide, at least a third of urban land is devoted to roads, parking lots, gasoline stations, and other automobile-related uses. In the United States, more land is now devoted to cars than to housing. Half the land in an average U.S. city is used for cars, prompting urban expert Lewis Mumford to suggest that the U.S. national flower should be the concrete cloverleaf.

In 1907 the average speed of horse-drawn vehicles through the borough of Manhattan was 18.5 kilometers (11.5 miles) per hour; today cars and trucks creep along Manhattan streets at an average speed of 5 kilometers (3 miles) per hour. If current trends continue, U.S. motorists will spend an average of 2 years of their lifetimes in traffic jams, imprisoned in metal boxes that were supposed to provide speed, freedom, and mobility. The U.S. economy loses at least $100 billion a year because of time lost in traffic delays. Building more roads is not the answer because, as economist Robert Samuelson put it, "Cars expand to fill available concrete."

The major hidden costs of driving include deaths and injuries from accidents, the value of time wasted in traffic jams, air pollution, increased threats from global warming, the drop in property values near roads because of noise and congestion, and the cost of maintaining a formidable military presence in the Middle East to ensure access to oil. Two recent estimates by economists put the mostly hidden harmful costs of driving in the United States at roughly $300–350 billion per year.

Environmentalists and a number of economists suggest that one way to break this increasingly destructive cycle is to make drivers pay directly for most of the true costs of auto use. This could be done by including the current harmful hidden costs of driving as a tax on gasoline and by phasing out government subsides for motor vehicle owners.

According to a study by the World Resources Institute, federal, state, and local government automobile subsidies in the United States amount to $300–600 billion a year (depending on the costs included), an average subsidy of $1,600–3,200 per vehicle. Taxpayers (drivers and nondrivers alike) foot this bill mostly unknowingly. If drivers had to pay these hidden costs directly in the form of a gasoline tax, the tax on each gallon would be about $5–7. Pollution emission fees, toll charges on roads (especially during peak traffic times), and higher parking fees are also ways to have motorists pay for the environmental and social costs of driving—a *user-pays* approach.

Deciding to include the hidden costs in the market prices of cars, trucks, and gasoline up front makes economic and environmental sense. However, this approach faces massive political opposition from the general public (mostly because they are unaware of the huge hidden costs they are already paying) and from the powerful transportation-related industries. In addition, such tools for encouraging more use of mass transportation will not work unless fast, efficient, reliable, and affordable transportation alternatives are available.

☸ Solutions: What Are Alternatives to the Car?

Cars are an important form of transportation, but some countries are encouraging alternatives. One alternative is the *bicycle*. Globally, bicycles outsell cars by almost 3 to 1 because most people can afford a bicycle whereas fewer than 10% can afford a car. Besides being inexpensive to buy and maintain, bicycles produce no pollution, are rarely a serious danger to pedestrians or cyclists, take few resources to make, and are the most energy-efficient form of transportation (including walking).

In 1996 a California firm developed a simple $15 bicycle that has no chain; the pedals are attached to the front wheel. This inexpensive design could greatly increase bicycle use in developing countries.

In urban traffic, cars and bicycles move at about the same average speed. Using separate bike paths or lanes running along roads, cyclists can make most trips shorter than 8 kilometers (5 miles) faster than drivers.

In China, at least 50% of urban trips are made by bicycle and the government gives subsidies to those who bicycle to work. In the Netherlands bicycle travel makes up 30% of all urban trips, and in Japan 15% of all commuters ride bicycles to work or to commuter-rail stations. Many train stations in Japan have high-rise parking towers with lifts for bicycles.

Only about 2% of Americans bicycle to work (the "no-pollute commute"), even though half of all U.S. commutes are under 8 kilometers (5 miles). However, according to recent polls, 20% of Americans say they would bicycle to work if safe bike lanes were available and if their employers provided secure bike storage and showers at work.

Another alternative to the car is *mass transit*. In the United States mass transit accounts for only 3% of all passenger travel, compared with 15% in Germany and 47% in Japan.

Rapid-rail, suburban train, and trolley systems have a number of advantages over highway and air transport. They are much more energy-efficient, produce less air pollution, cause fewer injuries and deaths, and take up less land. However, such train systems are efficient only where many people live along a narrow corridor and can easily reach properly spaced stations.

High-speed trains between cities can greatly reduce the need for travel by car and plane. In western Europe and Japan, *bullet* or supertrains travel on new or

---

A: 44% (73% in developed countries and 36% in developing countries) in 1998

upgraded tracks at speeds up to 330 kilometers (200 miles) per hour. They are ideal for trips of approximately 200–1,000 kilometers (120–600 miles). A high-speed train network could replace airplanes, buses, and private cars for most medium-distance travel between major American cities (Figure 3-15)

Bus systems are more flexible than rail systems; they can run throughout sprawling cities and be rerouted overnight if transportation patterns change. Bus systems also require less capital and have lower operating costs than heavy-rail systems. Because they must offer low fares to attract riders, however, bus systems often cost more to operate than they bring in. Because buses are cost-effective only when full, they are sometimes supplemented by more flexible car pools, van pools, and jitneys (small vans or minibuses that run along regular and stop-on-demand routes).

What Are the Major Resource and Environmental Problems of Urban Areas?

Most of today's urban areas don't even come close to being self-sustaining; they survive only by importing food, water, energy, minerals, and other resources from farms, forests, mines, and watersheds. They also produce enormous quantities of wastes that can pollute air, water, and land within and outside their boundaries (Figure 3-16)

However, urbanization has some environmental benefits. Recycling is more economically feasible because of the large concentration of recyclable materials. Concentrating people in urban areas also helps preserve biodiversity by reducing the stress on wildlife habitats.

The 44% of the world's people currently living in urban areas occupy only about 5% of the planet's land area. However, supplying these urban dwellers with resources is a major reason humans have disturbed about 73% of the earth's land area, if uninhabitable ice and rock areas are excluded. Furthermore, rural or wildlife areas downwind or downstream from urban areas are receptacles for much of the pollution produced in urban areas.

Most urban areas have several major resource and environmental problems. *Most cities have few trees, shrubs, or other natural vegetation* that absorbs air pollutants, gives off oxygen, helps cool the air (as water transpires from their leaves), muffles noise, provides wildlife habitats, and gives aesthetic pleasure. As one observer remarked, "Most cities are places where they cut down the trees and then name the streets after them." According to the American Forestry Association, one city tree provides over $57,000 worth of air conditioning, erosion and stormwater control, wildlife shelter, and air pollution control over a 50-year lifetime.

Most cities produce little of their own food. However, individuals can grow food by planting community gardens in unused lots and by using window-boxes,

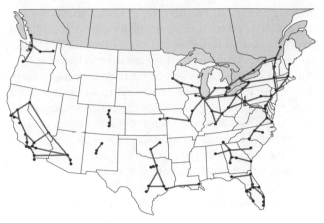

Figure 3-15 Potential routes for high-speed trains in the United States and parts of Canada. Such a system would allow rapid, comfortable, safe, and affordable travel between major cities in a region, and it would reduce dependence on cars, buses, and airplanes. (Data from High Speed Rail Association)

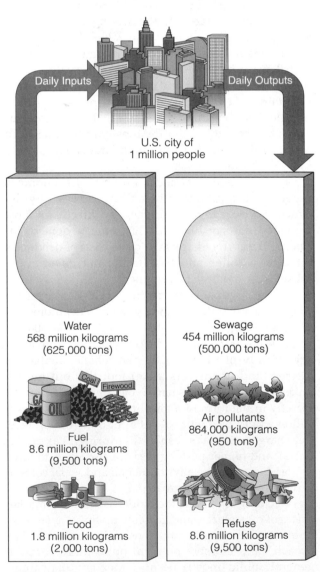

Figure 3-16 Typical daily inputs and outputs of matter and energy for a U.S. city of 1 million people.

Q: What percentage of urban children under age 5 live in extreme poverty?

balcony planters, and gardens or greenhouses on the roofs of apartment buildings. Urban gardens currently provide about 15% of the world's food and this proportion could be increased.

Cities are generally warmer, rainier, foggier, and cloudier than suburbs and nearby rural areas. The enormous amounts of heat generated by cars, factories, furnaces, lights, air conditioners, and people in cities create an **urban heat island** (Figure 3-17) surrounded by cooler suburban and rural areas. The dome of heat also traps pollutants, especially tiny solid particles (suspended particulate matter), creating a **dust dome** above urban areas. If wind speeds increase, the dust dome elongates downwind to form a **dust plume**, which can spread the city's pollutants for hundreds of kilometers. As urban areas grow and merge, individual heat islands also merge, which can affect the climate of a large area and keep polluted air from being diluted and cleansed.

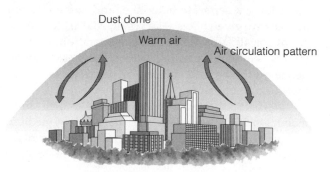

Figure 3-17 An urban heat island causes patterns of air circulation that create a dust dome over the city. Winds can elongate the dome toward downwind areas. A strong cold front can blow the dome away and lower local pollution levels, but increases pollution in downwind areas.

Many cities have water supply and flooding problems. As cities grow and their water demands increase, expensive reservoirs and canals must be built and deeper wells drilled. The transfer of water to urban areas deprives rural and wild areas of surface water and sometimes depletes groundwater faster than it is replenished. Covering land with buildings, asphalt, and concrete causes precipitation to run off quickly; it can overload sewers and storm drains, contributing to water pollution and flooding in cities and downstream areas.

Many of the world's largest cities are in coastal areas. If an enhanced greenhouse effect increases the average atmospheric temperature as projected (Section 9-2), a rise in average sea level of even a meter could flood many of these cities, perhaps sometime during the 21st century.

Urban areas also produce large quantities of air pollution (Chapter 9), *water pollution* (Chapter 10), *and garbage and other solid waste* (Chapter 11). According to the World Health Organization, more than 1.1 billion people—about one-fifth of humanity—live in urban areas where air pollution levels exceed healthful levels. The World Bank estimates that almost two-thirds of urban residents in developing countries don't have adequate sanitation facilities. In the developing world, it is estimated that 90% of all sewage is discharged directly into rivers, lakes, and coastal waters without treatment of any kind.

Most urban dwellers are also subjected to excessive noise (Figure 3-18). For example, every day one of every nine Americans lives, works, or plays around noise of sufficient duration and intensity to cause some permanent hearing loss, and that number is rising rapidly.

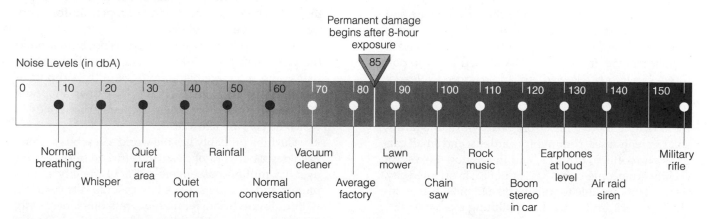

Figure 3-18 Noise levels (in decibel-A [dbA] sound pressure units) of some common sounds. You are being exposed to a sound level high enough to cause permanent hearing damage if you need to raise your voice to be heard above the racket, if a noise causes your ears to ring, or if nearby speech seems muffled. Prolonged exposure to lower noise levels and occasional loud sounds may not damage your hearing but can greatly increase internal stress. Noise pollution can be reduced by modifying noisy activities and devices to produce less noise, shielding noisy devices or processes, shielding workers or bystanders from noise, and using antinoise, a new technology that cancels out one noise with another.

A: About 50% (one-fifth of them street children with little or no family support)

Urban areas have beneficial and harmful effects on human health. Many aspects of urban life benefit human health, including better access to education, social services, and medical care. On the other hand, high-density city life increases the spread of infectious diseases (especially if adequate drinking water and sewage systems are not available), physical injuries (mostly from industrial and traffic accidents), and health problems caused by increased exposure to pollution and noise. According to the World Bank, at least 220 million people in cities in developing countries don't have safe drinking water.

Another problem is the *loss of rural land, fertile soil, and wildlife habitats as cities expand.* Once prime agricultural land or forestland is paved over or built on, it is lost for food production and habitat for most of its former wildlife. As a city expands, more energy is needed to transport food to its people; this in turn causes more pollution. In coastal areas, urban growth destroys or pollutes ecologically valuable wetlands.

Solutions: How Can We Make Cities More Sustainable?

An important goal in coming decades should be to make urban areas more self-reliant, sustainable, and enjoyable places to live. In a sustainable and ecologically healthy city, called an *ecocity* or *green city*, matter and energy resources are used efficiently. Far less pollution and waste are produced than in conventional cities. Emphasis is placed on pollution prevention, reuse, recycling, and efficient use of energy and matter resources. Per capita solid waste production is greatly reduced and at least 60% of what is produced is reused, recycled, or composted. An ecocity takes advantage of locally available energy sources and requires that all buildings, vehicles, and appliances meet high energy-efficiency standards.

Trees and plants adapted to the local climate and soils are planted throughout an ecocity to provide shade and beauty, to reduce pollution and noise, and to supply habitats for wildlife. Abandoned lots and polluted creeks are cleaned up and restored. Nearby forests, grasslands, wetlands, and farms are preserved instead of being devoured by urban sprawl. Much of the ecocity's food comes from nearby organic farms, solar greenhouses, community gardens, and small gardens on rooftops, in yards, and in window boxes. An ecocity is a people-oriented city, not a car-oriented city. Its residents are able to walk or bike to most places, including work, and to use low-polluting mass transit.

Case Study: Davis, California

The ecocity is not a futuristic dream. The citizens and elected officials of Davis, California—a city of about 54,000 people northeast of San Francisco—committed themselves in the early 1970s to making their city ecologically sustainable.

City building codes encourage the use of solar energy for water and space heating. All new homes must meet high standards of energy efficiency and when an existing home changes hands the buyer must bring it up to the energy conservation standards for new homes. In Davis's Village Homes Development, America's first solar neighborhood, houses are heated by solar energy.

Since 1975 Davis has cut its use of energy for heating and cooling in half. It has a solar power plant, and some of the electricity it produces is sold to the regional utility company. Eventually the city plans to generate all of its own electricity.

The city discourages the use of automobiles and encourages the use of bicycles by closing some streets to automobiles, building bike lanes on major streets, and building bicycle paths. Any new housing tract must have a separate bike lane, and some city employees are given bikes. As a result, 28,000 bicycles account for 40% of all in-city transportation, and less land is needed for parking spaces. This heavy dependence on the bicycle is aided by the city's warm climate and flat terrain.

Davis limits the type and rate of its growth, and it maintains a mix of homes for people with low, medium, and high incomes. Development of the fertile farmland surrounding the city for residential or commercial use is restricted.

Case Study: Curitiba, Brazil

One of the world's showcase ecocities is Curitiba, Brazil, with a population of 2.1 million. Trees are everywhere in Curitiba because city officials have given neighborhoods over 1.5 million trees to plant and care for. No tree in the city can be cut down without a permit; for every tree cut, two must be planted.

The air is clean because the city is not built around the car. There are 145 kilometers (90 miles) of bike paths, and more are being built. With the support of shopkeepers, many streets in the downtown shopping district have been converted to pedestrian zones in which no cars are allowed.

Curitiba probably has the world's best bus system. Each day a network of clean and efficient buses carries over 1.5 million passengers—75% of the city's commuters and shoppers—at a low cost (20–40¢ per ride, with unlimited transfers) on express bus lanes. Only high-rise apartment buildings are allowed near major bus routes, and each building must devote the bottom two floors to stores, which reduces the need for residents to travel. Since 1974 traffic has declined by 30% while the city's population doubled.

Q: What percentage of the world's cars and trucks are found in the United States?

The city recycles roughly 70% of its paper and 60% of its metal, glass, and plastic, which is sorted by households for collection. Recovered materials are sold mostly to the city's more than 340 major industries.

To equip the poor with basic technical training skills needed for jobs, the city set up old buses as roving technical training schools; courses cost the equivalent of two bus tokens. Each bus gives courses for 3 months in a particular area and then moves to another area.

The poor can swap sorted trash for food or bus tokens and receive free medical, dental, and child care; there are also 40 feeding centers for street children. As a result, the infant mortality rate has fallen by more than 60% since 1977.

All of these things have been accomplished despite Curitiba's enormous population growth, from 300,000 in 1950 to 2.2 million in 1997, as rural poor people have flocked to the city. Curitiba has slums, shantytowns, and most of the problems of other cities. But most of its citizens have a sense of vision, solidarity, pride, and hope, and they are committed to making their city even better.

One secret of Curitiba's success is the willingness of citizens to work together to create a better future. Another is city officials who genuinely care about providing a high quality of life for *all* of the city's inhabitants and who relish solving problems by using imagination, common sense, determination, and good planning. The entire program is the brainchild of architect and former college teacher Jaime Lerner, an energetic and charismatic leader who has served as the city's major three times since the 1970s.

We need the size of population in which human beings can fulfill their potentialities; in my opinion we are already overpopulated from that point of view; not just in places like India and China and Puerto Rico, but also in the United States and Western Europe.

GEORGE WALD (NOBEL LAUREATE, BIOLOGY)

CRITICAL THINKING

1. Why is it rational for a poor couple in India to have five or six children? What changes might induce such a couple to consider their behavior irrational?

2. Are there physical limits to population growth on the earth? Explain.

3. Are there social limits to population growth on the earth? Explain.

4. Explain why you agree or disagree with each of the following proposals:
   a. The number of legal immigrants and refugees allowed into the United States each year should be reduced sharply.
   b. The United States should adopt an official policy to stabilize its population and reduce unnecessary resource waste and consumption as rapidly as possible.
   c. Everyone should have the right to have as many children as they want.

5. Do you agree with China's current population control policies? Explain. What alternatives, if any, would you suggest?

6. Congratulations—you have just been put in charge of the world. List the five most important features of your population policy.

7. What conditions, if any, would encourage you to rely less on the automobile, and instead encourage you to travel to school or work by bicycle or motor scooter, on foot, by mass transit, or by a car or van pool?

8. Should squatters around cities of developing countries be given title to land they don't own? Explain. What are the alternatives?

9. Why do you think Curitiba has been so successful in its efforts to become an ecocity, compared to the generally unsustainable cities found in most more developed countries?

10. Congratulations—you have just been put in charge of the world. List the five most important features of your urban policy.

If the United States wants to save a lot of oil and money and increase national security, there are two simple ways to do it: Stop driving Petropigs and stop living in energy sieves.

AMORY B. LOVINS

4-1 EVALUATING ENERGY RESOURCES

What Types of Energy Do We Use? *Some 99% of the energy used to heat the earth and all of our buildings comes directly from the sun.* Without this input of essentially inexhaustible solar energy, the earth's average temperature would be –240°C (–400°F) and life as we know it would not exist. Solar energy also helps recycle the carbon, oxygen, water, and other chemicals we and other organisms need to stay alive and healthy. This direct input of solar energy also produces several forms of renewable energy: wind, falling and flowing water (hydropower), and biomass (solar energy converted to chemical energy stored in chemical bonds of organic compounds in trees and other plants).

The remaining 1%, the portion we generate to supplement the solar input, is *commercial energy* sold in the marketplace. Most commercial energy comes from extracting and burning mineral resources obtained from the earth's crust, primarily nonrenewable fossil fuels (Figure 4-1).

Developed countries and developing countries differ greatly in their sources of energy (Figure 4-2) and average per capita energy use. The most important supplemental source of energy for developing countries is potentially renewable biomass, especially fuelwood and charcoal made from fuelwood, the main sources of energy for heating and cooking for roughly half the world's population. Within a few decades one-fourth of the world's population in developed countries may face an oil shortage, but half the world's population in developing countries already faces a fuelwood shortage.

The United States is the world's largest user (and waster) of energy. With only 4.6% of the population, it uses 24% of the world's commercial energy, 93% from *nonrenewable* fossil fuels (85%) and nuclear energy (8%).

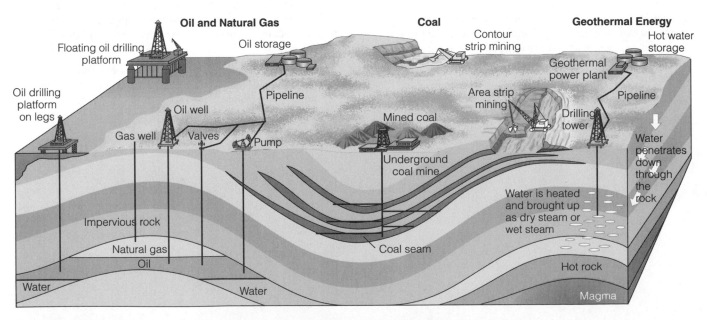

Figure 4-1 Important commercial energy resources from the earth's crust are geothermal energy, coal, oil, and natural gas. Uranium ore is also extracted from the crust and then processed to increase its concentration of uranium-235, which can be used as a fuel in nuclear reactors to produce electricity.

Campbell, C.J. 1998. "Running Out of Gas: This Time the Wolf is Coming." *The National Interest*, no. 51, 47(9).

In contrast, India, with 17% of the world's people, uses only about 3% of the world's commercial energy.

How Should We Evaluate Energy Resources?

The types of energy we use and how we use them are major factors determining our quality of life and our harmful environmental effects. Our current dependence on nonrenewable fossil fuels is the primary cause of air and water pollution, land disruption, and projected global warming. Moreover, affordable oil, the most widely used energy resource in developed countries, will probably be depleted within 40–80 years and will need to be replaced by other energy resources.

What is our best immediate energy option? The general consensus is to cut out unnecessary energy waste by improving energy efficiency. What is our next best energy option? There is disagreement about that. Some say we should get much more of the energy we need from the sun, wind, flowing water, biomass, heat stored in the earth's interior, and hydrogen gas by making the transition to a new *renewable energy* or *solar age*.

Others say we should burn more coal and synthetic liquid and gaseous fuels made from coal. Some believe natural gas is the answer, at least as a transition fuel to a new solar age built around improved energy efficiency and renewable energy. Others think nuclear power is the answer.

Experience shows that it usually takes at least 50 years and huge investments to phase in new energy alternatives (Figure 4-3). An exception is nuclear power, which after almost 50 years still provides only a small proportion of the world's commercial energy. Thus, we must plan for and begin the shift to a new mix of energy resources now. To do so involves answering the following questions for *each* energy alternative:

- How much of the energy source will be available in the near future (the next 15 years), intermediate future (the next 30 years), and long term (the next 50 years)?

- What is this source's net energy yield?

- How much will it cost to develop, phase in, and use this energy resource?

- How will extracting, transporting, and using the energy resource affect the environment?

- What will using this energy source do to help sustain the earth for us, future generations, and the other species living on this planet?

What Is Net Energy? The Only Energy That Really Counts

It takes energy to get energy. For example, oil must be found, pumped up from beneath the ground, transported to a refinery and converted to useful fuels (such as gasoline, diesel fuel, and heating oil), then transported to users, and finally burned in furnaces and cars before it is useful to us. Each of these steps uses energy, and the second law of energy (Section 2-2) tells us that each time we use energy to perform a task, some of it is always wasted and is degraded to low-quality energy.

The usable amount of high-quality energy available from a given quantity of an energy resource is its **net energy**: the total useful energy available from the resource

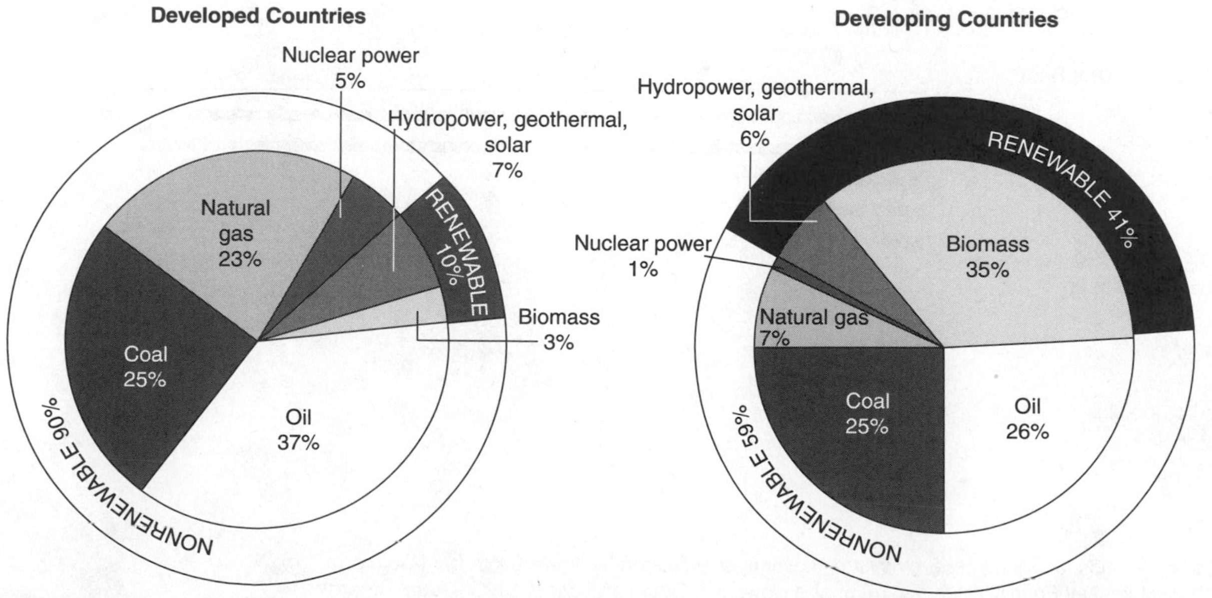

Figure 4-2 Commercial energy use by source for developed countries and developing countries. Commercial energy amounts to only 1% of the energy used in the world; the other 99% comes from the sun and is not sold in the marketplace. (Data from U.S. Department of Energy, British Petroleum, and Worldwatch Institute)

Hint: Enter the search term *energy crisis* using the Subject Guide.

CHAPTER 4   97

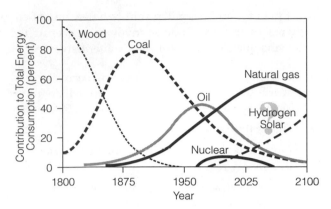

Figure 4-3 Shifts in the use of commercial energy resources in the United States since 1850, with projected changes to 2100. Shifts from wood to coal and then from coal to oil and natural gas have each taken about 50 years. Affordable oil is expected to be running out within 40–80 years, and burning fossil fuels is the primary cause of air pollution and projected warming of the atmosphere. For these reasons, most analysts believe we must make a new shift in energy resources over the next 50 years. Some believe that this shift should involve improved energy efficiency and greatly increased use of solar energy and hydrogen. (Data from U.S. Department of Energy)

over its lifetime minus the amount of energy used (the first law of energy), automatically wasted (the second law of energy), and unnecessarily wasted in finding, processing, concentrating, and transporting it to users. Net energy is like your net spendable income (your wages minus taxes and other deductions). For example, suppose that for each 10 units of energy in oil in the ground we have to use and waste 8 units of energy to find, extract, process, and transport the oil to users. Then we have only 2 units of useful energy available from each 10 units of energy in the oil.

We can look at this concept as the ratio of total energy produced to the energy used to produce it. In the example just given, the *net energy ratio* would be 10/8, or approximately 1.25. The higher the ratio, the greater the net energy yield. When the ratio is less than 1, there is a net energy loss.

Figure 4-4 shows estimated net energy ratios for various types of space heating, high-temperature heat for industrial processes, and transportation. Currently, oil has a high net energy ratio because much of it comes from large, accessible deposits such as those in the Middle East. When those sources are depleted, the net energy ratio of oil will decline and prices will rise.

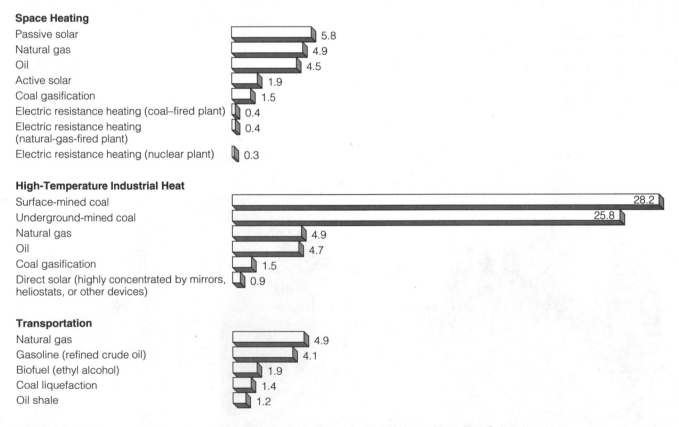

Figure 4-4 Net energy ratios for various energy systems over their estimated lifetimes. (Data from Colorado Energy Research Institute, *Net Energy Analysis*, 1976; and Howard T. Odum and Elisabeth C. Odum, *Energy Basis for Man and Nature*, 3d ed., New York: McGraw-Hill, 1981).

Q: What percentage of the cars carrying people to and from work in the United States have only one passenger?

Then more money and high-quality fossil-fuel energy will be needed to find, process, and deliver new oil from widely dispersed small deposits, deposits buried deep in the earth's crust, or deposits located in remote areas.

Conventional nuclear energy has a low net energy ratio because large amounts of energy are required to extract and process uranium ore, convert it into a usable nuclear fuel, and build and operate power plants. In addition, more energy is needed to dismantle the plants after their 15–40 years of useful life and to store the resulting highly radioactive wastes for thousands of years.

4-2 IMPROVING ENERGY EFFICIENCY

What Is Energy Efficiency? Doing More with Less You may be surprised to learn that *84% of all commercial energy used in the United States is wasted* (Figure 4-5). About 41% of this energy is wasted automatically because of the degradation of energy quality imposed by the second law of energy. More important, about 43% is wasted unnecessarily, mostly by using fuel-wasting motor vehicles, furnaces, and other devices, and by living and working in leaky, poorly insulated, and poorly designed buildings.

People in the United States unnecessarily waste as much energy as two-thirds of the world's population consumes. According to energy expert Amory Lovins, unnecessary energy waste in the United States amounts to about $300 billion per year—an average of $570,000 per minute! In 1998, this waste was more than the entire $268 billion military budget.

According to Lovins, the easiest, fastest, and cheapest way to get more energy with the least environmental impact is to eliminate much of this energy waste by making lifestyle changes that reduce energy consumption: walking or biking for short trips, using mass transit, putting on a sweater instead of turning up the thermostat, and turning off unneeded lights.

Another equally important way is to increase the efficiency of the energy conversion devices we use. **Energy efficiency** is the percentage of total energy input that does useful work (is not converted to low-quality, essentially useless heat) in an energy conversion system. The energy conversion devices we use vary in their energy efficiencies (Figure 4-6). We can save energy and money by buying the most energy-efficient home heating systems, water heaters, cars, air conditioners, refrigerators, computers, and other appliances that are available and by supporting research to invent even more energy-efficient devices. Some

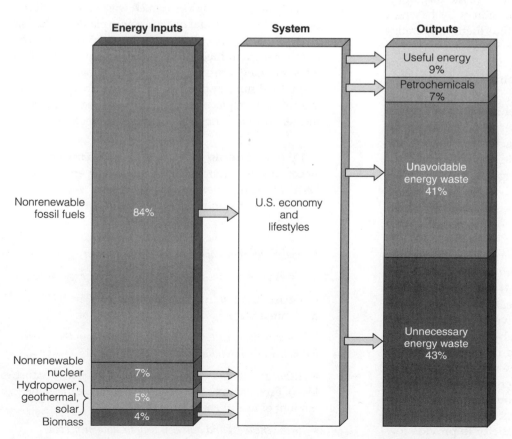

Figure 4-5 Flow of commercial energy through the U.S. economy. Note that only 16% of all commercial energy used in the United States ends up performing useful tasks or being converted to petrochemicals; the rest is either automatically and unavoidably wasted because of the second law of energy (41%) or wasted unnecessarily (43%).

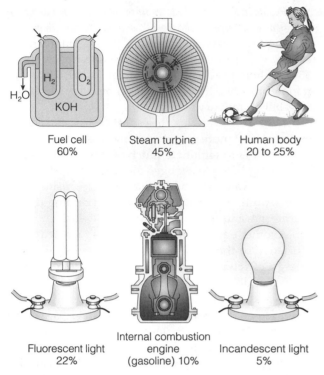

Fuel cell
60%

Steam turbine
45%

Human body
20 to 25%

Fluorescent light
22%

Internal combustion
engine
(gasoline) 10%

Incandescent light
5%

Figure 4-6 Energy efficiency of some common energy conversion devices.

energy-efficient models may cost more initially, but in the long run they usually save money by having a lower **life cycle cost**: initial cost plus lifetime operating costs. Although the United States has a long way to go, improvements in energy efficiency since the oil embargo in 1973 have cut the country's energy bills by a whopping $275 billion a year.

The net efficiency of the entire energy delivery process for a space heater, water heater, or car is determined by the efficiency of each step in the energy conversion process. For example, the sequence of energy-using (and energy-wasting) steps involved in using electricity produced from fossil or nuclear fuels is Extraction → Transportation → Processing → Transportation to power plant → Electric generation → Transmission → End use.

Figure 4-7 shows the net energy efficiency for heating two well-insulated homes: one with electricity produced at a nuclear power plant, transported by wire to the home, and converted to heat (electric resistance heating), the other heated passively, with an input of direct solar energy through high-efficiency windows facing the sun, with heat stored in rocks or water for slow release. This analysis shows that the process of converting the high-quality energy in nuclear fuel to high-quality heat at several thousand degrees in the power plant, converting this heat to high-quality electricity, and then using the electricity to provide low-quality heat for warming a house to only about 20°C

(68°F) is extremely wasteful of high-quality energy. Burning coal or any fossil fuel at a power plant to supply electricity for space heating is also inefficient. It is much less wasteful to collect solar energy from the environment, store the resulting heat in stone or water, and, if necessary, raise its temperature slightly to provide space heating or household hot water.

Physicist and energy expert Amory Lovins points out that using high-quality electrical energy to provide low-quality heating for living space or household water is like using a chain saw to cut butter or a sledgehammer to kill a fly. As a general rule, he suggests that we not use high-quality energy to do a job that can be done with lower-quality energy (Figure 2-6).

The logic of this point is illustrated by looking at the costs of providing heat using various fuels. In 1995, the average price of obtaining 250,000 kilocalories (1 million Btu) for heating space or water in the United States was $6.05 using natural gas, $7.56 using kerosene, $9.30 using oil, $9.74 using propane, and $24.15 using electricity. As these numbers suggest, if you don't mind throwing away hard-earned dollars, then use electricity to heat your house and bathwater. Yet almost one of every four homes in the United States is heated by electricity.

Perhaps the three least efficient energy-using devices in widespread use today are incandescent light bulbs (which waste 95% of the energy input), vehicles with internal combustion engines (which waste 86–90% of the energy in their fuel), and nuclear power plants producing electricity for space or water heating (which waste 86% of the energy in their nuclear fuel, and probably 92% when the energy needed to deal with radioactive wastes and retired nuclear plants is included). Energy experts call for us to replace these devices or greatly improve their energy efficiency over the next few decades.

Why Is Reducing Energy Waste So Important?
According to Amory Lovins and other energy analysts, reducing energy waste is one of the planet's best and most important economic and environmental bargains. It

- *Makes nonrenewable fossil fuels last longer*

- *Gives us more time to phase in renewable energy resources*

- *Decreases dependence on oil imports* (55% in 1997 in the United States)

- *Lessens the need for military intervention in the oil-rich but politically unstable Middle East*

- *Reduces local and global environmental damage* because less of each energy resource would provide the same amount of useful energy

- *Is the cheapest and quickest way to slow projected global warming*

Q: What percentage of Americans use public transportation to get to and from work?

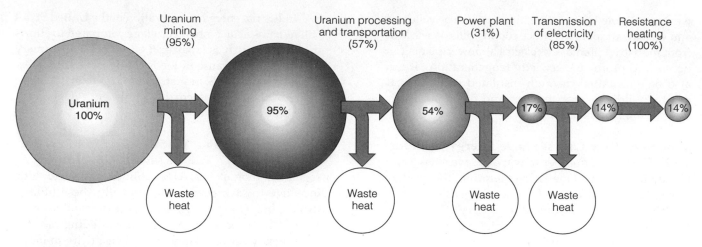

Electricity from Nuclear Power Plant

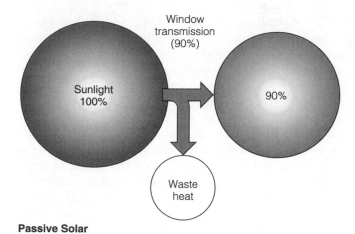

Passive Solar

Figure 4-7 Comparison of net energy efficiency for two types of space heating. The cumulative net efficiency is obtained by multiplying the percentage shown inside the circle for each step by the energy efficiency for that step (shown in parentheses). Because of the second law of thermodynamics, in most cases the greater the number of steps in an energy conversion process, the lower its net energy efficiency. About 86% of the energy used to provide space heating by electricity produced at a nuclear power plant is wasted. If the additional energy needed to deal with nuclear wastes and to retire highly radioactive nuclear plants after their useful life is included, then the net energy yield for a nuclear plant is only about 8% (or 92% waste). By contrast, with passive solar heating, only about 10% of incoming solar energy is wasted.

- *Saves more money, provides more jobs, improves productivity, and promotes more economic growth per unit of energy than other alternatives*

- *Improves competitiveness in the international marketplace*

According to Amory Lovins, if the world really got serious about improving energy efficiency, we could save $1 trillion per year. With such an impressive array of benefits, why isn't there more emphasis on improving energy efficiency? The primary reason is a glut of low-cost, underpriced fossil fuels. As long as such energy is artificially cheap because its harmful environmental costs are not included in its market prices, people are more likely to waste it and not make investments in improving energy efficiency.

Another cause is the use of huge government subsidies and tax breaks to help promote the use of energy-wasting power plants, motor vehicles, and buildings and few economic incentives for consumers to invest in improving energy efficiency even

if it saves them money. A third problem is that most consumers don't have adequate information about the availability of energy-saving devices and the amount of money such items can save them using *life cycle cost* analysis.

How Can We Use Waste Heat? Could we save energy by recycling energy? No. The second law of energy tells us that we cannot recycle energy. However, we can slow the rate at which waste heat flows into the environment when high-quality energy is degraded. For a house, the best way to do this is to insulate it thoroughly, eliminate air leaks, and equip it with an air-to-air heat exchanger to prevent buildup of indoor air pollutants. Many homes are so full of leaks that their heat loss in cold weather and heat gain in hot weather are equivalent to having a large window-size hole in the wall of the house.

In office buildings and stores, waste heat from lights, computers, and other machines can be collected and distributed to reduce heating bills during cold weather; during hot weather, this heat can be collected and vented outdoors to reduce cooling bills. Waste heat from industrial plants and electrical power plants can be used to produce electricity (cogeneration); it can also be distributed through insulated pipes to heat nearby buildings, greenhouses, and fish ponds, as is done in some parts of Europe.

Solutions: How Can We Save Energy in Industry? There are a number of ways to save energy and money in industry. One is **cogeneration**, the production of two useful forms of energy (such as steam and electricity) from the same fuel source. Waste heat from coal-fired and other industrial boilers can produce steam that spins turbines and generates electricity at roughly half the cost of buying it from a utility company. By using the electricity or selling it to the local power company for general use, a plant can save energy and money. Cogeneration allows up to 90% of the energy in fuel to be used productively—far more than the 33% efficiency for central power plants owned by U.S. utilities.

Cogeneration has been widely used in western Europe for years, and its use in the United States is growing. Within 8 years cogeneration could produce more electricity than all U.S. nuclear power plants, and much more cheaply. In Germany, small cogeneration units that run on natural gas or liquefied petroleum gas (LPG) supply restaurants, apartment buildings, and houses with all their energy. In 4–5 years they pay for themselves in saved fuel and electricity. In 1997, the U.S. company Allied Signal announced plans to market a small cogenerator (micro-powerplant).

Replacing energy-wasting electric motors is another strategy. About 60–70% of the electricity used in U.S. industry drives electric motors, most of which are run at full speed with their output throttled to match their task—somewhat like driving with the gas pedal to the floor and the brake engaged. Each year a heavily used electric motor consumes 10 times its purchase cost in electricity—equivalent to using $20,000 worth of gasoline each year to fuel a $20,000 car. According to Amory Lovins, it would be cost-effective to scrap virtually all such motors and replace them with adjustable-speed drives. The costs would be paid back in 1–3 years, depending on how the motors were used.

Energy can also be saved by switching to high-efficiency lighting. Additionally, computer-controlled energy management systems can turn off lighting and equipment in nonproduction areas and make adjustments during periods of low production. Recycling and reuse and making products that last longer and are easy to repair and recycle also saves energy compared to using virgin resources.

Industrial energy conservation in the United States still remains at an embryonic stage compared to efforts in countries such as Japan and Germany. The primary reason is that because of low energy costs, energy expenses for most U.S. industries are less than 2% of the value of their manufacturing shipments. Thus, they have little incentive to save energy.

Solutions: How Can We Save Energy in Producing Electricity? The Negawatt Revolution As regulated monopolies, U.S. utilities have had little incentive to save energy. Traditionally, these utilities make more money by increasing the demand for electricity. This process encourages the building of often unnecessary power plants to send electricity to inefficient appliances, heating and cooling systems, and industrial plants.

A small but growing number of utility companies in the United States and elsewhere are trying to reverse this wasteful process by reducing the demand for electricity. This new approach is known as *demand-side management* or the *negawatt revolution*. To reduce demand, utilities give customers cash rebates for buying efficient lights and appliances, free home-energy audits, low-interest loans for home weatherization or industrial retrofits, and lower rates to households or industries meeting certain energy-efficiency standards. To make demand-side management feasible, state and other utility regulators must allow utility investors to make a reasonable return on their money, based on the amount of energy the utilities save. Such a policy allows utility companies to profit by shifting their emphasis from producing megawatts to energy-saving negawatts. However, since 1993 such programs have been cut back sharply as the U.S. electric power industry is being restructured and deregulated to help increase competition.

🌀 Solutions: How Can We Save Energy in Transportation? The most important way to save energy (especially oil) and money in transportation is to *increase the fuel efficiency of motor vehicles*. Between 1973 and 1985, the average fuel efficiency doubled for new American cars and rose 37% for all passenger cars on the road because of government-mandated standards. However, since 1985 automakers have succeeded in opposing any raise in the government standards. Actually, the average fuel efficiency of new vehicles has fallen by about 5% in recent years because of the popularity of sport utility vehicles (SUVs), minivans, and light trucks (subject to much lower mileage standards than cars) and to larger, less efficient autos.

According to the U.S. Office of Technology Assessment, existing technology could raise current fuel efficiency of the entire U.S. automotive fleet to 15 kilo-

Q: What percentage of Americans walk or use a bicycle to get to and from work?

meters per liter (35 miles per gallon) by 2010. Doing this would save U.S. consumers about $65 billion a year (about $576 per household), cause a sharp drop in emissions of CO_2 and other air pollutants, cut oil imports in half (which would significantly lower the annual U.S. trade deficit), and create about 244,000 new jobs throughout the U.S. economy.

Since 1985, at least 10 automobile companies have made nimble and peppy prototype cars that meet or exceed current safety and pollution standards, with fuel efficiencies of 29–59 kilometers per liter (67–138 miles per gallon). If such cars were mass-produced, their slightly higher costs would be more than offset by their fuel savings. The problem is that there is little consumer interest in fuel-efficient cars when the inflation-adjusted price of gasoline today is the lowest it has been since 1920. This underpricing of gasoline encourages energy waste and pollution by failing to include in its market price the many harmful social and environmental costs of gasoline, which consumers ultimately end up paying anyway.

Conventional battery-powered *electric cars* might help reduce dependence on oil, especially for urban commuting and short trips. Electric vehicles are extremely quiet, need little maintenance, and can accelerate rapidly. The cars themselves, called zero-emission vehicles, produce no air pollution. However, using coal and nuclear power plants to produce the electricity needed to recharge their batteries daily does produce air pollution and nuclear wastes—something called *elsewhere pollution*. If solar cells or wind turbines could be used for recharging the car's batteries, CO_2 and other air pollution emissions would be virtually eliminated.

On the negative side, today's electric cars are not very efficient; they are equivalent to gasoline-powered cars that get about 7–11 kilometers per liter (16–25 miles per gallon). Current electric cars can travel only 81–161 kilometers (50–100 miles) vs. 480–640 kilometers (300–400 miles) for gasoline cars. Their batteries must be recharged for 3–8 hours (although a new device may reduce recharge time to 10–20 minutes at a cost of $8–15 per recharge) and must be replaced about every 48,000 kilometers (30,000 miles) at a cost of about at least $2,000. As a result, today's electric cars have twice the operating cost of gasoline-powered cars.

There is growing interest in developing hybrid electric–internal combustion *ecocars*, or *supercars* getting 64–128 kilometers per liter (150–300 miles per gallon). They get most of their power from a small hybrid electric/fuel engine that makes its own electricity and uses a small battery or flywheel to provide the extra energy needed for acceleration and hill climbing.

Amory Lovins and other researchers project that with proper financial incentives, low-polluting, ultralight, ultrasafe, ultraefficient hybrid electric–internal

combustion engine and fuel-cell *ecocars* could be designed, built, and sold within 10–20 years. With incentives such as tax credits or energy rebates, such cars could replace much of the existing car fleet within 10–12 years (the average road life of existing cars).

Lovins argues that such cars would sell "not just because they are more than 5 times as efficient and more than 100 times cleaner than current cars, but also because they will be markedly superior in comfort, quietness, performance, durability, beauty, and safety."

Another environmentally appealing type of ecocar is an electric vehicle that uses fuel cells powered by solar-produced hydrogen gas. Fuel cells consist of two electrodes immersed in a solution (electrolyte) that conducts electricity. They produce electricity by combining hydrogen and oxygen ions, typically from hydrogen and oxygen gas supplied as fuel for the cell or from gasoline. Fuel-cell cars running on hydrogen come close to being true zero-emission vehicles because they emit only water vapor and trace amounts of nitrogen oxides, easily controlled with existing technology.

In 1999, Toyota, which is spending $800 million a year on alternative-fuel research, began selling a sport utility vehicle retrofitted with a hydrogen-based fuel-cell system. Ford plans to build such a car by 2000.

Another way to save energy is to *shift to more energy-efficient ways to move people* (Figure 4-8) *and freight*. More freight could be shifted from trucks and planes to more energy-efficient trains and ships. The fuel efficiency of other means of transport can also be increased. With improved aerodynamic design, turbocharged diesel engines, and radial tires, new transport trucks can be 50% more fuel-efficient than today's conventional trucks.

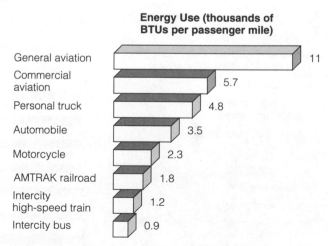

Energy Use (thousands of BTUs per passenger mile)

Type	Energy Use
General aviation	11
Commercial aviation	5.7
Personal truck	4.8
Automobile	3.5
Motorcycle	2.3
AMTRAK railroad	1.8
Intercity high-speed train	1.2
Intercity bus	0.9

Figure 4-8 Energy use of various types of domestic passenger transportation: The lower the energy use, the greater the efficiency.

A: About 2%

Solutions: How Can We Save Energy in Buildings?

Solutions: How Can We Save Energy in Buildings? There are a number of ways to improve the energy efficiency of buildings. One is to build more *superinsulated houses* (Figure 4-9). Such houses are so heavily insulated and so airtight that heat from direct sunlight, appliances, and human bodies warms them, with little or no need for a backup heating system. An air-to-air heat exchanger prevents buildup of indoor air pollution without wasting much heat. Although such houses typically cost 5% more to build than conventional houses of the same size, this extra cost is paid back by energy savings within 5 years and can save a homeowner $50,000–100,000 over a 40-year period.

Another way to save energy is to *use the most energy-efficient ways to heat houses* (Figure 4-10). The most energy-efficient ways to heat space are to build a superinsulated house, use passive solar heating, and use high-efficiency (85–98% efficient) natural gas furnaces. The most wasteful and expensive way is to use electric resistance heating with the electricity produced by a coal-fired or nuclear power plant.

Heat pumps can save energy and money for space heating in warm climates (where they aren't needed much), but not in cold climates; at low temperatures they automatically switch to wasteful, costly electric resistance heating (which is why utilities wanting to sell

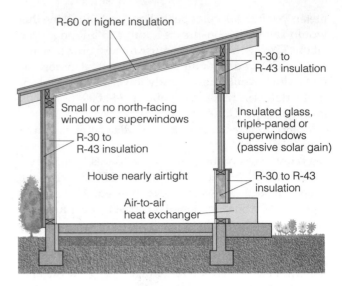

Figure 4-9 Major features of a superinsulated house, so heavily insulated and so airtight that heat from direct sunlight, appliances, and human bodies warms it, with little or no need for a backup heating system. An air-to-air heat exchanger prevents buildup of indoor air pollution without wasting much heat.

Net Energy Efficiency

Superinsulated house (100% of heat)(R-43)	98%
Passive solar (100% of heat)	90%
Passive solar (50% of heat) plus high-efficiency natural gas furnace (50% of heat)	87%
Natural gas with high-efficiency furnace	84%
Electric resistance heating (electricity from hydroelectric power plant)	82%
Natural gas with typical furnace	70%
Passive solar (50% of heat) plus high-efficiency wood stove (50% of heat)	65%
Oil furnace	53%
Electric heat pump (electricity from coal-fired power plant)	50%
High-efficiency wood stove	39%
Active solar	35%
Electric heat pump (electricity from nuclear plant)	30%
Typical wood stove	26%
Electric resistance heating (electricity from coal-fired power plant)	25%
Electric resistance heating (electricity from nuclear plant)	14%

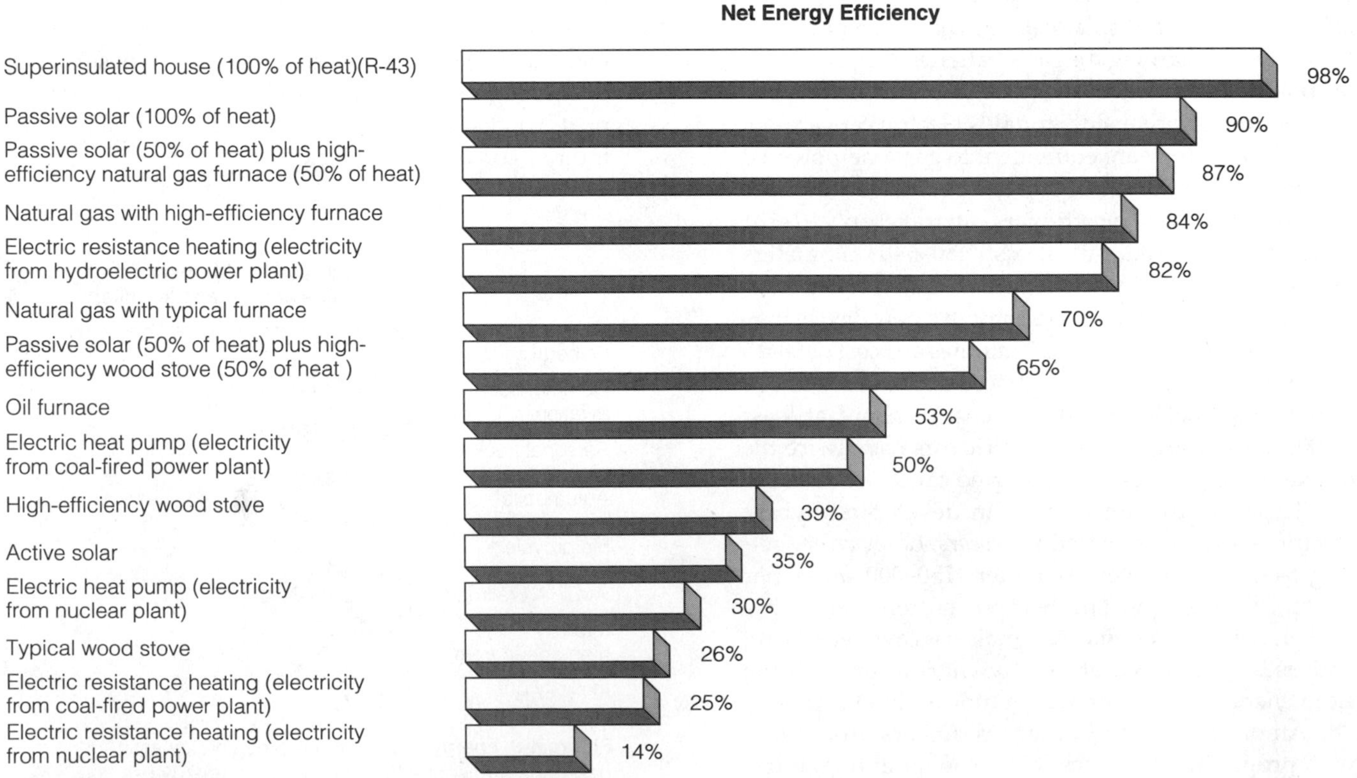

Figure 4-10 Net energy efficiencies for various ways to heat an enclosed space such as a house.

Q: How many people have been killed in motor vehicle accidents since the first automobile was built in 1885?

more electricity advocate their use). Some heat pumps in their air conditioning mode are also much less efficient than many individual air conditioning units. Most heat pumps also require expensive repair every few years. However, manufacturers have developed some improved models that produce warmer air and are more efficient than older models.

The energy efficiency of existing houses can be improved significantly by adding insulation, plugging leaks, and installing energy-saving windows. About one-third of heated air in U.S. homes and buildings escapes through closed windows and holes and cracks—equal to the energy in all the oil flowing through the Alaska pipeline every year. During hot weather these windows also let heat in, increasing the use of air conditioning.

We can also *use the most energy-efficient ways to heat household water.* An efficient method is to use tankless instant water heaters (about the size of bookcase loudspeakers) fired by natural gas or liquefied petroleum gas (LPG). These devices, widely used in many parts of Europe, heat the water instantly as it flows through a small burner chamber and provide hot water only when (and as long as) it is needed. A well-insulated, conventional natural gas or LPG water heater is also fairly efficient (although all conventional natural gas and electric resistance heaters waste energy by keeping a large tank of water hot all day and night and can run out after a long shower or two).

The most inefficient and expensive way to heat water for washing and bathing is to use electricity produced by any type of power plant. A $425 electric hot water heater can cost $5,900 in energy costs over its 15-year life, compared to about $2,900 for a comparable natural gas water heater over the same period.

Setting higher energy-efficiency standards for new buildings would also save energy. Building codes could require that all new houses use 60–80% less energy than conventional houses of the same size, as has been done in Davis, California (Case Study, p. 94). Because of tough energy-efficiency standards, the average Swedish home consumes about one-third as much energy as the average American home of the same size.

Another way to save energy is to *buy the most energy-efficient appliances and lights* (Figure 4-11).* If the most energy-efficient appliances and lights now available were installed in all U.S. homes over the next 20 years, the savings in energy would equal the estimated energy content

---

*Each year the American Council for an Energy-Efficient Economy (ACEEE) publishes a list of the most energy-efficient major appliances mass-produced for the U.S. market. A copy can be obtained from the council at 1001 Connecticut Ave. NW, Suite 801, Washington, DC 20036. Each year they also publish *A Consumer Guide to Home Energy Savings,* available in bookstores or from the ACEEE.

of Alaska's entire North Slope oil fields. Replacing a standard incandescent bulb with an energy-efficient compact fluorescent saves about $48–70 per bulb over its 10-year life. Thus, replacing 25 incandescent bulbs in a house with energy-efficient fluorescent bulbs saves $1,250–1,750. Energy-efficient lighting could save U.S. businesses $15–20 billion per year in electricity bills. The bad news is that by 1998 only 9% of U.S. homes used these bulbs.

Improvements in energy efficiency could be encouraged by giving rebates or tax credits for building energy-efficient buildings, improving the energy efficiency of existing buildings, and buying high-efficiency appliances and equipment.

Case Study: Saving Energy, Money, and Jobs in Osage, Iowa Osage, Iowa (population about 4,000), has become the energy-efficiency capital of the United States. Its transformation began in 1974 when Wes Birdsall, general manager of Osage Municipal Gas and Electric Company, started urging the townspeople to save energy and reduce their natural gas and electric bills. The utility would also save money by not having to add new electrical generating facilities.

Birdsall started his crusade by telling homeowners about the importance of insulating walls and ceilings and plugging leaky windows and doors. These repairs provided jobs and income for people selling and installing

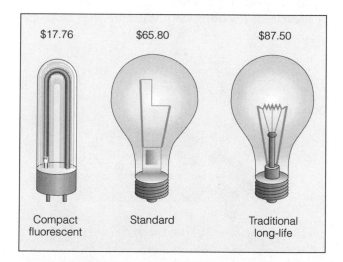

Figure 4-11 Cost of electricity for comparable light bulbs used for 10,000 hours. Because conventional incandescent bulbs are only 5% efficient and last only 750–1,500 hours, they waste enormous amounts of energy and money, and they add to the heat load of houses during hot weather. Socket-type fluorescent lights use one-fourth as much electricity as conventional bulbs. Although these new bulbs cost about $10–20 per bulb, they last 10–20 times longer than conventional incandescent bulbs used the same number of hours, and they save considerable money (compared with incandescent bulbs) over their long life. (Data from Electric Power Research Institute)

insulation, caulking, and energy-efficient windows. He also advised people to replace their incandescent light bulbs with more efficient fluorescent bulbs and to turn down the temperature on water heaters and wrap them with insulation—economic boons to the local hardware and lighting stores. The utility company even gave away free water heater blankets. Birdsall also suggested saving water and fuel by installing low-flow shower heads.

Birdsall then stepped up his efforts by offering to give every building in town a free thermogram—an infrared scan that shows where heat escapes. When people could see the energy (and their money) hemorrhaging out of their buildings, they took action to plug the leaks, again helping the local economy. Birdsall then stepped up his campaign even more, announcing that no new houses could be hooked up to the company's natural gas line unless they met minimum energy-efficiency standards.

Since 1974 the town has reduced its natural gas consumption by 45%, no mean feat in a locale with frigid winter temperatures. In addition, the utility company saved enough money to repay all its debt, accumulate a cash surplus, and cut inflation-adjusted electricity rates by a third (which attracted two new factories to the area). Furthermore, each household saves more than $1,000 per year; this money supports jobs, and most of it circulates in the local economy. Before the town's energy-efficiency revolution, about $1.2 million a year left town to buy energy.

Osage's success in making energy efficiency a way of life earned the town a National Environmental Achievement Award in 1991. What are your local util-

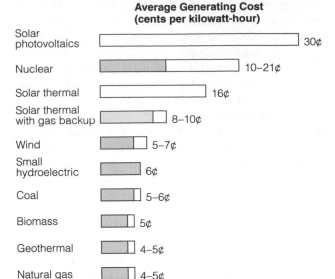

Figure 4-12 Generating costs of electricity per kilowatt-hour by various technologies in 1993. Between 2000 and 2005, costs per kilowatt-hour for wind are expected to decrease to 4–5¢, solar thermal with gas assistance 5–6¢, 4–5¢; solar photovoltaic 6–12¢, natural gas 3–4¢, and coal and nuclear power will remain at 10–21¢. Costs for the other technologies shown in this figure are expected to remain about the same. (Data from U.S. Department of Energy, Council for Renewable Energy Education, Investor Responsibility Research Center, California Energy Commission, and Worldwatch Institute)

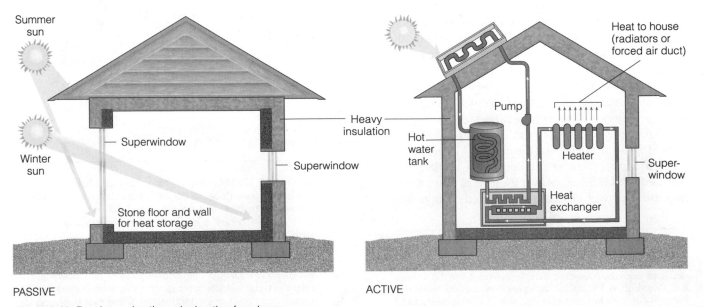

Figure 4-13 Passive and active solar heating for a home.

Q: Worldwide, how many people are killed each year by motor vehicles?

ity companies and community doing to save energy and stimulate the local economy?

4-3 RENEWABLE SOLAR ENERGY RESOURCES

🔲 **What Are the Advantages of Solar Energy? The Coming Renewable-Energy Age** About 92% of the known reserves and potentially available energy resources in the United States are renewable energy from the sun, wind, flowing water, biomass, and earth's internal heat. The other 8% of potentially available domestic energy resources are coal (5%), oil (2.5%), and uranium (0.5%).

Developing these mostly untapped renewable energy resources could meet 50–80% of projected U.S. energy needs by 2040 or sooner, and it could meet virtually all energy needs if coupled with improvements in energy efficiency. In 1994, Shell International Petroleum in London, which forecast the steep rise in oil prices in the 1970s, projected that renewable energy, especially solar, will dominate world energy production by 2050.

Developing renewable energy resources would save money, create two to five times more jobs per unit of electricity produced than coal and nuclear power plants, eliminate the need for oil imports, cause much less pollution and environmental damage per unit of energy used, and increase military, economic, and environmental security. In the United States, wind farms and power plants using geothermal, wood (biomass), hydropower, and high-temperature solar energy (with natural gas backup) can already produce electricity more cheaply than can new nuclear power plants, and with far fewer federal subsidies (Figure 4-12).

Solutions: How Can Solar Energy Be Used to Heat Houses and Water? Buildings and water can be heated by solar energy using two methods: passive and active (Figure 4-13). A **passive solar heating system** captures sunlight directly within a structure (Figures 4-13, left, and 4-14) and converts it into low-temperature heat for space heating. Energy-efficient windows, greenhouses, and sunspaces face the sun to collect solar energy by direct gain.

Engineer and builder Michael Sykes has designed a solar envelope house that is heated and cooled passively by solar energy and the slow storage and release of energy by massive timbers and the earth beneath the house (Figure 4-15). The front and back of this house are double walls of heavy timber impregnated with salt to increase the ability of the wood to store heat. The space between these two walls, plus the basement, forms a convection loop or envelope around the inner shell of the

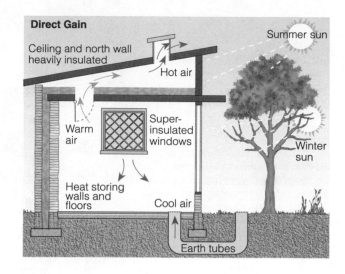

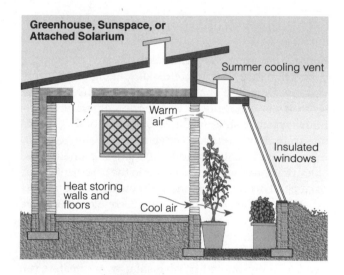

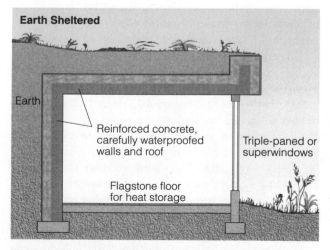

Figure 4-14 Three examples of passive solar design for houses.

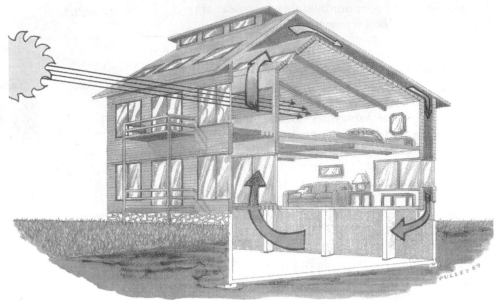

Figure 4-15 Solar envelope house that is heated and cooled passively by solar energy and the earth's thermal energy. This patented Enertia design developed by Michael Sykes needs no conventional heating or cooling system in most areas. It comes in a precut kit engineered and tailored to the buyer's design goals. Sykes plants 50 trees for each one used in his house kits and his design has received both the Department of Energy's Innovation Award and the North Carolina Governor's Energy Achievement Award. (Enertia Building Systems, Rt. 1, Box 67, Wake Forest, NC 27587)

© 1989 Enertia₈ Building Systems, Inc.   U.S. Patent No. 4,621,614

house. In summer, roof vents release heated air from the convection loop throughout the day; at night, these roof vents, with the aid of a fan, draw air into the loop, passively cooling the house. The interior temperature of the house typically stays within two degrees of 21°C (70°F) year-round, without any conventional cooling or heating system. In cold or cloudy climates, a small wood stove or vented natural gas heater in the basement can be used as a backup to heat the air in the convection loop.

With available and developing technologies, passive solar designs can provide at least 70% of a *residential* building's heating needs and at least 60% of its cooling needs, and up to 60% of a *commercial* building's energy needs. Roof-mounted passive solar water heaters can supply all or most of the hot water for a typical house. A good passive solar and superinsulated system usually adds 5–10% to the construction cost, but the life cycle cost of operating such a house is 30–40% lower. The typical payback time for passive solar features is 3–7 years.

In an **active solar heating system**, specially designed collectors absorb solar energy and a fan or a pump supplies part of a building's space-heating or water-heating needs (Figure 4-13, right). Several connected collectors are usually mounted on a portion of the roof with an unobstructed exposure to the sun. Some of the heat can be used directly, and the rest can be stored in insulated tanks containing rocks, water, or a heat-absorbing chemical for release later as needed. Active solar collectors can also supply hot water.

With current technology, active solar systems usually cost too much for heating most homes and small buildings because they use more materials to build, need more maintenance, and eventually deteriorate and must be replaced. In addition, some people consider active solar collectors sitting on rooftops or in yards to be ugly. However, retrofitting existing buildings is often easier with an active solar system. In Colombia, South America an ecopioneer has spent 30 years building a solar village in an inhospitable wet savanna infested with mosquitoes and with soil so toxic that it could support only tough grass (Individuals Matter, p. right).

Solar energy for low-temperature heating of buildings, whether collected actively or passively, is free, and the net energy yield is moderate (active) to high (passive). Both active and passive technology are well developed and can be installed quickly. No heat-trapping carbon dioxide is added to the atmosphere and environmental effects from air and water pollution are low. Land disturbance is also minimal because passive systems are built into structures and active solar collectors are usually placed on rooftops. However, owners of passive and active solar systems need *solar legal rights* to prevent others from building structures that block their access to sunlight.

Solutions: How Can We Cool Houses Naturally?
Superinsulation and superinsulated windows help make buildings cooler. Passive cooling can also be provided by blocking the high summer sun with deciduous trees, window overhangs, or awnings (Figure 4-14, top). Windows and fans can be used to take advantage of breezes and keep air moving. A reflective insulating foil sheet suspended in the attic can block heat from radiating down into the house. Solar-powered air conditioners have been developed but so far are too expensive for residential use.

Earth tubes can also be used for indoor cooling (Figure 4-14, top). At a depth of 3–6 meters (10–20 feet), the soil temperature stays at about 5–13°C (41–55°F) all

Q: How much of the air pollution in the United States is produced by motor vehicles?

year long in cold northern climates and about 19°C (67°F) in warm southern climates. Several earth tubes, simple plastic (PVC) plumbing pipes with a diameter of 10–15 centimeters (4–6 inches), buried about 0.6 meter (2 feet) apart and 3–6 meters (10–20 feet) deep, can pipe cool and partially dehumidified air into an energy-efficient house at a cost of a few dollars per summer.* People allergic to pollen and molds should add an air purification system, but this is also necessary with a conventional cooling system.

Solutions: How Can Solar Energy Be Used to Generate High-Temperature Heat and Electricity?
Several so-called *solar thermal* systems collect and transform radiant energy from the sun into thermal energy (heat) at high temperatures, which can be used directly or converted to electricity (Figure 4-16). In one such *central receiver system*, called a *power tower*, huge arrays of computer-controlled mirrors called *heliostats* track the sun and focus sunlight on a central heat-collection tower (Figure 4-16a). One government-subsidized power tower system began operating in the California desert in 1996.

In a *solar thermal plant* or *distributed receiver system*, sunlight is collected and focused on oil-filled pipes running through the middle of curved solar collectors (Figure 4-16b). This concentrated sunlight can generate temperatures high enough for industrial processes or for producing steam to run turbines and generate electricity. At night or on cloudy days, high-efficiency natural gas turbines supply backup electricity as needed. In California's Mojave Desert, solar thermal systems using parabolic trough collectors (Figure 4-16b) and a natural gas turbine backup system can produce power for 8–10¢ per kilowatt-hour—much cheaper than nuclear power plants.

Another type of distributed receiver system uses *parabolic dish collectors* (that look somewhat like TV satellite dishes) instead of parabolic troughs. These collectors can track the sun along two axes and are generally more efficient than troughs. A pilot plant is now being built in northern Australia. The U.S. Department of Energy projects that early in the 21st century parabolic dishes with gas turbine backup should be able to produce electrical power for 6¢ per kilowatt-hour.

Another promising approach to intensifying solar energy is a *nonimaging optical solar concentrator*. With this technology, the sun's rays are allowed to scramble instead of being focused on a particular point (Figure 4-16c). Experiments show that a nonimaging parabolic concentrator can intensify sunlight striking the earth 80,000 times. Because of its high efficiency and ability to generate extremely high temperatures, nonimaging

INDIVIDUALS MATTER

Gaviotas: A Solar Village in Colombia*

Thirty-three years ago ecopioneer and visionary Paolo Lugari began developing an ecologically sustainable solar village called Gaviotas in Colombia east of the Andes.

When he first saw this area in 1965 he thought that if people can live in this inhospitable area, they can live anywhere. With the help of professors and students from the universities in Bogota—16 hours away by jeep—he set out to build a sustainable ecovillage.

They invented cheap solar water heaters that could work even on cloudy days and helped finance development of the village by building and selling them to thousands of people in Bogota. They also produced electricity be designing ultra-light windmills to catch the area's mild but steady winds.

They built a village hospital heated by the sun and cooled by the wind for the area's then several hundred inhabitants. People in rural areas for hundreds of miles around came there for medical care and sent their children there to school.

The villagers found enough nontoxic soil along the riverbanks to grow mangoes, cassava, and cashews. They got protein from fish in the river and cattle that could eat the area's tough grass.

They planted hundreds of pines, which have grown into forests. The villagers tap gum oozing from the pines and use solar energy to distill it into turpentine and a valuable resin used in paints, cosmetics, and medicines. The huge market for these products provided a new industry for Gaviotas.

Dropping needles from the pines built up the soil, which then sprouted with hundreds of new kinds of plants. Now a rain forest that once grew in the area is returning from seeds carried by birds and roots emerging from the river edges.

These earth citizens believe that people who are willing to use their ingenuity and work hard to make their dreams reality can find solutions to their problems by learning to work with the earth. Their simple and affordable technologies, purposely unpatented, are spreading throughout the world.

---

*For more details see Alan Weisman, *Gaviotas: A Village to Reinvent the World* (White River Junction, Vt.: Chelsea Green Publishing, 1998).

---

*They work. I used them in a passively heated and cooled home/office for 15 years.

A: At least 50%, despite the world's strictest emission standards

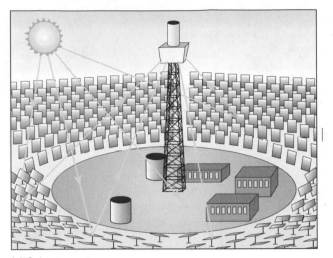

(a) Solar power tower

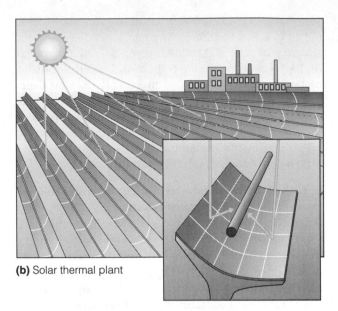

(b) Solar thermal plant

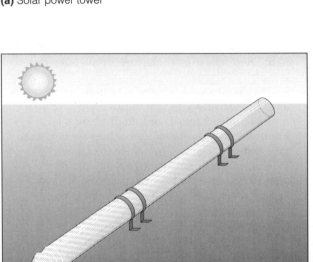

(c) Nonimaging optical solar concentrator

(d) Solar cooker

Figure 4-16 Several ways to collect and concentrate solar energy to produce high-temperature heat and electricity are in use. Today, solar plants are used mainly to supply reserve power for daytime peak electricity loads, especially in sunny areas with a high demand for air conditioning.

concentrators may make solar energy practical for widespread industrial and commercial use within a decade.

Inexpensive solar cookers can focus and concentrate sunlight and cook food, especially in rural villages in sunny developing countries. They can be made by fitting an insulated box big enough to hold three or four pots inside with a transparent, removable top (Figure 4-16d). Solar cookers reduce deforestation for fuelwood, the time and labor needed to collect firewood, and indoor air pollution from smoky fires.

The impact of solar thermal power plants on air and water is low. They can be built as large or as small as needed in 1–2 years, compared to 5–7 years for a coal-fired plant and 12–15 years for a nuclear power plant.

This saves investors millions in interest on construction loans. Distributed system solar thermal power plants with small turbines burning natural gas as backup can produce electricity at about half the cost of a nuclear power plant (Figure 4-12). According to the Worldwatch Institute, solar power plants installed on a total land area equal to that of Panama or South Carolina could provide as much electricity as the entire world now uses.

Solutions: How Can We Produce Electricity from Solar Cells? The PV Revolution Solar energy can be converted directly into electrical energy by **photovoltaic (PV)** cells, commonly called **solar cells** (Figure 4-17). Sunlight falling on a solar cell, a

Q: What percentage of the commercial energy used in the world comes from nonrenewable resources?

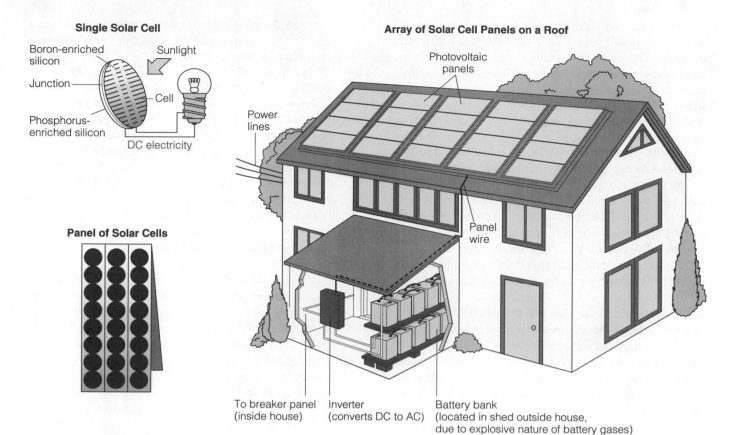

Single Solar Cell

Boron-enriched silicon

Sunlight

Junction

Cell

Phosphorus-enriched silicon

DC electricity

Panel of Solar Cells

Array of Solar Cell Panels on a Roof

Photovoltaic panels

Power lines

Panel wire

To breaker panel (inside house)

Inverter (converts DC to AC)

Battery bank (located in shed outside house, due to explosive nature of battery gases)

Figure 4-17 Photovoltaic (solar) cells can provide electricity for a house. Small and easily expandable arrays of such cells can provide electricity for urban villages throughout the world without large power plants or power lines. Massive banks of such cells can also produce electricity at a small power plant. Today, at least two dozen U.S. utility companies are using photovoltaic cells in their operations; as the price of such electricity drops, usage will increase dramatically. In 1990 a Florida builder began selling tract houses that get all of their electricity from roof-mounted solar cells. Although the solar-cell systems account for about one-third of the cost of each house, the savings in electric bills will pay this off over the term of a 30-year mortgage. Soon the racks of solar cell panels shown in this diagram will be incorporated in the roof, in the form of solar-cell roof shingles or PV panel roof systems that look like metal roofs.

transparent silicon wafer thinner than a sheet of paper, releases a flow of electrons when it strikes silicon atoms, creating an electrical current.

Solar cells are reliable and quiet, have no moving parts, and should last 20–30 years or more if encased in glass or plastic. The cells can be installed quickly and easily and expanded or moved as needed; maintenance consists of occasional washing to keep dirt from blocking the sun's rays. Arrays of cells can be located in deserts and marginal lands, along interstate highways, in yards, and on rooftops. Expandable banks of cells can also produce electricity at a small power plant, using efficient turbines burning natural gas to provide backup power when the sun isn't shining.

Solar cells produce no heat-trapping CO_2. Air and water pollution during operation is extremely low, air pollution from manufacture is low, land disturbance is very low for roof-mounted systems, and producing the materials doesn't require strip mining. The net energy

yield is fairly high and is increasing with new designs. Whereas solar thermal systems require direct sunlight, PV cells work in cloudy weather. Traditional-looking solar-cell roof shingles and photovoltaic panels that resemble metal roofs are now available, reducing the cost of solar cell installations by saving on roof costs.

With an aggressive program starting now, analysts project that solar cells could supply 17% of the world's electricity by 2010—as much as nuclear power does today—at a lower cost and much lower risk. With a strong push from governments and private investors, by 2050 solar cells could provide as much as 25% of the world's electricity (at least 35% in the United States). Solar cells are an ideal technology for providing electricity to 2 billion people in mostly sunny rural areas in most developing countries, which have not committed to expensive large-scale centralized nuclear and fossil power plants and electric-line transmission systems.

There are some drawbacks, however. The current costs of solar-cell systems are high (Figure 4-12), but they should become competitive in 5–15 years and are already cost-competitive in some situations. The manufacture of solar cells produces moderate levels of water pollution, which can be eliminated by effective pollution controls. Some people find racks of solar cells on rooftops or in yards to be unsightly, but new thin and flexible rolls of cells (and solar shingles and solar-cell metal roofs) should eliminate this problem.

According to some analysts, unless federal and private research efforts on photovoltaics are increased sharply, the United States may lose out on a huge global market ($1.4 billion in 1997 and at least $5 billion per year by 2010) and may have to import solar cells from Japan, Germany, Italy, and other countries that have invested heavily in this promising technology since 1980.

Solutions: How Can We Produce Electricity from Moving Water? In a *large-scale hydropower project*, a high dam is built across a large river to create a reservoir. Some of the water stored in the reservoir is allowed to flow through huge pipes at controlled rates, spinning turbines and producing electricity. Such projects have advantages and disadvantages (Figure 4-18). In a *small-scale hydropower project*, a low dam with no

reservoir (or only a small one) is built across a small stream. Output in small systems can vary with seasonal changes in stream flow.

Hydroelectric power supplies about 20% of the world's electricity (99% in Norway, 75% in New Zealand, 50% in developing countries, 25% in China, and 13% in the United States) and 6% of its total commercial energy. In 1998 the world's three largest producers of hydroelectric power were Canada, the United States, and Brazil. Any new large supplies of hydroelectric power in the United States will be imported from Canada, which gets more than 70% of its electricity from hydropower. Within a few years, China, with 10% of the world's hydropower potential, may become the largest producer of hydroelectricity.

Hydropower has a moderate to high net useful energy yield and fairly low operating and maintenance costs. Hydroelectric plants rarely need to be shut down, and they emit no heat-trapping carbon dioxide or other air pollutants during operation. They have life spans 2–10 times those of coal and nuclear plants. Large dams also help control flooding and supply a regulated flow of irrigation water to areas below the dam. However, hydropower—especially huge dams and reservoirs—has adverse effects on the environment (Figure 4-18).

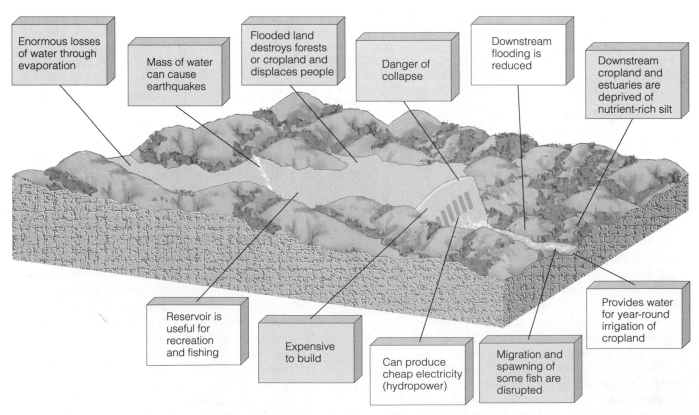

Figure 4-18 Advantages (white) and disadvantages (shaded) of large dams and reservoirs, which can be used to produce electricity.

Q: What percentage of the commercial energy used in the United States comes from nonrenewable resources?

Is Producing Electricity from Tides and Waves a Useful Option? Twice a day in high and low tides, water that flows into and out of coastal bays and estuaries can spin turbines to produce electricity (Figure 4-19). However, most analysts expect tidal power to make only a tiny contribution to world electricity supplies. There are few suitable sites, and construction costs are high.

The kinetic energy in ocean waves, created primarily by wind, is another potential source of electricity (Figure 4-19). Most analysts expect wave power to make little contribution to world electricity production, except in a few coastal areas with the right conditions (such as western England). Construction costs are moderate to high and the net energy yield is moderate, but equipment could be damaged or destroyed by saltwater corrosion and severe storms.

Solutions: How Can We Produce Electricity from Heat Stored in Water? Japan and the United States have been evaluating the use of the large temperature differences (between the cold, deep waters and the sun-warmed surface waters) of tropical oceans for producing electricity. If economically feasible, this would be done in *ocean thermal energy conversion* (OTEC) plants anchored to the bottom of tropical oceans in suitable sites (Figure 4-19). However, most energy analysts believe that the large-scale extraction of energy from ocean thermal gradients may never compete economically with other energy alternatives. Despite 50 years of work, the technology is still in the research-and-development stage.

Saline solar ponds, usually located near inland saline seas or lakes in areas with ample sunlight, can produce electricity from heat stored in layers of increasing concentrations of salt (Figure 4-19). Heat accumulated during the daytime in the more dense bottom layer can be used to produce steam that spins turbines, generating electricity. A small experimental saline solar pond power plant on the shore of the Israeli side of the Dead Sea was closed in 1989 because of high operating costs; smaller experimental systems have been built in California and Australia.

Freshwater solar ponds can be used to heat water and space (Figure 4-19). A shallow hole is dug and lined with concrete. A number of large, black plastic bags, each filled with several centimeters of water, are placed in the hole and then covered with fiberglass insulation panels. The panels let sunlight in but keep most of the heat stored in the water during the daytime from being lost to the atmosphere. When the water in the bags has reached its peak temperature in the afternoon, a computer turns on pumps to transfer hot water from the bags to large, insulated tanks for distribution.

Saline and freshwater solar ponds require no energy storage and backup systems. They emit no air pollution and have a moderate net energy yield. Freshwater solar ponds can be built in almost any sunny area and have moderate construction and operating costs. With adequate research and development support, proponents believe that freshwater solar ponds could supply 3–4% of U.S. electricity needs within 10 years.

Solutions: How Can We Produce Electricity from Wind? Since 1980 the use of wind to produce electricity has grown rapidly and is now the world's fastest growing energy resource (25% in 1998). In 1997 there were more than 27,000 wind turbines (Figure 4–20, left) worldwide, collectively producing about 7,000 megawatts of electricity (about 1% of the world's electricity).

Wind farms (clusters of 20–100 wind turbines, Figure 4-20, right), most of which are highly automated, now provide about 1% of California's electricity—enough to power 280,000 homes. Sizable wind farm projects are being planned in 12 other states. Denmark (which gets over 7% of its electricity from wind turbines) and Germany have pursued a different approach in which wind machines are installed one or two at a time across the rural landscape.

Wind power is a virtually unlimited source of energy at favorable sites; the global potential of wind power is about five times current world electricity use. Wind farms can be built in 6 months to a year and then easily expanded as needed. With a moderate to fairly high net energy yield, these systems emit no heat-trapping carbon dioxide or other air pollutants and need no water for cooling; manufacturing them produces little water pollution. The land under wind turbines can be used for grazing cattle or farming.

Wind power (with much lower government subsidies) has a significant cost advantage over nuclear power (Figure 4-12) and has become competitive with coal-fired power plants in many places. With new technological advances and mass production, projected cost declines should make wind power one of the world's cheapest ways to produce electricity within the next decade. In the long run, electricity from large wind farms in remote areas might be used to make hydrogen gas from water (Section 4-4) during off-peak periods. The hydrogen could then be fed into a pipeline and storage system.

One drawback of wind power is that it's economical primarily in areas with steady winds. When the wind dies down, backup electricity from a utility company or an energy storage system becomes necessary. Backup power could also come from linking wind farms with a solar-cell or hydropower plant, efficient natural-gas–burning turbines, or flywheels.

Other drawbacks to wind farms include visual pollution and noise, although these can be overcome by improving their design and locating them in isolated areas.

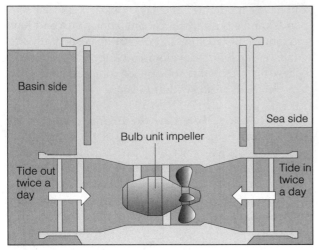

Tidal Power Plant

Basin side

Sea side

Bulb unit impeller

Tide out twice a day

Tide in twice a day

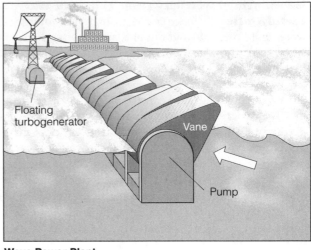

Wave Power Plant

Floating turbogenerator

Vane

Pump

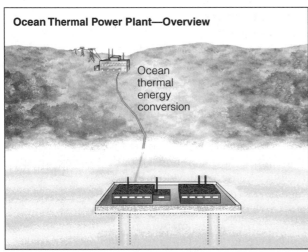

Ocean Thermal Power Plant—Overview

Ocean thermal energy conversion

Ocean Thermal Electric Plant

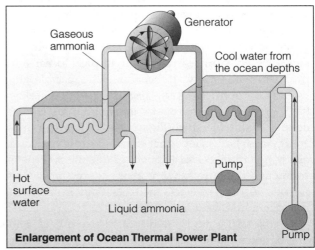

Gaseous ammonia

Generator

Cool water from the ocean depths

Pump

Hot surface water

Liquid ammonia

Pump

Enlargement of Ocean Thermal Power Plant

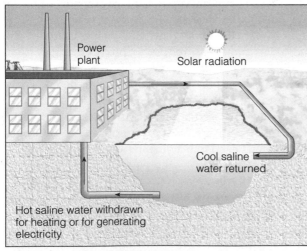

Power plant

Solar radiation

Cool saline water returned

Hot saline water withdrawn for heating or for generating electricity

Saline Water Solar Pond

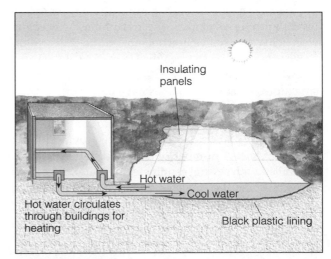

Insulating panels

Hot water

Cool water

Hot water circulates through buildings for heating

Black plastic lining

Freshwater Solar Pond

Figure 4-19 Ways to produce electricity from moving water and to tap into solar energy stored in water as heat. None of these systems are expected to be significant sources of energy in the near future.

Q: What percentage of the commercial energy used in the United States is wasted?

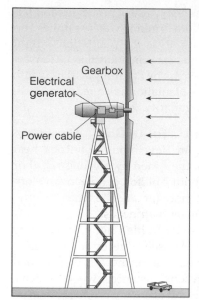 

Wind Turbine

Wind Farm

Figure 4-20 Wind turbines can be used to produce electricity individually or in clusters called wind farms. Wind farms, most of which are highly automated, now provide about 1% of California's electricity—enough to power 280,000 homes.

Large wind farms might also interfere with the flight patterns of migratory birds in certain areas, and they have killed large birds of prey (especially hawks, falcons, and eagles) that prefer to hunt along the ridge lines that are ideal for wind turbines. Researchers are evaluating how serious and widespread this problem is and are trying to find ways to deal with it.

Wind power experts project that by the middle of the 21st century wind power could supply more than 10–20% of the world's electricity and 10–25% of the electricity used in the United States. European governments are currently spending 10 times more for wind energy research and development than the U.S. government. Analysts warn that if this trend continues, U.S. consumers will be buying wind turbines from European countries (and eventually from China and India), and American companies will lose out on a rapidly growing global business ($2 billion in 1998).

What Are the Pros and Cons of Producing Energy from Plants? Many forms of *biomass*—organic matter in plants produced through photosynthesis—can be burned directly as a solid fuel or converted into gaseous or liquid *biofuels* (Figure 4-21). The burning of wood and manure to heat buildings and cook food supplies about 14% of the world's energy (4–5% in Canada and the United States) and about 35% of the energy used in developing countries.

Biomass is a potentially renewable energy resource as long as trees and plants are not harvested faster than they grow back, a requirement that is not being met in many places. No net increase in atmospheric levels of heat-trapping carbon dioxide occurs as long as the rate of removal and burning of trees and plants and the rate of loss of below-ground organic matter do not exceed the rate of replenishment. Burning biomass fuels adds much less air pollution to the atmosphere per unit of energy produced than does the uncontrolled burning of coal.

According to a 1992 UN study, by 2050 biomass could supply as much as 55% of the total energy used globally today. Whether such a projection becomes reality depends on the availability of large areas of productive land and adequate water and fertilizer—resources that may be in short supply in coming decades—and the ability to minimize the harmful environmental effects of such large-scale production of biomass. Currently, potentially renewable biomass is being exploited in nonrenewable and unsustainable ways, primarily because of deforestation, soil erosion, and the inefficient burning of wood in open fires and energy-wasting stoves.

Without effective land-use controls and replanting, widespread removal of trees and plants can deplete soil nutrients and cause excessive soil erosion, water pollution, flooding, and loss of wildlife habitat. Biomass resources also have a high moisture content (15–95%). The added weight of moisture makes collecting and hauling wood and other plant material fairly expensive, and the moisture also reduces the net energy yield. A 1998 analysis by David Pimentel and two other researchers concluded, "Large-scale biofuel production is not an

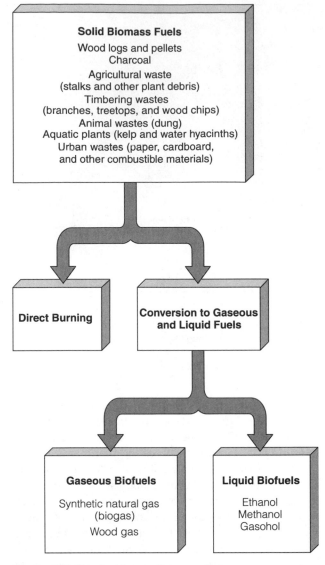

Solid Biomass Fuels

Wood logs and pellets
Charcoal
Agricultural waste
(stalks and other plant debris)
Timbering wastes
(branches, treetops, and wood chips)
Animal wastes (dung)
Aquatic plants (kelp and water hyacinths)
Urban wastes (paper, cardboard,
and other combustible materials)

Direct Burning

Conversion to Gaseous and Liquid Fuels

Gaseous Biofuels

Synthetic natural gas
(biogas)
Wood gas

Liquid Biofuels

Ethanol
Methanol
Gasohol

Figure 4-21 Principal types of biomass fuel.

alternative to the current use of oil and is not even an advisable option to cover a significant fraction of it.

Is Producing Energy from Biomass Plantations a Useful Option? One way to produce biomass fuel is to plant large numbers of fast-growing trees (especially cottonwoods, poplars, sycamores, and leucaenas), shrubs, and water hyacinths in *biomass plantations*. After harvest, these "Btu bushes" can be burned directly, converted into burnable gas, or fermented into a liquid alcohol fuel. Such plantations can be located on semi-arid land not needed for crops (although lack of water can limit productivity) and can be planted to reduce soil erosion and help restore degraded lands.

However, this industrialized approach to biomass production requires large areas of land (about 81 times as much land as solar cells to provide the same amount

of electricity) as well as heavy use of pesticides and fertilizers, which can pollute drinking water supplies and harm wildlife. In some areas the plantations might compete with food crops for prime farmland. Conversion of large forested areas or natural grasslands into single-species biomass plantations also reduces biodiversity and ecological integrity.

Is Burning Wood a Useful Option? Almost 70% of the people living in developing countries heat their homes and cook their food by burning wood or charcoal. However, about 2 billion people in developing countries cannot find (or are too poor to buy) enough fuelwood to meet their needs.

Wood has a moderate to high net useful energy yield when collected and burned directly and efficiently near its source. However, in urban areas, to which wood must be hauled from long distances, wood can cost homeowners more per unit of energy produced than oil or electricity. Harvesting wood can cause accidents (mostly from chain saws), and burning wood in poorly maintained or operated wood stoves can cause house fires. According to the EPA, air pollution from wood burning causes as many as 820 cancer deaths a year in the United States. Since 1990, the EPA has required all new wood stoves sold in the United States to emit at least 70% less particulate matter than earlier models.

Is Burning Agricultural and Urban Wastes a Useful Option? In agricultural areas, crop residues (such as sugarcane residues, rice husks, cotton stalks, and coconut shells) and animal manure can be collected and burned or converted into biofuels. The ash from biomass power plants can sometimes be used as fertilizer.

This approach makes sense when crop residues are burned in small power plants or burners located near areas where the residues are produced. Otherwise, it takes too much energy to collect, dry, and transport the residues to power plants. Some ecologists argue that it makes more sense to use animal manure as a fertilizer and to use crop residues to feed livestock, retard soil erosion, and fertilize the soil.

An increasing number of cities in Japan, western Europe, and the United States have built incinerators that burn trash and use the energy released to produce electricity or heat nearby buildings. However, this approach has been limited by opposition from citizens concerned about emissions of toxic gases and disposal of the resulting toxic ash. Some analysts argue that more energy is saved by composting or recycling paper and other organic wastes than by burning them. For example, recycling paper saves at least twice as much energy as is produced by burning the paper.

Is Producing Gaseous and Liquid Fuels from Solid Biomass a Useful Option? Various forms of

   Flavin, Christopher, and Molly O'Meara. 1998. "Solar Power Markets Boom." *World Watch*, vol. 11, no. 5, 23(5).

solid biomass can be converted by bacteria and various chemical processes into gaseous and liquid biofuels (Figure 4-21). Examples include *biogas*, a mixture of 60% methane (the principal component of natural gas) and 40% carbon dioxide); *liquid ethanol* (ethyl, or grain, alcohol); and *liquid methanol* (methyl, or wood alcohol).

In China, anaerobic bacteria in more than 6 million *biogas digesters* convert organic plant and animal wastes into methane fuel for heating and cooking. These simple devices can be built for about $50 including labor. After the biogas has been separated, the solid residue is used as fertilizer on food crops or, if contaminated, on trees. When they work, biogas digesters are very efficient. However, they are also slow and unpredictable, a problem that could be corrected with development of more reliable models.

Some analysts believe that liquid ethanol and methanol produced from biomass could replace gasoline and diesel fuel when oil becomes too scarce and expensive. *Ethanol* can be made from sugar and grain crops (sugarcane, sugar beets, sorghum, and corn) by fermentation and distillation. Gasoline mixed with 10–23% pure ethanol makes *gasohol*, which can be burned in conventional gasoline engines and is sold as super unleaded or ethanol-enriched gasoline.

Another alcohol, *methanol*, is made mostly from natural gas but can also be produced at a higher cost from wood, wood wastes, agricultural wastes (such as corncobs), sewage sludge, garbage, and coal. Table 4-1 gives the advantages and disadvantages of using ethanol, methanol, and several other fuels as alternatives to gasoline. Recently a new water-based gasoline fuel has entered the picture and is currently undergoing extensive testing by state and federal agencies (Individuals Matter, right).

4-4 THE SOLAR–HYDROGEN REVOLUTION

What Can We Use to Replace Oil? Good-Bye Oil and Smog, Hello Hydrogen When oil is gone (or when what's left costs too much to use) what will we use to fuel vehicles, industry, and buildings? Some scientists say the fuel of the future is hydrogen gas (H_2), a gaseous fuel that is easy to store and transport (Table 4-1).

There is very little hydrogen gas around, but we can get it from something we have plenty of: *water*. Water can be split by electricity into gaseous hydrogen and oxygen (Figure 4-22). If we can make the transition to an *energy-efficient solar–hydrogen age*, we could say goodbye to smog, oil spills, acid rain, and nuclear energy, and perhaps to the threat of global warming. The reason is simple: When hydrogen burns it combines with oxygen gas in the air and produces nonpolluting water vapor and some nitrogen oxides, thus eliminating most of the air pollution problems we face today.

A New Car Fuel That Is Mostly Water

INDIVIDUALS MATTER

It seems too good to be true, but a breakthrough fuel that is more than half water could power cars, trucks, train locomotives, generators, and aircraft running on gasoline and diesel fuel within a few years.

This new, milky-looking fuel, called A-55, was developed by Reno, Nevada, inventor Rudolf Gunnerman. It consists of 30–55% water and 45% naphtha (a cheap-to-produce by-product of petroleum distillation). In addition, it contains lubricants, antirust and antifreeze agents, and small amounts of a blending agent developed by Gunnerman (which attracts the normally repellent water and oily naphtha to one another). The water in A-55 allows the fuel to burn cooler and more efficiently, producing more power and less pollution. In 1994 Gunnerman and diesel equipment giant Caterpillar formed Advanced Fuels, a joint venture to test and market the new fuel.

So far in every test the new fuel tops conventional gasoline and diesel fuel as a clean, safe, and cheap fuel that can be used in almost any combustion engine. In November 1995, Nevada certified A-55 as a clean alternative fuel. That means that it meets federal and Nevada laws requiring clean fuels in fleets and other vehicles. A-55 is especially attractive to the trucking industry, facing federal crackdowns on dirty diesel engine emissions.

Tests so far show that using A-55 leads to a 60% drop in EPA-monitored emissions of air pollutants such as carbon monoxide, nitrous oxide, and hydrocarbons. Because it doesn't burn outside engine combustion chambers, the fuel can be stored in aboveground tanks without risking explosions.

Converting vehicles to use A-55 costs about $300 and requires only fairly simple adjustments to fuel injectors and timing. Once that's done, the vehicle could use either the water-based fuel or conventional gasoline or diesel fuels.

Car and diesel truck drivers could see their fuel costs cut in half. That's because the refining process for producing naphtha (currently used mostly as a hardener in road tars) is faster and cheaper than producing gasoline and diesel fuel. In addition to cutting costs, using naphtha instead of gasoline and diesel fuel would eliminate up to 90% of the air pollutants produced by oil refineries. Stay tuned to see whether this fuel passes its final tests.

Hint: Enter the search term *solar energy* using the Subject Guide.

Table 4-1 Evaluation of Alternatives to Gasoline

Advantages	Disadvantages
Compressed Natural Gas	
High octane	Large fuel tank required
Fairly abundant domestic and global supplies	Expensive engine modification required ($2,000)
Lower emissions of hydrocarbons, CO, and CO_2, and particulates	One-fourth the range
Currently inexpensive	New filling stations required
Vehicle development advanced	Nonrenewable resource
Reduced engine maintenance	CO_2 emissions (but less than gasoline)
Well suited for fleet vehicles	
Electricity	
Renewable if not generated from fossil fuels or nuclear power	Limited range and power
Zero vehicle emissions	Batteries expensive
Electric grid in place	Slow refueling (6–8 hours)
Efficient and quiet	Power-plant emissions if generated from coal or oil
	Radioactive wastes if produced by nuclear power
	High cost if produced by nuclear power
Reformulated Gasoline (Oxygenated Fuel)	
No new filling stations required	Nonrenewable resource
Low to moderate reduction of CO emissions	Dependence on imported oil perpetuated
No engine modification required	Possible high cost to modify refineries
Lower emission of CO and hydrocarbons	No emission reduction of CO_2
	Higher cost
	Water resources contaminated by leakage and spills
Methanol	
High octane	Large fuel tank required
Lower CO_2 emissions (total amount depends on method of production)	One-half the range of gasoline
Reduced total air pollution (30–40%)	Corrosive to metal, rubber, plastic
Potentially renewable (if made from biomass)	Increased emissions of potentially carcinogenic formaldehyde
	High CO_2 emissions if generated by coal
	Slightly higher ozone concentrations (smog)
	High capital cost to produce
	Hard to start in cold weather
	Nonrenewable (if made from coal)
Ethanol	
High octane	Large fuel tank required
Lower CO_2 (total amount depends on distillation process and efficiency of crop growing)	Much higher cost
Lower CO emissions	Corn supply limited
Potentially renewable	Competition with food growing for cropland
	Less range
	Higher ozone concentrations (smog)
	Higher emissions of nitrogen oxides (NO and NO_2)
	Corrosive
	Hard to start in cold weather
Solar-Hydrogen	
Renewable if produced using solar energy	Nonrenewable if generated by fossil fuels or nuclear power
Lower flammability than gasoline	Large fuel tank required
Virtually emission-free	No distribution system in place
Zero emissions of CO_2	Engine redesign required
Nontoxic	Currently expensive

Q: What percentage of the energy input of a screw-in fluorescent light bulb is converted to light?

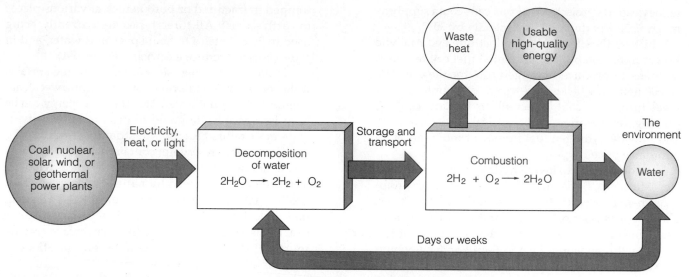

Figure 4-22 The hydrogen energy cycle. The production of hydrogen gas requires electricity, heat, or solar energy to decompose water, thus leading to a negative net energy yield. However, hydrogen is a clean-burning fuel that can replace oil, other fossil fuels, and nuclear energy. Using solar energy to produce hydrogen from water could also eliminate most air pollution and greatly reduce the threat of global warming.

What's the Catch? If you think using hydrogen as an energy source sounds too good to be true, you're right. Several problems must be solved to make hydrogen one of our primary energy resources, but scientists are making rapid progress in finding solutions.

One problem is that it takes energy to produce this marvelous fuel. We could use electricity from coal-burning and nuclear power plants to split water and produce hydrogen, but this subjects us to the harmful environmental effects associated with using these fuels (Sections 4-6 and 4-7) and it costs more than the hydrogen fuel is worth.

Most proponents of hydrogen gas believe that to get its very low pollution benefits, the energy to produce the gas from water must come from the sun, in the form of electricity generated by sources such as hydropower, solar thermal and solar cell power plants, and wind farms. If scientists and engineers can learn how to use sunlight to decompose water cheaply enough, they will set in motion a *solar–hydrogen revolution* over the next 50 years and change the world as much as the agricultural and industrial revolutions did. An important step in this direction occurred in 1998 when researchers at the U.S. National Renewable Energy Laboratory created a combined *photovoltaic–photoelectrochemical cell* that uses sunlight to split water into hydrogen and oxygen at an efficiency of 12.4%, a spectacular advance according to prominent researchers in this field.

Currently, using solar energy to produce hydrogen gas is too costly, but the costs of using solar energy to produce electricity are coming down. Moreover, if the health and environmental costs of using gasoline were included in its market prices (through gasoline taxes), it would cost about $1 per liter ($4 per gallon)—roughly the price in Japan and in many European countries. At this price, hydrogen produced by any form of solar energy would be competitive.

Hydrogen gas is easier to store than electricity. It can be stored in a pressurized tank or in metal powders or activated carbon, which absorb gaseous hydrogen and release it when heated for use as a fuel. Unlike gasoline, compounds and powders containing absorbed hydrogen will not explode or burn if a vehicle's tank is ruptured in an accident. If properly handled, hydrogen is a safer fuel than gasoline and natural gas. However, it's difficult to store enough hydrogen gas in a car as a compressed gas or as a metal powder for it to run very far—a problem similar to the one electric cars face. Scientists and engineers are seeking solutions to this problem.

Another possibility is to power a car with a *fuel cell* (Figure 4-6) in which hydrogen and oxygen gas combine to produce electrical current. Fuel cells produce no air pollution, have no moving parts, and have high energy efficiencies of up to 65% (Figure 4-6)—several times the efficiency of conventional gasoline-powered engines and electric cars. When fuel cells are used to cogenerate both electricity and heat, their total system energy efficiency will approach 90%. A number of prototype fuel-cell systems for cars, buses, homes, and buildings are being tested and evaluated.

Currently, fuel cells are expensive and heavy, and they produce energy at a cost equivalent to gasoline at $2.50 per gallon, but this could change with more research and mass production. In 1997, a panel of tech-

A: About 22% (over four times as much as an incandescent bulb)

nology experts projected that fuel cells could supply as much as 30% of the world's electricity by 2017.

In 1998 the Electric Power Institute and Analytic Power in Boston built a residential fuel cell that uses methane to produce hydrogen fuel for the cell. The $3,000 unit, now being field-tested, is about the size of a gas furnace. It could meet all the heating, cooling, cooking, refrigeration, and electrical needs of a home and provide hydrogen fuel for one or more cars at an affordable price.

We don't need to invent hydrogen-powered vehicles. Mercedes, BMW, and Mazda already have prototypes being tested on the roads. In the United States, a *hydrogen corridor* of research and production facilities is beginning to develop in the desert east of Los Angeles. It is anchored by a $2.5-million facility built by Xerox for using solar energy to convert water into hydrogen.

What's Holding Up the Solar–Hydrogen Revolution? Politics and economics, not a lack of promising technology, are the main factors holding up the transition to a solar–hydrogen age. Phasing out dependence on fossil fuels and phasing in new solar–hydrogen technologies over the next 40–50 years involves convincing investors and energy companies with strong vested interests in fossil fuels to risk a lot of capital on hydrogen. It also involves convincing governments to put up some of the money for developing hydrogen energy (as they have done for decades for fossil fuels and nuclear energy).

In the United States, large-scale government funding of hydrogen research is generally opposed by powerful U.S. oil companies, electric utilities, and automobile manufacturers, who see it as a serious threat to their profits. In contrast, the Japanese and German governments have been spending seven to eight times more on hydrogen research and development than the United States. Some analysts warn that without greatly increased government and private research and development, Americans may be buying solar–hydrogen equipment and fuel cells from Germany and Japan and may lose out on a huge global market and source of domestic jobs.

4-5 GEOTHERMAL ENERGY

How Can We Tap the Earth's Internal Heat? Heat contained in underground rocks and fluids is an important source of energy. Over millions of years, this geothermal energy from earth's mantle (Figure 2-8) has been transferred to underground concentrations of *dry steam* (steam with no water droplets), *wet steam* (a mixture of steam and water droplets), and *hot water*

trapped in fractured or porous rock at various places in earth's crust. All three types are currently being used in 20 countries to heat space and water, and in some cases to produce electricity (Figure 4-1).

If such geothermal sites are close to the surface, wells can be drilled to extract the dry steam, wet steam (Figure 4-23), or hot water. This thermal energy can be used for space heating and to produce electricity or high-temperature heat for industrial processes.

Geothermal reservoirs containing dry steam, wet steam, or hot water can be depleted if heat is removed faster than it is renewed by natural processes. Thus, geothermal resources are nonrenewable on a human time scale, but the potential supply is so vast that it is usually classified as a potentially renewable energy resource. However, easily accessible concentrations of geothermal energy are fairly scarce.

Currently, about 22 countries (most of them in the developing world) are extracting energy from geothermal

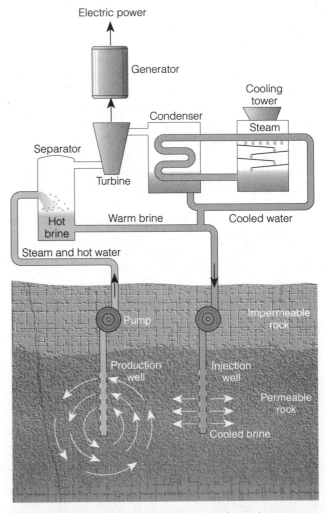

Figure 4-23 Tapping the earth's heat or geothermal energy in the form of wet steam to produce electricity.

Loeb, Penny. 1997. "Shear madness." *U.S. News & World Report*, vol. 123, no. 6, 26(9).

sites to produce less than 1% of the world's electricity. The United States accounts for 44% of the geothermal electricity generated worldwide, with most of the favorable sites in California, Hawaii, Nevada, and Utah.

Three other virtually nondepletable or potentially renewable sources of geothermal energy are *molten rock* (magma); *hot dry-rock zones*, where molten rock that has penetrated the earth's crust heats subsurface rock to high temperatures; and low- to moderate-temperature *warm-rock reservoir deposits*, which could be used to preheat water and run heat pumps for space heating and air conditioning. Research is being carried out to see whether hot dry-rock zones, which can be found almost anywhere if one can drill deep enough, can provide affordable geothermal energy.

The biggest advantages of geothermal energy include a vast, reliable, and sometimes renewable supply of energy for areas near reservoir sites, moderate net energy yields for large and easily accessible reservoir sites, about 96% fewer CO_2 emissions per unit of energy than fossil fuels, and a competitive cost of producing electricity (Figure 4-12).

A serious limitation of geothermal energy is the scarcity of easily accessible reservoir sites. Dry steam, wet steam, and hot water geothermal reservoirs must be carefully managed or they can be depleted within a few decades. Furthermore, geothermal development in some areas can destroy or degrade forests or other ecosystems.

Noise, odor, and local climate changes can also be problems. With proper controls, however, most experts consider the environmental effects of geothermal energy to be less (or no greater) than those of fossil fuel and nuclear power plants. Economics is the main barrier to greater use of geothermal energy. Currently, the cost of tapping geothermal energy is too high for all but the most concentrated and accessible sources.

4-6   NONRENEWABLE FOSSIL FUELS

What Are the Pros and Cons of Conventional Oil? **Petroleum,** or **crude oil** (oil as it comes out of the ground) is a fossil fuel produced by the decomposition of deeply buried dead organic matter from plants and animals under high temperatures and pressures over millions of years. Typically this gooey, smelly liquid consists mostly of hydrocarbons, with small amounts of sulfur, oxygen, and nitrogen impurities.

Crude oil and natural gas are often trapped together under a dome deep within the earth's crust (Figure 4-1). The crude oil is dispersed in pores and cracks in underground rock formations, like water in a sponge. A well can be drilled, and the crude oil that flows by gravity into the bottom of the well can be pumped out.

Most crude oil travels by pipeline to a *refinery* (Figure 4-24), where it is heated and distilled in gigantic columns to separate it into liquid components with different boiling points, such as naphtha (Individuals Matter, p. 117), diesel oil, heating oil, aviation fuel, and gasoline and solids such as grease, wax, and asphalt. Some of the resulting products, called **petrochemicals**, are used as raw materials in industrial organic chemicals, fertilizers, pesticides, plastics, synthetic fibers, paints, medicines, and many other products.

Oil is still cheap; when adjusted for inflation it costs about the same as it did in 1975. It is easily transported within and between countries, and when extracted from easily accessible deposits it has a high net energy

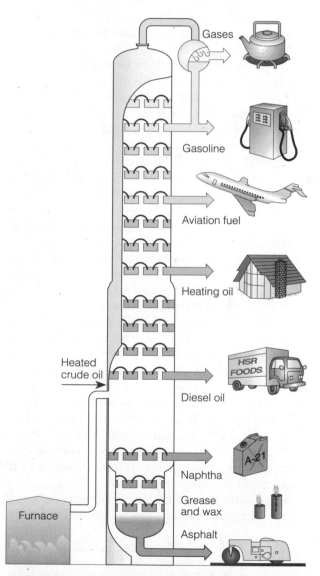

Figure 4-24 Refining crude oil. Components are removed at various levels, depending on their boiling points, in a giant distillation column. The most volatile components with the lowest boiling points are removed at the top of the column.

Hint: Enter the search terms *coal industry, environmental aspects* using the Subject Guide.

yield (Figure 4-4). Oil's low price has encouraged developed countries and developing countries alike to become heavily dependent on—indeed, addicted to—this important resource. Low prices have also encouraged waste of oil and discouraged improvements in energy efficiency and the switch to other sources of energy.

Oil is currently the lifeblood of the global economy, as evidenced by the amount used around the world (Figure 4-2). Oil *reserves* are identified deposits from which oil can be extracted profitably at current prices with current technology. The 13 countries that make up the Organization of Petroleum Exporting Countries (OPEC)* have 67% of the world's reserves, which explains why OPEC is expected to have long-term control over world oil supplies and prices. Saudi Arabia, with 26%, has the world's largest known crude oil reserves, followed by Iraq with 10%.

Currently, the United States has only 2.3% of the world's oil reserves but uses nearly 30% of the oil extracted worldwide each year, 65% of it for transportation. Despite an upsurge in exploration and test drilling, U.S. oil extraction has declined since 1985. Over half of the country's biggest oil fields are 80% depleted, and the net energy yield for small new domestic oil supplies is low and falling.

Mostly because of declining oil reserves and increased oil use, the United States imported 55% of the oil it used in 1997 (up from 36% in 1973); by 2010 it could be importing 70% or more of the oil it uses. This dependence on imported oil (about half of it from OPEC countries) and the likelihood of much higher oil prices within 10–20 years could drain the United States and other major oil-importing nations of vast amounts of money. This could lead to severe inflation and widespread economic recession, perhaps even a major depression.

Oil's fatal flaw is that its reserves may be 80% depleted within 35–84 years, depending on how rapidly it is used. At the current rate of consumption, global oil reserves will last at least 44 years. Undiscovered oil that is thought to exist might last another 20–40 years. Instead of remaining at the current level, however, global oil consumption is projected to increase by about 25% by 2010. This will hasten depletion of global oil reserves. At today's consumption rate, U.S. oil reserves will be depleted in about 24 years (less if consumption increases as projected); potential reserves might yield an additional 24 years of production.

Some analysts argue that rising oil prices (when oil consumption exceeds oil production) will stimulate

exploration and lead to enough new reserves to meet future demand through the next century. Other analysts argue that such optimistic projections about future oil supplies ignore the consequences of exponentially increasing consumption of oil.

Even assuming that we continue to use crude oil at the current rate, Saudi Arabia, with the largest known crude oil reserves, could supply all the world's oil needs for only 10 years; the estimated reserves under Alaska's North Slope (the largest ever found in North America) would meet world demand for only 6 months or U.S. demand for 3 years. In short, just to keep on using oil at the *current* rate and not run out, we must discover and add to global oil reserves the equivalent of a new Saudi Arabian supply *every 10 years*. According to the U.S. Geological Survey, global discovery of large oil fields peaked in 1962 and has been declining since.

Another major drawback of oil (and all fossil fuels) is that burning it releases heat-trapping carbon dioxide, which could alter global climate (Section 9-2), and it releases other air pollutants that harm people, crops, trees, fish, and other species (Section 9-6). Oil spills and leakage of toxic drilling muds pollute water, and the brine solution injected into oil wells can contaminate groundwater.

Suppose that all of the harmful environmental effects of using oil (Figure 1-11) were included in its market price (full-cost pricing) and that current government subsidies were phased out. Then analysts project that oil would become so expensive that it would probably be replaced by improved energy efficiency and by a variety of less harmful and cheaper renewable energy resources (also evaluated using full-cost pricing).

What Are the Pros and Cons of Using Heavy Oil Produced from Oil Shale and Tar Sand? Oil shale is a fine-grained rock that contains a solid, waxy mixture of hydrocarbon compounds called **kerogen**. After being removed by surface or subsurface mining, the shale is crushed and heated above ground in a retort to vaporize the kerogen (Figure 4-25). The kerogen vapor is condensed, forming heavy, slow-flowing, dark-brown **shale oil**.

The shale oil potentially recoverable from U.S. deposits, mostly on federal lands in Colorado, Utah, and Wyoming, could probably meet the country's crude oil demand for 41 years at current use levels. Canada, China, and several republics in the former Soviet Union also have large oil shale deposits. Indeed, according to energy expert John Harte, estimated potential global supplies of shale oil are 200 times larger than estimated global supplies of conventional oil.

However, there are problems with shale oil. It has a lower net energy yield than does conventional oil

---

*OPEC was formed in 1960 so that developing countries with much of the world's known and projected oil supplies could get a higher price for this resource. Today its members are Algeria, Ecuador, Gabon, Indonesia, Iran, Iraq, Kuwait, Libya, Nigeria, Qatar, Saudi Arabia, United Arab Emirates, and Venezuela.

Q: If the world really got serious about improving energy efficiency, how much money could be saved?

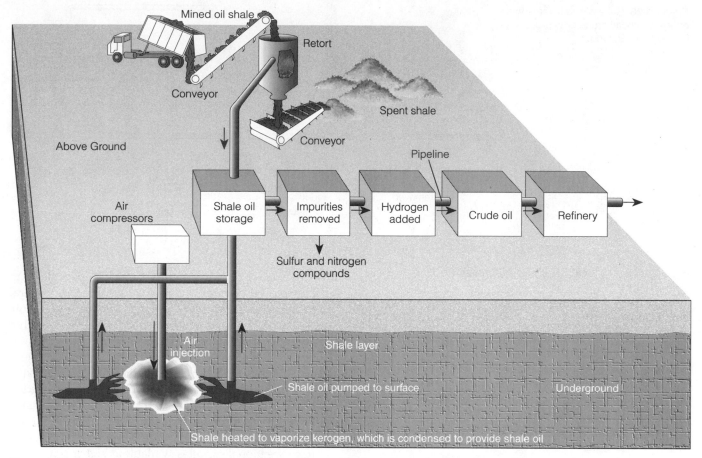

Figure 4-25 Aboveground and underground (in situ) methods for producing synthetic crude oil from oil shale.

because it takes the energy from almost half a barrel of conventional oil to extract, process, and upgrade one barrel of shale oil. Processing the oil requires large amounts of water, which is scarce in the semiarid locales where the richest deposits are located. Surface mining of shale oil tears up the land, leaving mountains of shale rock (which expands somewhat like popcorn when heated). In addition, salts, cancer-causing substances, and toxic metal compounds can be leached from the processed shale rock into nearby water supplies. Some of these problems can be reduced by extracting shale oil underground (Figure 4-25), but this method is too expensive and produces more sulfur dioxide air pollution than does surface processing.

Tar sand (or oil sand) is a mixture of clay, sand, water, and bitumen (a gooey, black, high-sulfur heavy oil). It is usually removed by surface mining. It is then heated with pressurized steam until the bitumen fluid softens and floats to the top. The bitumen is then purified and chemically upgraded into a synthetic crude oil suitable for refining (Figure 4-26).

The world's largest known deposits of tar sands, the Athabasca Tar Sands, lie in northern Alberta,

Canada. Currently, these deposits supply about 21% of Canada's oil needs. These deposits could supply all of Canada's projected oil needs for about 33 years at its current consumption rate, but they would last the world only about 2 years. Other large deposits of tar sands are in Venezuela, Colombia, and parts of the former Soviet Union.

Producing synthetic crude oil from tar sands has several disadvantages. The net energy yield is low because it takes the energy in almost one-half a barrel of conventional oil to extract and process one barrel of bitumen and upgrade it to synthetic crude oil. Processing requires large quantities of water, and upgrading bitumen to synthetic crude oil releases large quantities of air pollutants. The plants also create huge waste disposal ponds.

What Is the Future of Natural Gas? In its underground gaseous state, **natural gas** is a mixture of 50–90% by volume of methane (CH_4), the simplest hydrocarbon, and smaller amounts of heavier gaseous hydrocarbons such as ethane (C_2H_6), propane (C_3H_8), and butane (C_4H_{10}). *Conventional natural gas* lies above

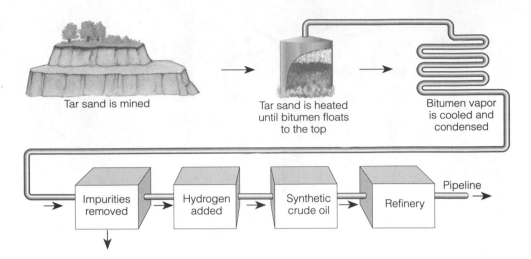

Figure 4-26 Generalized summary of how synthetic crude oil is produced from tar sand.

Tar sand is mined

Tar sand is heated until bitumen floats to the top

Bitumen vapor is cooled and condensed

Impurities removed → Hydrogen added → Synthetic crude oil → Refinery → Pipeline

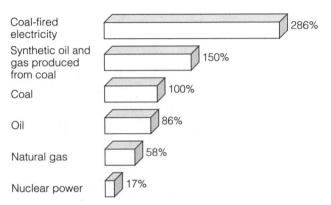

Coal-fired electricity — 286%

Synthetic oil and gas produced from coal — 150%

Coal — 100%

Oil — 86%

Natural gas — 58%

Nuclear power — 17%

Figure 4-27 Carbon dioxide emissions per unit of energy produced by various fuels, expressed as percentages of emissions produced by coal.

most reservoirs of crude oil (Figure 4-1). *Unconventional natural gas* is found by itself in other underground sources. It is not yet economically feasible to get natural gas from such unconventional sources, but the extraction technology is being developed rapidly.

When a natural gas field is tapped, propane and butane gases are liquefied and removed as **liquefied petroleum gas (LPG)**. LPG is stored in pressurized tanks for use mostly in rural areas not served by natural gas pipelines. The rest of the gas (mostly methane) is dried to remove water vapor, cleansed of poisonous hydrogen sulfide and other impurities, and pumped into pressurized pipelines for distribution. At a very low temperature of 184°C (–300°F), natural gas can be converted to **liquefied natural gas (LNG)**. This highly flammable liquid can then be shipped to other countries in refrigerated tanker ships.

Russia and Kazakhstan have almost 40% of the world's natural gas reserves. Other countries with large known natural gas reserves are Iran (15%), Qatar (5%), Saudi Arabia (4%), Algeria (4%), the United States (3%), Nigeria (3%), and Venezuela (3%). Geolo-

gists expect to find more natural gas, especially in unexplored developing countries. Most U.S. natural gas reserves are located in the same places as crude oil.

Natural gas has a number of advantages over other nonrenewable energy sources. To date, it is cheaper than oil. At the current consumption rate, known reserves and undiscovered, potential reserves of conventional natural gas in the United States are projected to last 65–80 years; world reserves are expected to last at least 125 years at the current consumption rate. It is estimated that conventional supplies of natural gas, plus unconventional supplies available at higher prices, will last at least 200 years at the current consumption rate and 80 years if usage rates rise 2% per year.

Natural gas can be transported easily over land by pipeline; it has a high net energy yield (Figure 4-4), burns hotter, and produces less air pollution than any other fossil fuel. Burning natural gas produces 43% less heat-trapping carbon dioxide per unit of energy than coal and 30% less than oil (Figure 4-27). Extracting natural gas also damages the environment much less than extracting either coal or uranium ore.

Natural gas is easier to process than oil and, because it can be transported by pipeline, it is less expensive to transport by pipeline than coal (which is usually moved by rail). It can also be used to power vehicles. In 1997, Honda designed an engine that runs on natural gas and is the cleanest internal combustion engine ever made.

New *combined-cycle natural gas systems* can produce electricity much more efficiently and cheaply (and with less pollution and no risk of emissions of radioactive wastes) than burning coal or oil or using nuclear power. In addition, natural gas can be burned cleanly and efficiently by *cogeneration* to produce both high-temperature heat and electricity, further improving the efficiency of this fuel. Natural gas can also be used in highly efficient fuel cells.

Because of its advantages over oil, coal, and nuclear energy, some analysts see natural gas as the best

Q: What is the most inefficient and costly way to produce electricity for heating an interior space or water?

fuel to help us make the transition to improved energy efficiency and more renewable energy over the next 50 years. Additionally, hydrogen gas produced from water by solar-generated electricity could be mixed with natural gas to help smooth the shift to a solar–hydrogen economy (Section 4-4).

🔋 What Are the Pros and Cons of Coal?
Coal is a solid fossil fuel formed in several stages as buried plant remains are subjected to intense heat and pressure over many millions of years (Figure 4-28). Currently coal provides about 25% of the world's commercial energy (22% in the United States). It is used to generate almost 64% of the world's electricity and to make 75% of its steel. About 66% of the world's proven coal reserves and 85% of the estimated undiscovered coal deposits are located in the United States, the former Soviet Union, and China (which gets 76% of its commercial energy from coal).

Coal is both the most abundant and the dirtiest fossil fuel. *Identified* world reserves of coal should last at least 220 years at current usage rates, but only 65 years if usage rises 2% per year. The world's *unidentified* coal reserves are projected to last about 900 years at the current consumption rate, and 149 years if the usage rate increases 2% per year. Identified U.S. coal reserves should last about 300 years at the current consumption rate; unidentified U.S. coal resources could extend those supplies for perhaps 100 years, at a much higher average cost. Coal also has a high net energy yield (Figure 4-4).

However, coal has a number of drawbacks, especially the harmful environmental effects associated with its extraction, processing, and use (Figure 1-11). Coal mining is dangerous because of accidents and black lung disease, a form of emphysema caused by prolonged breathing of coal dust and other particulate matter. Underground mining causes land to sink when a mine shaft collapses during or after mining. Surface mining of coal causes severe land disturbance and soil erosion, which can pollute nearby streams. Most surface-mined coal is removed by area strip mining or contour strip mining, depending on the terrain (Figure 4-1). Surface-mined land can be restored, but this is expensive and is not done in many countries; in arid and semiarid areas the land cannot be fully restored. Surface and subsurface coal mining can severely pollute nearby streams and groundwater with acids and toxic metal compounds.

Once coal is mined it is expensive to move from one place to another. In addition, coal cannot be used in solid form as a fuel for cars and trucks; it must be converted to liquid or gaseous fuels, with a significant drop in its net energy yield.

Without expensive air pollution control devices, burning coal produces more air pollution per unit of energy than any other fossil fuel. In addition to conventional air pollutants, burning coal releases thousands of times more radioactive particles into the atmosphere per unit of energy produced than does a normally operating nuclear power plant. Because coal produces more carbon dioxide per unit of energy than do other fossil fuels (Figure 4-27), burning more coal accelerates global warming (Section 9-2).

Burning coal is also one of the greatest threats to human health. Each year in the United States alone, air pollutants from coal burning kill thousands of people (with estimates ranging from 65,000 to 200,000), cause at least 50,000 cases of respiratory disease, and result in several billion dollars of property damage.

Suppose that all of coal's harmful environmental costs (Figure 1-11) were included in its market price and current government subsidies that make coal artificially cheap were phased out. According to some analysts, coal would become so expensive that it would probably

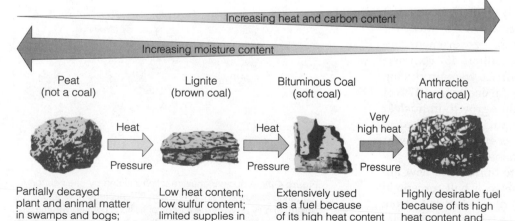

Figure 4-28 Stages in the formation of coal over millions of years. Three types of increasingly harder coal are formed: lignite, bituminous coal, and anthracite. Peat, a soil material made of moist, partially decomposed organic matter is not a coal. It is burned in some places but has a low heat content. Lignite and bituminous coal are sedimentary rocks, whereas anthracite is a metamorphic rock (Figure 2-27). Low-sulfur coal (lignite and anthracite) produces less sulfur dioxide when burned than does high-sulfur coal (bituminous). Anthracite is the most desirable type of coal because of its high heat content and low sulfur content.

Increasing heat and carbon content

Increasing moisture content

Peat
(not a coal)

Heat / Pressure

Lignite
(brown coal)

Heat / Pressure

Bituminous Coal
(soft coal)

Very high heat / Pressure

Anthracite
(hard coal)

Partially decayed plant and animal matter in swamps and bogs; low heat content

Low heat content; low sulfur content; limited supplies in most areas

Extensively used as a fuel because of its high heat content and large supplies; normally has a high sulfur content

Highly desirable fuel because of its high heat content and low sulfur content; supplies are limited in most areas

A: Nuclear power (only about 8–14% efficient)

be replaced for most uses by a combination of improved energy efficiency and cheaper and less environmentally harmful renewable energy resources.

What Are the Pros and Cons of Converting Solid Coal into Gaseous and Liquid Fuels? Solid coal can be converted into **synthetic natural gas (SNG)** by *coal gasification*, or into *hydrogen gas* (Section 4-4), or into a liquid fuel such as methanol or synthetic gasoline by *coal liquefaction* (Table 4-1). These *synfuels* can be transported by pipeline, and they produce much less air pollution than solid coal. They can be burned to produce high-temperature heat and electricity, to heat houses and water, and to propel vehicles.

However, coal gasification has a low net energy yield (Figure 4-4), as does coal liquefaction. A synfuel plant costs much more to build and run than an equivalent coal-fired power plant fully equipped with air-pollution control devices. In addition, the widespread use of synfuels would accelerate the depletion of world coal supplies because 30–40% of the energy content of coal is lost in the conversion process. It would also lead to greater land disruption from surface mining because producing a unit of energy from synfuels uses more coal than burning solid coal does.

Producing synfuels also requires huge amounts of water. Additionally, synfuels release more carbon dioxide per unit of energy than coal does (Figure 4-27). For these reasons, most analysts expect synfuels to play only a minor role as an energy resource in the next 30–50 years.

4-7 NONRENEWABLE NUCLEAR ENERGY

What Happened to Nuclear Power? In the 1950s, researchers predicted that by the end of the century 1,800 nuclear power plants would supply 21% of the world's commercial energy (25% in the United States) and most of the world's electricity. By 1996, after over 40 years of development, enormous government subsidies, and an investment of $2 trillion, 437 commercial nuclear reactors in 32 countries were producing only 6% of the world's commercial energy and 17% of its electricity. Little or no further growth in nuclear power is projected, and its capacity is expected to decline between 2000 and 2020 as existing plants wear out and are retired (decommissioned).

In western Europe, plans to build more new nuclear power plants have come to a halt, except in France, which gets about 80% of its electricity from nuclear power. However, France has more electricity than it needs (and the heavily subsidized government agency that builds and operates France's nuclear plants

has lost money for 25 years, accumulating a $30-billion debt—a serious burden to the French economy).

In the United States, no new nuclear power plants have been ordered since 1978, and all 120 plants ordered since 1973 have been canceled—more than the total nuclear capacity existing in the country today. In 1997, the 107 licensed commercial nuclear power plants in the United States generated about 22% of the country's electricity. This percentage is expected to decline over the next two decades as many of the current reactors reach the ends of their useful lives.

What happened to nuclear power? The answer is multibillion-dollar construction cost overruns, high operating costs, frequent malfunctions, false assurances and cover-ups by government and industry officials, inflated estimates of electricity use, poor management, the Chernobyl and Three Mile Island accidents, and public concerns about safety, costs, and radioactive waste disposal.

What Is Nuclear Fission? The source of energy for nuclear power is **nuclear fission**: a nuclear change in which nuclei of certain isotopes with large mass numbers (such as uranium-235; see Figure 2-3) are split apart into lighter nuclei when struck by neutrons; each fission releases two or three more neutrons and energy. Each of these neutrons, in turn, can cause an additional fission.

Multiple fissions within a critical mass form a **chain reaction**, which releases an enormous amount of energy (Figure 4-29). The rate at which this happens can

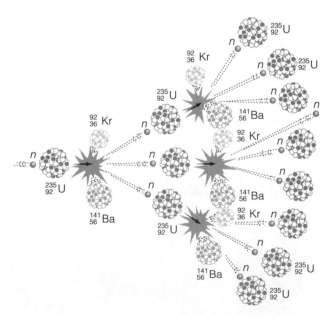

Figure 4-29 A nuclear chain reaction initiated by one neutron triggering fission in a single uranium-235 nucleus. This figure illustrates only a few of the trillions of fissions caused when a single uranium-235 nucleus is split within a critical mass of uranium-235 nuclei. The elements krypton (Kr) and barium (Ba), shown here as fission fragments, are only two of many possibilities.

Q: What are the two most energy-efficient ways to heat interior space?

be controlled in the nuclear fission reactor of a nuclear power plant, and the heat generated can be used to produce high-pressure steam, which spins turbines and thus generates electricity.

Nuclear fission produces radioactive fission fragments containing isotopes that spontaneously shoot out fast-moving particles (alpha and beta particles), gamma rays (a form of high-energy electromagnetic radiation, Figure 2-5), or both at a fixed rate. The unstable isotopes are called **radioactive isotopes** or **radioisotopes**. This spontaneous process is called *radioactive decay* and continues until the original isotope is changed into a new stable isotope that is not radioactive.

Exposure to alpha, beta, and gamma radiation and the high-speed neutrons emitted in nuclear fission (Figure 4-29) can harm cells in two ways. First,

harmful mutations of DNA molecules in genes and chromosomes can cause genetic defects in immediate offspring or several generations later. Second, tissue damage such as burns, miscarriages, eye cataracts, and cancers (bone, thyroid, breast, skin, and lung) can occur during the victim's lifetime.

How Does a Nuclear Fission Reactor Work? *Light-water reactors (LWRs)* like the one diagrammed in Figure 4-30 produce about 85% of the world's nuclear-generated electricity (100% in the United States). An LWR has several key parts. One is its *core*, which contains 35,000–40,000 long, thin fuel rods, each of which is packed with pellets of uranium oxide fuel. Each pellet is about one-third the size of a cigarette. About 97% of the uranium in each fuel pellet is uranium-238, a

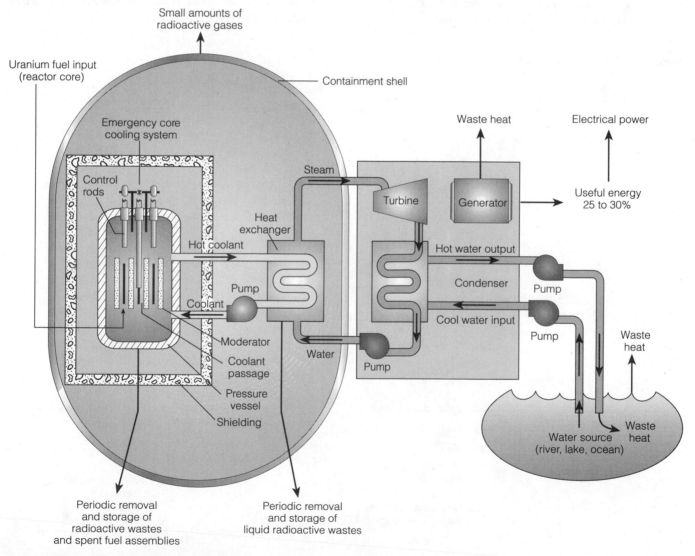

Figure 4-30 Light-water-moderated and -cooled nuclear power plant with a pressurized water reactor. The ill-fated Chernobyl nuclear reactor (Figure 4-32) in Ukraine was a graphite-moderated reactor with less extensive safety features than in most reactors in the rest of the world.

A: Passive solar and natural gas (Figure 4-10)

nonfissionable isotope; the other 3% (versus 0.7% in nature) is uranium-235, which is fissionable. The concentration of uranium-235 in the ore is increased (enriched) from 0.7% to 3% by removing some of the uranium-238 to create a suitable fuel.

Another important component is *control rods*, which are moved in and out of the reactor core to absorb neutrons and thus regulate the rate of fission and amount of power the reactor produces. The *moderator* slows down the neutrons emitted by the fission process so that the chain reaction can be kept going. A *coolant*, usually water, circulates through the reactor's core to remove heat (to keep fuel rods and other materials from melting) and to produce steam for generating electricity.

Nuclear power plants, each with one or more reactors, are only one part of the nuclear fuel cycle (Figure 4-31). In evaluating the safety and economic feasibility of nuclear power, we need to look at this entire cycle, not just the nuclear plant itself.

What Are the Advantages of Nuclear Power? Using nuclear power to produce electricity has some important advantages, especially when compared to mining, processing, and burning coal.

Nuclear plants don't emit air pollutants (as coal-fired plants do) as long as they are operating properly. The entire nuclear fuel cycle (Figure 4-31) adds about one-sixth as much heat-trapping carbon dioxide per unit of electricity as using coal does, thus making it more attractive than coal and other fossil fuels for reducing the threat of global warming.

Water pollution and disruption of land are low to moderate if the entire nuclear fuel cycle operates normally. Thick and strong reactor vessel walls, a steel-reinforced containment building, an emergency core cooling system, and multiple other safety systems with automatic backups greatly decrease the likelihood of a catastrophic accident releasing deadly radioactive material into the environment.

Figure 4-31 The nuclear fuel cycle.

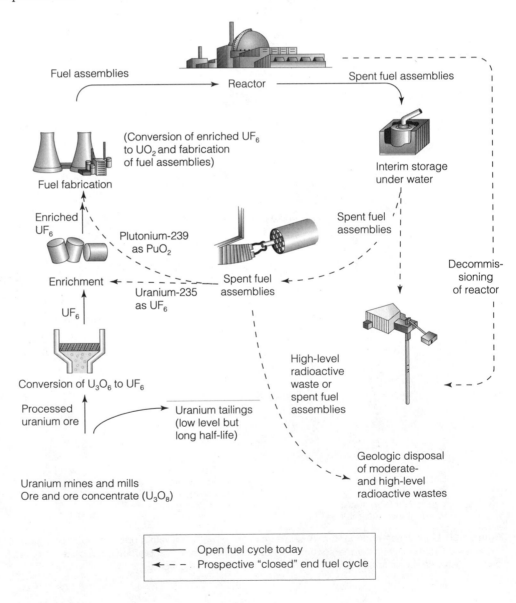

Q: What is the most energy-efficient fuel for powering a motor vehicle?

Chernobyl (Figure 4-32) may have caused the premature deaths of 32,000 people between 1986 and 1995, but *each year* air pollution from coal burning causes the premature death of 65,000 to 200,000 people in the United States alone. Globally the *annual* death toll from coal burning is in the millions.

How Safe Are Nuclear Power Plants and Other Nuclear Facilities? Because of the built-in safety features, the risk of exposure to radioactivity from nuclear power plants in the United States (and presumably in most other developed countries) is extremely low. However, a partial or complete meltdown or explosion is possible, as accidents at the Chernobyl nuclear power plant in Ukraine and the Three Mile Island plant in Pennsylvania have taught us.

On March 29, 1979, the number 2 reactor at the Three Mile Island (TMI) nuclear plant near Harrisburg, Pennsylvania, lost its coolant water because of a series of mechanical failures and human operator errors not anticipated in safety studies. The reactor's core became partially uncovered and about 50% of it melted and fell to the bottom of the reactor. Unknown amounts of radioactive materials escaped into the atmosphere, 50,000 people were evacuated, and another 50,000 fled the area on their own. Partial cleanup of the damaged TMI reactor, lawsuits, and payment of damage claims has cost $1.2 billion so far, almost twice the reactor's $700-million construction cost.

Most analysts say that the estimated radiation released was not enough to cause deaths or cancers. However, there is controversy over how much radiation was released and there has been an increase in certain cancers among residents who live near the plant. A 1991 study concluded that the excess cancers may have been caused by stress related to the accident. Stress is known to damage the immune system so that it may fail to prevent cancers. A 1997 study concluded that the increased cancer rates were caused by radiation released from the plant.

On April 26, 1986, a series of explosions in the Chernobyl nuclear power plant in Ukraine (then in the Soviet Union) blew the massive roof off the reactor building and flung radioactive debris and dust high into the atmosphere to encircle the planet. The accident happened when engineers turned off most of the reactor's automatic safety and warning systems to keep them from interfering with an unauthorized safety experiment (Figure 4-32). Here are some consequences of this disaster caused by poor reactor design and human error:

- In 1998 the Ukrainian Health Ministry put the official death toll from the accident at 3,576. However, Greenpeace Ukraine estimates that by 1995 the total death toll from the accident was about 32,000.

- According to the United Nations, almost 400,000 people have been forced to leave their homes, probably never to return. Most were not evacuated until 10 days or more after the accident.

- According to a recent UN report, some 160,000 square kilometers (62,000 square miles) of the former Soviet Union—almost the size of the state of Florida or the combined areas of Denmark and Greece—remain contaminated with radioactivity.

- Over half a million people were exposed to dangerous radioactivity, and some may suffer from cancers, thyroid tumors, and eye cataracts. Between 1986 and 1995 the rate of thyroid cancer (mostly among children) in Ukraine and nearby Belarus was more than 10 times the 1986 level and 30 times higher among those evacuated from the Chernobyl area. A 1996 study found that genetic mutations occur twice as often in children exposed to radioactive fallout from Chernobyl as in other families.

- Government officials say that the total cost of the accident will reach at least $358 billion—many times greater than the value of all the nuclear electricity that has ever been generated in the former Soviet Union.

The environmental refugees evacuated from the Chernobyl region had to leave their possessions behind and say good-bye to lush, green wheat fields and blossoming apple trees, to land their families had farmed for generations, to cows and goats that would be shot because the grass they ate was radioactive, and to their radioactivity-poisoned cats and dogs. They will not be able to return.

World-famous gymnast Olga Korbut gave this account in 1991:

> *I was . . . in Minsk when Chernobyl happened, and they didn't tell us for three or four days. . . . We were all outdoors, because it was close to the May 1 celebration, and we were planting gardens and enjoying the spring. . . . It has been five years . . . but people are still very frightened . . . and very angry. Our food and water supply is contaminated from radiation.*
>
> *When I went into the schools in Byelorussia, I learned that the first-graders have never been in the forest . . . because the trees were so contaminated. . . . When children want to see what nature used to be like, they go into a little courtyard inside the building, and the teacher says, "This is a bird and this is a tree," and they are plastic. Isn't that sad?*

Engineers fear that the hastily built, crumbling 20-story concrete shell (called a sarcophagus) built around the building to contain its radioactivity might collapse. This could allow potentially deadly radiation to escape into the atmosphere and increase cancer risks for millions of people worldwide. In 1997, the United States

Figure 4-32 Major events leading to the Chernobyl nuclear power-plant accident on April 26, 1986, in the former Soviet Union. The accident happened because engineers turned off most of the reactor's automatic safety and warning systems (to keep them from interfering with an unauthorized safety experiment) and because of inadequate safety design of the reactor (no secondary containment shell as in Western-style reactors, no emergency core cooling system to prevent reactor core meltdown, and a design flaw that leads to unstable operation at low power). After the reactor exploded, crews exposed themselves to lethal levels of radiation to put out fires and encase the shattered reactor in a hastily constructed concrete tomb. This tomb is now sagging and full of holes that allow water to seep in and radioactive dust to drift out. According to Ukrainian officials, the structure won't last another 10–15 years. Building a new tomb for the reactor will cost at least $1.5 billion—money the Ukrainian government doesn't have.

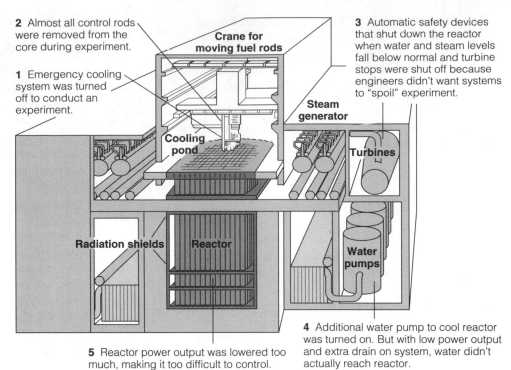

2 Almost all control rods were removed from the core during experiment.

1 Emergency cooling system was turned off to conduct an experiment.

3 Automatic safety devices that shut down the reactor when water and steam levels fall below normal and turbine stops were shut off because engineers didn't want systems to "spoil" experiment.

4 Additional water pump to cool reactor was turned on. But with low power output and extra drain on system, water didn't actually reach reactor.

5 Reactor power output was lowered too much, making it too difficult to control.

and six other developed countries agreed to provide $380 million to help stabilize the old shell by 2005 and build a new structure. It is a crucial race against time.

Chernobyl taught us that *a major nuclear accident anywhere is a nuclear accident everywhere.*

Nuclear scientists and government officials throughout the world are deeply concerned about 26 especially risky nuclear reactors in some republics of the former Soviet Union that have a Chernobyl-type design (15 reactors) or another flawed design (11 reactors). There is growing international consensus (including the Russian Academy of Sciences) that these 26 reactors, along with 10 other poorly designed and poorly operated nuclear plants in eastern Europe, should be shut down. However, without considerable government and private economic aid from the world's developed countries, it is extremely unlikely that these dangerous plants will be closed and replaced with safer nonnuclear alternatives.

A 1982 study by the Sandia National Laboratory estimated that a *worst-case accident* in a reactor near a large U.S. city might cause 50,000–100,000 immediate deaths, 10,000–40,000 subsequent deaths from cancer, and $100–150 billion in damages. Most citizens and businesses suffering injuries or property damage from a major nuclear accident would get little if any financial reimbursement because combined government and nuclear industry insurance covers only 7% of the estimated damage from such an accident. Critics of nuclear power contend that this 15-year-old study probably underestimates the deaths and damages. They also contend that plans for dealing with major nuclear accidents in the United States are inadequate.

Since 1986 government studies and once-secret documents have revealed that for decades most of the nuclear weapon production facilities supervised by the DOE have been operated with gross disregard for the safety of their workers and the people nearby. Since 1957 these facilities have released huge quantities of radioactive particles into the air and dumped tons of radioactive waste and toxic substances into flowing creeks and leaking pits without informing the public. As Senator John Glenn of Ohio summed up the situation, "We are poisoning our own people in the name of national security."

According to the nuclear power industry, nuclear power plants in the United States have not killed anyone. However, according to U.S. National Academy of Sciences estimates, U.S. nuclear plants cause 6,000 premature deaths and 3,700 serious genetic defects each year. If correct, this annual death toll is much smaller than the 65,000 to 200,000 deaths per year caused by coal-burning plants in the United States. However, critics point out the estimated annual deaths from both of these types of plants are unacceptable, given the much less harmful alternatives.

What Do We Do with Low-Level Radioactive Waste? Each part of the nuclear fuel cycle (Figure 4-31) produces solid, liquid, and gaseous radioactive wastes.

Q: What are the two best ways to save oil, slow ozone depletion and global warming, and reduce air pollution?

Wastes classified as *low-level radioactive wastes* give off small amounts of ionizing radiation and must be stored safely for 100–500 years before decaying to safe levels. From the 1940s to 1970, most low-level radioactive waste produced in the United States (and most other countries) was put into steel drums and dumped into the ocean; the United Kingdom and Pakistan still dispose of their low-level wastes in this way.

Since 1970, low-level radioactive wastes from U.S. military activities have been buried in commercial, government-run landfills. Today, low-level waste materials from U.S. commercial nuclear power plants, hospitals, universities, industries, and other producers are put in steel drums and shipped to two regional landfills run by federal and state governments (both scheduled for closure).

Attempts to build new regional dumps for low-level radioactive waste using improved technology have met with fierce public opposition. According to the EPA, even the best designed landfills eventually leak. Some environmentalists believe that low-level radioactive waste should be stored in carefully designed above-ground buildings that could capture and contain any leaks. They urge that all such buildings (or low-level waste dumps) be located at nuclear power-plant sites, which produce more than half of these wastes (as much as 80% in some states) and 80% of the radioactivity.

They point out that nuclear power plant sites have already been carefully evaluated to meet strict geological standards. They have the personnel with the expertise and equipment to manage the low-level wastes and are strictly regulated by the federal government. Locating low-level waste at plant sites would also reduce the need to transport most such wastes long distances. Such proposals are opposed by utility companies and the nuclear power industry because this would force the main producers of such wastes to be financially and legally responsible for the wastes they produce.

What Should We Do with High-Level Radioactive Waste? *High-level radioactive wastes* give off large amounts of ionizing radiation for a short time and small amounts for a long time. Such wastes must be stored safely for thousands of years—about 240,000 years if plutonium-239 is not removed by reprocessing. Most high-level radioactive wastes are spent fuel rods from commercial nuclear power plants and an assortment of wastes from plants that produce plutonium and tritium for nuclear weapons.

After 50 years of research, scientists still don't agree on whether there is any safe method of storing these wastes. Some scientists believe that the long-term safe storage or disposal of high-level radioactive wastes is technically possible. Others disagree, pointing out that it is impossible to demonstrate that any method will work for the 10,000–240,000 years of fail-safe storage needed for such wastes. Here are some of the proposed methods and their possible drawbacks:

- *Bury it deep underground.* This favored strategy is under study by all countries producing nuclear waste. To reduce storage time from hundreds of thousands of years to about 10,000 years, very long-lived radioactive isotopes such as plutonium-239 must be extracted from spent fuel rods. The remainder would be fused with glass or a ceramic material and sealed in metal canisters for burial in a deep underground salt, granite, or other stable geological formation that is earthquake resistant and waterproof. However, according to a 1990 report by the U.S. National Academy of Sciences, "Use of geological information—to pretend to be able to make very accurate predictions of long-term site behavior—is scientifically unsound."

- *Shoot it into space or into the sun.* Costs would be very high, and a launch accident, such as the explosion of the space shuttle *Challenger*, could disperse high-level radioactive wastes over large areas of the earth's surface. This strategy has been abandoned for now.

- *Bury it under the antarctic ice sheet or the Greenland ice cap.* The long-term stability of the ice sheets is not known. They could be destabilized by heat from the wastes, and retrieval of the wastes would be difficult or impossible if the method failed. This strategy is prohibited by international law and has been abandoned for now.

- *Dump it into descending subduction zones in the deep ocean.* Again, our geological knowledge is incomplete; wastes might eventually be spewed out somewhere else by volcanic activity. Also, waste containers might leak and contaminate the ocean before being carried downward, and retrieval would be impossible if the method did not work. This method is under active study by a consortium of 10 countries but current international agreements and U.S. laws forbid dumping radioactive waste beneath the seas.

- *Bury it in thick deposits of mud on the deep ocean floor in areas that tests show have been geologically stable for 65 million years.* The waste containers would eventually corrode and release their radioactive contents. Some geologists contend that gravity and the powerful adhesive qualities of the mud would prevent the wastes from migrating, but this hypothesis has not been verified. This approach is banned under current international agreements and U.S. law.

- *Change it into harmless or less harmful isotopes.* Currently there is no way to do this. Even if a method were developed, costs would probably be extremely high and the resulting toxic materials and low-level (but very long-lived) radioactive wastes would still need to be disposed of safely.

In 1985 the DOE announced plans to build the first repository for underground storage of high-level radioactive wastes from commercial nuclear reactors (Figure 4-33) on federal land in the Yucca Mountain desert region, 160 kilometers (100 miles) northwest of Las Vegas, Nevada. The facility, which is expected to cost at least $26 billion, was scheduled to open by 2003, but in 1990 its opening was postponed to at least 2010. The site may never open, mostly because of rock faults, a nearby active volcano, 36 active earthquake faults on the site itself, and leakage of rainwater through the site faster than previous studies had predicted.

Fed up with the delay, some utilities are taking matters into their own hands and are trying to develop a temporary, aboveground, monitored retrievable storage (MRS) facility for commercial high-level radioactive waste on the Mescalero Apache reservation in New Mexico. Because of the sovereignty of Native American nations, it is easier to site a radioactive storage facility on tribal lands than anywhere else in the United States. New Mexico officials have threatened to block the MRS facility, claiming that it would be too risky to transport spent fuel from 71 reactors in 31 states across the country to the facility. Even if a permit is granted, the site would not be open until 2003 at the earliest.

So far no country has come up with a scientifically and politically acceptable solution for the long-term storage of nuclear wastes. Regardless of the storage method, most citizens strongly oppose the location of a low-level or high-level nuclear waste disposal facility anywhere near them. Even if the problem is technically solvable, it may be politically unacceptable.

How Widespread Are Contaminated Radioactive Sites? In 1992 the EPA estimated that as many as 45,000 sites in the United States may be contaminated with radioactive materials, 20,000 of them belonging to the DOE and the Department of Defense. According to the DOE it will cost taxpayers at least $230 billion over the next 75 years to clean up these facilities (with some estimating that the price will range from $400 to $900 billion).

Some critics doubt that the necessary cleanup funds will be provided by Congress as the cleanup drags on and fades from serious public scrutiny. Others question spending so much money on a problem that is ranked by scientific advisers to the EPA as a low-risk ecological problem and not among the top high-risk health problems. However, the radioactive contamination situation in the United States pales in comparison to the post–Cold War legacy of nuclear waste and contamination in the republics of the former Soviet Union (Spotlight, p. right).

What Can We Do with Worn-Out Nuclear Plants? The useful operating life of today's nuclear power plants is supposed to be 40 years, but many plants are wearing out and becoming dangerous faster than anticipated.

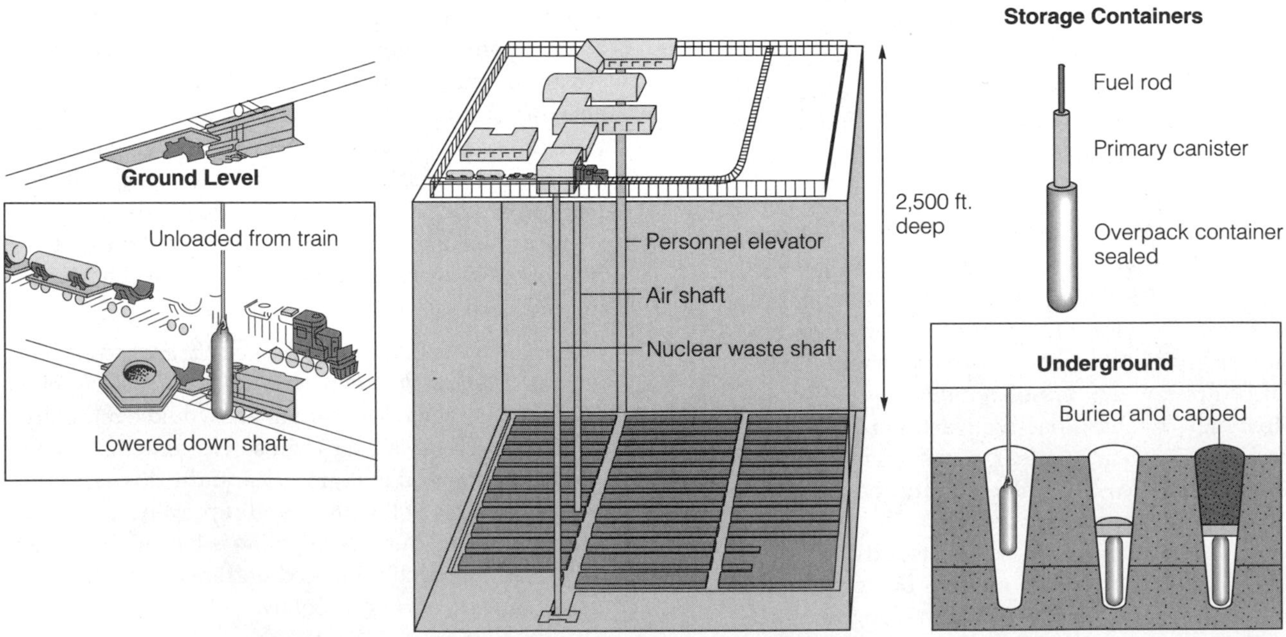

Figure 4-33 Proposed general design for deep-underground permanent storage of high-level radioactive wastes from commercial nuclear power plants in the United States. (Source: U.S. Department of Energy)

Q: What is the largest untapped energy source in the United States?

Worldwide, 81 reactors have been shut down after being in operation for an average of only 17 years. Because so many of its parts become radioactive, a nuclear plant cannot be abandoned or demolished by a wrecking ball the way a worn-out coal-fired power plant can.

Decommissioning nuclear power plants and nuclear weapon plants is the last step in the nuclear fuel cycle (Figure 4-31). Three methods have been proposed: **(1)** *immediate dismantling,* **(2)** *mothballing* for 30–100 years by putting up a barrier and setting up a 24-hour security system (costing about $15 million a year) and then dismantling it, and **(3)** *entombment* by covering the reactor with reinforced concrete and putting up a barrier to keep out intruders for several thousand years. Each method involves shutting down the plant, removing the spent fuel from the reactor core, draining all liquids, flushing all pipes, and sending all radioactive materials to an approved waste storage site yet to be built.

By 1995 more than 30 commercial reactors worldwide (12 in the United States) had been retired and awaited decommissioning. Another 228 large commercial reactors (20 in the United States) are scheduled for retirement between 2000 and 2012. By 2030 all U.S. reactors will have to be retired, based on the life of their current operating licenses, and many may be shut down early for safety or financial reasons.

Experience suggests that decommissioning old plants by dismantling and safely taking care of the resulting radioactive wastes sometimes exceeds the costs of building them in the first place. U.S. utilities have been setting aside funds for decommissioning more than 100 reactors, but to date these funds fall far short of the estimated costs. If sufficient funds are not available for decommissioning, the balance of the costs could be passed along to ratepayers and taxpayers.

What Is the Connection Between Nuclear Reactors and the Spread of Nuclear Weapons? Since 1958 the United States has been giving away and selling to other countries various forms of nuclear technology, mostly in the form of nuclear power plants and research reactors. Today the United States and at least 14 other countries sell nuclear power technology in the international marketplace. Information, components, and materials used to build and operate such reactors can be used to produce fissionable isotopes such as uranium-235 and plutonium-239 (the explosive materials in nuclear weapons).

We already live in a world with enough nuclear weapons to kill everyone on the earth 40 times over—20 times if current nuclear arms reduction agreements are carried out. By the end of this century, 60 countries—one of every three in the world—are expected to have nuclear weapons or the knowledge and ability to build them, with the fuel and knowledge coming mostly from research and com-

mercial nuclear reactors. Dismantling thousands of Russian and American nuclear warheads can increase the threat from the resulting huge amounts of bomb-grade plutonium that must be safeguarded.

So far no one has come up with truly effective solutions to the serious problems of nuclear weapon proliferation and what to do with retired weapon-grade plutonium. All we can do is try to slow the spread of such weapons and hope we can find a way to keep plutonium removed from weapons out of the hands of those who want nuclear arms.

A: Renewable energy

Can We Afford Nuclear Power? Experience has shown that nuclear power is an extremely expensive way to boil water to produce electricity, even when it is shielded partially from free-market competition with other energy sources by huge government subsidies. Thus, the major reason utility officials, investors, and most governments are shying away from nuclear power is not safety but the *extremely high cost* of making it a safe technology.

Some of these costs can be reduced by using standardized designs for all plants (which also cuts construction time in half), as France has done. Even so, the French government has run up a $30-billion debt subsidizing its government-run nuclear power industry.

Despite massive federal subsidies and tax breaks, the most modern nuclear power plants built in the United States produce electricity at an average of about 13.5¢ per kilowatt-hour, the equivalent of burning oil costing about $216 per barrel (compared to its current price of $15–20 per barrel) to produce electricity. All methods of producing electricity in the United States (except solar photovoltaic and solar thermal plants) have average costs below those of new nuclear power plants (Figure 4-12). By 2000–2005, even these methods (with few subsidies) are expected to be cheaper than nuclear power.

Banks and other lending institutions have become leery of financing new U.S. nuclear power plants. Abandoned reactor projects have cost U.S. utility investors over $100 billion since the mid-1970s, bankrupting U.S. utilities such as the Washington Public Power Supply System and the Public Service Company of New Hampshire. The Three Mile Island accident showed that utility companies could lose $1 billion worth of equipment in an hour and at least $1 billion more in cleanup costs, even without any established harmful effects on public health. A recent poll of U.S. utility executives found that only 2% would even consider ordering a new nuclear power plant. At the global level the World Bank said in 1995 that nuclear power is too costly and risky.

Forbes business magazine has called the failure of the U.S. nuclear power program "the largest managerial disaster in U.S. business history." The U.S. Department of Energy estimates that nearly one-fourth of the current 109 existing U.S. nuclear reactors may be closed prematurely for financial reasons by 2003 as utility companies write off their losses and move on to more profitable alternatives.

Since the Three Mile Island accident, the U.S. nuclear industry and utility companies have financed a $21-million-a-year public relations campaign by the U.S. Council for Energy Awareness to improve the industry's image and resell nuclear power to the American public.

Most of the council's ads advance the argument that the United States needs more nuclear power to reduce dependence on imported oil and to improve national security. But since 1979, only about 5% (3% in 1997) of the electricity in the United States has been produced by burning oil, and 95% of that is residual oil that can't be used for other purposes. The nuclear industry also does not point out that almost 75% of the uranium used for nuclear fuel in the United States is imported (most from Canada).

Nuclear industry officials also claim that nuclear power, unlike coal burning, adds no greenhouse-enhancing CO_2 to the atmosphere. Nuclear power plants themselves don't emit CO_2, but processing of uranium fuel (Figure 4-27) produces about one-sixth the CO_2 per unit of electricity as that from a coal-burning plant. To offset just 5% of current global CO_2 emissions would require nearly doubling the worldwide nuclear capacity at a cost of more than $1 trillion. According to energy expert Amory Lovins, if we want to reduce CO_2 emissions using the least-cost methods, then investing in energy efficiency and renewable energy resources are at the top of the list, and nuclear power is at the bottom.

The U.S. nuclear industry hopes to persuade the federal government and utility companies to build hundreds of new second-generation smaller plants using standardized designs with supposedly fail-safe features, which they claim are safer and can be built more quickly (in 3–5 years). However, according to *Nucleonics Week*, an important nuclear industry publication, "Experts are flatly unconvinced that safety has been achieved—or even substantially increased—by the new designs."

Furthermore, none of the new designs solves the problems of what to do with nuclear waste and worn-out nuclear plants and how to prevent the use of nuclear technology to build nuclear weapons. Indeed, these problems would become more serious if the number of nuclear plants increased from a few hundred to several thousand. None of these new designs changes the fact that nuclear power, even with huge government subsidies, is an incredibly expensive way to produce electricity compared to other alternatives. Most analysts agree that simple economics has proved to be the Achilles heel of nuclear power.

Should Conventional Nuclear Power Have a Future? Some analysts argue that we should continue some low-level government funding of research and development and pilot plant testing of new reactor designs to keep this option available for use in the future. There may be an urgent need to sharply reduce fossil-fuel greenhouse gas (because of serious global warming) or because improved energy efficiency and renewable energy options may fail to keep up with demands for electricity. What do you think?

Is Breeder Nuclear Fission a Feasible Alternative? Some nuclear power proponents urge the development and widespread use of **breeder nuclear fission reactors,**

Q: What percentage of U.S. and world energy needs could be provided by renewable energy resources by 2040 or sooner?

which generate more nuclear fuel than they consume by converting nonfissionable uranium-238 into fissionable plutonium-239. Because breeders would use over 99% of the uranium in ore deposits, the world's known uranium reserves would last at least 1,000 years, and perhaps several thousand years.

However, if the safety system of a breeder reactor fails, the reactor could lose some of its liquid sodium coolant, which ignites when exposed to air and reacts explosively if it comes into contact with water. This could cause a runaway fission chain reaction and perhaps a nuclear explosion powerful enough to blast open the containment building and release a cloud of highly radioactive gases and particulate matter. Leaks of flammable liquid sodium can also cause fires, as has happened with all experimental breeder reactors built so far.

In December 1986, France opened a commercial-size breeder reactor. Not only did it cost three times the original estimate to build, but the little electricity it produced was twice as expensive as that generated by France's conventional fission reactors. So far France has invested almost $13 billion in this breeder reactor, which after being shut down for expensive repairs was closed in 1999.

Tentative plans to build full-size commercial breeders in Germany, some republics of the former Soviet Union, the United Kingdom, and Japan have been abandoned because of the French experience and an excess of electricity-generating capacity and conventional uranium fuel. In addition, existing breeders produce plutonium fuel much too slowly. If this problem is not solved, it would take 100–200 years for breeders to begin producing enough plutonium to fuel a significant number of other breeder reactors. In 1994 U.S. Secretary of Energy Hazel O'Leary ended government-supported research for breeder technology after some $9 billion had been spent on it.

Is Nuclear Fusion a Feasible Alternative? Nuclear fusion is a nuclear change in which two isotopes of light elements, such as hydrogen (Figure 2-3), are forced together at extremely high temperatures until they fuse to form a heavier nucleus, releasing energy in the process (Figure 4-34). Scientists hope that someday controlled nuclear fusion will provide an almost limitless source of high-temperature heat and electricity. Research has focused on the D–T nuclear fusion reaction, in which two isotopes of hydrogen—deuterium (D) and tritium (T)—fuse at about 100 million °C (Figure 4-34).

After World War II the principle of *uncontrolled nuclear fusion* was used to develop extremely powerful hydrogen, or thermonuclear, weapons. These weapons use the D–T fusion reaction in which a hydrogen-2, or deuterium (D), nucleus and a hydrogen-3, or tritium

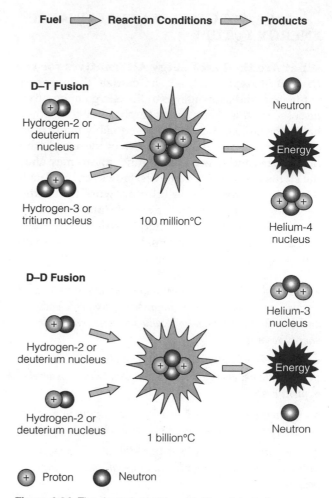

Figure 4-34 The deuterium–tritium (D–T) and deuterium–deuterium (D–D) nuclear fusion reactions, which take place at extremely high temperatures.

(T), nucleus are fused to form a larger, helium-4 nucleus, a neutron, and energy (Figure 4-34, top).

Despite 50 years of research and huge expenditures of mostly government funds, controlled nuclear fusion is still in the laboratory stage. If researchers can eventually get more energy out of nuclear fusion than they put in, the next step would be to build a small fusion reactor and then scale it up to commercial size—one of the most difficult engineering problems ever undertaken. The estimated cost of a commercial fusion reactor is several times that of a comparable conventional fission reactor.

Proponents contend that with greatly increased federal funding, a commercial nuclear fusion power plant might be built by 2030, but many energy experts don't expect nuclear fusion to be a significant energy source until 2100, if then. Meanwhile, experience shows that we can produce more electricity than we need using several quicker, cheaper, and safer methods.

What Are the Best Energy Alternatives for the United States? Table 4-2 summarizes the major advantages and disadvantages of the energy alternatives discussed in this chapter, with emphasis on their potential in the United States (and presumably many other countries). Energy experts argue over these and other projections, and new data and innovations may affect the status of certain alternatives. However, the data in the table do provide a useful framework for making decisions based on currently available information.

Many scientists and energy experts who have evaluated these energy alternatives have come to the following general conclusions:

■ *The best short-term, intermediate, and long-term alternatives are a combination of improved energy efficiency and greatly increased use of locally available renewable energy resources.*

■ *Future energy alternatives will probably have low to moderate net energy yields and moderate to high development costs.*

■ *Because there is not enough financial capital to develop all energy alternatives, projects must be chosen carefully.*

■ *We cannot and should not depend mostly on a single nonrenewable energy resource such as oil, coal, natural gas, or nuclear power.*

What Role Should Economics Play in Energy Resource Use? Cost is the biggest factor determining which commercial energy resources are widely used by consumers. Governments throughout the world use three basic economic and political strategies to stimulate or dampen the short-term and long-term use of a particular energy resource.

The first approach is *not attempting to control the price of an energy resource,* allowing all energy resources to compete in open, free-market competition. However, leaving energy pricing to the marketplace without any government interference is rarely politically feasible because of well-entrenched government intervention into the marketplace in the form of subsidies, taxes, and regulations. Furthermore, the free-market approach, with its emphasis on short-term gain, inhibits long-term development of new energy resources, which can rarely compete economically in their development stages without government support.

The second approach is *keeping energy prices artificially low,* which encourages use and development of a resource. In the United States (and most other countries), the energy marketplace is greatly distorted by huge government subsidies and tax breaks (such as depletion write-offs for fossil fuels). Such subsidies make the prices of fossil fuels and nuclear power artificially low and help perpetuate the use of these energy resources, even when better and less costly alternatives are available (Figure 4-12). At the same time, programs for improving energy efficiency and solar alternatives receive much lower subsidies and tax breaks. This creates an uneven economic playing field that encourages waste and rapid depletion of a nonrenewable energy resource and discourages the development of energy alternatives that are not getting at least the same level of subsidies and price control.

For example, in the U.S. the fossil-fuel industry gets almost $20 billion a year in taxpayer-supported government subsidies, compared to about $200 million for renewable energy. Worldwide government subsidies for fossil fuel and nuclear energy amount to an estimated $200 billion per year—more than half the value of all the crude oil produced each year.

The third approach is *keeping energy prices artificially high.* This can discourage development, use, and waste of a resource. Governments can raise the price of an energy resource by withdrawing existing tax breaks and other subsidies or by adding taxes on its use. This provides increased government revenues, encourages improvements in energy efficiency, reduces dependence on imported energy, and decreases use of an energy resource that has a limited future supply. However, increasing taxes on energy use can dampen economic growth and put a heavy economic burden on the poor and lower middle class unless some of the energy tax revenues are used to help offset their increased energy costs.

Solutions: How Can the United States Develop a More Sustainable Energy Future? Communities such as Osage, Iowa (p. 105), and Davis, California (p. 94), and individuals are taking energy matters into their own hands. At the same time, most environmentalists urge citizens to exert intense pressure on elected officials to develop a national energy policy based on much greater improvements in energy efficiency and a more rapid transition to a mix of renewable energy resources. A variety of analysts have suggested strategies, some of them highly controversial, to achieve such goals (Figure 4-35).

Government-based ways to improve energy efficiency include **(1)** *increasing fuel-efficiency standards for vehicles,* **(2)** *establishing energy-efficiency standards for buildings and appliances,* **(3)** *greatly increasing government-sponsored research and development to improve energy efficiency,* and **(4)** *giving tax credits and exemptions or government rebates for purchases of energy-efficient vehicles, houses, buildings, and appliances.* This last strategy could be accomplished with *freebates,* a self-financing mechanism that charges fees to purchasers of inefficient products and uses those funds to provide

Dunn, Seth. 1998. "Green Power Spreads to California." *World Watch,* vol. 11, no. 4, 7(1).

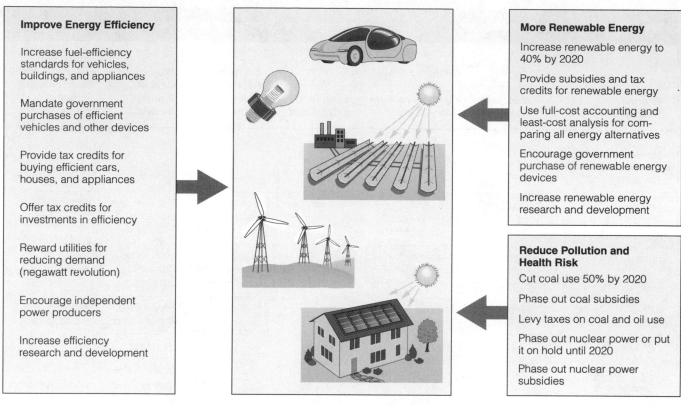

Improve Energy Efficiency

Increase fuel-efficiency standards for vehicles, buildings, and appliances

Mandate government purchases of efficient vehicles and other devices

Provide tax credits for buying efficient cars, houses, and appliances

Offer tax credits for investments in efficiency

Reward utilities for reducing demand (negawatt revolution)

Encourage independent power producers

Increase efficiency research and development

More Renewable Energy

Increase renewable energy to 40% by 2020

Provide subsidies and tax credits for renewable energy

Use full-cost accounting and least-cost analysis for comparing all energy alternatives

Encourage government purchase of renewable energy devices

Increase renewable energy research and development

Reduce Pollution and Health Risk

Cut coal use 50% by 2020

Phase out coal subsidies

Levy taxes on coal and oil use

Phase out nuclear power or put it on hold until 2020

Phase out nuclear power subsidies

Figure 4-35 Solutions. Some ways various analysts have suggested to attain a sustainable energy future.

rebates to buyers of energy-efficient products. This approach relies on market forces and requires no new government taxes or subsidies.

Another suggested government-based approach involves *taxing energy*. This would help include some of the harmful costs of using energy (which consumers are now paying indirectly) in the market prices of energy; the tax revenues would be used to improve energy efficiency, encourage use of renewable energy resources, and provide energy assistance to poor and lower-middle-class Americans. Some economists believe that the public might accept these higher taxes if income or other taxes were lowered as gasoline or other fossil fuel taxes were raised.

Governments can also improve energy efficiency by *modifying electric utility regulations*. These changes, which a few state utility commissions are now putting into effect, would require utilities to produce electricity on a least-cost basis, permit them to earn money for their shareholders by reducing electricity demand, and allow rate increases based primarily on improvements in energy efficiency.

Another suggested major goal would be to *rely more on renewable energy*. One way to encourage such a shift over the next 20–50 years would be to phase in *full-cost pricing*, in which environmental and health costs are included in the market price of energy. In other words, the prices of energy (and ideally all goods and services sold in the marketplace) should tell the ecological truth. This can provide consumers and power producers with more accurate information about the benefits and consequences of using various types of energy.

Two other ways to encourage more reliance on renewable energy are *to provide subsidies or tax credits for homeowners and businesses that switch to various forms of renewable energy* and *to greatly increase government funding of renewable energy research and development*.

Another innovation in the mostly monopolistic public utility industry is to *allow and encourage more competition from independent power producers*. A growing number of independent power companies will now build almost any kind of power plant anywhere in the world. Small decentralized devices for generating electricity and storing excess energy could allow developing countries to leapfrog into 21st-century power systems and bypass expensive central power plants and distribution systems.

Another suggested major goal is to *reduce the pollution and harmful health and ecological effects of relying on nonrenewable energy from coal and nuclear power*. This could be accomplished by phasing out most or all government subsidies and tax breaks for coal and nuclear power over a decade.

Energy experts estimate that implementing a carefully evaluated mix of such policies over the next two

Energy Resources	Estimated Availability			Estimated Net Useful Energy of Entire System	Projected Cost of Entire System	Actual or Potential Overall Environmental Impact of Entire System
	Short Term (2000–2010)	Intermediate Term (2010–2020)	Long Term (2020–2050)			
Nonrenewable Resources						
Fossil fuels						
Petroleum	High (with imports)	Moderate (with imports)	Probably low	High but decreasing	High for new domestic supplies	Moderate
Natural gas	High (with imports)	Moderate to high (with imports)	Moderate (with imports)	High but decreasing	High for new domestic supplies	Low
Coal	High	High	High	High but decreasing	Moderate but increasing	Very high
Oil shale	Low	Low to moderate	Low to moderate	Low to moderate	Very high	High
Tar sands	Low	Fair? (imports only)	Poor to fair (imports only)	Low	Very high	Moderate to high
Synthetic natural gas (SNG) from coal	Low	Low to moderate	Low to moderate	Low to moderate	High	High (increases use of coal)
Synthetic oil and alcohols from coal	Low	Moderate	High	Low to moderate	High	High (increases use of coal)
Nuclear energy						
Conventional fission (uranium)	Low to moderate	Low to moderate	Low to moderate	Low to moderate	Very high	Very high
Breeder fission (uranium and thorium)	None	None to low (if developed)	Moderate	Unknown, but probably moderate	Very high	Very high
Fusion (deuterium and tritium)	None	None	None to low (if developed)	Unknown, but may be high	Very high	Unknown (probably moderate to high)
Geothermal energy (some are renewable)	Low	Low	Moderate	Moderate	Moderate to high	Moderate to high
Renewable Resources						
Improving energy efficiency	High	High	High	Very high	Low	Decreases impact of other sources
Hydroelectric						
New large-scale dams and plants	Low	Low	Very low	Moderate to high	Moderate to very high	Low to moderate

Q: What percentage of the oil used in the United States is imported?

Energy Resources	Estimated Availability			Estimated Net Useful Energy of Entire System	Projected Cost of Entire System	Actual or Potential Overall Environmental Impact of Entire System
	Short Term (2000–2010)	Intermediate Term (2010–2020)	Long Term (2020–2050)			

Renewable Resources (continued)

Hydroelectric (continued)

Energy Resources	Short Term	Intermediate Term	Long Term	Net Useful Energy	Cost	Environmental Impact
Reopening abandoned small-scale plants	Moderate	Moderate	Low	Moderate	Moderate	Low to moderate
Tidal energy	Very low	Very low	Very low	Moderate	High	Low to moderate
Ocean thermal gradients	None	Low	Low to moderate (if developed)	Unknown (probably low to moderate)	High	Unknown (probably moderate to high)
Solar energy						
Low-temperature heating (for homes and water)	High	High	High	Moderate to high	Moderate	Low
High-temperature heating	Low	Moderate	Moderate to high	Moderate	High initially, but probably declining fairly rapidly	Low to moderate
Photovoltaic production of electricity	Low to moderate	Moderate	High	Fairly high	High initially but declining fairly rapidly	Low
Wind energy	Low	Moderate	Moderate to high	Fairly high	Moderate	Low
Geothermal energy (low heat flow)	Very low	Very low	Low to moderate	Low to moderate	Moderate to high	Moderate to high
Biomass (burning of wood and agricultural wastes)	Moderate	Moderate	Moderate to high	Moderate	Moderate	Moderate to high
Biomass (urban wastes for incineration)	Low	Moderate	Moderate	Low to fairly high	High	Moderate to high
Biofuels (alcohols and biogas from organic wastes)	Low to moderate	Moderate	Moderate to high	Low to fairly high	Moderate to high	Moderate to high
Hydrogen gas (from coal or water)	Very low	Low to moderate	Moderate to high	Variable but probably low to negative	Variable	Variable, but low if produced with solar energy

A: About 55% in 1997 (up from 36% in 1973)

decades could save money, create a net gain in jobs, improve competitiveness in the global marketplace, slow projected global warming, and sharply reduce air and water pollution. Some actions you can take toward achieving a sustainable energy future are listed in Appendix 4.

In the long run, humanity has no choice but to rely on renewable energy. No matter how abundant they seem today, eventually coal and uranium will run out.

DANIEL DEUDNEY AND CHRISTOPHER FLAVIN

CRITICAL THINKING

1. A homebuilder installs electric baseboard heat and claims that "it's the cheapest and cleanest way to go." Apply your understanding of the second law of energy (thermodynamics) to evaluate his claim.

2. Should the United States (or the country where you live) institute a crash program to develop solar photovoltaic cells and solar-produced hydrogen fuel? Explain.

3. Someone tells you that we can save energy by recycling it. How would you respond?

4. (a) Should air pollution emission standards for *all* new and existing coal-burning plants be tightened significantly? Explain. **(b)** Do you favor an energy strategy based on greatly increased use of coal-burning plants to produce electricity? Explain. What are the alternatives?

5. Explain why you agree or disagree with the following proposals by various energy analysts:
  a. Federal subsidies for all energy alternatives should be eliminated so that all energy choices can compete in a true free-market system.
  b. All government tax breaks and other subsidies for conventional fuels (oil, natural gas, coal), synthetic natural gas and oil, and nuclear power (fission and fusion) should be removed and replaced with subsidies and tax breaks for improving energy efficiency and developing solar, wind, geothermal, and biomass energy alternatives.
  c. Development of solar and wind energy should be left to private enterprise and receive little or no help from the federal government, but nuclear energy and fossil fuels should continue to receive large federal subsidies.
  d. To solve present and future U.S. energy problems, all we need to do is find and develop more domestic supplies of oil, natural gas, and coal, and increase dependence on nuclear power.
  e. A heavy federal tax should be placed on gasoline and imported oil used in the United States.
  f. Between 2000 and 2020, the United States should phase out all nuclear power plants.

6. Congratulations. You have just been put in charge of the world. List the five most important features of your energy policy.

5 BIODIVERSITY: SUSTAINING ECOSYSTEMS—FORESTS, RANGE-LANDS, PARKS, AND WILDERNESS

Forests precede civilizations, deserts follow them.

FRANÇOIS-AUGUSTE-RENÉ DE CHATEAUBRIAND

5-1 BIODIVERSITY, CONSERVATION BIOLOGY, AND ECOLOGICAL INTEGRITY

Why Is Biodiversity Loss Considered the Key Environmental Problem? Every country and region has three forms of wealth: material, cultural, and biological (biodiversity). The forests, rangelands, parks, wilderness, and aquatic systems where the genes, species, and ecosystems making up biodiversity are found are crucial sources of biological wealth that are coming under increasing pressure from population growth and economic development.

In Chapter 2 you learned that there are three components of the planet's biodiversity: *genetic diversity* (variability in the genetic makeup among individuals within a single species), *species diversity* (the variety of species on the earth and in different habitats on the earth), and *ecological diversity* (the variety of forests, deserts, grasslands, streams, lakes, wetlands, oceans, and other biological communities that interact with one another and with their nonliving environments).

Most ecologists and other environmental scientists consider the loss of biodiversity the key environmental problem. This revolution in awareness has led to a much deeper understanding and appreciation of *interrelationships or connections in nature*—a major theme of this book.

What Are Conservation Biology and Ecological Integrity? The science-based study of biodiversity has led to the development of **conservation biology**. It is a multidisciplinary science created in the late 1970s to deal with the crisis of maintaining the genes, species, communities, and ecosystems that make up the earth's *biological diversity*. Its goals are to investigate the human impacts on biodiversity and to develop practical approaches to preserving biodiversity. Conservation biology uses scientific data and concepts to find practical ways to protect critical ecosystems and biodiversity-rich areas and prevent the premature extinctions of species.

Since the mid-1980s conservation biologists and ecologists have increasingly emphasized the preservation of **ecological integrity**, the conditions and natural processes (such as the energy flow and matter cycling in ecosystems and species interactions) that generate and maintain biodiversity and allow evolutionary change as a key mechanism for adapting to changes in environmental conditions (Chapter 2). The **ecological health** of an area can be described in terms of the degree to which its biodiversity and ecological integrity remain intact.

Conservation biology rests on the following principles: **(1)** Biodiversity and ecological integrity are necessary to all life on earth and should not be reduced by human actions, **(2)** humans should not cause or hasten the premature extinction of populations and species by disrupting evolutionary processes and critical ecological processes, **(3)** the best way to preserve biodiversity and ecological integrity is to preserve habitats, niches, and ecological interactions, and **(4)** goals and strategies for preserving biodiversity and ecological integrity of an area must be based on a deep understanding of the ecological properties and processes of that system.

This *scientific approach* recognizes that saving wildlife means saving their habitats and not disrupting the complex interactions among species in an ecosystem (ecological integrity). It is also based on Aldo Leopold's ethical principle that something is right when it tends to maintain the earth's life-support systems for us and other species and wrong when it doesn't.

5-2 PUBLIC LANDS IN THE UNITED STATES

What Are the Major Types of U.S. Public Lands? No nation has set aside as much of its land—about 42%—for public use, enjoyment, and wildlife as has the United States. Almost one-third of the country's land belongs to every American and is managed for them by the federal government; 73% of this public land is in Alaska and another 22% is in the western states (where 60% of all land is public land). These public lands are classified as multiple-use lands, moderately restricted-use lands, and restricted-use lands.

Hint: Enter the search term island biogeography using the Subject Guide.

Multiple-Use Lands

- The 156 forests (Figure 5-1) and 20 grasslands of the *National Forest System* are managed by the U.S. Forest Service. These forests are used for logging (the dominant use in most cases), mining, livestock grazing, farming, oil and gas extraction, recreation, sport hunting, sport and commercial fishing, and conservation of watershed, soil, and wildlife resources. Off-road vehicles are usually restricted to designated routes.

- *National Resource Lands* in the western states and Alaska are managed by the Bureau of Land Management (BLM). Emphasis is on providing a secure domestic supply of energy and strategic minerals and on preserving rangelands for livestock grazing under a permit system.

Moderately Restricted-Use Lands

- The 508 *National Wildlife Refuges* (Figure 5-1) are managed by the U.S. Fish and Wildlife Service. Most refuges protect habitats and breeding areas for waterfowl and big game to provide a harvestable supply for hunters; a few protect endangered species from extinction. Sport hunting, trapping, sport and commercial fishing, oil and gas development, mining, logging, grazing, some military activities, and farming are permitted as long as the Department of the Interior finds such uses compatible with the purposes of each unit.

Restricted-Use Lands

- The 375 units of the *National Park System* include 54 major parks (mostly in the West) and 321 national recreation areas, monuments, memorials, battlefields, historic sites, parkways, trails, rivers, seashores, and lakeshores (Figure 5-1) are managed by the National Park Service. National parks may be used only for camping, hiking, sport fishing, and boating. Motor vehicles are permitted only on roads. In national recreation areas, these same activities, plus sport hunting, mining, and oil and gas drilling, are allowed.

- The 630 roadless areas of the *National Wilderness Preservation System*, which lie within the national parks, national wildlife refuges, and national forests, are managed by the National Park Service (42%), Forest Service (33%), Fish and Wildlife Service (20%), and BLM (5%). These areas are open only for recreational activities such as hiking, sport fishing, camping, nonmotorized boating, and, in some areas, sport hunting and horseback riding. Roads, logging, livestock grazing, mining, commercial activities, and buildings are banned, except when they predate the wilderness designation.

Between 1970 and 1996 the area within all public land systems (excluding national forests) increased significantly (2.9-fold in the National Park System, 3-fold in the National Wildlife Refuge System, and 9-fold in the National Wilderness Preservation System). Most of the additions, made by President Carter just before he left office in 1980, are in Alaska. Since then little land has been added. An exception was President Clinton's executive action in 1996 to add a new national monument area in Utah that is almost as large as Yellowstone National Park.

How Should Public Lands Be Managed? Because of the resources they contain, there has been intense controversy over how public lands should be used and managed. Economists, developers, and resource extractors tend to view areas of the earth's surface in terms of their usefulness in providing mineral and other resources, their potential for urban development, and their ability to increase short-term economic growth.

Ecologists and conservation biologists take a scientific approach and view the earth's ecosystems in terms of their usefulness in maintaining the planet's biodiversity and life-sustaining processes. To conservation biologists and ecologists, destroying or degrading most of the earth's remaining intact ecosystems for short-term economic gain will make them less useful to humans and other forms of life.

Most conservation biologists and many free-market economists believe that the following four principles should govern use of public land:

- Protection of biodiversity and the ecological integrity of vital public lands should be the primary goal.

- No one should be given subsidies or tax breaks for using or extracting resources on public lands.

- The American people deserve fair compensation for the use of their property.

- All users or extractors of resources on public lands should be fully responsible for any environmental damage they cause.

Do you agree or disagree with these principles? Why?

Since 1995, however, there have been increasing attempts by some members of the U.S. Congress (influenced by ranching, timber, mining, and development interests) to pass laws that would **(1)** sell public lands or their resources to corporations or individuals, usually at less than market value; **(2)** give public lands to the states (where these interests often have more influence); **(3)** slash federal funding for administration of public lands so that unlawful exploiters have less chance of getting

 Smith, M. B. 1998. "The Value of a Tree: Public Debates of John Muir and Gifford Pinchot." *The Historian*, vol. 60, no. 4, 757(22).

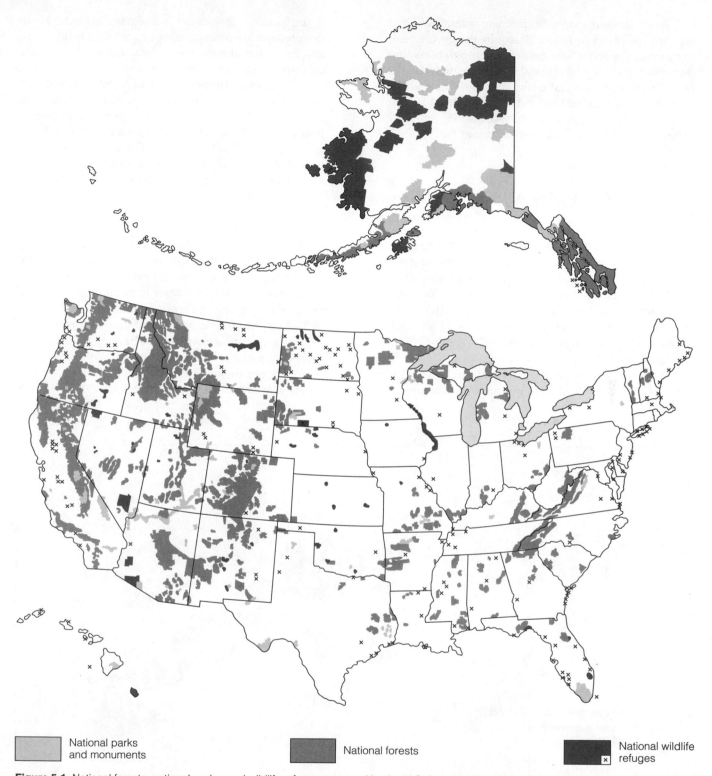

Figure 5-1 National forests, national parks, and wildlife refuges managed by the U.S. federal government. (Data from U.S. Geological Survey)

Legend:
- National parks and monuments
- National forests
- National wildlife refuges

Hint: Enter the search terms *environmental movement, history* using the Subject Guide.

CHAPTER 5   **143**

caught; **(4)** weaken, waive, or eliminate federal laws that regulate use of public lands; **(5)** require the government to reimburse private landowners for regulations that decrease the value of their lands; **(6)** allow highways to be built through national parks and wilderness areas; **(7)** redefine protected wetlands so that about half of them would no longer be protected; and **(8)** repeal or seriously weaken the Endangered Species Act.

5-3 MANAGING AND SUSTAINING FORESTS

What Are the Major Types of Forests?

There are three general types of forests, depending primarily on climate: tropical, temperate, and polar (Figure 2-30). Since agriculture began about 10,000 years ago, human activities have reduced the earth's forest cover by about one-quarter—from about 34% to 26% of the world's land area—and only 12% consists of intact forest ecosystems.

If the rate of cutting and degradation does not exceed the rate of regrowth, and if protecting biodiversity is emphasized, forests are renewable resources. However, forests are disappearing or are being fragmented and degraded almost everywhere, especially in tropical countries (Figure 5-2).

Old-growth forests are uncut forests and regenerated forests that have not been seriously disturbed for several hundred or thousands of years. Old-growth forests provide ecological niches for a variety of wildlife species (Figure 2-17). These forests also have large numbers of standing dead trees (snags) and fallen logs that are habitats for a variety of species. Decay of this dead vegetation returns plant nutrients to the soil and helps build fertile soil (Figure 2-13).

Second-growth forests are stands of trees resulting from secondary ecological succession after cutting (Figure 2-45). Most forests in the United States and other temperate areas are second-growth forests that grew back after virgin forests were cleared for timber or to create farms that were later abandoned. About 40% of tropical forests are second-growth forests.

Some second-growth stands have remained undisturbed long enough to become old-growth forests, but many are not diverse forests at all, but rather are **tree farms** or plantations: managed tracts with uniformly

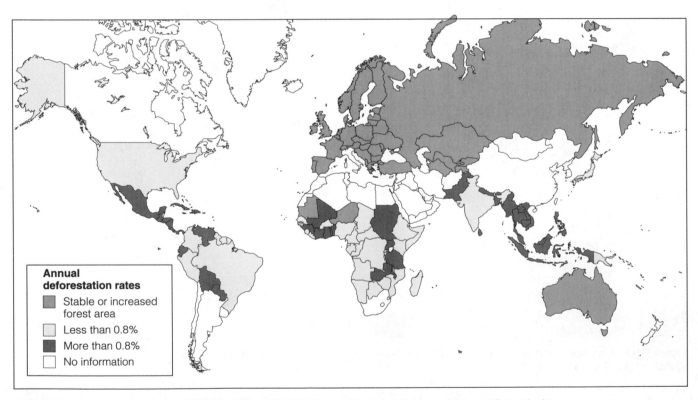

Figure 5-2 Estimated annual rates of deforestation, 1981–90. During this period the area of the world's tropical forest decreased by 8% (an average of 0.8% per year). At least 2 million square kilometers (770,000 square miles) of forest were lost between 1990 and 1995—an area three times larger than the state of Texas. Each year there is a net loss of another 160,000 square kilometers (62,000 square miles) of forest. This rate of deforestation has continued during the 1990s.(Data from UN Food and Agriculture Organization)

Noble, Ian R. 1997. "Forests As Human-Dominated Ecosystems." *Science*, vol. 275, no. 5325, 522(4).

aged trees of one species that are harvested by clear-cutting as soon as they become commercially valuable. Then they are replanted and clear-cut again on regular cycles. Studies of the effects of intensive tree plantation forestry in Germany show that in a forest where trees are repeatedly cut and removed, the soil becomes depleted of nutrients and the forests lose their resilience for coping with stresses such as drought, disease, and insect infestations.

🌐 What Is the Commercial and Ecological Importance of Forests?

Forests provide lumber for housing, biomass for fuelwood, pulp for paper, medicines, and many other products worth more than $300 billion a year. Many forestlands are also used for mining, grazing livestock, and recreation.

Forested watersheds act as giant sponges, slowing down runoff and holding water that recharges springs, streams, and groundwater. Thus, they regulate the flow of water from mountain highlands to croplands and urban areas, and they reduce the amount of sediment washing into streams, lakes, and reservoirs by reducing soil erosion.

Forests also influence climate. For example, 50–80% of the moisture in the air above tropical forests comes from trees via transpiration and evaporation. If large areas of these forests are cleared, average annual precipitation drops and the region's climate gets hotter and drier; then soils become depleted of already-scarce nutrients, and they are baked and washed away. This process can eventually convert a diverse tropical forest into a sparse grassland or even a desert.

Forests are vital to the global carbon cycle (Figure 2–23). They provide habitats for more wildlife species than any other biome, making them the planet's major reservoir of biodiversity. They also buffer us against noise, absorb air pollutants, and nourish the human spirit.

The ecological benefits of complex and diverse forests are undervalued in the marketplace. Until this situation is changed, most analysts believe that we will continue to sacrifice these forests and their long-term ecological services (earth capital, Figure 1-2) for short-term economic gain (Spotlight, p. 146).

What Are the Major Types of Forest Management?

The total volume of wood produced by a particular stand of forest varies as it goes through different stages of growth and ecological succession (Figure 5-3). If the goal is to produce fuelwood or fiber for paper production in the shortest time, the forest is usually harvested on a short rotation cycle, well before the volume of wood produced peaks (point A in Figure 5-3). Harvesting at point B in Figure 5-3 gives the

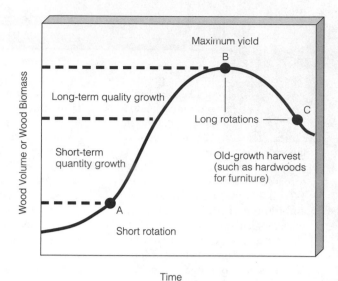

Figure 5-3 Changes in wood volume over various growth–harvest cycles in forest management. Forest management occurs over a cycle of decisions and events called a *rotation*. The most important steps in a rotation include taking an inventory of the site, developing a forest management plan, building roads into the site, preparing the site for harvest, harvesting timber, and regenerating and managing the site until the next harvest.

maximum yield of wood per unit of time. If the goal is high-quality wood for fine furniture or veneer, managers use longer rotations to develop larger, older-growth trees (point C in Figure 5-3), whose rate of growth has leveled off and is much lower than that of young trees.

There are two basic forest management systems: even-aged and uneven-aged. With **even-aged management**, trees in a given stand are maintained at about the same age and size. A major goal of even-aged management, sometimes called *industrial forestry*, is to grow and harvest trees using monoculture techniques. This is achieved by replacing a biologically diverse natural forest with a simplified tree farm of one or two fast-growing and economically desirable species that can be harvested every 10–100 years, depending on the species. Crossbreeding and genetic engineering can improve both the quality and the quantity of tree farm wood.

Even-aged management begins with one or two cuttings of all or most trees from an area; then the site is usually replanted with seedlings of one or more species of the same age. In a natural forest, dead and fallen trees are seen as vital wildlife habitats and integral parts of a natural cycle of decay and forest renewal. In most industrial forests, they are viewed as debris to be removed and burned to make way for the growth of planted tree seedlings.

Hint: Enter the search terms *forest ecology* using the Subject Guide.

How Much Is a Forest Worth?

In 1997 a group of 13 ecologists, economists, and geographers from 12 prestigious laboratories and universities attempted to estimate how much nature's ecological services (Figure 1-2) are worth. According to this ambitious and crude appraisal led by ecological economist Robert Costanza of the University of Maryland, the economic value of Mother Nature's bounty ranges from $16 trillion to $54 trillion. In other words, each year nature provides free goods and services worth somewhere between $2,800 and $9,000 per person.

Their medium estimate for nature's life-support services for humans is $33 trillion a year. This is nearly twice the $18-trillion-per-year economic output of the earth's 194 nations and almost five times the $7.0-trillion U.S. GDP in 1997.

To make these estimates the researchers divided the earth's surface into 16 biomes (Figure 2-30) and aquatic life zones such as various types of forests, grasslands, cropland, open oceans, estuaries, wetlands, and lakes and rivers (but omitted deserts and tundra because of a lack of data). Then they agreed on a list of 17 goods and services provided by nature (Figure 1-2) and sifted through more than 100 studies that attempted to put a dollar value on such services in the 16 different types of ecosystems.

Some analysts believe that such estimates are misleading and dangerous because they amount to putting a dollar value on ecosystem services that have an infinite value because they are irreplaceable. In the 1970s economist E. F. Schumacher warned, "To undertake to measure the immeasurable is absurd" and is a "pretense that everything has a price." Biologist and conservationist David Ehrenfield's response to this evaluation was, "I am afraid that I don't see much hope for a civilization so stupid that it demands a quantitative estimate of the value of its own umbilical cord." Those who cringe at the thought of putting a dollar value on nature's ecosystem services believe that they should be protected on *moral* grounds instead of being reduced to easily manipulated cost–benefit and risk-benefit analyses (Spotlight, p. 241).

The researchers admit that their estimates are full of many assumptions and omissions and could easily be too low by a factor of 10 to 1 million or more. For example, their calculations include only estimates of the ecosystem services themselves, not the natural or earth capital that generates them, and omitted the value of nonrenewable minerals and fuels. They also recognize that as the supply of ecosystem services declines their value will rise sharply and that such services can be viewed as having an infinite value.

However, they contend that their estimates are much more accurate than the *zero* value the market assigns to these ecosystem services. They hope such estimates will call people's attention to the fact that the earth's ecosystem services (Figure 1-2) are absolutely essential for all humans and their economies and that their economic value is huge. They believe that failure to estimate economic values for ecosystem services helps ensure that such vital life-support services will continue to be degraded for short-term economic gain.

Critical Thinking

Do you agree or disagree with the idea of trying to estimate the economic value of the earth's ecosystem services? Explain. What are the alternatives?

Even-aged management of an area leads to more even-aged management of that area. Once a diverse forest has been cleared and replaced with even-aged stands, the only economical thing for timber companies to do is to keep repeating the process of cutting down a stand of trees before it turns into a true forest. In even-aged management, forests are viewed primarily as lumber and fiber factories, and old-growth forests are seen as timber going to waste instead of as essential centers of the earth's biodiversity and ecological integrity.

In **uneven-aged management**, trees in a given stand are maintained at many ages and sizes to foster natural regeneration. Here the goals are biological diversity, long-term production of high-quality timber, a reasonable economic return, and multiple use. Mature trees are selectively cut; the removal of all trees is used only on small patches of species that benefit from such a practice.

How Are Trees Harvested? Before larger amounts of timber can be harvested, roads must be built for access and removal of timber. Even with careful design, logging roads have a number of harmful impacts (Figure 5-4). They cause erosion and increase sedimentation of waterways and cause severe habitat fragmentation and loss of biodiversity. They can expose forests to invasion by exotic pests, diseases, and introduced wildlife species. They can also open up once-inaccessible forests to farmers, miners, ranchers, hunters, and off-road vehicle users. In addition, logging roads on U.S. public lands disqualify the land for protection as wilderness.

Q: How long will the world's oil reserves last at the current consumption rate?

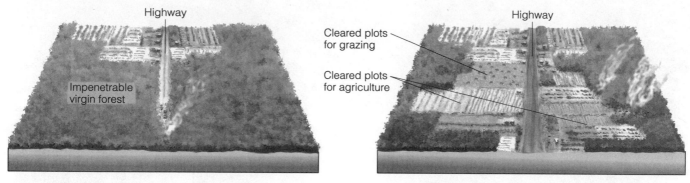

Figure 5-4 Building roads into previously inaccessible tropical forests paves the way to fragmentation, destruction, and degradation.

Once loggers can reach a forest, they use various methods for harvesting the trees (Figure 5-5). With **selective cutting**, intermediate-aged or mature trees in an uneven-aged forest are cut singly or in small groups (group selection), creating gaps no larger than the height of the standing trees (Figure 5-5a). Selective cutting reduces crowding, encourages the growth of younger trees, and maintains an uneven-aged stand of trees of different species. It also allows natural regeneration from the surrounding trees, thereby avoiding the financial and environmental costs of site preparation (usually by bulldozer or herbicides) and planting of trees after clearcutting. If done properly, selective cutting also helps protect the site from soil erosion and wind damage.

Many industry and government foresters contend that selective cutting is not profitable and that many of the trees that bring the best prices don't get enough sunlight for regeneration with selective cutting. But numerous private foresters and timber companies have been making satisfactory profits by using selective cutting of small and large areas of forests containing such species.

An unsound type of selective cutting is *high grading*, or *creaming*, which removes the most valuable trees. This practice, common in many tropical forests, ends up injuring one-third to two-thirds of the remaining trees when they are knocked over by logging equipment or by the large target trees when they are felled and removed.

Some tree species grow best in full or moderate sunlight in moderate to large clearings. Such sun-loving species are usually harvested by shelterwood cutting, seed-tree cutting, or clear-cutting. **Shelterwood cutting** removes all mature trees in two or typically three cuttings over a period of about 10 years (Figure 5-5b). The first cut removes most mature canopy trees, unwanted tree species, and diseased, defective, and dying trees. This opens the forest floor to light but leaves enough mature trees to cast seed and to shelter growing seedlings. Some years later, after enough seedlings take hold, a second cut removes more canopy trees, although some of the best mature trees are left to shelter the young trees. After these are well established, a third cut removes the remaining mature trees, and the even-aged stand of young trees then grows to maturity.

This method allows natural seeding from the best seed trees and keeps seedlings from being crowded out. It leaves a fairly natural-looking forest that can serve a variety of purposes; it also helps reduce soil erosion and provides a good habitat for wildlife. The danger is that loggers may take too many trees in the first cut.

Seed-tree cutting harvests nearly all of a stand's trees in one cutting, leaving a few uniformly distributed seed-producing trees to regenerate the stand (Figure 5-5c). After the new trees become established, the seed trees may be harvested. By allowing several species to grow at once, seed-tree cutting leaves an aesthetically pleasing forest that is useful for recreation, deer hunting, erosion control, and wildlife conservation. Leaving the best trees for seed can also lead to genetic improvement in the new stand.

Clear-cutting is the removal of all trees from an area in a single cutting. The clear-cut area may be a whole stand (Figure 5-5d), a strip, or a series of patches. After all trees are cut, the site is usually reforested; seed is released naturally, by the harvest, or artificially as foresters broadcast seed over the site or plant seedlings raised in a nursery. If clear-cut areas are small enough, seeding may occur from nearby uncut trees.

On the positive side, clear-cutting increases timber yield per hectare, permits reforesting with genetically improved stocks of fast-growing trees, and shortens the time needed to establish a new stand of trees. It also requires much less skill and planning than other harvesting methods, and it usually provides timber companies the maximum economic return in the shortest time.

However, clear-cutting leaves ugly, unnatural forest openings and eliminates any potential recreational

A: At least 44 years; undiscovered oil might add another 20–40 years

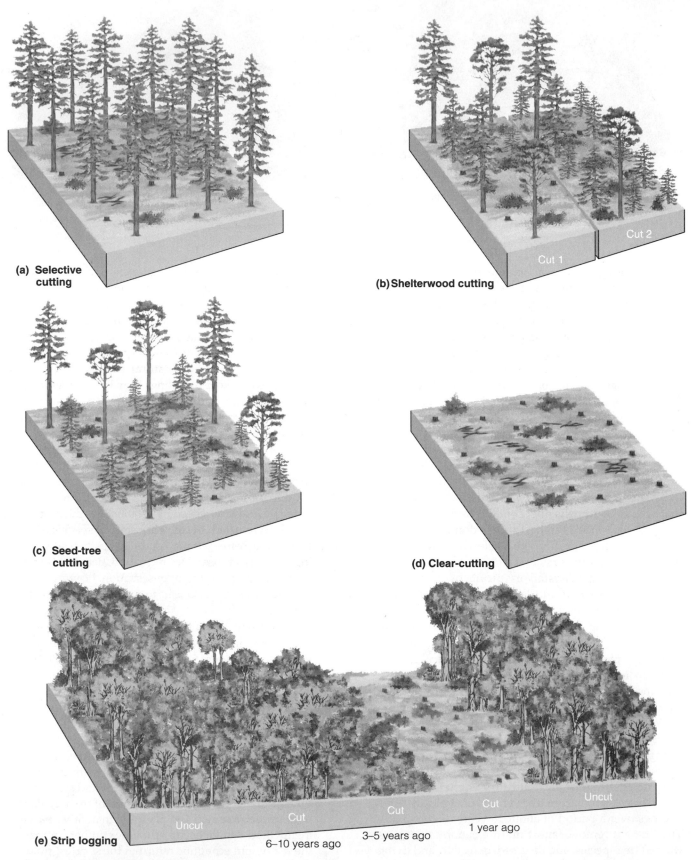

(a) Selective cutting

(b) Shelterwood cutting

Cut 1

Cut 2

(c) Seed-tree cutting

(d) Clear-cutting

(e) Strip logging

Uncut

Cut

6–10 years ago

Cut

3–5 years ago

Cut

1 year ago

Uncut

Figure 5-5 Tree harvesting methods.

value for several decades. It also destroys and fragments wildlife habitats and thus reduces biodiversity and disrupts ecological integrity. Environmental degradation from clear-cutting above ground is obvious, but equally serious damage occurs underground from the loss of fungi, worms, bacteria, and other microbes that condition soil and help protect plants from disease. Furthermore, trees in stands bordering clear-cut areas are more vulnerable to being blown down by windstorms. Large-scale clear-cutting on steep slopes leads to severe soil erosion, sediment water pollution, and flooding.

If done carefully and responsibly, clear-cutting is often the best way to harvest tree farms and stands of some tree species that require full or moderate sunlight for growth. However, for economic reasons it is often done irresponsibly and used on species that don't require this method.

A variation of clear-cutting that can allow a sustainable timber yield without widespread destruction is **strip cutting**. A strip of trees is clear-cut along the contour of the land, with the corridor narrow enough to allow natural regeneration within a few years (Figure 5-5e). After regeneration, another strip is cut above the first, and so on. This allows a forest to be clear-cut in narrow strips over several decades with minimal damage.

Is Industrial Forestry Sustainable? To timber companies, sustainable forestry means producing a maximum sustainable yield of commercial timber in as short a time as possible. This often means clearing diverse forests by using clear-cutting, seed-tree cutting, or shelterwood cutting (Figure 5-5b, c, and d) and replacing them with intensively managed tree farms as quickly as possible.

To conservation biologists and some foresters, such *maximum sustained yield forestry* does not result in long-term *sustainable forestry*. Research reveals that industrial or monoculture forestry has led to a decrease in the biodiversity of forests, elimination of competing species, suppression of ecologically important natural fires, draining of wetlands, and increased soil erosion and loss of soil nutrients. Trees have little resilience for coping with stresses such as drought, diseases, and pests. In other words, industrial forestry reduces overall biodiversity and disrupts the ecological integrity of forests.

Solutions: What Is Sustainable Forestry? Instead of conventional industrial forestry, conservation biologists and a growing number of foresters call for sustainable forest management that emphasizes the following:

- *Recycling more paper to reduce the harvest of pulpwood trees*

- *Using fibers from fast-growing plants such as kenaf to make tree-free paper, and reducing wood waste (p. 153)*

- *Growing more timber on long rotations, generally about 100–200 years (point C, Figure 5-3), depending on the species and soil quality*

- *Practicing selective cutting of individual trees or small groups of most tree species (Figure 5-5a)*

- *Minimizing fragmentation of remaining larger blocks of forest*

- *Using road building and logging methods that minimize soil erosion and compaction*

- *Practicing strip cutting (Figure 5-5e), instead of conventional clear-cutting, and banning clear-cutting (including seed-tree and shelterwood cutting) on land that slopes more than 15–20°*

- *Leaving standing dead trees (snags) and fallen timber to maintain diverse wildlife habitats and to be recycled as nutrients (Figure 2-13)*

- *Including the ecological and recreational services provided by trees and a forest in evaluating their economic value*

5-4 FOREST RESOURCES AND MANAGEMENT IN THE UNITED STATES

What Is the Status of Forests in the United States? Forests cover about one-third of the lower 48 United States and provide habitats for more than 80% of the country's wildlife species. Nearly two-thirds of this forestland is productive enough to grow commercially valuable trees. Public and private forests in the United States provide important wood and wood products. They are a prime setting for outdoor recreation and play vital ecological roles in helping sustain the country's biodiversity and ecological integrity.

Today American forests are generally bigger and often healthier than they were in 1900, when the country's population was below 100 million. This is an important accomplishment, but it can be misleading. Most of the virgin and old-growth forests in the lower 48 states have been cut (Figure 5-6), and much of what remains on private and public lands is threatened.

Most of the virgin forests that were cleared or partially cleared between 1800 and 1960 were allowed to grow back as fairly diverse second-growth forests. However, since the mid-1960s a rapidly increasing area of the nation's remaining old-growth and fairly diverse second-growth forests has been clear-cut and

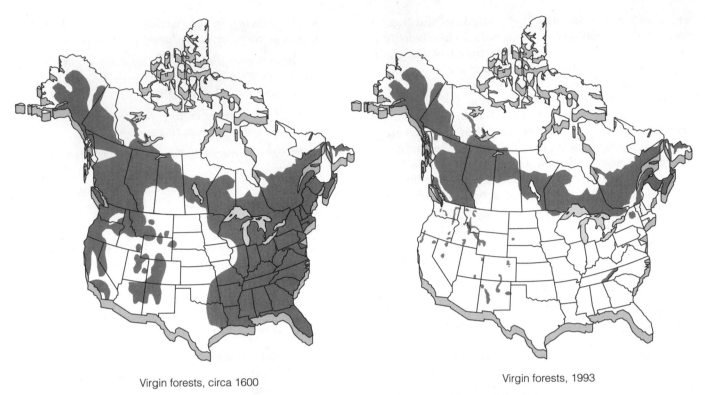

Virgin forests, circa 1600

Virgin forests, 1993

Figure 5-6 Vanishing old-growth forests in the United States and Canada. Since about 1600 most of the virgin forests that once covered much of the lower 48 states have been cleared away; most of the remaining old-growth forests in those states are on public lands (especially in the Pacific Northwest). About 60% of old-growth forests in western Canada have been cleared, and much of what remains is slated for cutting. (Data from the Wilderness Society, the U.S. Forest Service, and *Atlas Historique du Canada*, vol. 1)

replaced with tree plantations. Although this doesn't reduce the nation's overall tree cover, environmentalists argue that it does reduce the amount of eventually sustainable forestland and forest biodiversity and ecological integrity.

Case Study: How Are U.S. National Forests Managed? About 22% of the commercial forest acreage in the United States is located within the 156 national forests (Figure 5-1) managed by the U.S. Forest Service. These forests serve as grazing lands for more than 3 million cattle and sheep each year. They support multi–billion-dollar mining operations, supply about 4% of the nation's timber, contain a network of roads equal in area to the entire U.S. interstate highway system, and receive more recreational visits than any other federal public lands. The resulting networks of roads and clear-cuts severely fragment wildlife habitats, isolating many wildlife populations and making them vulnerable to genetic deterioration and predation.

The Forest Service is required by law to manage national forests according to the principles of *sustained yield* (which states that potentially renewable tree resources should not be harvested or used faster than they are replenished) and *multiple use* (which says that each of these forests should be managed simultaneously for a variety of uses such as sustainable timber harvesting, recreation, and wildlife conservation).

Environmentalists charge that especially since 1980 the Forest Service has allowed timber harvesting to become the dominant use in most national forests. To back up this charge, they point out that until 1998 about 70% of the Forest Service budget was devoted directly or indirectly to the sale of timber.

A major reason for this emphasis on timber cutting is that Congress allows the agency to keep most of the money it makes on timber sales, whereas any losses are passed on to taxpayers. Because logging increases its budget, the Forest Service has a powerful built-in incentive to encourage timber sales over other uses of the publicly owned lands. Local county officials also lobby Congress and encourage Forest Service officials to keep timber harvests high because counties get 25% of the gross receipts from national forests within their boundaries.

A drive through the 122 national forests open to logging, or a boat ride along streams in such forests, does not reveal the emphasis on timber cutting because the Forest Service leaves thin buffers of uncut trees, called *beauty strips*, along roads and streams. However, a flight over these forests reveals that many of the trees have been clear-cut and most of the forests have been highly fragmented.

Environmentalists and the U.S. General Accounting Office have accused the Forest Service of poor financial management of public forests. By law, the Forest Service must sell timber for no less than the cost of reforesting the land from which it was harvested. However, the cost of access roads is not included in this price but is provided as a subsidy to logging companies. Usually the companies also get the timber itself for less than they would pay a private landowner for an equivalent amount of timber.

Studies have shown that between 1978 and 1994, national forests lost at least $6.6 billion from below-cost timber sales. Each year typically about 5–25 of the 122 national forests open to logging make money. In effect, publicly owned national forests provide about 4% of the nation's timber at a *net loss* of about $75–500 million a year. In the early 1990s it was also discovered that timber companies were illegally cutting trees worth hundreds of millions of dollars per year on federal lands, sometimes with the knowledge of Forest Service agents.

For decades the Forest Service juggled its books to disguise timber sales losses by defraying its annual up-front costs of road building over 99 years. In 1998 this policy changed. Using more realistic bookkeeping the Forest Service announced that it lost $88.6 million in timber sales in 1997 and $230 million in 1995.

In 1998 Forest Service chief Mike Dombeck also announced a policy designed to shift the primary emphasis of the agency on maximizing timber harvests to placing greater value on overall forest health and multiple use. He pointed out that annual timber harvest levels in the national forests have dropped from 12 billion board feet in the 1980s to about 4 billion board feet in 1997. If this new policy remains in effect, much greater emphasis will be placed on the recreational use of national forests. This would recognize that in 1997 three times more people made recreational visits to national forests than they did to national parks.

Timber company officials argue that being able to get timber from federal lands fairly cheaply benefits taxpayers by keeping lumber and paper prices down.

Environmentalists respond that below-cost timber prices discourage investments to produce additional timber on private lands and put more pressure for timber cutting on national and state forests.

Congressional supporters of the timber industry threatened to cut the Forest Service budget if it moves away from logging as its top priority. In addition, Senator Larry Craig (R-Idaho) wants to revise the 1976 National Forest Management Act to **(1)** make timber harvesting the primary priority and make such harvests mandatory, **(2)** allow the Forest Service to impose fines up to $10,000 on individuals or groups who file appeals for an undefined improper purpose, **(3)** prevent forest projects and timber sales from being stopped in court, regardless of how much damage they do, **(4)** allow timber, mining, and grazing interests to meet in secret with officials of the Forest Service or the BLM to cut deals about extraction of forests and other resources, and **(5)** eliminate the requirement that the Forest Service and BLM consult with wildlife officials on the impacts of proposed logging on fish and wildlife.

Instead of providing economic stability for local industries and communities, logging in national forests tends to have the opposite effect. Communities that rely on proceeds from national forest timber sales experience boom-and-bust cycles as the timber is depleted and timber companies move to other areas. Analysts have proposed a number of reforms in Forest Service policy that would lead to more sustainable use and management of the national forests (Solutions, p. 152).

Case Study: Should Remaining Old-Growth Forests on U.S. Public Lands Be Cut or Preserved?

To officials of timber companies, the giant living trees and rotting dead trees in the Pacific Northwest's old-growth forests are valuable resources that could provide jobs and should be harvested for profit, not left alone to please environmentalists. These officials point out that the timber industry annually pumps millions of dollars into the region's economy and provides jobs for about 100,000 loggers and millworkers. Timber officials claim that protecting large areas of remaining old-growth forests on public lands will cost as many as 53,000 jobs and hurt the economies of logging and milling towns in such areas.

To conservation biologists, the remaining ancient forests on the nation's public lands are a treasure whose ecological, scientific, aesthetic, and recreational values far exceed the economic value of cutting them down for short-term economic gain. They argue that the fate of these forests is a national issue because these forests are owned by all U.S. citizens. They also

SOLUTIONS

Foresters and conservation biologists have proposed a number of management reforms that would allow national forests in the United States to be used more sustainably without seriously impairing their important ecological functions. These suggestions include the following:

- *Make sustaining biodiversity and ecological integrity the first priority of national forest management policy.*

- *Use full-cost accounting to include the ecological services provided by old-growth forests in the market price of timber harvested from national forests.*

- *Prohibit timber harvesting on at least half of remaining old-growth forests on public lands.*

- *Ban all timber cutting in national forests and fund the Forest Service completely from recreational user fees.* The Forest Service estimates that recreational user fees (as low as $3 per day) would generate three times what it earns from timber sales.

- *Sharply reduce all new road building in national forests.*

- *Allow individuals or groups to buy conservation easements that prevent timber harvesting on designated areas of public old-growth forests. In such conservation-for-tax-relief swaps, purchasers would be allowed tax breaks for the funds they put up.*

- *Require that timber from national forests be sold at a price that includes the costs of road building, site preparation, and site regeneration, and that all timber sales in national forests yield a profit for taxpayers.*

- *Do not use money from timber sales in national forests to supplement the Forest Service budget because it encourages overexploitation of timber resources.*

- *Eliminate the provision that returns 25% of gross receipts from national forests to counties containing the forests, or base such returns on recreational user-fee receipts only.*

- *Strictly enforce and close loopholes in the current federal ban on exporting raw logs harvested from public lands.*

- *Tax exports of raw logs but not finished wood products (which provide and create domestic milling and other jobs).*

- *Establish federally funded reforestation and ecological restoration projects on degraded public lands. In addition to the ecological benefits, this will provide jobs for displaced forest harvesting and wood products workers.*

- *Provide increased aid and job retraining for displaced workers in the forest harvesting and products industry.*

- *Require use of sustainable forestry methods (p. 149).*

Understandably, timber company officials, most county officials, and many Forest Service officials oppose such policies and have prevented such reforms.

Critical Thinking

Explain why you agree or disagree with each of these proposals for reforming federal forest management in the United States.

---

view this as a global issue because these forests are important reservoirs of the earth's biodiversity and ecological integrity.

Conservation biologists and economists point out that the major causes of past and projected job losses in the timber industry in this region are automation, export of logs overseas (depriving U.S. millworkers of jobs while providing jobs for millworkers in other countries), and timber imports from Canada. Loggers, millworkers, and store owners living in logging communities are caught in the middle of a highly controversial situation.

The threatened northern spotted owl lives almost exclusively in 200-year-old Douglas fir forests in western Oregon, Washington, and northern California (mostly in 17 national forests and 5 BLM parcels). The species is vulnerable to extinction because of its low reproductive rate and the low survival rate of juveniles through their first five years. Only 2,000–3,600 pairs remain. However, logging company officials contend that the owls do not require old-growth forest and can instead adapt to younger second-growth forests.

In July 1990 the U.S. Fish and Wildlife Service added the northern spotted owl to the federal list of threatened species. This requires that its habitat be protected from logging or other practices that would decrease its chances of survival. The timber industry hopes to persuade Congress to revise the Endangered Species Act to allow for economic considerations that will reverse this protection.

The owl and other threatened species are symbols of the broader clash between timber companies who want to clear-cut most remaining old-growth stands in the national forests and conservation biologists who want to protect their priceless biodiversity and ecological integrity, or at least allow only sustainable harvesting in some areas. The question is whether the biodiversity and ecological integrity of the last 5–8% of remaining ancient forests on public lands in the Northwest should be preserved to protect their biodi-

Manser-Fonds, Bruno. 1998. "Why Are Forests Burning?" *The Ecologist*, vol. 28, no. 1, 8(1).

versity and ecological integrity or cut to save a final decade or so of logging jobs. What do you think?

Solutions: How Can We Reduce Wood Waste and Use Other Fibers to Make Paper? According to the Worldwatch Institute and forestry analysts, up to 60% of the wood consumed in the United States is unnecessarily wasted. Because only 4% of the total U.S. timber production comes from the national forests, reducing the waste of wood and paper products by only 4% could eliminate the need to remove any timber from the national forests. Then these public lands could be used primarily for recreation and protection of biodiversity and ecological integrity.

Ways to reduce such waste include **(1)** reducing construction waste, **(2)** using laminated boards instead of solid wood, **(3)** using less packaging, **(4)** reducing junk mail, **(5)** recycling more paper with the highest possible content of postconsumer waste, and **(6)** using paper made from fibers obtained from rapidly growing plants other than trees.

Some have suggested that wood can be saved by switching to nonwood home building and remodeling materials such as steel studs and beams, aluminum and vinyl siding, concrete slabs instead of floor joists and plywood, and plastics instead of paper products. However, this involves shifting from a potentially renewable resource to nonrenewable resources such as limestone, iron ore and other mineral deposits, and crude oil used to make these products and plastics. Analyses show that switching to these nonrenewable resources usually requires more energy for transportation and manufacture and produces more pollution than use of wood and wood products.

Currently, most of the small amount of treeless paper produced in the United States is made from the fibers of a woody annual plant called *kenaf* (pronounced kuh-NAHF). The tall, bamboolike stalks of this annual woody plant grow to heights of 4 meters (12 feet) in only 4–5 months, compared to 7–60 years for pine trees.

Kenaf thrives in tropical and temperate climates and herbicides are rarely needed because it grows faster than most weeds; few insecticides are needed because its outer fibrous covering is nearly insect-proof. Because kenaf is a nitrogen fixer, cultivating it does not deplete soil nitrogen. It also takes fewer chemicals and less energy to break kenaf down into fibers used for making paper, so less toxic wastewater is produced and water treatment costs are lower than when pulpwood is used. The U.S. Department of Agriculture considers kenaf the best renewable fiber to replace trees in paper making and

predicts that the market for kenaf should expand rapidly in coming years.

There are some drawbacks, but they can be overcome. Kenaf paper currently costs three to five times more than either virgin or recycled paper stock, which is why this book is not printed on tree-free kenaf paper. However, as demand increases and producers cut production costs, the price of kenaf should come down. The basic problem is getting investors to gamble money on converting or constructing mills to produce rice, wheat, or kenaf paper without an established market.

5-5 TROPICAL DEFORESTATION AND THE FUELWOOD CRISIS

How Fast Are Tropical Forests Being Cleared and Degraded? Tropical forests cover about 6% of the earth's land area (roughly the area of the lower 48 states of the United States) and grow in equatorial Latin America, Africa, and Asia (Figure 2-28). Tropical forests come in several varieties: rain forests (which receive rainfall almost daily), tropical deciduous forests (with one or two dry seasons each year), dry and very dry deciduous forests, and forests on hills and mountains.

Climatic and biological data suggest that mature tropical forests once covered at least twice as much area as they do today, with most of the destruction occurring since 1950. Satellite scans and ground-level surveys used to estimate forest destruction indicate that large areas of tropical forests are being cut or degraded (Figure 5-2).

There is considerable debate over the current rates of tropical deforestation and degradation because of difficulties in interpreting satellite images, different ways of defining deforestation and forest degradation, and political and economic factors that cause countries to hide or exaggerate deforestation. The lowest estimated rate of loss and degradation of remaining tropical forests is 62,000 square kilometers (24,000 square miles) per year. This is equivalent to a loss and degradation of *14 city blocks of tropical forest per minute*. The highest estimated rate of destruction and degradation is 308,000 square kilometers (118,000 square miles) per year—equivalent to a loss and degradation of *68 city blocks of tropical forest per minute*.

Whether the best estimate of tropical deforestation and degradation is 14 or 68 city blocks per minute (or somewhere in between), it is clear that these oases of biodiversity are being lost and degraded at a high rate. Scientists estimate that this annual rate of destruction and degradation could well double within another decade.

Hint: Enter the search terms *rain forests, environmental aspects* using the Subject Guide.

About 40% of current tropical deforestation is taking place in South America (especially in the vast Amazon Basin). However, the *rates* of such deforestation in Southeast Asia and in Central America are about 2.7 times higher than those in South America. Haiti has lost 99% of its original forest cover, the Philippines 97%, and Madagascar 84%.

🌐 Why Should We Care About Tropical Forests?

Biologists consider the plight of tropical forests to be one of the world's most serious environmental problems, primarily because these forests are home to 50–90% of the earth's terrestrial species, most of them still unknown and unnamed. Biologist Edward O. Wilson estimates that by 2022 at least 20% of tropical forest species could be gone, and as many as 50% by 2042, if current rates of tropical deforestation and degradation continue. If these estimates are correct, no extinction of this size has occurred for 65 million years.

Tropical forests touch the daily lives of everyone on the earth through the products and ecological services they provide. These forests supply half of the world's annual harvest of hardwood, hundreds of food products (including coffee, tea, cocoa, spices, nuts, chocolate, and tropical fruits), and materials such as natural latex rubber, resins, dyes, and essential oils that can be harvested sustainably. A 1988 study by a team of scientists showed that sustainable harvesting of such nonwood products as nuts, fruits, herbs, spices, oils, medicines, and latex rubber in Amazon rain forests over 50 years would generate twice as much revenue per hectare as timber production and three times as much as cutting them and converting them to grassland for cattle ranching.

The active ingredients for 25% of the world's prescription drugs are derived from plants, most of which grow in tropical rain forests. Commercial sales of drugs with active ingredients derived from tropical forests total an estimated $100 billion per year worldwide and $15.5 billion per year in the United States. Of the 3,000 plants identified by the National Cancer Institute (NCI) as sources of cancer-fighting chemicals, 70% come from tropical forests.

Most of the original strains of rice, wheat, and corn that supply more than half of the world's food were developed from wild tropical plants. Botanists believe that tens of thousands of plant strains with potential food value await discovery in tropical forests.

Despite this immense potential, fewer than 1% of the estimated 125,000 flowering plant species in tropical forests (and less than 3% of all the world's 240,000 such species) have been examined closely for possible use as human resources. Biologist Edward Wilson warns that destroying these forests and the species they contain for short-term economic gain is like burning down an ancient library before reading the books.

What Causes Tropical Deforestation?

Tropical deforestation results from a number of interconnected causes, all of which are related to population growth, poverty, and certain government policies that encourage deforestation (Figure 5-7). Population growth and poverty combine to drive subsistence farmers and the landless poor to tropical forests, where they try to grow enough food to survive.

Government subsidies can accelerate deforestation by making timber or other resources cheap relative to their full ecological value and by encouraging the poor to colonize tropical forests by giving them title to land they clear (as is done in Brazil, Indonesia, and Mexico). International lending agencies encourage developing countries to borrow huge sums of money from developed countries to finance projects such as roads, mines, logging operations, oil drilling, and dams in tropical forests. To stimulate economic development (and in some cases to pay the interest on loans from developed countries), these countries often sell off some of their timber and other natural resources, mostly to developed countries.

The process of degrading a tropical forest begins with a road (Figure 5-4), usually cut by logging companies. Once the forest becomes accessible, it can be cleared and increasingly degraded by a number of factors. One is unsustainable forms of small-scale farming. Hordes of poor people follow logging roads into the forest to build homes and plant crops on small cleared plots. Instead of practicing various methods of traditional and potentially sustainable shifting cultivation (Figure 5-8), these inexperienced *shifted cultivators* often practice unsustainable farming that depletes soils and destroys large tracts of forests.

According to environmental scientist Norman Myers, these forest newcomers are responsible for well over half of all tropical forest destruction. Driven by rapidly growing population and poverty, their numbers and impact are rising rapidly.

Cattle ranching also degrades tropical forests. Cattle ranches, often aided by government subsidies, are often established on cropland that has been exhausted by small-scale farmers. Some farmers simply abandon their plots to ranchers; others seed their plots with grass and then sell them to ranchers. When torrential rains and overgrazing turn the thin, nutrient-poor soils into eroded wastelands, ranchers move to another area and repeat the destructive process known as *shifting ranching*.

Bromeliad

Primary Causes:
Rapid population growth
Poverty
Exploitive government policies
Exports to developed countries
**Failure to include ecological services
in evaluating forest resources**

Toucan

Scarlet
macaw

Golden lion
marmoset

Orchid

Secondary Causes:
Roads
Logging
Unsustainable peasant farming
Cash crops
Cattle ranching
Tree plantations
Flooding from dams
Mining
Oil drilling

Blue morpho
butterfly

Figure 5-7 Major interconnected causes of the destruction and degradation of tropical forests, all of which are ultimately related to population growth, poverty, and government policies that encourage deforestation.

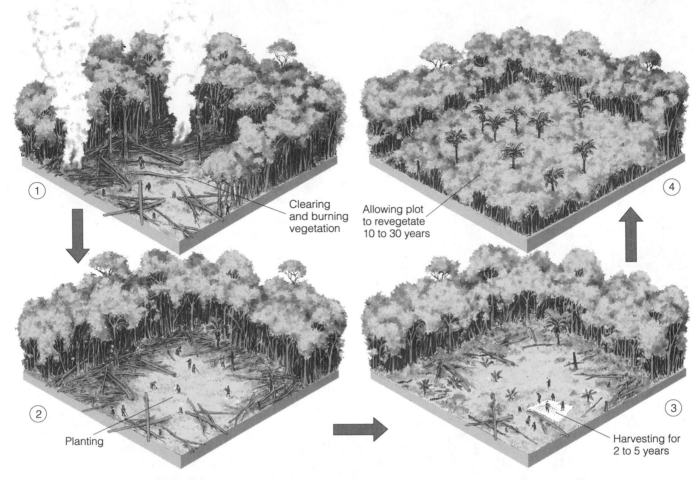

Figure 5-8 The first crop-growing technique may have been a combination of slash-and-burn and shifting cultivation in tropical forests. This method is sustainable only if small plots of the forest are cleared, cultivated for no more than 5 years, and then allowed to regenerate for 10–30 years to renew soil fertility. Indigenous cultures have developed many variations of this technique and have found ways to make some nondestructive uses of former plots while they are being regenerated.

Clearing large areas of tropical forest for raising cash crops severely degrades these reservoirs of biodiversity. Immense plantations grow crops such as sugarcane, bananas, pineapples, peppers, strawberries, cotton, tea, and coffee, mostly for export to developed countries. Mining and oil drilling also degrade tropical forests and dams built on rivers flood large areas.

As forests become opened up by roads, logging, agricultural plots, and cattle ranches, they become drier and more susceptible to fires. This phenomenon became apparent when huge areas of forests burned in Indonesia, Brazil, Guatemala, Nicaragua, and Mexico in 1997 and 1998. Most of these fires were caused by farmers using fire to prepare fields for planting or cattle-grazing. Tens of millions of people were sickened by the highly polluted air and hundreds died; the resulting damage and loss of

timber from the fires amounted to billions of dollars. In coming years more fires are expected in Latin America, Africa, and Asia, destroying large areas of forest, reducing biodiversity and ecological integrity, and adding to greenhouse gases. Some conservationists warn that mostly because of human actions coupled with prolonged droughts (also made more likely by large-scale land clearing) we may be entering an *age of fire*.

Commercial logging also degrades tropical forests. Since 1950 the consumption of tropical timber has risen 15-fold. Japan alone accounts for 53% of the world's tropical timber imports, followed by Europe (32%) and the United States (15%). Currently, almost three-fourths of exported tropical timber comes from Southeast Asia. As tropical timber in Asia is depleted in the 1990s, cutting is shifting to Latin America and Africa.

Q: How long will the world's proven reserves of coal last at current consumption rates?

Although timber exports to developed countries contribute significantly to tropical forest depletion and degradation, over 80% of the trees cut in developing countries are used at home. Loggers typically take only the best large and medium trees, but they also topple up to 17 other trees for each one they remove. The falling trees damage many others, with up to 70% of them eventually dying from their injuries.

Logging is directly responsible for only a small portion of tropical deforestation and degradation when compared to small-scale agriculture. However, the construction of logging and other resource access roads opens up these forests to small-scale farmers, ranchers, and miners.

Solutions: How Can We Reduce Tropical Deforestation and Degradation?
A number of analysts have suggested ways to protect tropical forests, to use them more sustainably, and to use and restore degraded areas of tropical forests. An important first step is to make a detailed survey to determine how much of the world is covered with tropical (and other) forests, how much has been deforested or degraded, and where. This could be done by a combination of remote-sensing satellites and ground-level evaluations, at a cost roughly equivalent to what the world spends for military purposes *every 3 minutes*. Conservation biologists also urge us to identify and move rapidly to protect areas of tropical forests that are rich in unique species and in imminent danger—so called *hot spots* or *critical ecosystems*.

Environmentalists urge governments to reduce the flow of the landless poor to tropical forests by slowing population growth and discouraging the poor from migrating to undisturbed tropical forests. They also call for establishing programs to help new settlers in tropical forests learn how to practice small-scale sustainable agriculture and forestry. The Lacandon Maya Indians of Chipas, Mexico, for example, have developed a multilayered system of agroforestry that allows them to cultivate up to 75 crop species on 1-hectare (2.5-acre) plots for up to 7 consecutive years. After that another plot must be planted to allow regeneration of the soil in the original plot.

Another suggestion is to use economic policies to protect and sustain tropical forests. One approach is to phase in *full-cost pricing* over the next two decades. As long as tropical forests and other forms of the earth capital are valued only for the resources that can be removed from them, not for the ecological services they provide, we will continue to destroy and degrade them.

Governments can also phase out subsidies that encourage unsustainable deforestation and phase in tariffs, use taxes, user fees, and subsidies that favor sustainable forestry and protection of biodiversity. Another economic approach is to pressure banks and international lending agencies (controlled by developed countries) not to lend money for environmentally destructive projects, especially road building and dams, involving old-growth tropical forests.

Debt-for-nature swaps and conservation easements can be used to encourage countries to protect tropical forests. In a *debt-for-nature swap*, participating countries act as custodians for protected forest reserves in return for foreign aid or debt relief; in *conservation easements*, a private organization, country, or group of countries compensates other countries for protecting selected forest areas. Currently less than 5% of the world's tropical forests are part of parks and preserves, and many of these are protected in name only.

In 1998, tiny Suriname (a former Dutch colony on South America's northern Caribbean coast with a population of 400,000) announced that it was setting aside 12% of its land as a nature reserve instead of turning it over to Asian loggers. Suriname's rain forest is larger than all those in Central America combined. Conservation International, which brokered the decision, hopes it will be a model for other developing countries.

Tropical timber-cutting regulations and practices can be reformed. New logging contracts could charge more for timber-cutting concessions and require companies to post adequate bonds for reforestation and restoration. However, such restrictions are hard to enact and enforce because of the economic and political power of timber companies, whose executives often see bribes and fines as a very minor business cost.

Gentler methods could be used for harvesting trees. For example, cutting canopy vines (lianas) before felling a tree can reduce damage to neighboring trees by 20–40%, and using the least obstructed paths to remove the logs can halve the damage to other trees.

Pressure for clearing old-growth tropical forests can be reduced by concentrating peasant farming, tree and crop plantations, and ranching activities on already-cleared tropical forest areas that are in various stages of secondary ecological succession. Conservationists can also work with local people to protect forests (Spotlight, p.158). In addition, global efforts can be mounted to reforest and rehabilitate degraded tropical forests and watersheds.

Finally, lowering the waste and overconsumption of forest products by consumers would play a major role in reducing the pressures on the world's forests. For example, 20% of the world's people living in the United States, Japan, and Western Europe consume more than half of the world's industrial timber and 70% of the paper. Reducing consumption and waste in these countries by even a small fraction would greatly reduce pressures on the world's forests.

A: About 220 years (65 years if use increases by 2% a year)

How Farmers and Loud Monkeys Saved a Forest

It's early morning in a tropical forest in the Central American country Belize. Suddenly, loud roars that trail off into wheezing moans—territorial calls of black howler monkeys—wake up everyone in or near the wildlife sanctuary by the Belize River. These long-tailed primates, who live in small troops headed by a dominant male, travel slowly among the treetops, feeding on leaves, flowers, and fruits.

This species is the centerpiece of an experiment recruiting peasant farmers to preserve nearby tropical forests and wildlife. The project is the brainchild of an American biologist, Robert Horwich, who in 1985 met with villagers and suggested that they establish a sanctuary that would benefit the local black howlers and themselves.

He proposed that the farmers leave thin strips of forest along the edges of their fields to provide the howlers with food and a travel route through the sanctuary's patchwork of active garden plots and young and mature forest. Leaving strips of forest along the river, he noted, would also reduce soil erosion and river silting, yielding more fish for the villagers.

To date, more than 100 farmers have participated, and the 47-square-kilometer (18-square-mile) sanctuary is now home for an estimated 1,100 black howlers. The idea has spread to seven other villages.

Now, as many as 6,000 ecotourists visit the sanctuary each year to catch glimpses of its loud monkeys and other wildlife. Villagers serve as tour guides, cook meals for the visitors, and lodge tourists overnight in their spare rooms. As long as ecotourism doesn't get so big that it disturbs the sanctuary ecosystem, this experiment can demonstrate the benefits of integrating ecology and economics, allowing local villagers to make money by helping sustain the forest and its wildlife.

Critical Thinking

How could projects such as this be used in the country where you live?

Case Study: Debt-for-Nature Swap in Bolivia

In 1984, biologist Thomas Lovejoy suggested that debtor nations should be rewarded for protecting their natural resources. In such *debt-for-nature swaps*, Lovejoy proposed that a certain amount of foreign debt be canceled in exchange for spending a certain sum on better natural resource management.

Typically, a conservation organization buys a specified amount of a country's debt from a bank at a discounted rate and negotiates the swap. A government or private agency must then agree to enact and supervise the conservation program.

In 1987, Conservation International, a private U.S. banking consortium, purchased $650,000 of Bolivia's $5.7-billion national debt from a Citibank affiliate in Bolivia for $100,000. In exchange for not having to pay back this part of its debt, the Bolivian government agreed to expand and protect from harmful forms of development on 1.5 million hectares (3.7 million acres) of tropical forest around its existing Beni Biosphere Reserve in the Amazon Basin, which contains some of the world's largest reserves of mahogany and cedar. The government was to establish maximum legal protection for the reserve and create a $250,000 fund, with the interest to be used to manage the reserve.

The plan was intended to be a model of how conservation of forest and wildlife resources could be compatible with sustainable economic development. Central to the plan was a virgin tropical forest to be set aside as a biological reserve, surrounded by a protective buffer of savanna used for sustainable grazing of livestock (Figure 5-9). Controlled commercial logging, as well as hunting and fishing by local inhabitants, would be permitted in some parts of the forest but not in the area above the tract, which was to be set aside to protect the area's watershed and prevent erosion.

In 1998, 11 years after the agreement was signed, the Bolivian government still had not provided legal protection for the reserve. Meanwhile, timber companies (with government approval) have cut thousands of mahogany trees from the area; most of this lumber has been exported to the United States. The area's 5,000 native inhabitants were not consulted about the swap plan, even though they are involved in a land ownership dispute with logging companies.

One lesson from this first debt-for-nature swap is that *legislative and budget requirements must be met before the swap is executed.* Another is that such *swaps must be carefully monitored by environmental organizations* to ensure that proposals for sustainable development are not disguises for unsustainable development.

Solutions: Forest Protection and Restoration in Costa Rica

Costa Rica, smaller in area than West Virginia, was once almost completely covered with tropical forests. Between 1963 and 1983, politically powerful ranching families cleared much of the country's forests to graze cattle, and they exported most of the beef to the United States and western Europe. By 1983, only 17% of the country's original tropical forest remained, and soil erosion was rampant.

Q: How long will proven reserves of coal in the United States last at current consumption rates?

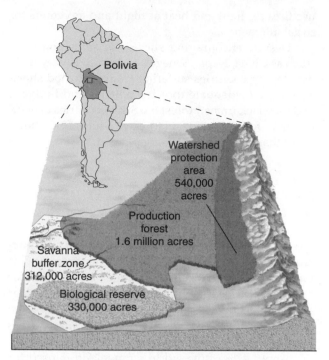

Figure 5-9 Blending economic development and conservation in a 1.5-million-hectare (3.7-million-acre) tract in Bolivia. A U.S. conservation organization arranged a debt-for-nature swap to help protect this land from destructive development.

Despite widespread forest loss, tiny Costa Rica is a superpower of biodiversity, with an estimated 500,000 species of plants and animals. A single park in Costa Rica is home for more bird species than all of North America.

In the mid-1970s, Costa Rica established a system of national parks and conservation reserves that by 1997 included 12% of its land (6% of it in reserves for indigenous peoples), compared to only 1.8% set aside as reserves in the lower 48 United States. As a result, Costa Rica now has a larger proportion of land devoted to biodiversity conservation than any other country. The government has also established a 15% national fuel tax, with part of the income used for reforestation.

This biodiversity conservation strategy has paid off. Today, revenue from tourism (almost two-thirds of it from ecotourists) is the country's largest source of outside income. However, legal and illegal deforestation threatens this plan because of the country's population growth and poverty, which still affects 10% of its people. If Costa Rica's attempts to protect and use biodiversity in sustainable ways are successful, it could be a model for the entire world (Individuals Matter, right).

How Serious Is the Fuelwood Crisis in Developing Countries? In 1998 about 2.2 billion people in 63 developing countries either could not get enough fuelwood to meet their basic needs or were forced to meet

Biocultural Restoration in Costa Rica

Costa Rica is the site of one of the world's largest and most innovative ecological restoration projects. In the lowlands of the country's Guanacaste National Park, a small tropical seasonal forest is being restored and relinked to the rain forest on adjacent mountain slopes.

Daniel Janzen, professor of biology at the University of Pennsylvania and a leader in the growing field of rehabilitation and restoration of degraded ecosystems, has helped galvanize international support and has raised more than $10 million for this restoration project.

Janzen's vision is to make the nearly 40,000 people who live near the park an essential part of the restoration of 740 square kilometers (285 square miles) of degraded forest—a concept he calls *biocultural restoration*. By actively participating in the project, local residents will reap enormous educational, economic, and environmental benefits. Local farmers have been hired to sow large areas with tree seeds and to plant seedlings started in Janzen's lab.

Students in grade schools, high schools, and universities study the ecology of the park in the classroom and then go on field trips to the park itself. Educational programs for civic groups and tourists from Costa Rica and elsewhere will stimulate the local economy.

The project will also serve as a training ground in tropical forest restoration for scientists from all over the world. Research scientists working on the project give guest classroom lectures and lead some of the field trips.

Janzen recognizes that in 20–40 years today's children will be running the park and the local political system. If they understand the importance of their local environment, they are more likely to protect and sustain its biological resources. He believes that education, awareness, and involvement—not guards and fences—are the best ways to protect ecosystems from unsustainable use.

Critical Thinking

What three things would you do to encourage such projects throughout the world?

their needs by using wood faster than it was being replenished. The UN Food and Agriculture Organization projects that by the end of this century this already serious shortage of fuelwood will affect 2.7 billion people in 77 developing countries. As burning wood (or charcoal derived from wood) to boil water becomes an unaffordable luxury, waterborne infectious diseases and death will spread.

Besides local deforestation and accelerated soil erosion, fuelwood scarcity has other harmful effects. It places an additional burden on the rural poor, especially women and children, who often must walk long distances searching for firewood. Buying fuelwood or charcoal can take 40% of a poor family's meager income.

An estimated 800 million poor people who can't get enough fuelwood burn dried animal dung and crop residues for cooking and heating. These natural fertilizers thus never reach the soil, cropland productivity is reduced, the land is degraded still further, and hunger and malnutrition increase.

Solutions: What Can Be Done About the Fuelwood Crisis? Developing countries can reduce the severity of the fuelwood crisis by planting more fast-growing fuelwood trees or shrubs, burning wood more efficiently, and switching to other fuels. However, fast-growing tree and shrub species used to establish fuelwood plantations must be selected carefully to prevent harm to local ecosystems.

Experience shows that planting projects are most successful when local people participate in their planning and implementation. Programs work best that give village farmers incentives, such as ownership of the land or ownership of any trees they grow on village land. Emphasis should be placed on community woodlots, which are easy to tend and harvest, rather than on creating large fuelwood plantations located far from where the wood is needed.

Another promising method is to encourage villagers to use the sun-dried roots of various gourds and squashes as cooking fuel. These rootfuel plants, which regenerate themselves each year, produce large quantities of burnable biomass per unit area on dry deforested lands. They also help reduce soil erosion and produce an edible seed with a high protein content.

The traditional three-stone fire is a very inefficient way to burn wood, typically wasting about 94% of the wood's energy content. New, cheap, more efficient, less polluting stoves can provide both heat and light while reducing indoor air pollution, a major health threat. However, such stoves won't be widely used if they don't take into account local needs, resources, and cultural practices. The stoves must be easy and cheap to build, and the materials used to make them must be easily accessible locally.

Cheap and easily made solar ovens (Figure 4-16d) that capture sunlight can also be used to reduce wood use and air pollution in sunny and warm areas; however, they are often not accepted by cultures that also use fires for light and heat at night and as centers for social interaction.

Despite encouraging success in some countries (such as China, Nepal, Senegal, and South Korea), most developing countries suffering from fuelwood shortages have inadequate forestry policies and budgets, and they lack trained foresters. Such countries are cutting trees for fuelwood and forest products 10–20 times faster than new trees are being planted.

5-6 MANAGING AND SUSTAINING RANGELANDS

What Is Rangeland, and Why Is It Important? Almost half of the earth's ice-free land is rangeland: land that supplies forage or vegetation for grazing (grass-eating) and browsing (shrub-eating) animals and that is not intensively managed. Most rangelands are grasslands in arid and semiarid areas too dry for nonirrigated crops.

About 42% of the world's rangeland is used for grazing livestock; much of the rest is too dry, cold, or remote from population centers to be grazed by large numbers of livestock. About 34% of the total U.S. land area is rangeland, most of it short-grass prairie in the arid and semiarid western half of the country. About 2% of the cattle and 10% of the sheep in the United States graze on public rangelands.

Most rangeland grasses have deep and complex root systems that help anchor the plants. Blades of rangeland grass grow from the base, not the tip. Thus, as long as only its upper half is eaten and its lower half remains, rangeland grass is a renewable resource that can be grazed again and again.

Rangeland has a number of important ecological functions. It provides forage for large numbers of wild herbivores and essential habitats for a variety of wildlife and plant species. Rangelands also act as crucial watersheds that help replenish surface and groundwater resources by absorbing and slowly releasing rainfall.

What Is Overgrazing? Overgrazing occurs when too many animals graze for too long and exceed the carrying capacity of a grassland area. It lowers the productivity of vegetation and changes the number and types of plants in an area. Large populations of wild ruminants can overgraze rangeland during prolonged dry periods, but most overgrazing is caused by excessive numbers of domestic livestock feeding for too long in a particular area.

Heavy overgrazing compacts the soil, which diminishes its capacity to hold water and regenerate itself. Overgrazing also converts continuous grass cover into

Q: How many people in the United States die prematurely each year from coal burning?

patches of grass and thus exposes the soil to erosion, especially by wind; then woody shrubs such as mesquite and prickly pear cactus take over. Overgrazing is the major cause of desertification in arid and semiarid lands (Figure 5-10 and p. 162).

What Is the Condition of the World's Rangelands? *Range condition* is usually classified as excellent (containing more than 75% of its potential forage production), good (51–75%), fair (26–50%), and poor (0–25%). No comprehensive survey of rangeland conditions has been done except in North America. However, data from surveys in various countries indicate that most of the world's rangelands have been degraded to some degree, mostly by desertification (Figure 5-10).

In 1990, 68% of nonarctic U.S. public rangeland was rated by the BLM and the General Accounting Office as being in unsatisfactory (fair or poor) condition, compared with 84% in 1936. On the other hand, since 1936 U.S. public range area in excellent or good condition

has doubled while the area rated as poor has shrunk by half. This is a considerable improvement, but there's still a long way to go.

What Are Riparian Zones, and Why Are They Important? According to some conservation biologists and rangeland experts, overall estimates of rangeland condition obscure severe damage to certain heavily grazed areas, especially vital **riparian zones**: thin strips of lush vegetation along streams. These zones help prevent floods (and help keep streams from drying out during droughts) by storing and releasing water slowly from spring runoff and summer storms. These centers of biodiversity also provide habitats, food, water, and shade for wildlife in the arid and semiarid western lands. Studies indicate that 65–75% of the wildlife in the western United States is totally dependent on riparian habitats.

Because cattle need lots of water, they tend to congregate near riparian zones and feed there until the grass and shrubs are gone. As a result, riparian vegetation

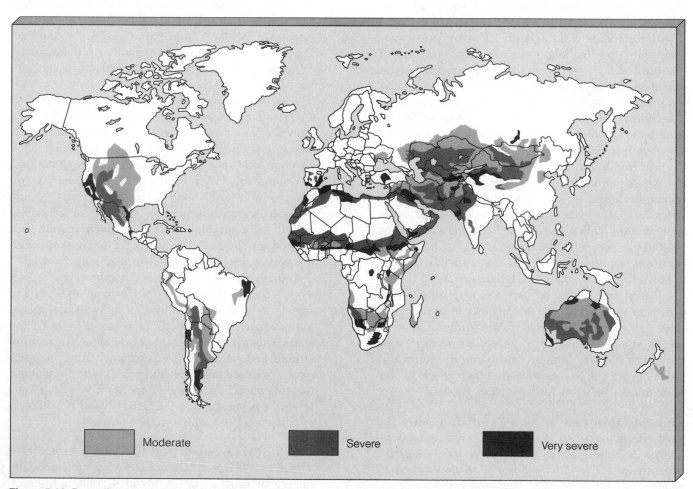

Figure 5-10 Desertification of arid and semiarid lands. (Data from UN Environmental Programme and Harold E. Dregnue)

A: 65,000 to 200,000

is destroyed by trampling and overgrazing. It takes only a few head of cattle to degrade a riparian zone.

Riparian areas can be restored by using fencing to restrict access to degraded areas and by developing off-stream watering sites for livestock. Sometimes protected areas can recover in a few years. Despite the threat to soil and wildlife, little has been done to protect or repair riparian zones on U.S. public land, mostly because of political opposition by ranchers and lack of awareness of the problem by the general public.

What Is Desertification, and How Serious Is This Problem? Desertification is a process whereby the productive potential of arid or semiarid land falls by 10% or more; this phenomenon results most-ly from human activities. *Moderate desertification* is a 10–25% drop in productivity, *severe desertification* is a 25–50% drop, and *very severe desertification* is a drop of 50% or more, usually creating huge gullies and sand dunes. Desertification is a serious and growing problem in many parts of the world (Figure 5-10).

Practices that leave topsoil vulnerable to desertification include (1) overgrazing on fragile arid and semi-arid rangelands, (2) deforestation without reforestation, (3) surface mining without land reclamation, (4) irrigation techniques that lead to increased erosion, (5) salt buildup and waterlogged soil, (6) farming on land with unsuitable terrain or soils, and (7) soil compaction by farm machinery and cattle hoofs. The consequences of desertification include worsening drought, famine, declining living standards, and swelling numbers of environmental refugees whose land is too eroded to grow crops or feed livestock.

An estimated 8.1 million square kilometers (3.1 million square miles)—an area the size of Brazil and 12 times the size of Texas—have become desertified in the past 50 years. This threatens the livelihoods of at least 900 million people in 100 countries. If current trends continue, within a few years desertification could threaten the livelihoods of 1.2 billion people.

Every year, low to moderate new desertification occurs on an estimated 60,000 square kilometers (23,000 square miles, an area the size of West Virginia); another 210,000 square kilometers (81,000 square miles, an area the size of Kansas) undergo severe desertification and lose so much soil and fertility that they are no longer economically valuable for farming or grazing.

Solutions: How Can We Slow Desertification? The most effective way to slow desertification is to drastically reduce overgrazing, deforestation, and the destructive forms of planting, irrigation, and mining that are to blame. In addition, planting trees and grasses will anchor soil and hold water while slowing desertification and reducing the threat of global warming (Section 9-2).

Such prevention and rehabilitation would cost $10–22 billion annually for the next 20 years. This expenditure is considerably less than the estimated $42 billion annual loss in agricultural productivity from desertified land; once this potential productivity is restored, the cost of the program could be recouped quickly. So far, less than $1 billion per year is spent globally to halt this form of land degradation.

Solutions: How Can Rangelands Be Managed? The primary goal of rangeland management is to maximize livestock productivity without overgrazing rangeland vegetation. The most widely used method to prevent overgrazing is controlling the *stocking rate*—the number of each kind of animal placed in a given area—so that it doesn't exceed carrying capacity. But determining the carrying capacity of a range site is difficult and costly, and carrying capacity varies with factors such as climatic conditions (especially drought), past grazing use, soil type, invasions by new species, kinds of grazing animals, and intensity of grazing.

Both the numbers and distribution of livestock on a rangeland must be controlled to prevent overgrazing. Ranchers can control livestock distribution by fencing off damaged rangeland and riparian zones, rotating livestock from one grazing area to another, providing supplemental feed at selected sites, and situating water holes and salt blocks in strategic places.

A more expensive and less widely used method of rangeland management involves suppressing the growth of unwanted plants (mostly invaders) by herbicide spraying, mechanical removal, or controlled burning. A cheaper and less harmful way to discourage unwanted vegetation is controlled, short-term trampling by large numbers of livestock.

Seeding and applying fertilizer can increase growth of desirable vegetation. Even though this method usually costs too much, reseeding is an excellent way to restore severely degraded rangeland.

Another aspect of rangeland management for livestock is control of predators such as wolves, bears, and coyotes. For decades hundreds of thousands of predators have been shot, trapped, and poisoned in the United States by ranchers, farmers, and federal predator control officials. Federally subsidized predator control programs have so reduced gray wolf and grizzly bear populations that they are now endangered species. Today additional funds are being spent by the U.S. Fish and Wildlife Service to protect and replenish their numbers.

Case Study: Livestock and U.S. Public Rangeland About 29,000 U.S. ranchers hold what are essentially lifetime permits to graze about 4 million livestock (3 million of them cattle) on BLM and Forest Service rangelands in 16 western states. About 10% of these

Q: What would happen if coal's harmful effects were included in its market price and government subsidies were removed?

permits are held by small livestock operators; the other 90% belong to ranchers with large livestock operations, including four U.S. billionaires, and to corporations such as Metropolitan Life Insurance, Union Oil, Getty Oil, and the Vail Ski Corporation.

Permit holders pay the federal government a grazing fee for this privilege. Since 1981, grazing fees on public rangeland have been set by Congress at only one-fourth to one-eighth the going rate for comparable private land. This means that taxpayers give the roughly 1% of U.S. ranchers with federal grazing permits subsidies amounting to $60–200 million a year, depending on how the costs of federal rangeland management are calculated.

The economic value of a permit is a part of the overall worth of a ranch. It can serve as collateral for a loan and is usually renewed automatically every 10 years, allowing permit holders to treat federal land as part of their ranches. It's not surprising that politically influential permit holders have fought so hard to block any change in this system.

The public subsidy does not end with low grazing fees, however. When other costs such as water pipelines, stock ponds, weed control, livestock predator control, clearing of undesirable vegetation, planting of grass for livestock, erosion, and loss of biodiversity are added, it's estimated that taxpayers are providing 29,000 ranchers with an annual subsidy of about $2 billion (an average of $69,000 per rancher) to produce only 2% of the country's beef. In 1991 one rancher alone received subsidies worth almost $950,000, which explains why many critics charge that the public lands grazing programs are little more than rancher welfare, mostly for the well-to-do ranchers who hold 90% of the permits.

However, ranchers with permits say that grazing fees on public lands should be low because most private rangeland is more productive than public rangeland. Environmentalists contend that this varies with the land involved and that even when this is true, current grazing fees on public rangelands are much too low.

Economic studies show that federal expenditures on grazing programs exceed not only grazing fee receipts but also the profits ranchers make from grazing their livestock on public lands. Thus, some analysts suggest that the government could save money by paying ranchers with permits not to graze their livestock on public lands.

Solutions: How Can U.S. Public Rangeland Be Managed More Sustainably?

Some environmentalists believe that all commercial grazing of livestock on western public lands should be phased out over the next 10–15 years. They contend that the water-poor

The Eco-Rancher

INDIVIDUALS MATTER

Wyoming rancher Jack Turnell is one of a new breed of cowpuncher who gets along with environmentalists. He talks about riparian ecology and biodiversity as fluently as he talks about cattle: "I guess I have learned how to bridge the gap between the environmentalists, the bureaucracies, and the ranching industry."

Turnell grazes cattle on his 32,000-hectare (80,000-acre) ranch south of Cody, Wyoming, and on 16,000 hectares (40,000 acres) of Forest Service land on which he has grazing rights. For the first decade after he took over the ranch, he raised cows the conventional way. Since then, he's made some changes.

Turnell disagrees with the proposals by environmentalists to raise grazing fees and remove sheep and cattle from public rangeland. He believes that if ranchers are kicked off the public range, ranches like his will be sold to developers and chopped up into vacation sites and homes (ranchettes), irreversibly destroying the range for wildlife and livestock alike.

At the same time, he believes that ranches can be operated in more ecologically sustainable ways. To demonstrate this, Turnell began systematically rotating his cows away from the riparian areas, gave up most uses of fertilizers and pesticides, and crossed his Hereford and Angus cows with a French breed that tends to congregate less around water. Most of his ranching decisions are made in consultation with range and wildlife scientists, and changes in range condition are carefully monitored with photographs.

The results have been impressive. Riparian areas on the ranch and Forest Service land are lined with willows and other plant life, providing lush habitat for an expanding population of wildlife (including pronghorn antelope, deer, moose, elk, bear, and mountain lions). And this *eco-rancher* now makes more money because the higher-quality grass puts more meat on his cattle.

western range is not a very good place to raise cattle and sheep, which require a lot of water.

However, some ranchers have demonstrated that western rangeland can be grazed sustainably; they make the case that such practices should not be prohibited on public rangeland (Individuals Matter, p. above). Some environmentalists agree that with proper management, ranching on western rangeland is a potentially sustainable operation and that encouraging sustainable ranching practices keeps the land from being developed and destroyed. They believe that the following measures can promote more sustainable use of public rangeland:

- *Allow no or only limited grazing on riparian areas.*

- *Ban grazing on rangeland in poor condition.*

- *Use competitive bidding for all grazing permits.* The existing noncompetitive system gives ranchers with essentially lifetime permits an unfair economic advantage over ranchers who can't get such permits, which amount to government subsidies.

- *Allow individuals and environmental groups to purchase grazing permits and not use the land for grazing.*

- *Until a competitive bidding system is implemented, raise grazing fees to fair market value.*

- *Give ranchers with small holdings grazing fee discounts to reduce the financial pressure for them to sell their ranch lands to developers.*

- *Abolish rancher-dominated grazing advisory boards.*

The problem is that despite their small numbers, western ranchers with grazing permits wield enough political power to see that measures that promote sustainability are not enacted by elected officials. There are also efforts in the U.S. Congress to turn management of public rangelands over almost completely to ranchers and exempt them from most environmental laws and any legal challenges.

5-7 MANAGING AND SUSTAINING NATIONAL PARKS

How Popular Is the Idea of National Parks? Today, over 1,100 national parks larger than 1,000 hectares (2,500 acres) each are located in more than 120 countries, together covering an area equal to that of Alaska, Texas, and California combined. This important achievement in global conservation was spurred by the creation of the first national park system in the United States in 1912.

The U.S. National Park System is dominated by 54 national parks, most of them in the West (Figure 5-1). These repositories of majestic beauty and biodiversity, sometimes called America's crown jewels, are supplemented by state, county, and city parks. Most state parks are located near urban areas and thus are more heavily used than national parks. Nature walks, guided tours, and other educational services offered by U.S. Park Service employees have given many visitors a better appreciation for and understanding of nature, but recently these services have diminished because of budget cutbacks.

How Are Parks Being Threatened? Parks everywhere are under siege. In developing countries parks are often invaded by local people who desperately need wood, cropland, and other resources. Poachers kill animals to obtain and sell rhino horns, elephant tusks, and furs. Park services in developing countries typically have too little money and too few personnel to fight these invasions, either by force or by education. In addition, most of the world's national parks are too small to sustain many of the larger animal species.

Popularity is one of the biggest problems of national and state parks in the United States (and other developed countries). Because of increased numbers of roads, cars, and affluent people, annual recreational visits to major national parks increased fourfold (and visits to state parks sevenfold) between 1950 and 1997; visits are projected to increase from 271 million in 1997 to 500 million by 2010.

During the summer, the most popular U.S. national and state parks are often choked with cars and trailers and hour-long entrance backups. They are also plagued by noise, traffic jams, litter, vandalism, poaching, deteriorating trails, polluted water, rundown visitor centers, garbage piles, and crime. Many visitors to heavily used parks leave the city to commune with nature, only to find the parks as noisy, congested, and stressful as where they came from.

U.S. Park Service rangers now spend an increasing amount of their time on law enforcement instead of resource conservation, management, and education. Since 1976, the number of federal park rangers has not changed while the number of visitors to parks has risen by 85 million; the number of visitors is expected to rise by another 162 million in the next 20 years. Currently, there is one ranger for every 84,200 visitors to the major national parks. Because overworked rangers earn an average salary of less than $28,000, many leave for better-paying jobs.

Wolves, bears, and other large predators in and near various parks have all but vanished because of excessive hunting, poisoning by ranchers and federal officials, and the limited size of most parks. As a result, populations of species they once controlled have exploded, destroying vegetation and crowding out other species. An attempt is being made to reintroduce the gray wolf into the Yellowstone National Park area (Case Study, right).

Nonnative species have moved into or been introduced into many parks. Wild boars (imported to North Carolina in 1912 for hunting) are threatening vegetation in part of the Great Smoky Mountains National Park. The Brazilian pepper tree has invaded Florida's Everglades National Park. Mountain goats in Washington's Olympic National Park trample native vegetation and accelerate soil erosion. At the same time that some species have moved into parks, other native species

Q: How many new nuclear power plants have been ordered in the United States since 1978?

CASE STUDY

At one time the gray wolf ranged over most of North America, but between 1850 and 1900 some 2 million wolves were systematically shot, trapped, and poisoned by ranchers, hunters, and government employees. The idea was to make the West and the Great Plains safe for livestock and for big game animals prized by hunters.

This strategy worked. The species is now listed as endangered in all 48 lower states except Minnesota (which has an estimated 2,000 wolves) where it is listed as threatened.

Ecologists now recognize the important role these predators once played in parts of the West and the Great Plains by culling herds of bison, elk, caribou, and mule deer of weaker animals, thereby strengthening the genetic pool of the survivors. In recent years, these herds have proliferated, devastating some of the area's vegetation, increasing erosion, and threatening the niches of other wildlife species.

In 1987, the U.S. Fish and Wildlife Service proposed that gray wolves be reintroduced into the Yellowstone ecosystem. This proposal to reestablish ecological connections that humans had eliminated brought outraged protests.

Some objections came from ranchers who feared the wolves would attack their cattle and sheep; one enraged rancher said that the idea was "like reintroducing smallpox." Other protests came from hunters who feared that the wolves would kill too many big game animals and from miners and loggers who worried that the government would force them to cease operations on wolf-populated federal lands.

National Park Service officials promised to trap or shoot any wolves that killed livestock outside designated areas and to reimburse ranchers from a private fund established by Defenders of Wildlife for any livestock verified to have been killed by wolves. However, these promises fell on deaf ears, and many ranchers and hunters worked hard to unsuccessfully delay or defeat the plan in Congress.

Since 1995 federal wildlife officials have been catching about 30–40 gray wolves per year in Canada and relocating them in Yellowstone National Park and northern Idaho. The goal is to allow the wolf populations to increase to sustainable levels by 2002. By 1998 there were almost 100 gray wolves in the Yellowstone ecosystem.

Some ranchers and hunters say that either way they'll take care of the wolves quietly—what they call the "shoot, shovel, and shut up" solution. Meanwhile, there are continuing efforts in Congress to kill the program or its funding.

In 1997 the recovery program received a serious setback when a Wyoming federal appeals court judge found the wolf recovery program illegal. He said that the U.S. Fish and Wildlife Service had violated the law by declaring the wolves an "experimental population." That designation denies them the protection of the Endangered Species Act. The judge ordered that the Canadian wolves reintroduced into Yellowstone National Park and central Idaho be removed. The case has been appealed by several wildlife conservation organizations.

Critical Thinking

Do you agree or disagree with the program to reestablish populations of the gray wolf in the Yellowstone ecosystem? Explain. What are the major drawbacks of this program? Could the money be better spent on other wildlife programs? If so, what programs would you suggest?

---

of animals and plants (including many threatened or endangered species) are being killed or removed illegally in almost half of U.S. national parks.

Some wildlife biologists point to another internal danger to the ecological health of U.S. national parks: the population explosion of some native grazing animals whose numbers are no longer controlled by natural predators or culling (killing) of a certain number of individuals each year by park service officials. For example, studies indicate that the National Park Service's hands-off management policy of *natural regulation* has been the key factor in allowing the elk population in Yellowstone National Park to increase from 3,100 in 1968 to about 20,000 in 1995. Research has shown that this large elk population is unraveling the park's ecosystem as stands of aspens and willows die, woody shrubs disappear, populations of once abundant species from beavers to birds dwindle or disappear, and stream banks erode.

Under mounting criticism, the National Park Service has hinted that it may reexamine its long-standing natural regulation policy. However, this may not be politically feasible because polls indicate that the general public (mostly unaware of the ecological decline of the park) favors natural regulation of the park over the park's old policy of allowing hunters to cull (kill) thousands of elk each winter.

Nearby human activities outside many national parks also threaten the ecological health of such parks.

A: None, and all 120 plants ordered since 1973 have been canceled

Wildlife and recreational values are threatened by mining, logging, livestock grazing, coal-burning power plants, water diversion, and urban development. Polluted air drifts hundreds of kilometers, killing ancient trees in California's Sequoia National Park and blurring the awesome vistas at Arizona's Grand Canyon. According to the National Park Service, air pollution affects scenic views in national parks more than 90% of the time.

That's not all. Mountains of trash wash ashore daily at Padre Island National Seashore in Texas. Water use in Las Vegas threatens to shut down geysers in the Death Valley National Monument. Unless a massive ecological restoration project is successful, Florida's Everglades National Park may dry up.

Solutions: How Can Park Management in Developing Countries Be Improved? Some park managers, especially in developing countries, are developing integrated management plans that combine conservation practices with sustainable development of resources in the park and surrounding areas. In such plans, the inner core and vulnerable areas of each park are to be protected from development and treated as wilderness. Restricted numbers of people are allowed to use these areas for hiking, nature study, ecological research, and other nondestructive recreational and educational activities.

In buffer areas surrounding the core, controlled commercial logging, sustainable grazing by livestock, and sustainable hunting and fishing by local people are allowed. Money spent by park visitors adds to local income. By involving nearby residents in developing park management and restoration plans, managers and conservationists seek to help them see the park as a vital resource to protect and sustain rather than ruin (Individuals Matter, p. 163).

Integrated park management plans look good on paper, but they cannot be carried out without adequate funding and the support of nearby landowners and users. Moreover, in some cases the protected inner core may be too small to sustain the park's larger animal species.

Solutions: How Can Management of U.S. National Parks Be Improved? The 54 major U.S. national parks are managed under the principle of natural regulation, as if they were wilderness ecosystems that would sustain themselves if left alone. Many ecologists consider this a misguided policy. Most parks are far too small to even come close to sustaining themselves. Even the biggest ones, such as Yellowstone, cannot be isolated from the harmful effects caused by activities in nearby areas and from degradation from within by exploding populations of some plant-eating species (such as elk).

The U.S. National Park Service has two goals that increasingly conflict: to preserve nature in parks and to make nature more available to the public. The Park Service must accomplish these goals with a $1.5-billion annual budget and a $6-billion backlog of maintenance, repairs, and high-priority construction projects at a time when park usage and external threats to the parks are increasing. Private developers continually pressure Congress and the Park Service to turn national parks into high-tech, heavily developed theme or amusement parks instead of the repositories of natural ecosystems and centers for learning about nature they were intended to be.

In 1988, the Wilderness Society and the National Parks and Conservation Association made a number of suggestions for sustaining and expanding the national park system, including the following:

- *Have all user and entrance fees for national park units be used for the management, upkeep, and repair of the national parks.* Currently, such fees go into the general treasury.

- *Require integrated management plans for parks and other nearby federal lands.*

- *Increase the budget for adding new parkland near the most threatened parks.*

- *Increase the budget for buying private lands inside parks.*

- *Locate all new and some existing commercial facilities and visitor parking areas outside parks.*

- *Require private concessionaires who provide campgrounds, restaurants, hotels, and other services for park visitors to compete for contracts, raise their fees to 22% of their gross (not net) receipts, and pour receipts back into parks.* In 1996, private concessionaires in national parks paid only 6.7% of their gross receipts in franchise fees to the government. Many large concessionaires have long-term contracts by which they pay the government as little as 0.75% of their gross receipts.

- *Allow concessionaires to lease but not own facilities inside parks.*

- *Provide more funds for park system maintenance and repairs.*

- *Raise fees for park visitors and pour receipts back into parks.*

- *Restrict the number of visitors to crowded park areas.*

- *Increase the number and pay of park rangers.*

- *Survey condition and types of wildlife species in parks.*

- *Encourage volunteers to give tours and lectures to visitors.*

- *Encourage individuals and corporations to donate money for park maintenance and repair.*

Q: What percentage of current U.S. nuclear power generation are expected to be retired by 2015?

5-8 PROTECTING AND MANAGING WILDERNESS AND OTHER BIODIVERSITY SANCTUARIES

How Much of the Ecosphere Should We Protect from Exploitation? Most wildlife and conservation biologists believe that the best way to preserve biodiversity and ecological integrity—or wildness—is through a worldwide network of reserves, parks, wildlife sanctuaries, wilderness, and other protected areas. According to Aldo Leopold, "a species must be saved *in many places* if it is to be saved at all."

Currently, about 6% of the world's land area is either strictly or partially protected in more than 8,600 nature reserves, parks, wildlife refuges, and other areas around the globe. This is an important beginning, but conservation biologists say that to keep biodiversity and ecological integrity from being depleted, a minimum of 10–12% of the globe's land area should be protected. Moreover, many existing reserves are too small to provide any real protection for the wild species that live on them. Many existing preserves also receive little protection.

In 1981 the UN Educational, Scientific, and Cultural Organization (UNESCO) proposed that at least one (and ideally five or more) *biosphere reserves* be set up in each of the earth's 193 biogeographical zones. Each reserve should be large enough to prevent gradual species loss and should combine conservation with sustainable use of natural resources. To date, more than 300 biosphere reserves have been established in 76 countries.

A well-designed biosphere reserve has three zones (Figure 5-11): **(1)** a *core area* containing an important ecosystem that has had little or no disturbance from human activities, **(2)** a *buffer zone* where activities and uses are managed in ways that help protect the core, and **(3)** a second buffer or *transition zone*, which combines conservation and sustainable forestry, grazing, agriculture, and recreation. Buffer zones can also be used for education and research.

So far most biosphere reserves fall short of the ideal (Figure 5-11) and too little funding has been provided for their protection and management. An international fund to help developing countries protect and manage bioregions and biosphere reserves would cost $100 million per year—about what the world's nations spend on arms every 90 minutes.

Because there won't be enough money to protect most of the world's biodiversity, environmentalists believe that efforts should be focused on the most biodiverse countries (Figure 5-12) and threatened species-rich areas within such countries. In the United States, conservation biologists urge Congress to pass an Endangered Ecosystems or Biodiversity Protection Act

as an important step toward surveying and preserving the country's biodiversity.

However, most economists, developers, and resource extractors disagree with protecting even the current 6% of the earth's remaining undisturbed ecosystems. They contend that such areas contain valuable resources that would add to economic growth. They view protected areas as "useless" and "empty" because they are not being used to provide mineral and other resources and land for development for humans.

Why Preserve Wilderness? Protecting biodiversity and ecological integrity means protecting *wildness*. One way to do this is to protect undeveloped lands from exploitation by setting them aside as wilderness. According to the U.S. Wilderness Act of 1964, **wilderness** consists of areas "where the earth and its community of life are untrammeled by man, where man himself is a visitor who does not remain." President Theodore Roosevelt summarized what we should do with wilderness: "Leave it as it is. You cannot improve it."

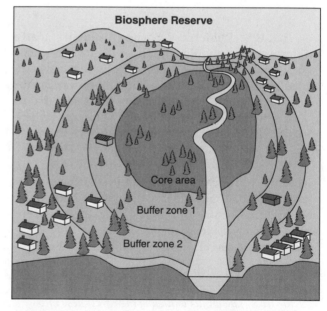

Human settlements   Research station

Tourism and education center

Figure 5-11 Design of a model biosphere reserve. In traditional parks and wildlife reserves, well-defined boundaries keep people out and wildlife in. By contrast, biosphere reserves recognize people's needs for access to sustainable use of various resources in parts of the reserve.

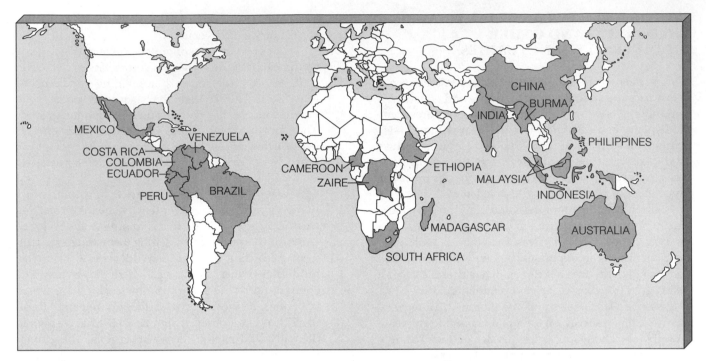

Figure 5-12 The earth's most biodiverse countries. Conservation biologists believe that efforts to preserve repositories of biodiversity as protected wilderness areas should be concentrated in these species-rich countries. (Data from Conservation International and World Wildlife Fund)

The U.S. Wilderness Society estimates that a wilderness area should contain at least 4,000 square kilometers (1,500 square miles); otherwise it can be affected by air, water, and noise pollution from nearby human activities. Conservation biologists urge that remaining wilderness be protected by law everywhere, focusing first on the most endangered spots in wilderness- and species-rich countries (Figure 5-12).

According to wilderness supporters, we need wild places where we can experience the beauty of nature and observe natural biological diversity, where we can enhance our mental and physical health by getting away from noise, stress, and large numbers of people. Wilderness preservationist John Muir advised us to

Climb the mountains and get their good tidings. Nature's peace will flow into you as the sunshine into the trees. The winds will blow their freshness into you, and the storms their energy, while cares will drop off like autumn leaves.

Even those who never use the wilderness may want to know it is there, a feeling expressed by novelist Wallace Stegner:

Save a piece of country . . . and it does not matter in the slightest that only a few people every year will go into it. This is precisely its value. . . . We simply need

that wild country available to us, even if we never do more than drive to its edge and look in. For it can be a means of reassuring ourselves of our sanity as creatures, a part of the geography of hope.

Recently some critics have argued that protecting wilderness for its scenic and recreational value for a small number of people is an outmoded concept that keeps some areas of the planet from being economically useful to humans. They also contend that wilderness protection diverts money, energy, and talent from learning how to better manage urban and other landscapes disturbed by human activities.

Scientists argue that this idea is outmoded because it fails to take into account the research that has revealed the ecological importance of wilderness areas for all people and all life. To most biologists, *the most important reason for protecting wilderness and other areas from exploitation and degradation is to preserve the biodiversity and ecological integrity they contribute as a vital part of earth capital* (Figure 1-2).

Wilderness areas provide mostly undisturbed habitats for wild plants and animals, protect diverse biomes from damage, and provide a laboratory in which we can discover more about how nature works. In other words, wilderness is a biodiversity and wildness bank and an ecoinsurance policy.

Q: How many people in the United States die prematurely each year from U.S. nuclear power plants?

Conservation biologists and other environmental scientists agree that we need much more research and emphasis on learning how to manage and restore land disturbed by human activities. However, they don't believe that this should be done by decreasing or unraveling wilderness protection, which will lead to even more areas being disturbed by human activities.

To protect more of the earth's biodiversity and ecological integrity, conservation biologists and other environmental scientists believe the designation of future wilderness (and other protected) areas should focus on prairies, lowland forests, and wetlands that are largely absent from the current system. This will bring wilderness advocates into increasing conflict with developers, farmers, ranchers, and resource extractors who view remaining lands in these categories as prime sites for human use.

Some analysts believe that wilderness should be preserved because the wild species it contains have a right to exist (or struggle to exist) and to play their roles in the earth's ongoing saga of biological evolution and ecological processes, without human interference.

Case Study: The U.S. National Wilderness Preservation System In the United States, preservationists have been trying to save wild areas from development since 1900. On the whole, they have fought a losing battle. Not until 1964 did Congress pass the Wilderness Act, which allowed the government to protect undeveloped tracts of public land from development as part of the National Wilderness Preservation System.

Only about 4% of U.S. land area is protected as wilderness, with almost three-fourths of it in Alaska. Only 1.8% of the land area of the lower 48 states is protected, most of it in the West.

Of the 413 wilderness areas in the lower 48 states, only 4 are larger than 4,000 square kilometers. Furthermore, the present wilderness preservation system includes only 81 of the country's 233 distinct ecosystems. Like the national parks, most wilderness areas in the lower 48 states are threatened habitat islands in a sea of development.

There remain almost 400,000 square kilometers (150,000 square miles) of scattered blocks of public lands that could qualify for designation as wilderness. Conservation biologists who would like to see all this land protected as wilderness also advocate rehabilitating other lands to enlarge existing wilderness areas. Such efforts are strongly opposed by timber, mining, ranching, energy, and other interests who want either to extract resources from these and most other public lands or convert them to private ownership.

Some conservation biologists also urge that *wilderness recovery areas* be created by closing and obliterat-ing nonessential roads in large areas of public lands, restoring habitats, allowing natural fires to burn, and reintroducing species that have been driven from such areas. However, resource developers lobby elected officials and government agencies to build roads into national forests (and other areas being evaluated for inclusion in the wilderness system) so that they can't be designated as wilderness; they also strongly oppose the idea of wilderness recovery areas.

How Can Wilderness Be Protected? To protect the most popular areas from damage, wilderness managers must designate sites where camping is allowed and limit the number of people using these sites at any one time. Managers have also increased the number of wilderness rangers to patrol vulnerable areas, and they have enlisted volunteers to pick up trash discarded by thoughtless users.

Environmental historian and wilderness expert Roderick Nash suggests that wilderness areas be divided into three categories. The easily accessible, popular areas would be intensively managed and have trails, bridges, hiker's huts, outhouses, assigned campsites, and extensive ranger patrols. Large, remote wilderness areas would be used only by people who get a permit by demonstrating their wilderness skills. The third category, biologically unique areas, would be left undisturbed as gene pools of plant and animal species, with no human entry allowed.

Why Is It Difficult to Protect Marine Biodiversity? One reason it is difficult to protect marine biodiversity is that shore-hugging species are adversely affected by coastal development and the accompanying massive inputs of sediment and other wastes from land into coastal waters. This poses a severe threat to biologically diverse and highly productive coastal ecosystems such as coral reefs, marshes, and mangrove swamps (Section 2-7).

Protecting marine biodiversity is also difficult because much of the damage is not visible to most people. In addition, the seas are viewed by many as an inexhaustible resource, capable of absorbing an almost infinite amount of waste and pollution. Finally, most of the world's ocean area lies outside the legal jurisdiction of any country and is thus an open-access resource, subject to overexploitation because of the tragedy of the commons.

Protecting marine biodiversity requires countries to enact and enforce tough regulations to protect coral reefs, mangrove swamps, and other coastal ecosystems from unsustainable use. Furthermore, much more effective international agreements are needed to protect biodiversity in the open seas.

In this chapter we have seen that the task of protecting the earth's biodiversity and ecological integrity and restoring lands we have damaged is enormous. However, conservation biologists believe that it can be done if enough of us become involved in this vital earth-sustaining activity.

We abuse land because we regard it as a commodity belonging to us. When we see land as a community to which we belong, we may begin to use it with love and respect.

ALDO LEOPOLD

CRITICAL THINKING

1. Explain why you agree or disagree with the four principles given on p. 142, which are advocated by most environmentalists and many free-market economists as guidelines for the use of public land in the United States.

2. What difference could the loss of essentially all the remaining old-growth tropical forests and old-growth forests in North America have on your life and on the lives of your descendants?

3. Explain why you agree or disagree with each of the proposals given on p.149 for sustainable forestry.

4. Explain why you agree or disagree with each of the proposals listed on p. 157 for protection of the world's tropical forests.

5. In the name of progress, the policies of various governments are virtually eliminating the cultures of remaining hunter–gatherers and other indigenous peoples by taking over their lands for growing crops, cutting timber, grazing cattle, mining, building hydroelectric dams and reservoirs, and encouraging other forms of economic development. Many observers argue that the world's remaining hunter–gatherer societies should be given title to the land on which they and their ancestors have lived for centuries and should be left alone by modern civilization. We have created protected reserves for endangered wild species, so why not create reserves for these endangered human cultures? What do you think? Explain.

6. Explain why you agree or disagree with the proposals given on p.164 for making more sustainable use of public rangeland in the United States.

7. Should cattle be banned from grazing on public rangeland? Explain. If not, what new restrictions (if any) would you place on ranchers allowed to graze livestock on U.S. public lands?

8. Should trail bikes, dune buggies, and other off-road vehicles be banned from public rangeland in the United States to reduce damage to vegetation and soil? Explain. Should such a ban also include national forests, national wildlife refuges, and national parks? Explain.

9. Explain why you agree or disagree with each of the proposals given on p. 166 concerning the U.S. national park system.

10. Should more wilderness areas and wild and scenic rivers be preserved in the United States, especially in the lower 48 states? Explain. Should this be done by reclassifying a large portion of U.S. Forest Service and Bureau of Land Management land as wilderness, closing it off for many uses, and allowing it to undergo natural restoration as wildlife habitat? Explain.

6 BIODIVERSITY: SUSTAINING WILD SPECIES

The last word in ignorance is the person who says of an animal or plant: "What good is it?" . . . If the land mechanism as a whole is good, then every part of it is good, whether we understand it or not. . . . Harmony with land is like harmony with a friend; you cannot cherish his right hand and chop off his left.

ALDO LEOPOLD

6-1 WHY PRESERVE WILD SPECIES?

Case Study: The Passenger Pigeon In the early 1800s, bird expert Alexander Wilson watched a single migrating flock of passenger pigeons darken the sky for over 4 hours. He estimated that this flock was more than 2 billion birds strong, 386 kilometers (240 miles) long, and 1.6 kilometers (1 mile) wide.

By 1914 the passenger pigeon had disappeared forever. How could a species that was once the most common bird in North America become extinct in only a few decades?

The answer is humans. The main reasons for the extinction of this species were uncontrolled commercial hunting and loss of the bird's habitat and food supply as forests were cleared to make room for farms and cities.

Passenger pigeons were good to eat, their feathers made good pillows, and their bones were widely used for fertilizer. They were easy to kill because they flew in gigantic flocks and nested in long, narrow colonies.

Commercial hunters would capture one pigeon alive, sew its eyes shut, and tie it to a perch called a stool. Soon a curious flock would land beside this "stool pigeon," and the birds would then be shot or ensnared by nets that might trap more than 1,000 birds at once.

Beginning in 1858, passenger pigeon hunting became a big business. Shotguns, traps, artillery, and even dynamite were used. Birds were sometimes suffocated by burning grass or sulfur below their roosts. Live birds were even used as targets in shooting galleries. In 1878 one professional pigeon trapper made $60,000 by killing 3 million birds at their nesting grounds near Petoskey, Michigan.

By the early 1880s commercial hunting had ceased because only a few thousand birds were left. At that point, recovery of the species was doomed because the females laid only one egg per nest. On March 24, 1900, a young boy in Ohio shot the last known wild passenger pigeon. The last passenger pigeon on earth, a hen named Martha after Martha Washington, died in the Cincinnati Zoo in 1914. Her stuffed body is now on view at the National Museum of Natural History in Washington, D.C.

Eventually, all species become extinct or evolve into new species, but humans have become the primary factor in the premature extinction of more and more species. Conservation biologists estimate that every day at least 10 (and perhaps as many as 200) species become extinct because of human activities, and studies indicate that the rate of this loss of biodiversity is increasing rapidly.

If all species eventually become extinct, why should we worry about losing a few more because of our activities? Does it matter that the passenger pigeon, the great auk, or some unknown plant or insect in a tropical forest becomes prematurely extinct because of human activities (Figure 6-1)?

Biologists contend that the answer is yes because of the economic, medical, scientific, ecological, aesthetic, and recreational value of all species. Some environmentalists go further and contend that each species has an inherent right to play its role in the ongoing evolution of life on earth until it becomes extinct without interference by humans.

What Is the Economic and Medical Importance of Wild Species? Some 90% of today's food crops were domesticated from wild tropical plants. Moreover, agricultural scientists and genetic engineers need existing wild plant species to derive today's crop strains and to develop the new crop strains of tomorrow.

Wild plants and plants domesticated from wild species supply rubber, oils, dyes, fiber, paper, lumber, and other useful products. Nitrogen-fixing microbes in the soil and in plants' root nodules supply nitrogen to grow food crops worth almost $50 billion per year worldwide ($7 billion in the United States alone). Pollination by birds and insects is essential to many food crops, including 40 U.S. crops valued at approximately $30 billion per year.

Figure 6-1 Some species that have become extinct largely because of human activities, mostly habitat destruction and overhunting.

Passenger pigeon Great auk Dodo Dusky seaside sparrow Aepyornis (Madagascar)

About 80% of the world's population relies on plants or plant extracts for medicines. At least 40% of all medicines, drugs, and pharmaceuticals, worth $100 billion per year, owe their existence to the genetic resources of wild plants, mostly from tropical developing countries. Plant-derived anticancer drugs save an estimated 30,000 lives per year in the United States. Over 3,000 antibiotics, including penicillin and tetracycline, are derived from microorganisms. Only about 5,000 of the 250,000 known plant species have been studied thoroughly for their possible medical uses.

What Is the Scientific and Ecological Importance of Wild Species? Every species can help scientists understand how life has evolved and functions and how it will continue to evolve on this planet. Wild species also provide many of the ecological services that make up earth capital (Figure 1-2) and thus are key factors in sustaining the earth's biodiversity and ecological integrity.

They supply us (and other species) with food, recycle nutrients essential to agriculture, and help generate and maintain soils. They also produce oxygen and other gases in the atmosphere, absorb pollution, moderate the earth's climate, help regulate local climates and water supplies, reduce erosion and flooding, and store solar energy. Moreover, they detoxify poisonous substances, break down organic wastes, control potential crop pests and disease carriers, and make up a vast gene pool for future evolutionary processes.

What Is the Aesthetic and Recreational Importance of Wildlife? Wild plants and animals are a source of beauty, wonder, joy, and recreational pleasure for many people. Wildlife tourism, sometimes called *ecotourism*, is the fastest growing segment of the global travel industry and generates an estimated $30 billion in revenues each year. Wildlife biologist Michael Soulé estimates that one male lion living to age 7 generates

$515,000 in tourist dollars in Kenya; by contrast, if killed for its skin the lion would bring only about $1,000. Similarly, over a lifetime of 60 years, a Kenyan elephant is worth close to $1 million in ecotourist revenue. Florida's coral reefs are worth an estimated $1.6 billion a year in tourism revenue. However, care must be taken to ensure that ecotourists do not damage or disturb wildlife and ecosystems and disrupt local cultures.

Why Is It Ethically Important to Preserve Wild Species? Some people believe that each wild species has an inherent right to exist, or to struggle to exist. According to this view, it is wrong for us to hasten the extinction of any species, and we have an ethical responsibility to protect species from becoming prematurely extinct as a result of human activities. Biologist Edward O. Wilson believes most people feel obligated to protect other species and the earth's biodiversity because of the natural affinity for nature built into our genes (Spotlight, right).

Some people distinguish between the survival rights of plants and those of animals, mostly for what they consider practical reasons. Poet Alan Watts once said that he was a vegetarian "because cows scream louder than carrots."

Other people distinguish among various types of species. They might think little about killing a fly, mosquito, cockroach (Spotlight, p. 35), or rat or ridding the world of disease-causing bacteria. Unless they are strict vegetarians, they might also see no harm in having others kill domesticated animals in slaughterhouses to provide them with meat, leather, and other products. However, these same people might deplore the killing of wild animals such as deer, squirrels, or rabbits.

Some proponents go further and assert that each individual organism, not just each species, has a right to survive without human interference, just as each human being has the right to survive. Others emphasize

Q: What percentage of current U.S. nuclear power generation will be retired by 2015?

the importance of preserving the whole spectrum of biodiversity by protecting entire ecosystems rather than individual species or organisms, as discussed in Chapter 5. This view is based on the principle that humans have an ethical obligation to prevent premature extinction of wildlife by saving their habitats and not disrupting the complex ecological interactions that sustain all life.

6-2 THE RISE AND FALL OF SPECIES

How Does Background Extinction Differ from Mass Extinction? Extinction is a natural process and eventually all species become extinct. David Raup and several other evolutionary biologists estimate that more than 99.9% of all the species that have ever existed are now extinct because of a combination of background and mass extinctions.

Each year, a small number of species become extinct naturally at a low rate, a phenomenon called the *natural, or background, rate of extinction*. In contrast, *mass extinction* is an abrupt rise in extinction rates above the background level. It is a catastrophic, widespread (often global) event in which large groups of existing species (perhaps 25–70%) are wiped out. Most mass extinctions are believed to result from one or a combination of global climate changes that kill many species and leave behind those able to adapt to the new conditions.

Fossils and geological evidence indicate that the earth's species have experienced five great mass extinctions (20–60 million years apart) during the past 500 million years in which large numbers of species became extinct each year for tens of thousands to millions of years (Figure 2-43). Evidence also shows that these mass extinctions were followed by other periods, called *adaptive radiations*, when the diversity of life increased and spread for 10 million years or more (Figure 2-43). The last mass extinction took place about 65 million years ago, when the dinosaurs became extinct, for reasons that are hotly debated, after thriving for 140 million years.

Why Do Conservation Biologists Believe There Is a New Mass Extinction Crisis? Imagine that you have built a two-story house using wood as the basic structural material and that termites are slowly destroying various parts of the structure. How long will it be before they destroy enough of the structure for parts of the house to collapse?

Conservation biologists believe that this urgent question, when applied to the earth, is one that we as a species should be asking ourselves. As we tinker with the only home for us and other species, we are rapidly removing parts of the earth's natural biodiversity and ecological integrity on which we and other species depend in ways we know little about. We are not heeding

Biophilia

SPOTLIGHT

Biologist Edward O. Wilson contends that because of the billions of years of biological connections leading to the evolution of the human species (Figure 2-41), we have an inherent affinity for the natural world— a phenomenon he calls *biophilia* (love of life). He points out that we cannot erase this evolutionary imprint in our genes because of a few generations of urban living.

Evidence of this natural affinity for life is seen in the preference people have for almost any natural scene over one from an urban environment. Given a choice, people prefer to live in an area where they can see water, grassland, or a forest. More people visit zoos and aquariums than attend all professional sporting events combined.

In the 1970s I was touring the space center at Cape Canaveral in Florida. During our bus ride the tour guide was pointing out each of the abandoned multimillion-dollar launch sites and giving a brief history of each launch. Most of us were utterly bored. All of a sudden people started rushing to the front of the bus and staring out the window with great excitement. What they were looking at was a baby alligator—a dramatic example of how *biophilia* can triumph over *technophilia*.

Critical Thinking

Do you have a built-in affinity for wildlife and wild ecosystems (biophilia)? If so, how do you display this inherent love of wildlife in your daily actions? What aspects of your pattern of consumption help destroy and degrade wildlife?

Aldo Leopold's warning: "To keep every cog and wheel is the first precaution of intelligent tinkering."

It's difficult to document extinctions, so most go unrecorded. However, some conservation biologists estimate that currently 18,000–73,000 species become extinct each year (on average, 50–200 species per day)— thousands of times the estimated natural background extinction rate of 3–30 species per year. Even if this estimate of the erosion of biodiversity is 100 times too high, the current extinction rate is many times the estimated rate of background extinction (Spotlight, p. 174).

Scientists expect this rate of extinction to accelerate as the human population grows and takes over more of the planet's wildlife habitats and net primary productivity. They warn that within the next few decades we could easily lose at least 1 million of earth's species, most of

Some social scientists and a few biologists question the existence of an extinction crisis caused by our activities. They point to several problems in estimating species loss. First, we don't know how many species there are; estimates range between 5 million and 100 million.

Second, it is difficult to observe species extinction, especially for species we know little or nothing about. A species is generally considered extinct when it has not been seen for at least 50 years or when the last of a few closely monitored individuals dies. This means that species extinctions recorded today really pertain to the 1940s rather than the 1990s.

As a result, biologists have to use models and field data to make estimates of current and future extinction rates. Using such approaches, the annual loss of tropical forest habitat is estimated at about 1.8% per year. Edward O. Wilson and several tropical biologists who have counted species in patches of tropical forest before and after destruction or degradation estimate that this 1.8% loss in habitat results in roughly a 0.5% loss of species. However, biomes vary in the number of species they contain (species diversity), and the ratio of habitat loss to species loss varies in different biomes.

Do such estimates add up to an extinction crisis? Let us assume, as Wilson and many other biologists do, that a loss of 1 million species over several decades represents an extinction crisis. If we assume the global decline in species to be 0.5% per year, then we will lose 25,000 species per year if there are 5 million species, 100,000 per year if there are 20 million species, and 500,000 per year if there are 100 million species. If these assumptions are correct, we will lose 1 million species in 40 years if there are 5 million species, in 10 years with 20 million species, and in only 2 years with 100 million species.

Let's assume, however, that the estimate of 0.5% species loss per year is too high for the earth as a whole. If it is 0.25% per year, then we will lose 1 million species in 80 years with 5 million species, in 20 years with 20 million species, and in 4 years with 100 million species. Even if we halve the estimated species loss again, to 0.125% per year, we can still lose 1 million species within 8–160 years, enough to easily qualify the situation as an extinction crisis.

According to a 1998 survey, 70% of the biologists polled believe that we are in the midst of a mass extinction and that this loss of species will pose a major threat to human existence in the next century. Biologists don't contend that their estimates of extinction rates are precise enough to make a firm prediction. Instead, they argue that there is ample evidence that we are destroying and degrading wildlife habitats at an increasing rate and that our actions certainly lead to a significant loss of species, even though the number and rate vary in different parts of the world.

Critical Thinking

Do you believe that we are in the midst of a sixth mass extinction caused mostly by human activities? Explain. If so, list five ways in which you contribute to this loss of biodiversity.

---

them in tropical regions. According to biodiversity expert Edward O. Wilson, "Clearly, we are in the midst of one of the great extinction spasms of geological history."

Mass extinctions occurred long before humans evolved (Figure 2-43), but there are three important differences between the current mass extinction most biologists believe we are bringing about and those of the past. First, *the current extinction crisis is the first to be caused by a single species: our own.* By using or wasting approximately 40% of the earth's terrestrial net primary productivity (25% of the world's total net primary productivity), we are crowding out other terrestrial species. What will happen to wildlife and the services they provide for humans and other species if our population doubles in the next 90 years?

Second, *the current mass wildlife extinction is taking place in only a few decades, rather than over thousands to millions of years.* Such rapid extinction cannot be balanced by speciation because it takes 2,000–100,000 generations for new species to evolve. Fossil and other evidence related to past mass extinctions indicates that it takes millions of years to recover biodiversity through adaptive radiations (Figure 2-43). Thus, repercussions for humans and other species from the current human-caused mass extinction will affect the future course of evolution for 5–10 million years and could lead to the premature extinction of the human species.

Third, *besides killing off species, we are eliminating or degrading many biologically diverse environments such as tropical forests, tropical coral reefs, wetlands, and estuaries that in the past have served as evolutionary centers for the recovery of biodiversity after a mass extinction.* However, a few analysts question whether there is an extinction crisis (Spotlight, above).

Even without extinctions, ecosystems may lose their ability to support many forms of life because of the

Q: What countries have the most sites contaminated with radioactive materials?

Table 6-1 Characteristics of Extinction-Prone Species

Characteristic	Examples
Low reproduction rate	Blue whale, polar bear, California condor, Andean condor, passenger pigeon, giant panda, whooping crane
Specialized feeding habits	Everglades kite (eats apple snail of southern Florida), blue whale (krill in polar upwellings), black-footed ferret (prairie dogs and pocket gophers), giant panda (bamboo), koala (certain eucalyptus leaves)
Feed at high trophic levels	Bengal tiger, bald eagle, Andean condor, timber wolf
Large size	Bengal tiger, lion, elephant, Javan rhinoceros, American bison, giant panda, grizzly bear
Limited or specialized nesting or breeding areas	Kirtland's warbler (6- to 15-year-old jack pine trees), whooping crane (marshes), orangutan (only on Sumatra and Borneo), green sea turtle (lays eggs on only a few beaches), bald eagle (prefers forested shorelines), nightingale wren (only on Barro Colorado Island, Panama)
Found in only one place or region	Woodland caribou, elephant seal, Cooke's kokio, many unique island species
Fixed migratory patterns	Blue whale, Kirtland's warbler, Bachman's warbler, whooping crane
Preys on livestock or people	Timber wolf (livestock), some crocodiles (people)
Behavioral patterns	Passenger pigeon and white-crowned pigeon (nest in large colonies), redheaded woodpecker (flies in front of cars), Carolina parakeet (when one bird is shot, rest of flock hovers over body), key deer (forages for cigarette butts along highways—it's a "nicotine addict")

disappearance of local populations of key organisms. According to a 1997 study by biologist Jennifer Hughes and her colleagues at Stanford University, every year an estimated 16 million local plant and animal populations—an average of 1,800 populations every hour—disappear from the world's tropical forests.

🐾 What Are Endangered and Threatened Species? Biologists distinguish among three levels of extinction: **(1)** *Local extinction* occurs when a species is no longer found in an area it once inhabited but is still found elsewhere in the world, **(2)** *ecological extinction* occurs when there are so few members of a species left that it can no longer play its ecological roles in the biological communities where it is found, and **(3)** *biological extinction* occurs when a species is no longer found anywhere on the earth. Biological extinction is forever.

Species heading toward biological extinction are classified as either *endangered* or *threatened* (Figure 6-2). An **endangered species** has so few individual survivors that the species could soon become extinct over all or most of its natural range. Examples are the California condor in the United States (40 in the wild), the whooping crane in North America (about 288 left), the giant panda in central China (about 1,000 left), the snow leopard in central Asia (about 2,500 left), and the black rhinoceros in Africa (about 2,400 left).

A **threatened species** is still abundant in its natural range but is declining in numbers and is likely to become endangered. Examples are the grizzly bear, the southern sea otter, and the American alligator. Endangered and threatened species are ecological smoke alarms.

Some species have characteristics that make them more vulnerable than others to premature extinction (Table 6-1). In general, species more vulnerable to extinction are found in a limited area and have a small population size, low population density, large body size, specialized niche, and low reproductive rate. Such species also tend to undergo fairly fixed migrations, feed atop long food chains or webs, and have high economic value to people.

According to a 1996 joint study by the International Union for the Conservation and Conservation International, more than 5,200 known animal species are at risk of extinction. They include 34% of the world's fish, 25% of amphibians, 25% of mammals, 20% of reptiles, and 11% of the bird species. The study also estimated that at least one of every eight plant species in the world—and nearly one of three in the United States—is under threat of extinction. According to biologists, such figures are probably underestimated because of lack of data from many areas, including many species-rich tropical countries (Figure 5-12), where areas are being rapidly cleared (Figure 5-2).

According to a 1995 study by the Nature Conservancy, about 32% of the 20,500 U.S. animal and plant species studied by scientists are vulnerable to premature extinction, mostly because of human activities. The animals that are most at risk are those that depend on aquatic ecosystems, such as freshwater fish and amphibians; a large number of flowering plants are also in trouble.

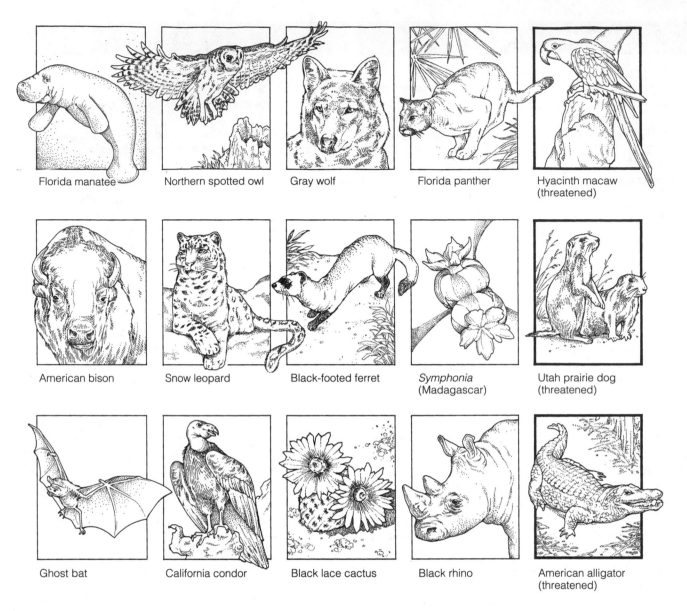

Figure 6-2 Species that are endangered or threatened largely because of human activities. Populations of some of the species are recovering, such that some endangered species may be relisted as threatened and some threatened species (such as the bald eagle and peregrine falcon) may be delisted.

The populations of many wild species that are not yet in danger of extinction have diminished locally or regionally to the point at which they are locally or ecologically extinct. Such species may be a better indicator of the condition of entire ecosystems than endangered and threatened species. These *indicator species* can serve as early warnings so that we can prevent species extinction rather than respond to emergencies, often with little chance of success.

🐾 **Case Study: Bats Are Getting a Bad Rap** Despite their variety (950 known species) and worldwide distribution, bats—the only mammals that can fly—have certain traits that place them at risk because of human activities. Bats reproduce slowly, and many bat species that live in huge colonies in caves and abandoned mines become vulnerable to destruction when people block the passageways or disturb their hibernation.

Because of unwarranted fears of bats and lack of knowledge about their vital ecological roles, several species have been driven to extinction. Currently, 26% of the world's bat species are listed as endangered or threatened.

Conservation biologists urge us not to kill bats but to protect them because of their important ecological roles. About 70% of all bat species feed on crop-

Q: Does using nuclear power add carbon dioxide to the atmosphere?

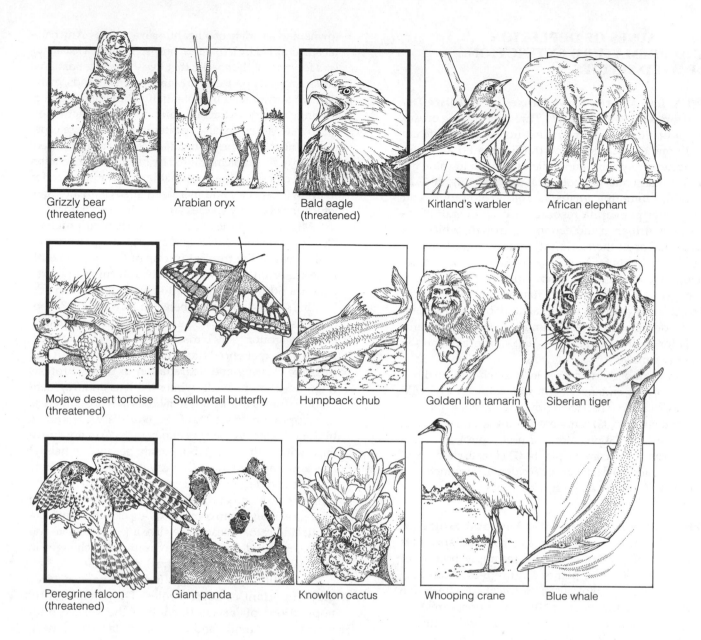

Grizzly bear
(threatened)

Arabian oryx

Bald eagle
(threatened)

Kirtland's warbler

African elephant

Mojave desert tortoise
(threatened)

Swallowtail butterfly

Humpback chub

Golden lion tamarin

Siberian tiger

Peregrine falcon
(threatened)

Giant panda

Knowlton cactus

Whooping crane

Blue whale

damaging nocturnal insects and other pest species such as mosquitoes, making them the primary control agents for such insects.

In some tropical forests and on many tropical islands, pollen-eating bats pollinate flowers; fruit-eating bats distribute plants throughout tropical forests by excreting undigested seeds. As keystone species, they are vital for maintaining plant biodiversity and for regenerating large areas of tropical forest cleared by human activities. If you enjoy bananas, cashews, dates, figs, avocados, or mangos, you can thank bats.

Many people mistakenly view bats as fearsome, filthy, aggressive, rabies-carrying bloodsuckers. But

most bat species are harmless to people, livestock, and crops. In the United States, only 10 people have died of bat-transmitted disease in four decades of recordkeeping; more Americans die each year from falling coconuts.

One way to protect bats from human disturbance and killing is to prevent human access to caves and mines that are major bat roosts and hibernation sites. *Bat gates*—vandalproof metal grids that allow bats to enter and leave but keep out people—are one option. Another is to educate people about the importance of bats to humans and other species. We need to see bats as valuable allies, not as enemies.

A: Yes, but it produces only one-sixth as much CO_2 per unit of electricity as a coal-burning plant

6-3 CAUSES OF DEPLETION AND PREMATURE EXTINCTION OF WILD SPECIES

✦ What Are the Root Causes of Wildlife Depletion and Extinction? Three underlying causes of population reduction and extinction of wildlife are (1) human population growth (Figures 1-1 and 1-6), (2) economic systems and policies that fail to value the environment and its ecological services (Figure 1-2), thereby promoting unsustainable exploitation, and (3) greater per capita resource use as a result of increasing affluence and economic growth, which is a prime factor in the exploitation and degradation of wildlife habitats for human uses (Figure 1-15). In developing countries, the combination of rapid population growth and poverty pushes the poor to cut forests, grow crops on marginal land, overgraze grasslands, deplete fish species, and kill endangered animals for their valuable furs, tusks, or other parts in order to survive.

These underlying causes lead to other more direct causes of the endangerment and extinction of wild species, such as (1) habitat loss and degradation, (2) habitat fragmentation, (3) commercial hunting and poaching, (4) overfishing, (5) predator and pest control, (6) sale of exotic pets and decorative plants, (7) climate change and pollution, and (8) deliberate or accidental introduction of nonnative (exotic or alien) species into ecosystems.

What Is the Role of Habitat Loss and Fragmentation? The greatest threat to most wild species is reduction of habitats as we increasingly occupy or degrade more of the planet (Figure 6-3). According to conservation biologists, tropical deforestation (Figure 5-2) is the greatest eliminator of species, followed by destruction of coral reefs and wetlands (Section 2-7), plowing of grasslands, and pollution of freshwater and marine habitats (Section 10-5).

In the lower 48 states of the United States, 98% of the tall-grass prairies have been plowed, half of the wetlands drained, and 85–95% of old-growth forests cut (Figure 5-6). Overall forest cover has been reduced by 33%. At least 500 native species have been driven to extinction, and others to near-extinction (Figure 6-3), mostly because of habitat loss and fragmentation.

Island species, many of them *endemic species* found nowhere else on earth, are especially vulnerable to extinction. Most have no other place to go and few have evolved defenses against predators or diseases accidentally or deliberately introduced onto islands.

In addition to habitat loss, an increasing amount of the planet's remaining wildlife habitat is being fragmented into vulnerable patches by roads, fences, fields, towns, and a variety of other human activities. Any habitat surrounded by a different one is, in effect, a *habitat island* for most of the species that live there. Most national parks and other protected areas are habitat islands, many of them surrounded by potentially damaging logging, mining, energy extraction, and industrial activities. Freshwater lakes are also habitat islands that are especially vulnerable when nonnative species are introduced. The two main problems caused by habitat fragmentation are a decrease in the sustainable population size for many wild species and increased surface area (edge), which makes many species vulnerable to predators.

Migrating species face a double habitat problem. Nearly half of the 700 U.S. bird species spend two-thirds of the year in the tropical forests of Central or South America or on Caribbean islands, returning to North America during the summer to breed. A U.S. Fish and Wildlife study showed that between 1978 and 1987, populations of 44 of the 62 surveyed species of insect-eating, migratory songbirds in North America declined; 20 species experienced drops of 25–45%. The main culprits are clearing and degradation of tropical forests (Figure 5-2) in the birds' winter habitats and fragmentation of their summer forest habitats in North America.

Approximately 69% of the world's 9,600 known bird species are declining in numbers (58%) or are threatened with extinction (11%), mostly because of habitat loss and fragmentation. Conservation biologists view this loss and decline of bird species as an early warning of the widespread and greater loss of biodiversity and ecological integrity to come. Birds are excellent environmental indicators because they live in every climate and biome, respond quickly to environmental changes in their habitats, and are easy to track and count.

In addition to serving as indicator species, birds play important ecological roles: They help control populations of insects (including the spruce budworm, gypsy moth, and tent caterpillar, which decimate many tree species) and rodents, pollinate a wide variety of flowering plants, and spread plants throughout their habitats by consuming plant seeds and excreting them in their droppings.

What Is the Role of Commercial Hunting and Poaching? The international trade in wild plants and animals is big business, bringing in up to $12 billion a year worldwide, with almost half of this trade involving the illegal sale of endangered and threatened species or their parts. Organized crime has moved into wildlife smuggling because of the huge profits involved (second only to drug smuggling). An estimated 60–80% of all live animals smuggled around the world die in transit.

Worldwide, some 622 species of animals and plants face extinction, mostly because of illegal trade. A live

Q: Nuclear weapons existing today could kill everyone in the world how many times over?

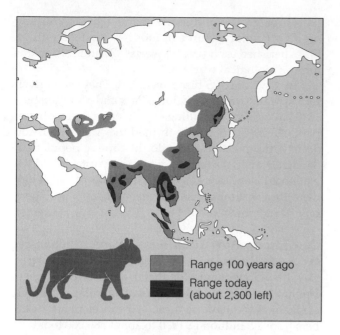

Indian Tiger

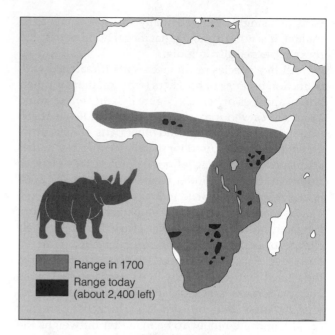

Black Rhino

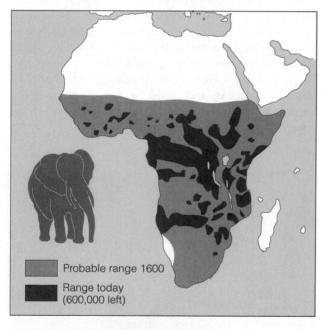

African Elephant

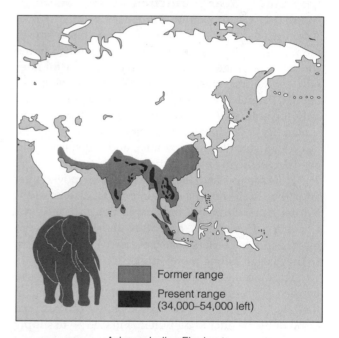

Asian or Indian Elephant

Figure 6-3 Reductions in the ranges of four wildlife species, mostly the result of a combination of habitat loss and hunting. What will happen to these and millions of other species when the global human population doubles in the next few decades? (Data from International Union for the Conservation of Nature and World Wildlife Fund)

mountain gorilla is worth $150,000, a live chimpanzee $50,000, and a live imperial Amazon macaw $30,000. Bengal tigers are at risk because a tiger fur sells for $100,000 in Tokyo. Rhinoceros horn sells for as much as $28,600 per kilogram.

Despite international protection, the world's total tiger population has dwindled to 4,600–8,000, mostly because of habitat loss and poaching for their furs and bones. Without emergency action, there may be few or no tigers left in the wild within 20 years. With the body

A: About 40 (20 if current arms reduction proposals are carried out)

parts of a single tiger potentially worth as much as $5 million, it is not surprising that illegal hunting has skyrocketed, especially in India.

All five species of rhinoceros are threatened with extinction because of poachers (who kill them for their horns) and loss of habitat (Figure 6-3). In Yemen, rhino horns are carved into ornate dagger handles that sell for $500–12,000. In China and other parts of Asia, powdered rhino horn is used for medicinal purposes and as an alleged aphrodisiac. Efforts to protect these species from extinction are under way, but only about 12,000 rhinos remain in the wild today.

As more species become endangered, the demand for them on the black market soars, hastening their chances of extinction. Poaching of endangered or threatened species (many for markets in Asia) is increasing in the United States, especially in western national parks and wilderness areas covered by only 200 federal wildlife protection officers. Most poachers are not caught, and the money to be made far outweighs the risk of fines and the much smaller risk of imprisonment.

Case Study: Near Extinction of the American Bison In 1500, before Europeans settled North America, 60–125 million North American bison grazed the plains, prairies, and woodlands over much of the continent. A single herd on the move might thunder past for hours. Several Native American tribes depended heavily on bison, and typically they killed only the animals they needed for food, clothing, and shelter. By 1906, however, the once-vast range of the bison had shrunk to a tiny area, and the species had been driven nearly to extinction (Figure 6-4).

How did this happen? First, as settlers moved west after the Civil War, the sustainable balance between Native Americans and bison was upset. The Sioux, Cheyenne, Comanches, and other plains tribes traded bison skins to settlers for steel knives and firearms, so they began killing more bison.

The most relentless slaughter, however, was caused by the new settlers. As railroads spread westward in the late 1860s, railroad companies hired professional bison hunters—including Buffalo Bill Cody—to supply construction crews with meat. Passengers also gunned down bison from train windows for sport, leaving the carcasses to rot. Commercial hunters shot millions of bison for their hides and tongues (considered a delicacy), leaving most of the meat to rot. "Bone pickers" collected the bleached bones that whitened the prairies and shipped them east to be ground up as fertilizer.

Farmers shot bison because they damaged crops, fences, telegraph poles, and sod houses. Ranchers killed them because they competed with cattle and sheep for grass. The U.S. Army killed bison as part of its campaign to subdue the plains tribes by killing off their primary source of food. At least 2.5 million bison perished each year between 1870 and 1875 in this form of biological warfare.

By 1892 only 85 bison were left. They were given refuge in Yellowstone National Park and protected by a 1893 law against the killing of wild animals in national parks. In 1905, 16 people formed the American Bison Society to protect and rebuild the captive population. Soon thereafter, the federal government established the National Bison Range near Missoula, Montana. Today there are an estimated 200,000 bison, about 97% of them on privately owned ranches.

Case Study: How Should We Protect Elephants from Extinction? For decades poachers have slaughtered elephants for their valuable ivory tusks. Habitat loss (Figure 6-3) and legal and illegal trade in elephant ivory have reduced African elephant numbers from 2.5 million in 1970 to about 580,000 today.

Since 1989 this decline has been slowed by an international ban on the sale of ivory from African elephants, but things are not quite that simple. Increases in elephant populations in areas where their habitat has shrunk has resulted in widespread destruction of vegetation by these animals. This in turn reduces the niches available for other wild species. Some analysts also argue that sustainable legal harvesting of elephants for their ivory, meat, and hides, with the profits going mostly to local people, will be more successful in the long run than banning all ivory sales.

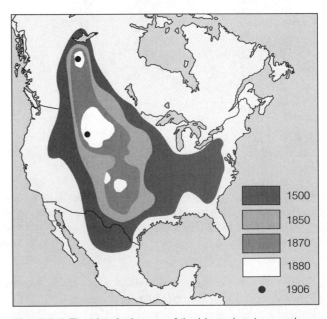

■	1500
▨	1850
▤	1870
☐	1880
●	1906

Figure 6-4 The historical range of the bison shrank severely between 1500 and 1906.

Q: Can switching to increased use of nuclear power in the United States save much oil?

In several southern African nations in which elephant populations were high, the governments allowed a certain number of elephants to be killed each year. Income from this sustainable harvesting of elephants encouraged local people to protect elephants from unsustainable and illegal poaching, and it also provided governments with funds to help pay for conservation of elephants and other wildlife species. With the ban on ivory trade, such income dried up.

To restore these important sources of income, some wildlife conservationists and leaders of several southern African nations have called for a partial lifting of the elephant ivory ban in areas in which the elephant populations are not endangered. Before being sold in the international marketplace, ivory from the sustainable culling of elephants in these areas would be marketed in a way that certifies that it was obtained legally; most of the profits would go to local people.

In most Western countries elephants are viewed as benign. However, according to David Western, head of the Kenya Wildlife Service, in some countries (such as Kenya) elephants are highly destructive of ecosystems and some wildlife preserves (such as Kenya's Amboseli National Park) where elephant populations are too high relative to available vegetation and space. Elephants also destroy fences, crops, and village houses and sometimes kill people.

The ban on elephant ivory sales has increased the killing of other species with ivory parts such as bull walruses (for their ivory tusks, worth $500–1500 a pair) and hippos (for their ivory teeth, worth $70 per kilogram)—another example of the principle that we can never do only one thing. Eastern Zaire's hippo population, once one of the world's largest, has dropped from about 23,000 to 11,000 since the 1989 elephant ban, mostly because of poaching for hippo teeth.

Others argue that opening up the legal elephant ivory market again, even partially, would lead to a renewal of massive poaching of elephants to supply a new ivory black market driven by increased consumption of ivory. They see continuing the ban as a way to help prevent African elephants from becoming threatened or endangered by keeping the price of ivory down.

In 1997 the Convention on International Trade in Endangered Species voted to let Zimbabwe, Botswana, and Namibia sell a total of almost 55 metric tons (60 tons) of stockpiled ivory under strict monitoring. Profits from the sale will be used for elephant conservation programs. What do you think about this policy?

What Is the Role of Overfishing in Reducing Aquatic Biodiversity? Concentrations of particular aquatic species suitable for commercial harvesting

in a given ocean area or inland body of water are called **fisheries**. Various methods for harvesting fish are shown in Figure 6-5.

Some fishing boats, called *trawlers*, catch bottom-dwelling fish and shellfish by dragging a funnel-shaped net held open at the neck along the ocean bottom. Newer trawling nets are large enough to swallow 12 jumbo jets in a single gulp, and even larger ones are on the way. The large mesh of the net allows most small fish to escape but can capture and kill other species such as seals and endangered and threatened sea turtles. Only the large fish are kept, with most of the fish and other aquatic species thrown back into the ocean either dead or dying.

Surface-dwelling species such as tuna, which feed in schools near the surface or in shallow areas, are often caught by *purse-seine fishing* (Figure 6-5). After a school is found, it's surrounded by a large purse-seine net. The net is then closed like a drawstring purse to trap the fish, and the catch is hauled aboard. Nets used to capture yellowfin tuna in the eastern tropical Pacific Ocean have also killed large numbers of dolphins, which swim on the surface above schools of the tuna.

Another increasingly used method for catching open-ocean fish species such as swordfish, tuna, and sharks (Case Study, p. 38) is *longlining*, in which fishing vessels put out lines up to 130 kilometers (80 miles) long, hung with thousands of baited hooks. This practice has helped cause such a severe drop in the population of Atlantic swordfish that many restaurant owners and chefs have agreed to stop serving this fish and many consumers are refusing to order it. Long lines also hook pilot whales, dolphins, endangered sea turtles, and sea-feeding albatross birds.

One of the greatest threats to marine biodiversity is *drift-net fishing* (Figure 6-5). These monster nets drift in the water and catch fish when their gills become entangled in the nylon mesh. Each net descends as much as 15 meters (50 feet) below the surface and is up to 65 kilometers (40 miles) long. Almost anything that comes in contact with these nearly invisible "curtains of death" becomes entangled.

In 1990 the UN General Assembly declared a moratorium on the use of drift nets longer than 2.5 kilometers (1.6 miles) in international waters after December 31, 1992, to help reduce overfishing and reduction of marine biodiversity. However, compliance is voluntary and the ban has numerous loopholes. There is no effective mechanism for monitoring fleets' activities over vast ocean areas and no structure for enforcement or punishment of violators. The financial rewards of illegal harvesting using drift nets far outweigh the very slim chances of being caught and having to pay small fines.

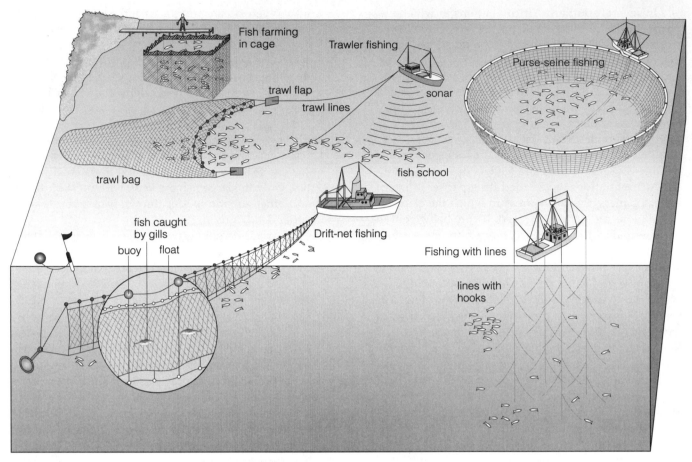

Figure 6-5 Major commercial fishing methods used to harvest or raise marine species.

How Serious Are the Threats to the Biodiversity of Freshwater and Marine Systems?
According to a number of aquatic scientists, the economic pressures to catch more and more fish pose a serious threat to freshwater and marine diversity. Most people are unaware how vulnerable freshwater environments are to environmental degradation and depletion. Currently, about 34% of the known fish species are at risk of becoming extinct. According to a recent study by the World Resources Institute, coastal developments threaten marine biodiversity along roughly half of the world's coasts.

The major causes of declines in fish populations are habitat loss and degradation (mostly from pollution), invasions by nonnative species, and overfishing. Overfishing leads in most cases to *commercial extinction*, which is usually only a temporary depletion of fish stocks, as long as depleted areas are allowed to recover. However, large-scale fish harvesting can lead to serious depletion

and extinction of species such as sea turtles and other marine mammals that are unintentionally caught.

What Is the Role of Predator and Pest Control?
People try to exterminate species that compete with them for food and game animals. For example, U.S. fruit farmers exterminated the Carolina parakeet around 1914 because it fed on fruit crops. The species was easy prey because when one member of a flock was shot, the rest of the birds hovered over its body, making themselves easy targets (Table 6-1).

African farmers kill large numbers of elephants to keep them from trampling and eating food crops. Ranchers, farmers, and hunters in the United States support the killing of coyotes, wolves, and other species that can prey on livestock and on species prized by game hunters. Since 1929 U.S. ranchers and government agencies have poisoned 99% of North America's prairie dogs because horses and cattle sometimes step

Q: What would happen if nuclear power's harmful effects were included in its market price and all subsidies were removed?

into the burrows and break their legs. This has also nearly wiped out the endangered black-footed ferret (Figure 6-2), which preyed on the prairie dog.

What Is the Role of the Market for Exotic Pets and Decorative Plants? The global legal and illegal trade in wild species for use as pets is a huge and very lucrative business. However, for every live animal captured and sold in the pet market an estimated 50 other animals are killed.

Worldwide, over 5 million live wild birds are captured and sold legally each year, and an estimated 2.5 million more are captured and sold illegally, primarily in Europe, Japan, and the United States. Over 40 species, mostly parrots, are endangered or threatened because of this wild-bird trade. Collectors of exotic birds may pay $10,000 for a threatened hyacinth macaw (Figure 6-2) smuggled out of Brazil; however, during its lifetime a single macaw left in the wild might yield as much as $165,000 in tourist income.

About 25 million U.S. households have exotic birds as pets, 85% of them imported. For every wild bird that reaches a pet shop legally or illegally, as many as 10 others die during capture or transport. Keeping pet birds may also be hazardous to human health. A 1992 study suggested that keeping indoor pet birds for more than 10 years doubles a person's chances of getting lung cancer.

Some exotic plants, especially orchids and cacti such as the black lace cactus (Figure 6-2), are also endangered because they are gathered (often illegally), sold to collectors, and used to decorate houses, offices, and landscapes. A collector may pay $5,000 for a single rare orchid. A single rare mature crested saguaro cactus can earn cactus rustlers as much as $15,000.

What Are the Roles of Climate Change and Pollution? A potential problem for many species is the possibility of fairly rapid (50–100 years) changes in climate, accelerated by deforestation and emissions of heat-trapping gases into the atmosphere (Section 9-2). Wildlife in even the best-protected and best-managed reserves could be depleted in a few decades if such changes in climate take place.

Toxic chemicals degrade wildlife habitats, including wildlife refuges, and kill some plants and animals. Slowly degradable pesticides, such as DDT, can accumulate in fatty tissue and be biologically magnified to very high concentrations in food chains and webs (Figure 6-6). The resulting high concentrations of DDT or other slowly biodegraded, fat-soluble organic chemicals can kill some organisms, reduce their ability to reproduce, or make them more vulnerable to diseases, parasites, and predators.

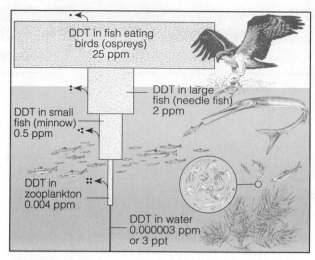

Figure 6-6 Bioaccumulation and biomagnification. DDT is a fat-soluble chemical that can bioaccumulate in the fatty tissues of animals. In a food chain or food web the accumulated concentrations of DDT can be biologically magnified in the bodies of animals at each higher trophic level. This diagram shows that the concentration of DDT in the fatty tissues of organisms was biomagnified about 10 million times in this food chain in an estuary near Long Island Sound. If each phytoplankton organism in such a food chain takes up from the water and retains one unit of DDT, a small fish eating thousands of zooplankton (which feed on the phytoplankton) will store thousands of units of DDT in its fatty tissue. Then each large fish that eats 10 of the smaller fish will ingest and store tens of thousands of units, and each bird (or human) that eats several large fish will ingest hundreds of thousands of units. Dots represent DDT, and arrows show small losses of DDT through respiration and excretion.

During the 1950s and 1960s, populations of fish-eating birds such as the osprey, cormorant, brown pelican, and bald eagle plummeted. Research indicated that a chemical derived from DDT, when biologically amplified in food webs (Figure 6-6), made the birds' eggshells so fragile that they could not successfully reproduce. Also hard-hit were such predatory birds as the prairie falcon, sparrow hawk, and peregrine falcon (Figure 6-2), which help control rabbits, ground squirrels, and other crop-eaters.

Since the U.S. ban on DDT in 1972, most of these species have made a comeback. In 1980, however, DDT levels were again rising in the peregrine falcon and the osprey. Scientists believe that the birds may be picking up DDT and other banned pesticides in Latin America, where they winter and where use of such chemicals is still legal. Illegal use of banned pesticides in the United States, as well as airborne drift of long-lived pesticides from developing countries onto U.S. land and surface water, may also play a role. According to the U.S. Fish and Wildlife Service, pesticides menace about 20% of the endangered and threatened species in the United States.

Table 6-2 Damage Caused by Plants and Animals Imported into the United States

Name	Origin	Mode of Transport	Type of Damage
Mammals			
European wild boar	Russia	Intentionally imported (1912), escaped captivity	Destroys habitat by rooting; damages crops
Nutria (cat-sized rodent)	Argentina	Intentionally imported, escaped captivity (1940)	Alters marsh ecology; damages levees and earth dams; destroys crops
Birds			
European starling	Europe	Intentionally released (1890)	Competes with native songbirds; damages crops; transmits swine diseases; causes airport nuisance
House sparrow	England	Intentionally released by Brooklyn Institute (1853)	Damages crops; displaces native songbirds
Fish			
Carp	Germany	Intentionally released (1877)	Displaces native fish; uproots water plants; lowers waterfowl populations
Sea lamprey	North Atlantic Ocean	Entered Great Lakes via Welland Canal (1829)	Wiped out lake trout, lake whitefish, and sturgeon in Great Lakes
Walking catfish	Thailand	Imported into Florida	Destroys bass, bluegill, and other fish
Insects			
Argentine fire ant	Argentina	Probably entered via coffee shipments from Brazil (1918)	Damages crops; destroys native ant species
Camphor scale insect	Japan	Accidentally imported on nursery stock (1920s)	Damaged nearly 200 plant species in Louisiana, Texas, and Alabama
Japanese beetle	Japan	Accidentally imported on irises or azaleas (1911)	Defoliates more than 250 species of trees and other plants, including many of commercial importance
Plants			
Water hyacinth	Central America	Intentionally introduced (1884)	Clogs waterways; shades out other aquatic vegetation
Chestnut blight (fungus)	Asia	Accidentally imported on nursery plants (1900)	Killed nearly all eastern U.S. chestnut trees; disturbed forest ecology
Dutch elm disease (fungus)	Europe	Accidentally imported on infected elm timber used for veneers (1930)	Killed millions of elms; disturbed forest ecology

From *Biological Conservation* by David W. Ehrenfeld. Copyright © 1970 by Holt, Rinehart & Winston, Inc. Modified and reprinted by permission.

What Is the Role of Introduced Species?

Deliberate or accidental introduction of nonnative or alien species into habitats where they are not native is the second-ranked cause of animal extinctions (after habitat alteration and degradation).

Travelers sometimes collect plants and animals and introduce them intentionally or accidentally into new geographical regions. Many of these introduced species ultimately provide food, game, and aesthetic beauty and may even help control pests in their new environments.

However, some introduced species have no natural predators, competitors, parasites, or pathogens to control their numbers in their new habitats. This can allow them to dominate their new ecosystem and reduce or wipe out the populations of many native species, making them a form of *biological pollution* (Table 6-2).

The United States is home to at least 4,000 nonnative plant species and 2,300 alien animal species. Damages and pest control costs for these unwanted species amount to an estimated $122 million per year—an average of $14,000 per

Tuxill, J., and C. Bright. 1998 "Protecting Nature's Diversity: Mending Strands in the Web of Life." *The Futurist*, v. 32, no.5, 46(6).

minute—and future losses are expected to be much higher. About 49% of the species in the United States on the official list of endangered and threatened species are there in part because of population declines caused by nonnative species. In Hawaii, more than 95% of the 282 imperiled plant and bird species are threatened by alien species.

An example of the effects of the *accidental introduction* of a nonnative species is the fast-growing water hyacinth, which is native to Central and South America. In 1884, a woman took one from a New Orleans exhibition and planted it in her backyard, from which it spread to Florida's waterways.

With no natural enemies and the ability to double their population every 2 weeks, the plants took advantage of nutrient-rich waters and spread rapidly. Within 10 years water hyacinths had smothered and displaced many native plants and fish and clogged many ponds, streams, and canals, first in Florida and later elsewhere in the southeastern United States.

Mechanical harvesters and herbicides have failed to keep the plant in check. Grazing Florida manatees (Figure 6-2) found in the coastal rivers of Florida and southern Georgia can control water hyacinths better than mechanical or chemical methods. However, these gentle and playful herbivores are threatened with extinction because of a combination of pollution, habitat loss, and slashing by powerboat propellers.

To help control its spread, scientists have introduced water hyacinth-eaters, including a weevil from Argentina, a water snail from Puerto Rico, and the grass carp, a fish from the former Soviet Union. These species can help, but water snails and grass carp also feed on other, desirable aquatic plants.

The good news is that water hyacinths can provide several benefits. They absorb toxic chemicals in sewage treatment lagoons; they can be fermented into a biogas fuel similar to natural gas, added as a mineral and protein supplement to cattle feed, and applied to the soil as fertilizer; and they can be used to clean up polluted ponds and lakes if their growth can be kept under control.

An example of the impact of *deliberately introduced* species on ecosystems is wild African bees, which were imported to Brazil in 1957 to help increase honey production. Instead, these bees have displaced domestic honey bees and reduced the honey supply. Since then these nonnative bee species, popularly known as "killer bees," have moved northward into Central America (killing 150 people in Mexico since 1986). Since 1994 they have become established in Texas (one death in 1994), Arizona (one death in 1993), New Mexico, and Puerto Rico. They are now heading north at 240 kilometers (150 miles) per year, although they may be stopped eventually by cold winters in the central United States. Although they are not the killer bees portrayed in some horror movies, these bees are aggressive and unpredictable.

6-4 SOLUTIONS: PROTECTING WILD SPECIES FROM DEPLETION AND EXTINCTION

How Can We Protect Wildlife and Biodiversity?
There are three basic approaches to managing wildlife and protecting biodiversity. The *ecosystem approach*, discussed in Chapter 5, aims to preserve balanced populations of species in their native habitats, establish legally protected wilderness areas and wildlife reserves, and eliminate nonnative species.

Biologists consider protecting ecosystems to be the best way to preserve biological diversity and ecological integrity. The problem is that fully or partially protected wildlife sanctuaries make up only about 6% of the world's land area, and the human population is expected to double in 49 years.

The *species approach* is based on protecting endangered species by identifying them, giving them legal protection, preserving and managing their crucial habitats, propagating them in captivity, and reintroducing them into suitable habitats. The *wildlife management approach* manages game species for sustained yield by using laws to regulate hunting, establishing harvest quotas, developing population management plans, and using international treaties to protect migrating game species such as waterfowl. These two approaches are discussed in the remainder of this chapter.

How Can International Treaties Help Protect Endangered Species? Several international treaties and conventions help protect endangered or threatened wild species. One of the most far-reaching is the 1975 Convention on International Trade in Endangered Species (CITES). This treaty, now signed by 136 countries, lists almost 700 species that cannot be commercially traded as live specimens or wildlife products because they are endangered or threatened.

However, enforcement of this treaty is spotty, convicted violators often pay only small fines, and member countries can exempt themselves from protecting any listed species. Furthermore, much of the estimated $6 billion annual illegal trade in wildlife and wildlife products goes on in countries that have not signed the treaty. Centers of illegal animal trade are Singapore, Argentina, Indonesia, Spain, Taiwan, and Thailand.

How Can Laws Help Protect Endangered Species?
The United States controls imports and exports of endangered wildlife and wildlife products through two important laws. The Lacey Act of 1900 prohibits transporting live or dead wild animals or their parts across state borders without a federal permit. The Endangered Species Act of 1973 (amended in 1982 and 1988)

Hint: Enter the search terms *extinction, prevention* using the Subject Guide.

makes it illegal for Americans to import or trade in any product made from an endangered or threatened species unless it is used for an approved scientific purpose or to enhance the survival of the species.

The Endangered Species Act of 1973 is one of the world's toughest environmental laws. It authorizes the National Marine Fisheries Service (NMFS) to identify and list endangered and threatened ocean species; the U.S. Fish and Wildlife Service (USFWS) identifies and lists all other endangered and threatened species. These species cannot be hunted, killed, collected, or injured in the United States.

Any decision by either agency to add or remove a species from the list must be based on biology only, not on economic or political considerations. However, economic factors can be used in deciding whether and how to protect endangered habitat and in developing recovery plans for listed species. The act also forbids federal agencies to carry out, fund, or authorize projects that would either jeopardize an endangered or threatened species or destroy or modify the critical habitat it needs to survive.

Between 1973 and 1998, the number of U.S. species included in the official endangered and threatened list increased from 92 to over 1,100 species (of which approximately 60% are plants and 40% animals). Each year about 85 species are added to the list.

Getting listed is only half the battle. Next, the USFWS or NMFS is supposed to prepare a plan to help the species recover. However, because of a lack of funds, final recovery plans have been developed and approved for only about two-thirds of the endangered or threatened U.S. species, and half of those plans exist only on paper.

The act requires that all commercial shipments of wildlife and wildlife products enter or leave the country through one of nine designated ports. Few illegal shipments are confiscated because the 60 USFWS inspectors can examine only about one-fourth of the 90,000 shipments that enter and leave the United States each year. Even if caught, many violators are not prosecuted, and convicted violators often pay only a small fine.

In 1998, Secretary of Interior Bruce Babbitt announced that the U.S. Fish and Wildlife would remove (delist) 17 animals and 12 plants from the endangered and threatened species list over 2 years, subject to final approval by federal biologists. Some would be downgraded from endangered to threatened and others removed from the law's protection. Candidates for this delisting include well-known species such as the bald eagle, peregrine falcon, brown pelican, and gray wolf, as well as lesser-known species such as Pahrump poolfish and the Missouri bladder-pod plant.

Since 1995 there have been intense efforts to seriously weaken the Endangered Species Act by (1) making the protection of endangered species on private land voluntary, (2) having the government pay landowners if it forces them to stop using part of their land to protect endangered species, (3) making it harder and more expensive to list newly endangered species by requiring government wildlife officials to navigate through a series of hearings and peer-review panels, (4) giving the secretary of interior the power to permit a listed species to become extinct without trying to save it, (5) allowing the secretary of interior to give any state, county, or landowner permanent exemption from the law, with no requirement for public notification or comment, (6) allowing landowners to lock in long-term endangered species management plans—known as *habitat conservation plans* (HCPs)—that exempt the owners from further obligations for 100 years or more, and (7) prohibiting the public from commenting on or bringing lawsuits on any changes in habitat conservation plans for endangered species.

Before his death California Representative Sonny Bono joked that the best way to deal with endangered species is to "give them all a designated area and then blow it up." Washington Senator Slade Gordon suggests that all endangered species be removed from the wild and bred in zoos as "a way to preserve animals without blocking economic development."

Some of these critics have spread horror stories about how environmentalists have used endangered species (many of them small and largely unknown) to block development and resource extraction, violate private property rights, and waste tax dollars trying to save useless creatures that are on the verge of extinction anyway. However, careful investigation has revealed that most of these widely circulated anecdotal stories were either false or misleading.

Should the Endangered Species Act Be Strengthened? Most conservation biologists and wildlife scientists contend that the Endangered Species Act has not been a failure (Spotlight, right). They also refute the charge that the act has caused severe economic losses.

The truth is that the Endangered Species Act has had virtually no impact on the nation's overall economic development. For example, between 1989 and 1992 only 55 of more than 118,000 projects evaluated by the USFWS were blocked or withdrawn as a result of the Endangered Species Act.

Moreover, *the act does allow for economic concerns.* By law, a decision to list a species must be based solely on science. But once a species is listed, economic considerations can be weighed against species protection in protecting critical habitat and in designing and

Q: What do most scientists believe are our best energy options?

implementing recovery plans. The act also allows a special Cabinet-level panel, called the "God Squad," to exempt any federal project from having to comply with the act if the economic costs are too high. In addition, the act allows the government to issue permits and exemptions to landowners with listed species living on their property.

Most biologists and wildlife conservationists believe that we should develop a new system to protect and sustain biological diversity and ecological integrity based on three principles: **(1)** Find out what species and ecosystems we have, **(2)** locate and protect the most endangered ecosystems and species, and **(3)** give private landowners financial carrots (tax breaks and write-offs) for helping protect endangered species and ecosystems.

Should We Try to Protect All Endangered and Threatened Species? Because of limited funds and trained personnel, only a few endangered and threatened species can be saved. Many wildlife experts suggest that the limited funds available for preserving threatened and endangered wildlife be concentrated on species that **(1)** have the best chance for survival, **(2)** have the most ecological value to an ecosystem, and **(3)** are potentially useful for agriculture, medicine, or industry.

Others oppose such ideas on ethical grounds or contend that currently we don't have enough biological information to make such evaluations. Proponents argue that, in effect, we are already deciding by default which species to save and that, despite limited knowledge, this approach is more effective and a better use of limited funds than the current one. What do you think?

How Can Wildlife Refuges and Other Protected Areas Help Protect Endangered Species? Since 1903, when President Theodore Roosevelt established the first U.S. federal wildlife refuge at Pelican Island, Florida, the National Wildlife Refuge System has grown to 508 refuges (Figure 5-1). About 85% of the area included in these refuges is in Alaska.

Over three-fourths of the refuges are wetlands for protection of migratory waterfowl. About 20% of the species on the U.S. endangered and threatened list have habitats in the refuge system, and some refuges have been set aside for specific endangered species. These have helped Florida's key deer, the brown pelican, and the trumpeter swan to recover. Conservation biologists urge the establishment of more refuges for endangered plants. They are also urging Congress and state legislatures to allow abandoned military lands that contain significant wildlife habitat to become national or state wildlife refuges.

So far Congress has not established guidelines (such as multiple use or sustained yield) for management

SPOTLIGHT

Has the Endangered Species Act Been a Failure?

Critics of the Endangered Species Act call it an expensive failure because only a few species have been removed from the endangered list. Most of these critics are ranchers, developers, and officials of timber and mining companies who want more access to resources on public lands.

Most biologists strongly disagree that the act has been a failure, for several reasons. *First*, species are listed only when they are already in serious danger of extinction. This is like setting up a poorly funded hospital emergency room that takes only the most desperate cases, often with little hope for recovery, and then saying it should be shut down because it has not saved enough patients.

Second, it takes decades for most species to become endangered or threatened. Thus, it should not be surprising that it usually takes decades to bring a species in critical condition back to the point where it can be removed from the list. Expecting the Endangered Species Act (which has been in existence only since 1973) to quickly repair the biological depletion of centuries is unrealistic. The most important measure of the law's success is that the conditions of almost 40% of the listed species are stable or improving. A hospital emergency room taking only the most desperate cases yet stabilizing or improving the condition of 40% of its patients would be considered an astounding success.

Third, the federal endangered species budget was only $93 million in 1999 (up from $23 million in 1993). This funding is equal to about what it takes to build 3.2 kilometers (1.5 miles) of urban interstate highway, or about 36¢ a year per U.S. citizen—a pittance to help save some of the country's irreplaceable biodiversity. One C-17 transport airplane costs $300 million—almost three times what was spent on protecting endangered species in the United States during 1999.

To most biologists it's amazing that so much has been accomplished in stabilizing or improving the condition of almost 40% of nearly terminal species on a shoestring budget.

Critical Thinking

1. Do you believe that the Endangered Species act should be weakened or strengthened? Explain.

2. Do you agree or disagree with each of the proposals made to **(a)** weaken the act given on p. 186 and **(b)** strengthen the act listed on p. 187. Defend each of your answers.

CASE STUDY

Should Oil and Gas Development Be Allowed in the Arctic National Wildlife Refuge?

The Arctic National Wildlife Refuge on Alaska's North Slope (Figure 6-7) contains more than one-fifth of all the land in the National Wildlife Refuge System and has been called the crown jewel of the system. The refuge's coastal plain, its most biologically productive part, is the only stretch of Alaska's arctic coastline not open to oil and gas development.

For years, U.S. oil companies have been working to change this because they believe that this coastal area *might* contain oil and natural gas deposits. They argue that such exploration is needed to increase U.S. oil and natural gas supplies and to reduce dependence on oil imports. However, oil companies plan to export most of any oil they find in the reserve to Japan.

Conservation biologists oppose this proposal and urge Congress to designate the entire coastal plain as wilderness. They cite Department of Interior estimates that there is only a 19% chance of finding as much oil there as the United States consumes every 6 months. Even if the oil does exist, environmentalists do not believe the potential degradation of any portion of this irreplaceable wilderness area would be worth it, especially when improvements in energy efficiency would save far more oil at a much lower cost (Section 4-2).

Oil company officials claim they have developed Alaska's Prudhoe Bay oil fields without significant harm to wildlife; they also contend that the area they seek to open to oil and gas development is less than 1.5% of the entire coastal plain region. However, the huge 1989 oil spill from the tanker *Exxon Valdez* in Alaska's Prince William Sound casts serious doubt on such claims.

Moreover, a study leaked from the U.S. Fish and Wildlife Service in 1988 revealed that oil drilling at Prudhoe Bay has caused much more air and water pollution than was anticipated before drilling began in 1972. According to this study, oil development in the coastal plain could cause the loss of 20–40% of the area's 180,000-head caribou herd, 25–50% of the remaining musk oxen, and 50% or more of the wolverines that live there part of the year. A 1988 EPA study also found that "violations of state and federal environmental regulations and laws are occurring at an unacceptable rate" in portions of the Prudhoe Bay area in which oil fields and facilities have been developed.

Critical Thinking

Do you believe that exploration and removal of oil from this wildlife refuge should be allowed? Why?

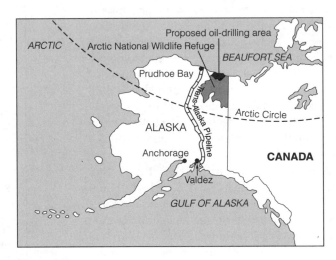

Figure 6-7
Proposed oil-drilling area in Alaska's Arctic National Wildlife Refuge. (Courtesy U.S. Fish and Wildlife Service)

within the National Wildlife Refuge System, as it has for other public lands. A 1990 report by the General Accounting Office found that activities considered harmful to wildlife occur in nearly 60% of the nation's wildlife refuges. There is much controversy over whether to allow oil and gas development in Alaska's Arctic National Wildlife Refuge (Case Study, above).

Private groups also play an important role in establishing wildlife refuges and other protected areas. For example, since 1951 the Nature Conservancy (with over 800,000 members) has preserved over 38,000 square kilometers (15,800 square miles) of areas in the United States with unique ecological or aesthetic significance.

Can Gene Banks and Botanical Gardens Help Save Most Endangered Species? Botanists preserve genetic information and endangered plant species by storing their seeds in gene or seed banks: refrigerated, low-humidity environments. Scientists urge

Gayton, Don. 1997. "Terms of Endangerment." *Canadian Geographic*, vol. 117, no. 3, 30(12).

that many more such banks be established, especially in developing countries. But some species can't be preserved in gene banks, and maintaining the banks is very expensive.

The world's 1,500 botanical gardens and arboreta maintain at least 40,000 plant species. However, these sanctuaries have too little storage capacity and too little funding to preserve most of the world's rare and threatened plants.

⚜ Can Zoos Help Protect Most Endangered Species? Worldwide, 1,000 zoos house about 540,000 terrestrial vertebrate animals from 3,000 species of mammals, amphibians, birds, and reptiles. Many of them are neither threatened nor endangered. However, zoos and animal research centers are increasingly being used to preserve some individuals of critically endangered animal species, with the long-term goal of reintroducing the species into protected wild habitats.

Two techniques for preserving such species are egg pulling and captive breeding. *Egg pulling* involves collecting wild eggs laid by critically endangered bird species and then hatching them in zoos or research centers. In *captive breeding*, some or all wild individuals of a critically endangered species are captured for breeding in captivity, with the aim of reintroducing the offspring into the wild.

Captive breeding programs at zoos in Phoenix, San Diego, and Los Angeles saved the nearly extinct Arabian oryx (Figure 6-2), a large antelope that once lived throughout the Middle East. Since 1980, a few oryx bred in captivity have been returned to the wild in protected habitats in the Middle East; the wild population is now about 120. Endangered species now being bred in captivity in the United States and returned to the wild include the peregrine falcon (Figure 6-2), the black-footed ferret (Figure 6-2), and the golden lion tamarin (Figure 6-2). However, most reintroductions fail because of a lack of suitable habitat and the inability of individuals bred in captivity to survive in the wild.

Efforts to maintain populations of endangered species in zoos and research centers are limited by lack of space and money. The captive population of each species must number 100 to 500 individuals to avoid extinction through accident, disease, or loss of genetic diversity through inbreeding. Recent genetic research indicates that 10,000 or more individuals are needed for an endangered species to maintain its capacity for biological evolution.

Moreover, caring for and breeding captive animals is very expensive. It is estimated that today's zoos and research centers have space to preserve healthy and sustainable populations of only 925 of the 2,000 large vertebrate species that could vanish from the planet. It is doubtful that the more than $6 billion needed to care for these animals for 20 years will become available.

Some critics see zoos as prisons for wild animals. They also contend that zoos foster the notion that we don't need to preserve large numbers of wild species in their natural habitats.

Whether one agrees or disagrees with this position, it is clear that zoos and botanical gardens are not a biologically or economically feasible solution for most of the world's current endangered species and the much larger number expected to become endangered over the next few decades.

6-5 SOLUTIONS: WILDLIFE MANAGEMENT

How Can Wildlife Populations Be Managed? Wildlife management entails manipulating wildlife populations (especially game species) and their habitats for their welfare and for human benefit. It includes preserving endangered and threatened wild species and enforcing wildlife laws.

In the United States, funds for state game management programs come from the sale of hunting and fishing licenses and from federal taxes on hunting and fishing equipment. Two-thirds of the states also have provisions on state income tax returns that allow individuals to contribute money to state wildlife programs. Only 10% of all government wildlife dollars are spent to study or benefit nongame species, which make up nearly 90% of the country's wildlife species.

The first step in wildlife management is to decide which species are to be managed in a particular area, a decision that is a source of much controversy. Ecologists and conservation biologists emphasize preserving biodiversity and ecological integrity, wildlife conservationists are concerned about endangered species, bird-watchers want the greatest diversity of bird species, and hunters want sufficiently large populations of game species for harvest during hunting season. In the United States and other developed countries, most wildlife management is devoted to producing surpluses of game animals and birds for hunters.

After goals have been set, the wildlife manager must develop a management plan. Ideally, this is based on principles of ecological succession (Figures 2-44 and 2-45), wildlife population dynamics (Figure 2-40), and an understanding of the cover, food, water, space, and other habitat requirements of each species to be managed. The manager must also consider the number of potential hunters, their success rates, and the regulations available to prevent excessive harvesting.

Hint: Enter the search terms *extinction, prevention* using the Subject Guide.

This information is difficult, expensive, and time-consuming to obtain. It involves much educated guesswork and trial and error, which is why wildlife management is as much an art as a science. Management plans must also be sensitive to political pressures from conflicting groups and to budget constraints.

How Can Vegetation and Water Supplies Be Manipulated to Manage Wildlife?

Wildlife managers can encourage the growth of plant species that are the preferred food and cover for a particular animal species in a given area by controlling the ecological succession of the vegetation in that area (Figure 6-8).

Various types of habitat management can be used to attract a desired species and encourage its population growth. Examples are planting seeds, transplanting certain types of vegetation, building artificial nests, and deliberately setting controlled, low-level ground fires to help control vegetation. Wildlife managers often create or improve ponds and lakes in wildlife refuges to provide water, food, and habitat for waterfowl and other wild animals.

How Useful Is Sport Hunting in Managing Wildlife Populations?

Most developed countries use sport hunting laws to manage populations of game animals. Licensed hunters are allowed to hunt only during certain portions of the year so as to protect animals during their mating season. Limits are set on the size, number, and sex of animals that can be killed, as well as on the number of hunters allowed in a given area.

Close control of sport hunting is difficult. Accurate data on game populations may not exist and may cost too much to get. People in communities near hunting areas, who benefit from money spent by hunters, may seek to have hunting quotas raised.

How Can Populations of Migratory Waterfowl Be Managed?

Migratory birds—including ducks, geese, swans, and many songbirds—make north–south journeys from one habitat to another each year, usually to find food, suitable climate, and other conditions necessary for reproduction. Such bird species use many different north–south routes called **flyways**, but only about 15 are considered major routes (Figure 6-9). Some countries along such flyways have entered into agreements and treaties to protect crucial habitats needed by such species, both along their migration routes and at each end of their journeys. However, the populations of migrating snow geese have risen to the point where they are causing serious ecological damage (Connections, p. 192).

Wildlife officials manage waterfowl by regulating hunting, protecting existing habitats, and developing new habitats, including artificial nesting sites, ponds, and nesting islands. More than 75% of the federal wildlife refuges in the United States are wetlands used by migratory birds. Local and state agencies and private conservation groups such as Ducks Unlimited, the Audubon Society, and the Nature Conservancy have also established waterfowl refuges.

Since 1934 the Migratory Bird Hunting and Conservation Stamp Act has required waterfowl hunters to buy a duck stamp each season they hunt. Revenue from these sales goes into a fund to buy land and easements for the benefit of waterfowl.

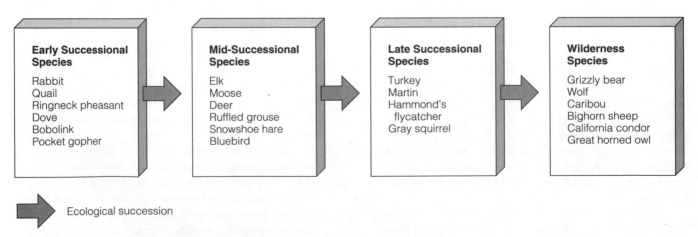

Figure 6-8 Examples of wildlife species typically found in areas at different stages of ecological succession in areas of the United States with a temperate climate.

 Botsford, Louis W. 1997. "The Management of Fisheries and Marine Ecosystems." *Science*, vol. 275, no. 5325, 509(7).

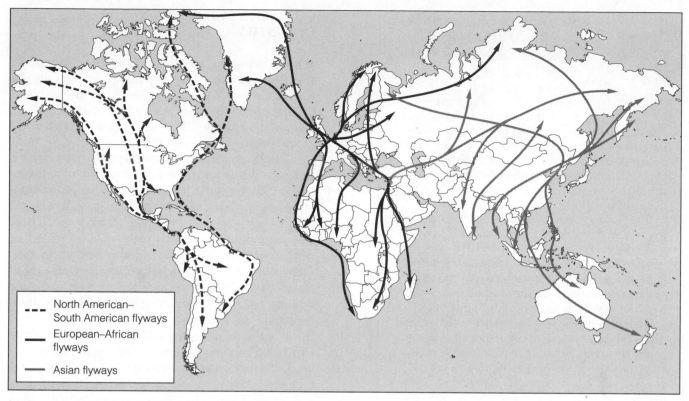

Figure 6-9 Major flyways used by migratory birds, mostly waterfowl. Each route has a number of subroutes.

Legend:
- – – – North American–South American flyways
- —— European–African flyways
- —— Asian flyways

6-6 SOLUTIONS: FISHERY MANAGEMENT

How Can Freshwater Fisheries Be Managed and Sustained? Managing freshwater fish involves encouraging populations of commercial and sport fish species and reducing or eliminating populations of less desirable species. This is usually done by regulating time and length of fishing seasons and the number and size of fish that can be taken.

Other techniques include building reservoirs and farm ponds and stocking them with fish, fertilizing nutrient-poor lakes and ponds, and protecting and creating spawning sites. Habitats can also be protected from buildup of sediment and other forms of pollution, and debris can be removed. Excessive growth of aquatic plants from cultural eutrophication (excessive inputs of plant nutrients from human activities) can be prevented, and small dams can be built to control water flow.

Predators, parasites, and diseases can be controlled by improving habitats, breeding genetically resistant fish varieties, and using antibiotics and disinfectants judiciously. Hatcheries can be used to restock ponds, lakes, and streams with prized species such as trout and salmon, and entire river basins can be managed to protect such valued species as salmon.

How Can Marine Fisheries Be Managed? By international law, a country's offshore fishing zone extends to 370 kilometers (200 nautical miles, or 230 statute miles) from its shores. Foreign fishing vessels can take certain quotas of fish within such zones, called exclusive economic zones, but only with a government's permission. Ocean areas beyond the legal jurisdiction of any country are known as the high seas; any limits on the use of the living and mineral common-property resources in these areas are set by international maritime law and international treaties.

Managers of marine fisheries use several techniques to help prevent overfishing and commercial extinction and allow depleted stocks to recover. Fishery commissions, councils, and advisory bodies (with representatives from countries or states using a fishery) can set annual quotas and establish rules for dividing the allowable catch among the participating countries or states, limiting fishing seasons, and regulating the type of fishing gear that can be used to harvest a particular species.

Hint: Enter the search term *fisheries management* using the Subject Guide.

CONNECTIONS

What Should Be Done About Snow Geese?

The population of the lesser snow goose has grown so large that it threatens its arctic tundra and wetland habitats. Over the past three decades, the population has risen from about 800,000 to at least 3 million. Hunters kill about 500,000 of the birds annually, but their population is still growing at a rate of about 250,000 birds per year.

Their massive numbers have converted vast tracts of arctic tundra in Canada and Alaska to highly saline, bare soil where few plants can grow, creating what has been called a "botanical desert." Snow geese have also destroyed about one-third of the salt marshes along the Hudson and James bays and have heavily damaged another third.

The snow geese population has thrived because each year they can gorge themselves off of the abundant grains in the U.S. agricultural heartland from the Great Plains to the Gulf Coast. As a result, they arrive at their winter breeding grounds in better condition, leading to higher rates of reproduction.

There are several options for dealing with this problem. Eventually the geese will destroy their winter breeding ground and their population will crash. However, this may take decades and lead to widespread ecosystem degradation.

Another strategy is to encourage hunters to kill up to 3 million of these birds each year—about 6 times the current hunting rate. A third suggestion is to have U.S. and Canadian troops to gather and destroy about 2 million snow goose eggs a year in the arctic tundra.

This problem, which connects high agricultural productivity with overpopulation of a species and ecosystem degradation, is another reminder that we can never do one thing in nature.

Critical Thinking

What do you think should be done about the snow goose problem? Explain.

As voluntary associations, however, fishery commissions have no legal authority to compel their members to follow their rules. Furthermore, it is very difficult to estimate the sustainable yields of various marine species. Experience has shown that many fishery commissions, most of them dominated by fishing industry representatives, have not prevented overfishing.

6-7 CASE STUDY: THE WHALING INDUSTRY

Why Is Whaling an Example of the Tragedy of the Commons? *Cetaceans* are an order of mostly marine mammals ranging in size from the 0.9-meter (3-foot) porpoise to the giant 15- to 30-meter (50- to 100-foot) blue whale. They are divided into two major groups: toothed whales and baleen whales.

Toothed whales, such as the porpoise, sperm whale, and killer whale (orca), bite and chew their food; they feed mostly on squid, octopus, and other marine animals. *Baleen whales*, such as the blue, gray, humpback, and finback, are filter feeders. Instead of teeth, they have several hundred horny plates made of baleen, or whalebone, that hang down from their upper jaw. These plates filter plankton from the seawater, especially shrimplike krill, which are smaller than your thumb (Figure 2-19). Baleen whales are the most abundant group of cetaceans.

Whales are fairly easy to kill because of their large size and their need to come to the surface to breathe. Mass slaughter has become especially efficient since the advent of fast propeller-driven ships, harpoon guns, and inflation lances (which pump dead whales full of air and make them float).

Whale harvesting—mostly in international waters, a commons open to all resource extractors—has followed the classic pattern of a tragedy of the commons. Between 1925 and 1975, whalers killed an estimated 1.5 million whales; this overharvesting drove the populations of 8 of the 11 major species to commercial extinction, the point where it no longer paid to hunt and kill them.

Case Study: Near Extinction of the Blue Whale In addition to commercial extinction, the populations of some prized species were reduced to the brink of biological extinction. A prime example is the endangered blue whale (Figures 2-19 and 6-2), the world's largest animal. Fully grown, it's more than 30 meters (100 feet) long—longer than three train boxcars—and weighs more than 25 elephants. The adult has a heart the size of a Volkswagen Beetle car, and some of its arteries are so big that a child could swim through them.

Blue whales spend about 8 months of the year in antarctic waters. There they find an abundant supply of krill, which they filter daily by the trillions from seawater. During the winter, they migrate to warmer waters, where their young are born.

Before commercial whaling began, an estimated 200,000 blue whales roamed the Antarctic Ocean. Today, the species has been hunted to near biological extinction for its oil, meat, and bone. Its decline was

Q: What percentage of all U.S. land consists of public lands?

caused by a combination of prolonged overharvesting and certain natural characteristics of blue whales. Their huge size made them easy to spot. They were caught in large numbers because they grouped together in their Antarctic feeding grounds. They also take 25 years to mature sexually and have only one offspring every 2–5 years, a reproductive rate that makes it difficult for the species to recover once its population falls beneath a certain threshold.

Blue whales haven't been hunted commercially since 1964 and have been classified as an endangered species since 1975. Despite this protection, some marine biologists believe that too few blue whales—an estimated 3,000–10,000—remain for the species to recover and avoid extinction.

How Have Whale Populations Been Managed?
In 1946, the International Whaling Commission (IWC) was established to regulate the whaling industry by setting annual quotas to prevent overharvesting and commercial extinction. However, these quotas often were based on inadequate data or were ignored by whaling countries. Without any powers of enforcement, the IWC has been unable to stop the decline of most whale species.

In 1970, the United States stopped all commercial whaling and banned all imports of whale products. Under intense pressure from environmentalists and governments of many countries, led by the United States, the IWC has imposed a moratorium on commercial whaling since 1986.

Some whaling has continued because Japan, Norway, Peru, and the former Soviet Union have exempted themselves from the moratorium. Since 1985, Norway, Iceland, and Japan have each killed several hundred whales a year by exploiting a loophole in the IWC treaty that allows whales to be killed for scientific research. Opponents of whaling call this a sham. With whale meat fetching more than $600 per kilogram ($273 per pound) in Japan, there is a great incentive to cheat.

Should the International Ban on Whaling Be Lifted?
Whaling is a traditional part of the economies and cultures of countries such as Japan, Iceland, and Norway. To many people in these cultures hunting whales is no more immoral than hunting deer or elk. These and other whaling nations believe that the international ban on commercial whaling should be lifted, arguing that the moratorium is based on emotion, not science.

Most environmentalists disagree. Some argue that whales are peaceful, intelligent, sensitive, and highly social mammals that pose no threat to humans and should not be killed; others fear that opening the door

to any commercial whaling may eventually lead to widespread harvests of whales by weakening current international disapproval and economic sanctions against commercial whaling. They cite the earlier failure of the IWC to enforce quotas that prevent commercial extinction of most whale species. They also contend that comparing whale populations to deer and elk populations is misleading because populations of deer and elk in the United States are large and in no danger of becoming endangered or extinct.

In 1994 the IWC voted 23 to 1 to establish a permanent whale sanctuary in the Antarctic Ocean. However, it will be almost impossible to monitor and prevent illegal whaling in this huge sanctuary. Moreover, Japan and Russia filed objections to the IWC decision, and Japan announced its intentions to continue commercial whaling in the sanctuary.

During our short time on this planet we have gained immense power over which species, including our own, live or die. We named ourselves the wise (sapiens) species. Most conservation biologists believe that in the next few decades, we will learn whether we are indeed a wise species—whether we have the wisdom to learn from and work with nature to protect ourselves and other species. Some actions you can take to help protect wildlife and preserve biodiversity are listed in Appendix 4.

A greening of the human mind must precede the greening of the earth. A green mind is one that cares, saves, and shares. These are the qualities essential for conserving biological diversity now and forever.

M. S. SWAMINATHAN

CRITICAL THINKING

1. Discuss your gut-level reaction to the following statement: "It doesn't really matter that the passenger pigeon is extinct and that the blue whale, the whooping crane, the California condor, and the world's remaining species of rhinoceros and tigers are endangered mostly because of human activities, because eventually all species become extinct anyway." Be honest about your reaction and give arguments for your position.

2. Make a log of your own consumption of all products for a single day. Relate your level and types of consumption to the decline of wildlife species and the increased destruction and degradation of wildlife habitats in the United States, in tropical forests, and in aquatic ecosystems.

3. **(a)** Do you accept the ethical position that each *species* has the inherent right to survive without human interference, regardless of whether it serves any useful purpose for humans? Explain. **(b)** Do you believe that each *individual* of an animal species has an inherent right to

survive? Explain. Would you extend such rights to individual plants and microorganisms? Explain.

4. Do you believe that the use of animals to test new drugs and vaccines and the toxicity of chemicals should be banned? Should the use of animals to test cosmetics be banned? Explain. What are the alternatives?

5. Should the U.S. government impose an economic boycott on products imported from any nation that illegally resumes whaling? Explain.

6. Which of the following statements best describes your feelings toward wildlife: **(a)** As long as it stays in its space, wildlife is OK; **(b)** as long as I don't need its space, wildlife is OK; **(c)** I have the right to use wildlife habitat to meet my own needs; **(d)** when you've seen one redwood tree, fox, elephant, or some other form of wildlife you've seen them all, so lock up a few of each species in a zoo or wildlife park and don't worry about protecting the rest; **(e)** wildlife should be protected.

7. List your three favorite species. Examine why they are your favorites. Are they cute and cuddly-looking, like the giant panda and the koala? Do they have human-like qualities, like apes or penguins that walk upright? Are they large, like elephants or blue whales? Are they beautiful, like tigers and monarch butterflies? Are any of them plants? Are any of them species such as bats, sharks, snakes, or spiders that most people are afraid of? Are any of them microorganisms that help keep you alive? Reflect on what your choice of favorite species tells you about your attitudes toward most wildlife.

7 BIODIVERSITY: SUSTAINING SOILS AND PRODUCING FOOD

Below that thin layer comprising the delicate organism known as the soil is a planet as lifeless as the moon.

G. Y. JACKS AND R. O. WHYTE

7-1 SOIL AND SOIL EROSION

What Are the Major Layers Found in Mature Soils? The material we call **soil** is a complex mixture of eroded rock, mineral nutrients, decaying organic matter, water, air, and billions of living organisms, most of them microscopic decomposers. Although soil is a potentially renewable resource, it is produced very slowly by the weathering of rock, deposit of sediments by erosion, and decomposition of organic matter in dead organisms.

Our lives and the lives of most other organisms depend on soil, especially topsoil. To a large extent all flesh is soil nutrients. In addition to food, soil indirectly provides us with wood, paper, fiber, and medicines, and it helps to purify the water we drink and to decompose and recycle biodegradable wastes. Yet since the beginning of agriculture human activities have led to rapid soil erosion, which can convert this potentially renewable resource into a nonrenewable resource. Entire civilizations have collapsed because they mismanaged the topsoil that supported their populations.

Mature soils are arranged in a series of zones called **soil horizons**, each with a distinct texture and composition that varies with different types of soils. A cross-sectional view of the horizons in a soil is called a **soil profile.** Most mature soils have at least three of the possible horizons (Figure 7-1).

The top layer, the *surface-litter layer,* or *O horizon,* consists mostly of freshly fallen and partially decomposed leaves, twigs, animal waste, fungi, and other organic materials. Normally, it is brown or black. The *topsoil layer,* or *A horizon,* is a porous mixture of partially decomposed organic matter, called **humus,** and some inorganic mineral particles. It is usually darker and looser than deeper layers. The roots of most plants and most of a soil's organic matter are concentrated in these two upper layers. As long as these layers are anchored by vegetation, soil stores water and releases it in a nourishing trickle instead of a devastating flood.

The two top layers of most well-developed soils teem with bacteria, fungi, earthworms, and small insects that interact in complex food webs (Figure 7-2). Bacteria and other decomposer microorganisms found by the billions in every handful of topsoil recycle the nutrients we and other land organisms need. They break down some complex organic compounds into simpler inorganic compounds soluble in water. Soil moisture carrying these dissolved nutrients is drawn up by the roots of plants and transported through stems and into leaves.

Some organic litter in the two top layers is broken down into a sticky, brown residue of partially decomposed organic material (humus). Because this humus is only slightly soluble in water, most of it stays in the topsoil layer. A fertile soil that produces high crop yields has a thick topsoil layer with lots of humus, which helps topsoil hold water and nutrients taken up by plant roots.

The color of its topsoil tells us a lot about how useful a soil is for growing crops. For example, dark-brown or black topsoil is nitrogen-rich and high in organic matter. Gray, bright yellow, or red topsoils are low in organic matter and need nitrogen enrichment to support most crops.

The *B horizon (subsoil)* and the *C horizon (parent material)* contain most of a soil's inorganic matter, mostly broken-down rock consisting of varying mixtures of sand, silt, clay, and gravel. The C horizon lies on a base of unweathered parent rock called *bedrock.*

The spaces, or pores, between the solid organic and inorganic particles in the upper and lower soil layers contain varying amounts of air (mostly nitrogen and oxygen gas) and water. Plant roots need oxygen for cellular respiration.

Some of the precipitation that reaches the soil percolates through the soil layers and occupies many of the soil's open spaces or pores. This downward movement of water through soil is called **infiltration.** As the water seeps down, it dissolves various soil components in upper layers and carries them to lower layers in a process called **leaching.**

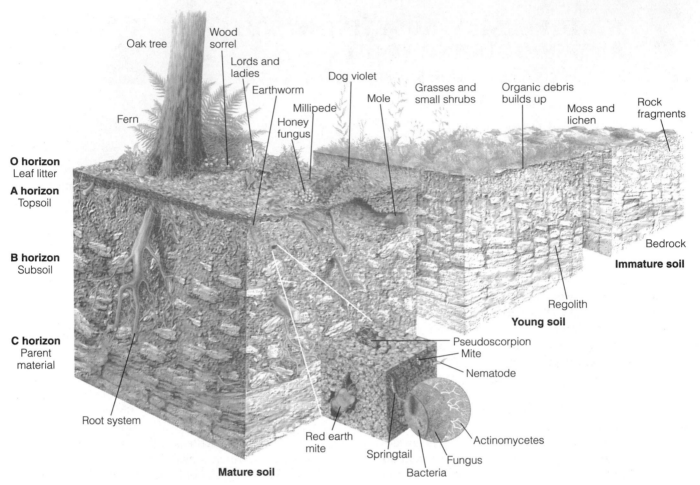

O horizon
Leaf litter

A horizon
Topsoil

B horizon
Subsoil

C horizon
Parent material

Oak tree

Wood sorrel

Lords and ladies

Earthworm

Fern

Millipede

Honey fungus

Dog violet

Mole

Grasses and small shrubs

Organic debris builds up

Moss and lichen

Rock fragments

Bedrock

Immature soil

Regolith

Young soil

Root system

Pseudoscorpion

Mite

Nematode

Red earth mite

Springtail

Bacteria

Fungus

Actinomycetes

Mature soil

Figure 7-1 Formation and generalized profile of soils. Horizons, or layers, vary in number, composition, and thickness, depending on the type of soil. (Used by permission of Macmillan Publishing Company from Derek Elsom, *Earth*, New York: Macmillan, 1992. Copyright ©1992 by Marshall Editions Developments Limited)

Soils develop and mature slowly. The earth's current mature soils vary widely from biome to biome in color, content, pore space, acidity, and depth. Five important soil types, each with a distinct profile, are shown in Figure 7-3. Most of the world's crops are grown on soils exposed when grasslands and deciduous forests are cleared.

How Do Soils Differ in Texture, Porosity, and Acidity?
Soils vary in their content of *clay* (very fine particles), *silt* (fine particles), *sand* (medium-size particles), and *gravel* (coarse to very coarse particles). The relative amounts of the different sizes and types of mineral particles determine **soil texture**, as depicted in Figure 7-4. Soils with roughly equal mixtures of clay, sand, silt, and humus are called **loams**.

To get an idea of a soil's texture, take a small amount of topsoil, moisten it, and rub it between your fingers and thumb. A gritty feel means that it contains a lot of sand. A sticky feel means a high clay content, and you should be able to roll it into a clump. Silt-laden soil feels smooth, like flour. A loam topsoil, which is best suited for plant growth, has a texture between these extremes— a crumbly, spongy feeling—with many of its particles clumped loosely together.

Soil texture helps determine **soil porosity**, a measure of the volume of pores or spaces per volume of soil and of the average distances between those spaces. A porous soil has many pores and can hold more water and air than a less porous soil. The average size of the spaces or pores in a soil determines **soil**

Q: Where is most of the federally owned and managed land in the United States?

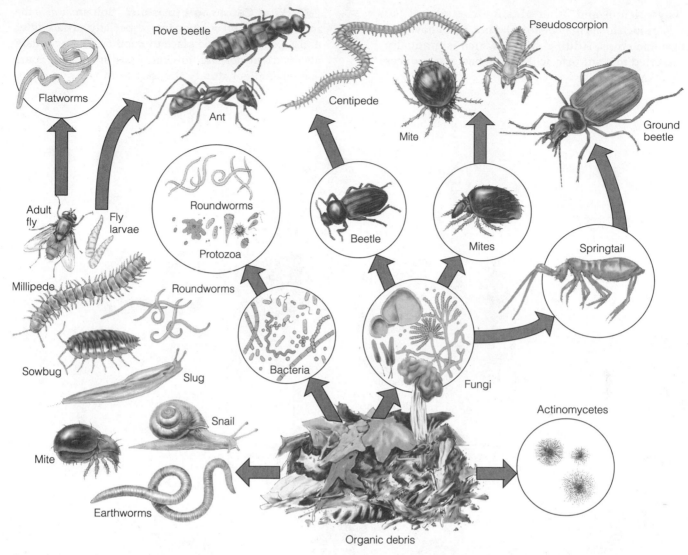

Figure 7-2 Greatly simplified food web of living organisms found in soil.

permeability: the rate at which water and air move from upper to lower soil layers. Soil porosity is also influenced by **soil structure**: how soil particles are organized and clumped together.

Loams are the best soils for growing most crops because they hold lots of water, but not too tightly for plant roots to absorb. Sandy soils are easy to work, but water flows rapidly through them. They are useful for growing irrigated crops or those with low water requirements, such as peanuts and strawberries.

The particles in clay soils are very small and easily compacted. When these soils get wet, they form large, dense clumps, which is why wet clay can be molded into bricks and pottery. Clay soils are more porous and have a greater water-holding capacity than sandy soils, but the pore spaces are so small that these soils have a low permeability. Because little water can infiltrate to lower levels, the upper layers can easily become too waterlogged for most crops.

A numerical scale of **pH** values is used to compare the acidity and alkalinity in water solutions (Figure 7-5). The pH of a soil influences the uptake of soil nutrients by plants, which vary in the pH ranges they can tolerate.

When soils are too acidic, the acids can be partially neutralized by an alkaline substance such as lime. Because lime speeds up the decomposition of organic matter in the soil, however, manure or another organic fertilizer should also be added to maintain soil fertility.

In dry regions such as much of the western and southwestern United States, rain does not leach away calcium and other alkaline compounds, so soils in such areas may be too alkaline (pH above 7.5) for some crops. Adding sulfur, which is gradually converted into sulfuric acid by soil bacteria, reduces soil alkalinity.

What Causes Soil Erosion? Soil erosion is the movement of soil components, especially surface litter and topsoil, from one place to another. The two main agents of erosion are flowing water and wind. Some soil erosion is natural—the long-term wearing down

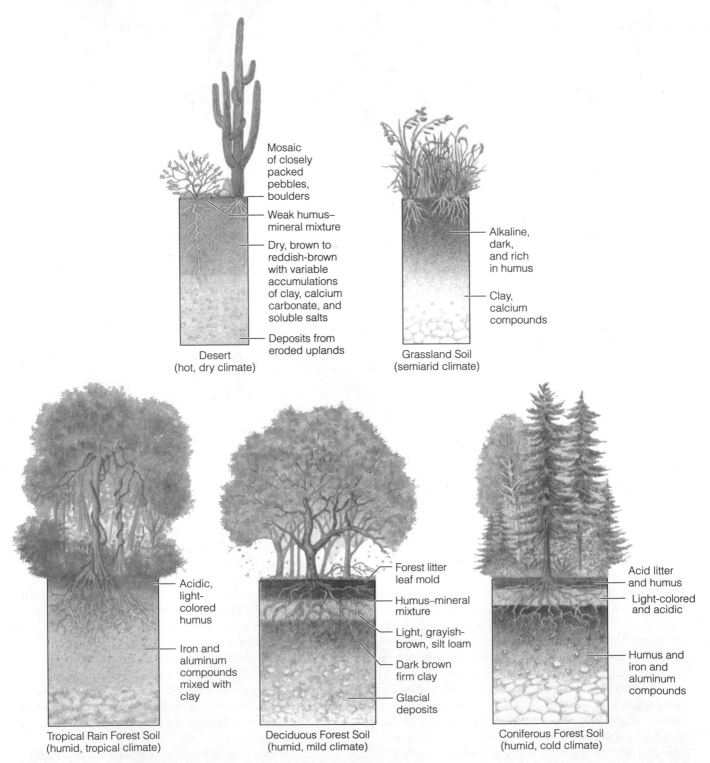

Desert
(hot, dry climate)

Mosaic of closely packed pebbles, boulders
Weak humus–mineral mixture
Dry, brown to reddish-brown with variable accumulations of clay, calcium carbonate, and soluble salts
Deposits from eroded uplands

Grassland Soil
(semiarid climate)

Alkaline, dark, and rich in humus
Clay, calcium compounds

Tropical Rain Forest Soil
(humid, tropical climate)

Acidic, light-colored humus
Iron and aluminum compounds mixed with clay

Deciduous Forest Soil
(humid, mild climate)

Forest litter leaf mold
Humus–mineral mixture
Light, grayish-brown, silt loam
Dark brown firm clay
Glacial deposits

Coniferous Forest Soil
(humid, cold climate)

Acid litter and humus
Light-colored and acidic
Humus and iron and aluminum compounds

Figure 7-3 Profiles of the principal soil types typically found in five different biomes.

Q: What percentage of earth's land area is covered by tropical forests?

of mountains and building up of plains and deltas by the combined action of physical, chemical, and biological forces. In undisturbed vegetated ecosystems, the roots of plants help anchor the soil, and usually soil is not lost faster than it forms.

However, farming, logging, construction, overgrazing by livestock, off-road vehicles, deliberate burning of vegetation, and other activities that destroy plant cover leave soil vulnerable to erosion. Such human activities can speed up erosion and destroy in a few decades what nature took hundreds to thousands of years to produce. In 1937, U.S. President Franklin D. Roosevelt sent a letter to the governors of the states in which he said, *"The nation that destroys its soil destroys itself."*

Losing topsoil makes a soil less fertile and less able to hold water. The resulting sediment—the largest source of water pollution—clogs irrigation ditches, boat channels, reservoirs, and lakes. The sediment-laden water is cloudy and tastes bad, fish die, and flood risk increases.

Soil, especially topsoil, is classified as a potentially renewable resource because it is continuously regenerated by natural processes. However, in tropical and temperate areas it takes 200–1,000 years (depending on climate and soil type) for 2.54 centimeters (1 inch) of new topsoil to form. If topsoil erodes faster than it

forms on a piece of land, the soil becomes a nonrenewable resource. Annual erosion rates for farmlands throughout the world are 7–100 times the natural renewal rate. Soil erosion is milder on forestland and rangeland than on cropland, but forest soil takes two to three times longer to restore itself than does cropland. Construction sites usually have the highest erosion rates by far.

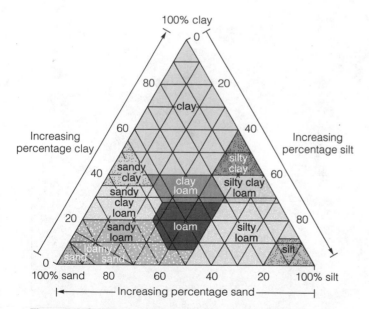

Figure 7-4 Soil texture depends on the proportions of clay, silt, and sand particles in the soil. Soil texture affects soil porosity, the average number and spacing of pores in a given volume of soil. Loams—roughly equal mixtures of clay, sand, silt, and humus—are the best soils for growing most crops. (Data from Soil Conservation Service)

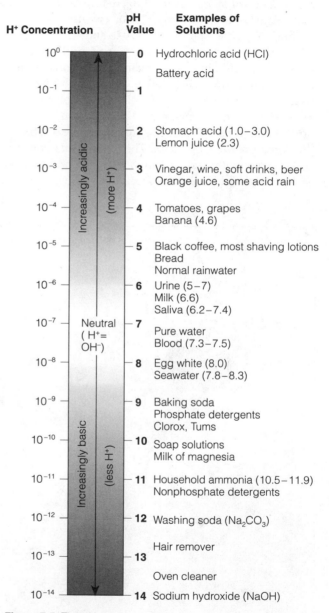

Figure 7-5 The pH scale, used to measure acidity and alkalinity of water solutions. Values shown are approximate. A neutral solution has a pH of 7, one with a pH greater than 7 is basic, or alkaline, and one with a pH less than 7 is acidic. The lower the pH below 7, the more acidic the solution. Each whole-number drop in pH represents a tenfold increase in acidity.

A: 6% (about 2% by tropical rain forests)

How Serious Is Global Soil Erosion? According to a UN Environment Programme survey, topsoil is eroding faster than it forms on about one-third of the world's cropland, causing an estimated 85% of the world's land degradation from human activities.

A 1992 study by the World Resources Institute and the UN Environment Programme found that soil on an area the size of China and India combined had been seriously eroded since 1945 (Figure 7-6). The study also found that about 15% of land scattered across the globe was too eroded to grow crops anymore because of a combination of overgrazing (35%), deforestation (30%), and unsustainable farming (28%). Two-thirds of the seriously degraded lands are in Asia and Africa.

Each year we must feed about 84 million more people with an estimated 24 billion metric tons (26 billion tons) less topsoil, eroded as a result of human activities. A 1995 study by David Pimentel and his associates estimated that global soil erosion is 75 billion metric tons (82.5 billion tons) per year—three times the previous estimate. The topsoil that washes and blows into the world's streams, lakes, and oceans each year would fill a train of freight cars long enough to encircle the planet at least 150 times—450 times if the larger estimate is correct. At that rate, the world is losing about 7–21% of its topsoil from actual or potential cropland each decade.

The situation is worsening as many poor farmers in some developing countries plow up marginal (easily erodible) lands to survive. In 1995, soil expert David Pimentel estimated that soil erosion causes nearly $400 billion per year worldwide in direct damage to agricultural lands and indirect damage to waterways, infrastructure, and human health—an average of $46 million in damages per hour.

Case Study: Soil Erosion in the United States According to the Soil Conservation Services, (now called the Natural Resource Conservation Service) about one-third of the nation's original prime topsoil has been washed or blown into streams, lakes, and oceans, mostly as a result of overcultivation, overgrazing, and deforestation.

In the 1930s, Americans learned a harsh environmental lesson when much of the topsoil in several windy and dry Midwestern states was lost through a

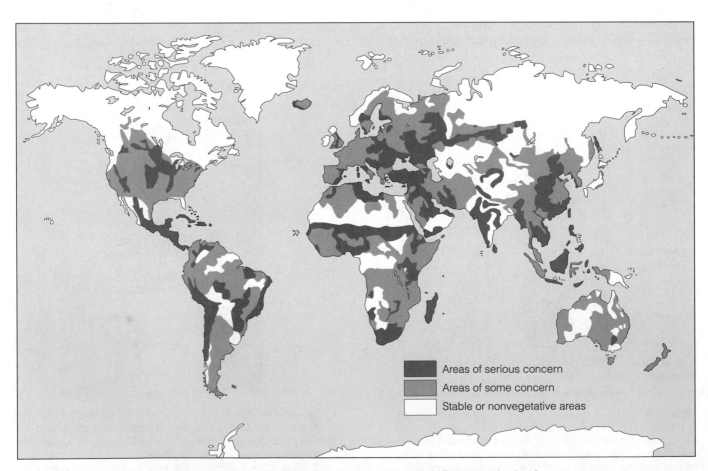

Figure 7-6 Global soil erosion. (Data from UN Environmental Programme and the World Resources Institute)

Legend:
- Areas of serious concern
- Areas of some concern
- Stable or nonvegetative areas

Q: What percentage of the world's species are found in tropical forests?

combination of poor cultivation practices and prolonged drought. Before settlers began grazing livestock and planting crops there in the 1870s, the deep and tangled root systems of native prairie grasses anchored the fertile topsoil firmly in place (Figure 7-3). Plowing the prairie tore up these roots, and the annual crops the settlers planted in their place had less extensive root systems.

After each harvest, the land was plowed and left bare for several months, exposing it to the plains winds. Overgrazing also destroyed large expanses of grass, denuding the ground. The stage was set for severe wind erosion and crop failures; all that was needed was a long drought.

Such a drought occurred between 1926 and 1934. In the 1930s, dust clouds created by hot, dry windstorms darkened the sky at midday in some areas; rabbits and birds choked to death on the dust. During May 1934, the entire eastern United States was blanketed by a cloud of topsoil blown off the Great Plains, some 2,400 kilometers (1,500 miles) away. Journalists gave the Great Plains a new name: the *Dust Bowl* (Figure 7-7).

During the so-called Dirty Thirties, cropland equal in area to Connecticut and Maryland combined was stripped of topsoil, and an area the size of New Mexico was severely eroded. Thousands of displaced farm families from Oklahoma, Texas, Kansas, and Colorado migrated to California or to the industrial cities of the Midwest and East. Most found no jobs because the country was in the midst of the Great Depression.

In May 1934, Hugh Bennett of the U.S. Department of Agriculture (USDA) went before a congressional hearing in Washington to plead for new programs to protect the country's topsoil. Lawmakers took action when Great Plains dust began seeping into the hearing room.

In 1935, the United States passed the Soil Erosion Act, which established the Soil Conservation Service (SCS) as part of the USDA. With Bennett as its first head, the SCS began promoting sound conservation practices, first in the Great Plains states and later elsewhere. Soil conservation districts were formed throughout the country, and farmers and ranchers were given technical assistance in setting up soil conservation programs.

Unfortunately, these heroic efforts have not yet stopped human-accelerated erosion in the Great Plains. The basic problem is that much of the region is better suited for moderate grazing than for farming. If the earth warms as projected, the region could become even drier and farming might have to be abandoned.

Today, soil on cultivated land in the United States is eroding about 16 times faster than it can form. Erosion rates are even higher in heavily farmed regions, including the Great Plains, which has lost one-third or

Figure 7-7 The Dust Bowl of the Great Plains, where a combination of extreme drought and poor soil conservation practices led to severe wind erosion of topsoil in the 1930s.

more of its topsoil in the 150 years since it was first plowed. Some of the country's most productive agricultural lands, such as those in Iowa, have lost about half their topsoil. California's soil is eroding about 80 times faster than it can be formed.

The estimated amount of topsoil that erodes away each day in the United States would fill a line of dump trucks 5,600 kilometers (3,500 miles) long. In 1995, soil expert David Pimentel estimated that the direct and indirect costs of soil erosion and runoff in the United States exceed $44 billion per year—an average loss of $5 million per hour.

Critics say that estimates of soil erosion and damages from such erosion are exaggerated. They point to studies by several soil scientists concluding that if current rates of cropland erosion in the United States continue for 100 years, crop yields will be only 3–10% less than they would be without such erosion.

However, David Pimentel and others point out that this estimate does not include all of the ecological effects caused by soil erosion. Such effects include reductions in soil depth, availability of soil water for crops, and soil organic matter and nutrients. When such effects are included, soil scientists and ecologists estimate that soil erosion causes a 15–30% reduction in crop productivity. This requires costly use of inorganic fertilizers to help replace lost soil nutrients. However, because fertilizers are not a substitute for fertile soil, there is a limit to the amount of fertilizer that can be applied before crop yields level off and then begin to decline.

How Do Excess Salts and Water Degrade Soils?

Approximately 16% of the world's cropland is now irrigated, producing about one-third of the world's food. Irrigated land can produce crop yields that are two to three times greater than those from rain watering, but

A: At least 50% (some biologists say 90%)

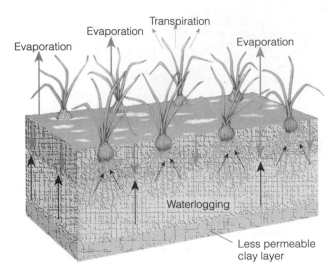

Evaporation Evaporation Transpiration Evaporation

Waterlogging

Less permeable
clay layer

Salinization

1. Irrigation water contains small amounts of dissolved salts.

2. Evaporation and transpiration leave salts behind.

3. Salt builds up in soil.

Waterlogging

1. Precipitation and irrigation water percolate downward.

2. Water table rises.

Figure 7-8 Salinization and waterlogging of soil on irrigated land without adequate drainage lead to decreased crop yields.

irrigation has its downside. Most irrigation water is a dilute solution of various salts, picked up as the water flows over or through soil and rocks.

Irrigation water not absorbed into the soil evaporates, leaving behind a thin crust of dissolved salts (such as sodium chloride) in the topsoil. The accumulation of these salts, called **salinization** (Figure 7-8), stunts crop growth, lowers yields, and eventually kills plants and ruins the land. According to a 1995 study, severe salinization has reduced yields on 20% of the world's irrigated cropland and another 30% has been moderately salinized.

Precipitation can desalinate soil, but this takes thousands of years in the arid and semiarid areas where irrigation is used. Salts can be flushed out of soil by applying much more irrigation water than is needed for crop growth, but this practice increases pumping and crop-production costs, wastes enormous amounts of water, and waterlogs plants if the water table rises close to the surface.

Heavily salinized soil can also be renewed by taking the land out of production for 2 to 5 years, installing an underground network of perforated drainage pipes, and flushing the soil with large quantities of low-salt water. However, this costly scheme only slows the salt buildup; it does not stop the process. Flushing salts from the soil also makes downstream irrigation water saltier unless the saline water can be drained into evaporation ponds rather than returned to the stream or canal.

Another problem with irrigation is **waterlogging** (Figure 7-8). Farmers often apply large amounts of irrigation water to leach salts deeper into the soil. Without adequate drainage, however, water accumulates underground, gradually raising the water table. Saline water then envelops the roots of plants, lowering their productivity and killing them after prolonged exposure. At least one-tenth of all irrigated land worldwide suffers from waterlogging, and the problem is getting worse.

7-2 SOLUTIONS: SOIL CONSERVATION

How Can Conservation Tillage Reduce Soil Erosion? Soil conservation involves reducing soil erosion and restoring soil fertility. For hundreds of years, farmers have used various methods to reduce soil erosion, most of which involve keeping the soil covered with vegetation.

In **conventional-tillage farming** the land is plowed and then the soil is broken up and smoothed to make a planting surface. In areas such as the Midwestern United States, harsh winters prevent plowing just before the spring growing season. Thus, cropfields are often plowed in the fall, baring the soil during the winter and early spring and leaving it vulnerable to erosion.

To reduce erosion, many U.S. farmers are using **conservation-tillage farming** (either *minimum-tillage* or *no-till farming*). The idea is to disturb the soil as little as possible while planting crops. With minimum-tillage farming, special tillers break up and loosen the subsurface soil without turning over the topsoil, previous crop residues, and any cover vegetation. In no-till farming, special planting machines inject seeds, fertilizers, and weed-killers (herbicides) into slits made in the unplowed soil.

Besides reducing soil erosion, conservation tillage saves fuel, cuts costs, holds more water in the soil, keeps the soil from getting packed down, and allows more crops to be grown during a season (multiple cropping). Yields are at least as high as those from conventional tillage. At first, conservation tillage was thought to require more herbicides, but a 1990 USDA study of corn production in the United States found no real difference in levels of herbicide use between conventional and conservation tillage systems. However, no-till cultivation of corn does leave stalks, which can serve as habitats for the corn borer; this can potentially increase the use of pesticides.

Willis, K.J., et al. 1998. "Prehistoric Land Degradation in Hungary: Who, How and Why?" *Antiquity*, vol. 72, no. 275, 101(13).

By 1997 conservation tillage was used on about 37% of U.S. croplands and is projected to be used on over half of it by 2005. The USDA estimates that using conservation tillage on 80% of U.S. cropland would reduce soil erosion by at least half. So far, the practice is not widely used in other parts of the world.

How Can Terracing, Contour Farming, Strip Cropping, and Alley Cropping Reduce Soil Erosion? Terracing can reduce soil erosion on steep slopes, each of which is converted into a series of broad, nearly level terraces that run across the land contour (Figure 7-9a). Terracing retains water for crops at each level and reduces soil erosion by controlling runoff.

Soil erosion can be reduced by 30–50% on gently sloping land by means of **contour farming**: plowing and planting crops in rows across, rather than up and down, the sloped contour of the land (Figure 7-9b). Each row planted along the contour of the land acts as a small dam to help hold soil and slow water runoff.

In **strip cropping**, a row crop such as corn alternates in strips with another crop (such as a grass or a grass–legume mixture) that completely covers the soil and thus reduces erosion (Figure 7-9c). The strips of the cover crop trap soil that erodes from the row crop. They also catch and reduce water runoff and help prevent the spread of pests and plant diseases. Nitrogen-fixing legumes such as soybeans or alfalfa planted in some of the strips help restore soil fertility.

Erosion can also be reduced by **alley cropping**, or **agroforestry**, a form of *intercropping* in which several crops are planted together in strips or alleys between trees and shrubs that can provide fruit or fuelwood (Figure 7-9d). The trees provide shade (which reduces water loss by evaporation) and help to retain and slowly release soil moisture. The tree and shrub trimmings can be used as mulch (green manure) for the crops and as fodder for livestock.

How Can Gully Reclamation, Windbreaks, and Land Classification Reduce Soil Erosion? Gully **reclamation** can restore sloping bare land on which water runoff quickly creates gullies (Figure 7-9e). Small gullies can be seeded with quick-growing plants such as oats, barley, and wheat for the first season, whereas deeper gullies can be dammed to collect silt and gradually fill in the channels. Fast-growing shrubs, vines, and trees can also be planted to stabilize the soil, and channels can be built to divert water from the gully and prevent further erosion.

Windbreaks, or **shelterbelts**, can reduce wind erosion. Long rows of trees are planted to partially block the wind (Figure 7-9f). Windbreaks, which are especially effective if uncultivated land is kept covered with vegetation, also help retain soil moisture, supply some

wood for fuel, and provide habitats for birds, pest-eating and pollinating insects, and other animals. Unfortunately, many of the windbreaks planted in the upper Great Plains after the 1930s Dust Bowl disaster have been cut down to make way for large irrigation systems and modern farm machinery.

Land can be evaluated with the goal of identifying easily erodible (marginal) land that should be neither planted with crops nor cleared of vegetation. In the United States, the SCS has set up a classification system to identify types of land that are suitable or unsuitable for cultivation.

The SCS relies on voluntary compliance with its guidelines in the almost 3,000 local and state soil and water conservation districts it has established, and it provides technical and economic assistance through local district offices. Such efforts and recent farm legislation have helped reduce soil erosion in the United States (Case Study, p. 205).

How Can We Maintain and Restore Soil Fertility? Fertilizers partially restore plant nutrients lost by erosion, crop harvesting, and leaching. Farmers can use either **organic fertilizer** from plant and animal materials or **commercial inorganic fertilizer** produced from various minerals.

Three basic types of *organic fertilizer* are animal manure, green manure, and compost. **Animal manure** includes the dung and urine of cattle, horses, poultry, and other farm animals. It improves soil structure, adds organic nitrogen, and stimulates beneficial soil bacteria and fungi. Despite its effectiveness, the use of animal manure in the United States has decreased. One reason is that separate farms for growing crops and raising animals have replaced most mixed animal-raising and crop-farming operations. Animal manure is available at feedlots near urban areas, but transporting it to distant rural crop-growing areas usually costs too much. Thus, much of this valuable resource is wasted and can end up polluting nearby bodies of water. In addition, tractors and other motorized farm machinery have replaced horses and other draft animals that naturally added manure to the soil.

Green manure is fresh or growing green vegetation plowed into the soil to increase the organic matter and humus available to the next crop. **Compost** is a rich natural fertilizer and soil conditioner that aerates soil, improves its ability to retain water and nutrients, helps prevent erosion, and prevents nutrients from being wasted in landfills. Farmers, homeowners, and communities produce compost by piling up alternating layers of nitrogen-rich wastes (such as grass clippings, weeds, animal manure, and vegetable kitchen scraps), carbon-rich plant wastes (dead leaves, hay, straw, sawdust), and topsoil. Composting also reduces the amount of waste

Hint: Enter the search term *soil degradation* using the Subject Guide.

CHAPTER 7   203

(a) Terracing

(b) Contour farming

(c) Strip cropping

(d) Alley cropping or agroforestry

(e) Gully reclamation

(f) Windbreaks or shelterbelts

Figure 7-9 Solutions: soil conservation methods.

Q: At current loss rates, how long will it take to destroy most remaining tropical forests?

Slowing Soil Erosion in the United States

Of the world's major food-producing countries, only the United States is reducing some of its soil losses through conservation tillage and government-sponsored soil conservation programs.

The 1985 Farm Act established a strategy for reducing soil erosion in the United States. In the first phase of this program, farmers are given a subsidy for highly erodible land they take out of production and replant with soil-saving grass or trees for 10 years. The land in such a *conservation reserve* cannot be farmed, grazed, or cut for hay. Farmers who violate their contracts must pay back all subsidies plus interest. Between 1985 and 1995, this program has cut soil losses on cropland in the United States by about 60%— a shining example of good news—

and could eventually cut such losses as much as 80%.

The second phase of the program required all farmers with highly erodible land to develop SCS-approved 5-year soil conservation plans for their entire farms by the end of 1990. A third provision of the Farm Act authorizes the government to forgive all or part of farmers' debts to the Farmers Home Administration if they agree not to farm highly erodible cropland or wetlands for 50 years. The farmers must plant trees or grass on this land or restore it to wetland.

In 1987, however, the SCS eased the standards that farmers' soil conservation plans must meet to keep them eligible for other subsidies. Environmentalists have accused the SCS of laxity in enforcing the Farm Act's "swampbuster" provisions, which deny federal funds to farm-

ers who drain or destroy wetlands on their property. In 1996 Congress weakened the act by allowing farmers to end their contracts without USDA approval. Despite some weaknesses, the 1985 Farm Act makes the United States the first major food-producing country to make soil conservation a national priority.

Even though these efforts to slow soil erosion are an important step, effective soil conservation is practiced on only about half of all U.S. agricultural land and on less than half of the country's most erodible cropland.

Critical Thinking

Do you believe that U.S. tax dollars should be used to pay farmers for taking highly erodible land out of production? Explain. What are the alternatives?

---

taken to landfills and incinerators. It can be done easily, with little labor by individuals and communities.

Another method for conserving soil nutrients is **crop rotation**. Corn, tobacco, and cotton can deplete the topsoil of nutrients (especially nitrogen) if planted on the same land several years in a row. Farmers using crop rotation may plant areas or strips with such nutrient-depleting crops one year; the next year, however, they plant the same areas with legumes—whose root nodules add nitrogen to the soil—or with crops such as soybeans, oats, barley, rye, or sorghum. This method helps restore soil nutrients and reduces erosion by keeping the soil covered with vegetation. It also helps reduce crop losses to insects by presenting them with a changing target.

Will Inorganic Fertilizers Save the Soil? Today, many farmers (especially in developed countries) rely on *commercial inorganic fertilizers* containing nitrogen (as ammonium ions, nitrate ions, or urea), phosphorus (as phosphate ions), and potassium (as potassium ions). Other plant nutrients may also be present in low or trace amounts.

Inorganic commercial fertilizers are easily transported, stored, and applied. Worldwide, their use

increased about tenfold between 1950 and 1989, but declined by 12% between 1990 and 1996. Today, the additional food they help produce feeds one of every three people in the world; without them, world food output would plummet an estimated 40%.

Commercial inorganic fertilizers have some disadvantages, however. They do not add humus to the soil. Unless animal manure and green manure are also added, the soil's content of organic matter, and thus its ability to hold water, will decrease, and the soil will become compacted and less suitable for crop growth. By decreasing the soil's porosity, inorganic fertilizers also lower its oxygen content and keep added fertilizer from being taken up as efficiently. In addition, most commercial fertilizers supply only 2 or 3 of the 20-odd nutrients needed by plants. Moreover, producing, transporting, and applying inorganic fertilizers require large amounts of energy and have significant environmental impacts.

The widespread use of commercial inorganic fertilizers, especially on sloped land near streams and lakes, also causes water pollution as some fertilizer nutrients are washed into nearby bodies of water; the resulting plant-nutrient enrichment causes algae blooms that use up oxygen dissolved in the water, thereby killing fish.

Rainwater seeping through the soil can also leach nitrates in commercial fertilizers into groundwater. Drinking water drawn from wells containing high levels of nitrate ions can be toxic, especially for infants.

Environmental historian Donald Worster reminds us that fertilizers are not a substitute for fertile soil:

> We can no more manufacture a soil with a tank of chemicals than we can invent a rain forest or produce a single bird. We may enhance the soil by helping its processes along, but we can never recreate what we destroy. The soil is a resource for which there is no substitute.

According to soil scientists, responsibility for reducing soil erosion should not be limited to farmers. Timber cutting, overgrazing, mining, and urban development carried out without proper regard for soil conservation cause at least 40% of soil erosion in the United States. Each of us has a role in seeing that these vital soil resources are used sustainably. Some things you can do to reduce soil erosion are listed in Appendix 4.

7-3 HOW FOOD IS PRODUCED

What Plants and Animals Feed the World?
Global food production has increased substantially over the past two decades, but producing food and other agricultural products by conventional means uses more soil, water, plant, animal, and energy resources—and causes more pollution and environmental damage—than any other human activity. To feed the 8 billion people projected by 2025, we must produce and equitably distribute as much food during the next 25 years as was produced since agriculture began about 10,000 years ago.

The multitude of species of plants and animals (species diversity) and the varieties of plants and animals (genetic diversity) that provide us with food are an important part of the planet's biodiversity. Biologists estimate that even though the earth has perhaps 30,000 plant species with parts that people can eat, only 15 plant and 8 animal species supply 90% of our food.

Four crops—wheat, rice, corn, and potato—make up more of the world's total food production than all other crops combined. These four, and most of our other food crops, are *annuals*, whose seeds must be replanted each year.

Two out of three of the world's people survive on grains (mainly rice, wheat, and corn), mostly because they can't afford meat. As incomes rise people consume even more grain, but indirectly in the form of meat (mostly beef, pork, and chicken), eggs, milk, cheese, and other products of grain-eating domesticated livestock.

What Are the Major Types of Food Production?
There are two major types of agricultural systems: industrialized and traditional. **Industrialized agriculture**, or high-input agriculture, uses large amounts of fossil fuel energy, water, commercial fertilizers, and pesticides to produce huge quantities of single crops (monocultures) or livestock animals for sale. Practiced on about 25% of all cropland, mostly in developed countries (Figure 7-10), industrialized agriculture has spread since the mid-1960s to some developing countries. **Plantation agriculture**, a form of industrialized agriculture practiced primarily in tropical developing countries, grows cash crops such as bananas, coffee, and cacao, mostly for sale in developed countries.

Traditional agriculture consists of two main types, which together are practiced by about 2.7 billion people in developing countries—almost half the people on the earth. **Traditional subsistence agriculture** typically produces only enough crops or livestock for a farm family's survival; in good years there may be a surplus to sell or to put aside for hard times. Subsistence farmers use primarily human labor and draft animals. Examples of this type of agriculture include numerous forms of shifting cultivation in tropical forests (Figure 5-8) and nomadic livestock herding. In **traditional intensive agriculture**, farmers increase their inputs of human and draft labor, fertilizer, and water to get a higher yield per area of cultivated land to produce enough food to feed their families and to sell for income.

How Have Green Revolutions Increased Food Production? Farmers can produce more food either by farming more land or by getting higher yields per unit of area from existing cropland. Since 1950, most of the increase in global food production has resulted from increased yields per unit of area of cropland in a process called the **green revolution**.

This process involves three steps: **(1)** developing and planting monocultures of selectively bred or genetically engineered high-yield varieties of key crops such as rice, wheat, and corn; **(2)** lavishing fertilizer, pesticides, and water on crops to produce high yields; and **(3)** often increasing the intensity and frequency of cropping. This approach dramatically increased crop yields in most developed countries between 1950 and 1970 in what is considered the *first green revolution* (Figure 7-11).

A *second green revolution* has been taking place since 1967 (Figure 7-11), when fast-growing dwarf varieties of rice and wheat, specially bred for tropical and subtropical climates, were introduced into several developing countries. With sufficient fertile soil and enough fertilizer, water, and pesticides, yields of these new plants can be two to five times those of traditional wheat and rice varieties. The fast growth also allows

Q: What percentage of tropical forest plants have been studied for their possible use as human resources?

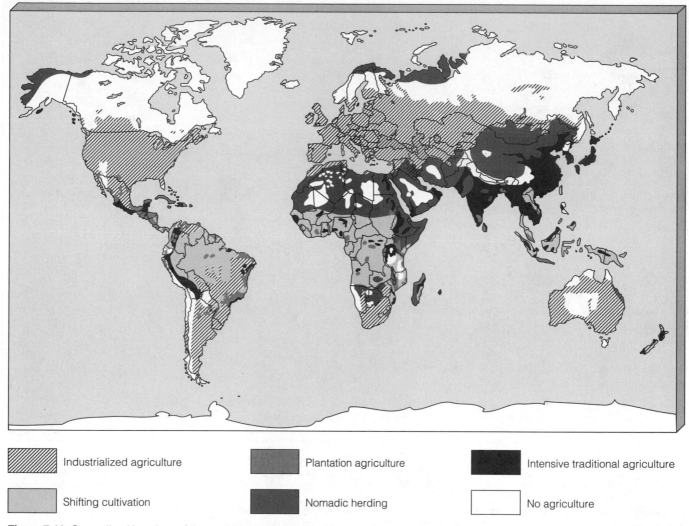

| Industrialized agriculture | Plantation agriculture | Intensive traditional agriculture |
| Shifting cultivation | Nomadic herding | No agriculture |

Figure 7-10 Generalized locations of the world's principal types of food production.

farmers to grow two or even three crops a year (multiple cropping) on the same land.

Producing more food on less land is also an important way to protect biodiversity by saving large areas of forests, grasslands, wetlands, and mountain terrain from being used to grow food. According to Dennis Avery, without the two major green revolutions the world would have lost wild land equal to the combined land area of the United States, Europe, and Brazil.

These yield increases depend not only on having fertile soil and ample water, but also on extensive use of fossil fuels to run machinery, produce and apply inorganic fertilizers and pesticides, and pump water for irrigation. All told, green-revolution agriculture uses about 8% of the world's oil output.

These high inputs of energy, water, fertilizer, and pesticides on high-yield crop varieties have yielded

dramatic results. At some point, however, additional inputs become useless because no more output can be squeezed from the land and the crop varieties—the principle of diminishing returns in action. In fact, yields may even start dropping, for a number of reasons: The soil erodes, loses fertility, and becomes salty and waterlogged (Figure 7-8); underground and surface water supplies become depleted and polluted with pesticides and nitrates from fertilizers; and populations of rapidly breeding pests develop genetic immunity to widely used pesticides.

Case Study: Food Production in the United States
Since 1940 U.S. farmers have more than doubled crop production without cultivating more land, a result of industrialized agriculture using green-revolution techniques in a favorable climate on some of the world's

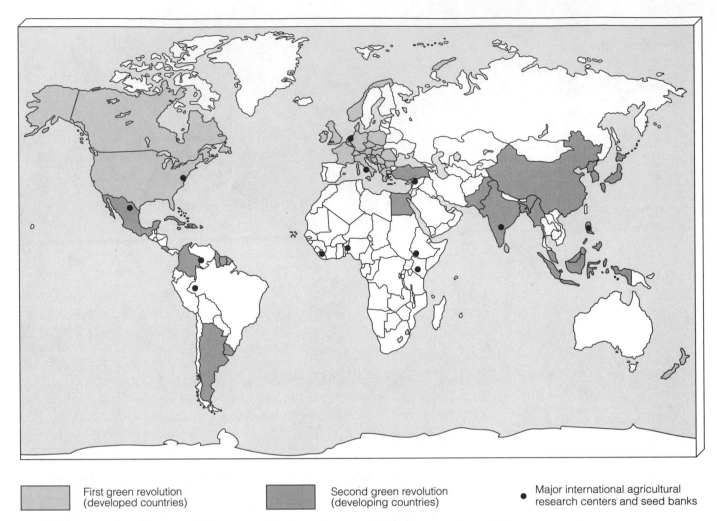

Figure 7-11 Countries whose crop yields per unit of land area increased during the two green revolutions. The first took place in developed countries between 1950 and 1970; the second has occurred since 1967 in developing countries with enough rainfall or irrigation capacity. Several agricultural research centers and gene or seed banks play a key role in developing high-yield crop varieties.

most fertile and productive soils. This has also kept large areas of forests, grasslands, wetlands, and easily erodible land from being converted to farmland.

Farming has become *agribusiness* as big companies and larger family-owned farms have taken control of most U.S. food production. Only about 650,000 Americans are full-time farmers. However, about 9% of the population is involved in the U.S. agricultural system, from growing and processing food to distributing it and selling it at the supermarket.

In terms of total annual sales, agriculture is the biggest industry in the United States, bigger than the automotive, steel, and housing industries combined. It generates about 18% of the country's gross national product and 19% of all jobs in the private sector, employing more people than any other industry. The U.S. agricultural system is highly productive. Currently,

each U.S. farmer fed and clothed 140 people (105 at home and 35 abroad), up from 58 in 1976.

The industrialization of agriculture was made possible by the availability of cheap energy, most of it from oil. Agriculture consumes about 17% of all commercial energy in the United States each year (Figure 7-12). Most plant crops in the United States provide more food energy than the energy used to grow them. However, if we include livestock as well as crops, the U.S. food production system currently uses about three units of fossil fuel energy to produce one unit of food energy.

Energy efficiency is much lower if we look at the whole U.S. food system. Considering the energy used to grow, store, process, package, transport, refrigerate, and cook all plant and animal food, *about 10 units of nonrenewable fossil fuel energy are needed to put 1 unit of food energy on the table.* By comparison, every unit of energy from human

Brown, Lester. 1998. "Food Scarcity: An Environmental Wakeup Call." *The Futurist*, vol. 32, no. 1, 34(5).

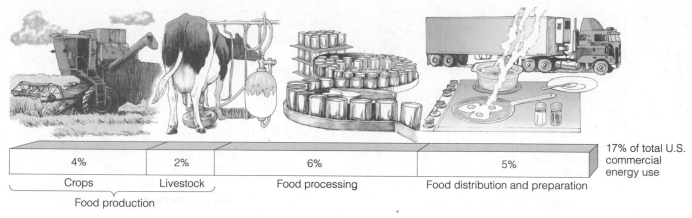

4%	2%	6%	5%	17% of total U.S. commercial energy use
Crops	Livestock	Food processing	Food distribution and preparation	

Food production

Figure 7-12 In the United States, industrialized agriculture uses about 17% of all commercial energy.

labor in subsistence farming provides at least 1 unit of food energy; with traditional intensive farming, each unit of energy provides up to 10 units of food energy.

What Is Traditional Agriculture? Agrodiversity in Action Traditional farmers in developing countries today grow about 20% of the world's food on about 75% of its cultivated land. Many traditional farmers simultaneously grow several crops on the same plot, a practice known as *interplanting*. Such crop diversity reduces the chance of losing most or all of their year's food supply to pests and other misfortunes.

Common interplanting strategies found throughout the world, mostly in developing countries, include the following:

- *Polyvarietal cultivation*, in which a plot is planted with several varieties of the same crop.

- *Intercropping*, in which two or more different crops are grown at the same time on a plot (for example, a carbohydrate-rich grain that uses soil nitrogen alongside a protein-rich legume that puts it back).

- *Agroforestry*, or *alley cropping*, in which crops and trees are planted together (Figure 7-9d). For example, a grain or legume crop can be planted around fruit-bearing orchard trees or in rows between fast-growing trees or shrubs that can be used for fuelwood or for adding nitrogen to the soil.

- *Polyculture*, a more complex form of intercropping in which many different plants maturing at various times are planted together. If cultivated properly, these plots can provide food, medicines, fuel, and natural pesticides and fertilizers on a sustainable basis.

Recent ecological research on crop yields of 14 artificial ecosystems found that on average, polyculture (with four or five different crop species) produces higher yields per unit of area than high-input monoculture. This important finding has major implications for development of high-yield sustainable agriculture in developing countries, which can combine the techniques of traditional high-yield interplanting with modern inputs of organic or inorganic fertilizer and irrigation.

7-4 WORLD FOOD PROBLEMS AND CHALLENGES

How Much Has Food Production Increased? Figure 7-13 shows the success story of global agriculture. Between 1950 and 1990, world grain production almost tripled (Figure 7-13, left), and per capita production rose by about 36% (Figure 7-13, right), helping reduce global hunger and malnutrition. During the same period, average food prices adjusted for inflation dropped by 25% and the amount of food traded in the world market quadrupled.

Despite these impressive achievements in food production, population growth is outstripping food production and distribution in areas that support 2 billion people. Since 1978 grain production has lagged behind population growth in 88 developing countries. Food production in Africa has been rising steadily since 1961, but not fast enough to keep up with population growth; between 1974 and 1996 per capita production fell by 20%.

Since 1950 global grain production has been rising but its rate of growth has slowed (Figure 7-13, left), and the rate of per capita grain production has declined slightly since 1985 (Figure 7-13, right). More than 100 countries regularly import food from the United States, Canada, Australia, Argentina, western Europe, New Zealand, Thailand, and a few other countries.

Unless death rates rise sharply, we seem destined to have a population of around 8 billion people by 2025.

Hint: Enter the search term *food supply, environmental aspects* using the Subject Guide.

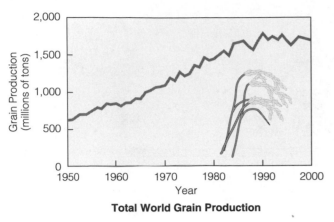

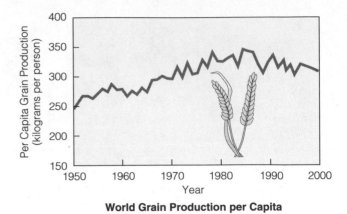

Figure 7-13 Total worldwide grain production of wheat, corn, and rice, and per capita grain production, 1950–96. (Data from U.S. Department of Agriculture and Worldwatch Institute)

To provide this many people with even a meatless subsistence diet will require doubling food production and distribution between 1990 and 2025. Actually, food production will have to be increased even more because projected increasing affluence in many developing countries will raise meat consumption and require higher amounts of grain for animal feed.

🦠 How Serious Are Undernutrition, Malnutrition, and Overnutrition?

People who cannot grow or buy enough food to meet their basic energy needs suffer from **undernutrition**. To maintain good health and resist disease, people need not only a certain number of calories, but also food with the proper amounts of protein (from animal or plant sources), carbohydrates, fats, vitamins, and minerals.

People who are forced to live on a low-protein, high-carbohydrate diet consisting only of grains such as wheat, rice, or corn often suffer from **malnutrition**: deficiencies of protein and other key nutrients. Many of the world's desperately poor people, especially children, suffer from both undernutrition and malnutrition.

Here's some good news. *Between 1970 and 1995 the worldwide proportion of people suffering from chronic undernutrition fell from 36% to 14%.* Also, despite population growth, *the estimated number of chronically malnourished people fell from 940 million in 1970 to 840 million in 1995.*

Despite this progress, about one of every five people in developing countries (including one of every three children below age 5) is chronically undernourished or malnourished; 87% of them live in Asia and Africa. Such people are disease prone, and adults are too weak to work productively or think clearly. As a result, their children also tend to be underfed and malnourished. If these children survive to adulthood, many are locked in a tragic malnutrition–poverty cycle (Figure 7-14) that can be perpetuated for generations.

It is estimated that each year at least 10 million people, half of them children under age 5, die prematurely from undernutrition, malnutrition, or normally nonfatal diseases such as measles and diarrhea worsened by malnutrition. Some researchers put this annual death toll at 18 million. However, children don't have to die prematurely because of undernutrition and malnutrition (Solutions, p. 213).

Each of us must have a small daily intake of vitamins that cannot be made in the human body. Although balanced diets, vitamin-fortified foods, and vitamin supplements have slashed the number of vitamin-deficiency diseases in developed countries, millions of cases occur each year in developing countries. Each year up to 500,000 children go blind because their diet lacks vitamin A.

Other nutritional-deficiency diseases are caused by the lack of certain minerals. For example, too little iron (a component of hemoglobin that transports oxygen in the blood) causes anemia. This mineral deficiency causes fatigue, makes infection more likely, increases a woman's chances of dying in childbirth, and increases an infant's chances of dying from infection during its first year of life. In tropical regions of Asia, Africa, and Latin America, iron-deficiency anemia affects about 350 million people.

Whereas an estimated 17% of the people in developing countries suffer from undernutrition and malnutrition, about 15% of the people in developed countries (32% in the United States) suffer from **overnutrition**: an excessive intake of food, especially fats, that can cause obesity (excess body fat).

Overnutrition is associated with at least two-thirds of the deaths in the United States each year. A study of thousands of Chinese villagers indicates that the healthiest diet for humans is largely vegetarian, with only 10–15% of calories coming from fat—in contrast to the typical meat-based diet, in which 40% of the calories come from fat.

Q: How many people cannot find or buy enough fuelwood to meet their basic needs?

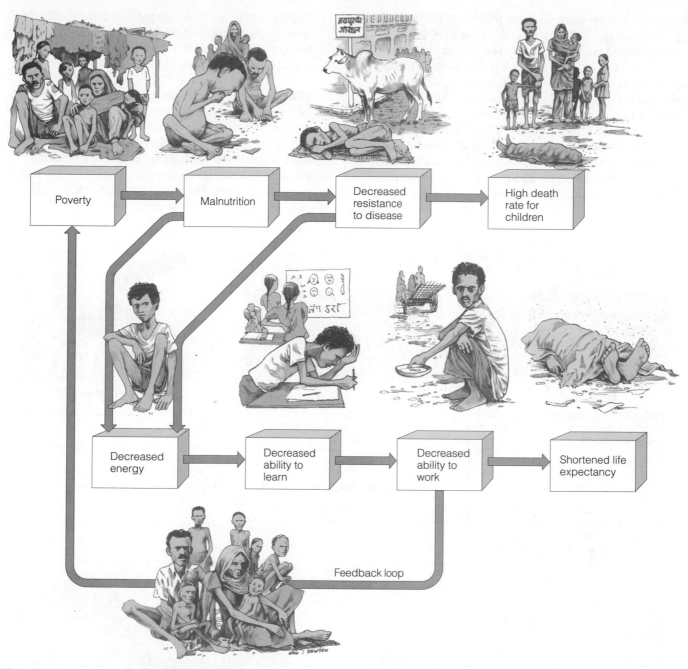

Figure 7-14 Interactions among poverty, malnutrition, and disease form a tragic cycle that tends to perpetuate such conditions in succeeding generations of families.

Do We Produce Enough Food to Feed the World's People? The *good news* is that we produce more than enough food to meet the basic nutritional needs of every person on the earth today. If distributed equally, the grain currently produced worldwide is enough to give everyone a *meatless subsistence diet.*

The *bad news* for those not getting enough food is that food is not distributed equally among the world's people because of differences in soil, climate, political and eco-

nomic power, and average per capita income throughout the world. Most agricultural experts agree that *the principal cause of hunger and malnutrition is and will continue to be poverty,* which prevents poor people from growing or buying enough food regardless of how much is available.

If everyone ate the diet typical of a person in a developed country, with 30–40% of the calories coming from animal products, estimates suggest that the current world agricultural system would support

A: About 2.2 billion in 1998, about one of every three people on earth

only 2.5 billion people—less than half the current population and only one-fourth of the 10 billion people projected sometime in the next century.

Moreover, increases in global and per capita food production often hide large differences in food supply and quality among and within countries. For example, despite impressive gains in total and per capita food production since 1970, roughly 40% of India's population suffers from malnutrition because they are too poor to buy or grow enough food to meet their basic needs.

Developed countries also have pockets of poverty, hunger, and malnutrition. According to a 1997 U.S. Department of Agriculture study, as many as 11 million Americans (excluding 600,000 homeless people) go hungry each year and 2 million of them suffer preventable severe hunger.

What Are the Environmental Effects of Producing Food? *Agriculture has a greater harmful impact on air, soil, water, and biodiversity resources than any other human activity* (Figure 7-15). Biologist David Pimentel has estimated that the harmful costs not included in the prices of food in the United States are $150–200 billion per year. According to a 1990 UN study, degradation of

Biodiversity Loss

Loss and degradation of habitat from clearing grasslands and forests and draining wetlands

Fish kills from pesticide runoff

Killing of wild predators to protect livestock

Loss of genetic diversity from replacing thousands of wild crop strains with a few monoculture strains

Soil

Erosion

Loss of fertility

Salinization

Waterlogging

Desertification

Air Pollution

Greenhouse gas emissions from fossil fuel use

Other air pollutants from fossil fuel use

Pollution from pesticide sprays

Water

Aquifer depletion	Surface and groundwater pollution from pesticides and fertilizers
Increased runoff and flooding from land cleared to grow crops	Overfertilization of lakes and slow-moving rivers from runoff of nitrates and phosphates from fertilizers, livestock wastes, and food processing wastes
Sediment pollution from erosion	
Fish kills from pesticide runoff	

Human Health

Nitrates in drinking water

Pesticide residues in drinking water, food, and air

Contamination of drinking and swimming water with disease organisms from livestock wastes

Bacterial contamination of meat

Figure 7-15 Major environmental effects of food production.

Q: Over the last 50 years what has happened to the total area of forest and woodlands in the United States?

irrigated cropland, rain-fed cropland, and rangeland now costs the world more than $42 billion a year in lost crop and livestock output; this loss is roughly equal to the annual value of the entire U.S. grain harvest.

Many analysts believe that it is possible to produce enough food to feed the 8 billion people projected by 2025 through new advances in agricultural technology and by spreading the use of existing green-revolution or high-yield techniques. Other analysts disagree. They have serious doubts about the ability of new food production technologies and food distribution systems to keep up with current levels of population growth, mostly because the harmful environmental effects of agriculture will reduce yields (Case Study, p. 214).

According to Worldwatch Institute estimates, between 1945 and 1990, erosion, salinization, waterlogging, and other forms of environmental degradation eliminated an area of land from food production equal to the cropland of two Canadas. This trend is expected to accelerate as modern industrialized farming and environmentally unsound subsistence farming increase in coming decades.

7-5   SOLUTIONS TO WORLD FOOD PROBLEMS

Is Increasing Crop Yields the Answer? Agricultural experts expect most future increases in food yields per hectare on existing cropland to result from improved strains of plants and from expansion of green-revolution technology to new parts of the world. For example, a new strain of rice developed by the International Rice Research Institute and expected to be commercially available by 2000 could increase rice yields by as much as 20%.

Scientists are working to create new green revolutions—actually *gene revolutions*—by using genetic engineering and other forms of biotechnology. Over the next 20–40 years they hope to breed high-yield plant strains that are more resistant to insects and disease, thrive on less fertilizer, make their own nitrogen fertilizer, do well in slightly salty soils, can withstand drought, and use solar energy more efficiently during photosynthesis. But according to Donald Duvick, former director of research at Pioneer HiBred International (one of the world's largest seed producers), "No breakthroughs are in sight. Biotechnology, while essential to progress, will not produce sharp upward swings in yield potential except for isolated crops in certain situations."

Several factors have limited the success of the green and gene revolutions to date, and may continue to do so. Without huge amounts of fertilizer and water, most green-revolution crop varieties produce yields that are no higher (and are sometimes lower) than those

from traditional strains; this is why the second green revolution has not spread to many arid and semiarid areas (Figure 7-11). Without ample water, good soil, and favorable weather, new genetically engineered crop strains could fail. Furthermore, the cost of genetically engineered crop strains is too high for most of the world's subsistence farmers in developing countries.

Continuing to increase inputs of fertilizer, water, and pesticides eventually produces no additional increase in crop yields; the *J*-shaped curve of crop productivity slows down, reaches its limits, levels off, and becomes an *S*-shaped curve. At that point, the yield potential for any particular grain in a country depends mostly on that country's soil moisture (from rainfall or ability to irrigate), day length or latitude, temperatures, and solar intensity—factors that are difficult to alter. Grain yields per hectare are still increasing in many parts of the world, but at a much slower rate.

Moreover, Indian economist Vandana Shiva contends that overall gains in crop yields from new green- and gene-revolution varieties may be much lower

Can China's Population Be Fed?

CASE STUDY

Since 1970 China has made significant progress in feeding its people and slowing its rate of population growth (Section 3-4). But with its economy doubling in size every 8 years and 12 million more people each year, there is growing concern that crop yields may not be able to keep up with demand. A basic problem is that with 21% of the world's people, China has only 7% of the world's cropland and fresh water, 3% of its forests, and 2% of its oil.

China needs to expand its area of cropland. Instead, China is experiencing a drop in available farmland, mostly because of a lack of suitable cropland, massive conversion of existing cropland to nonfarm uses as the country industrializes, and environmental degradation (especially erosion, salinization, and waterlogging of soil and severe air pollution as a result of increased burning of coal).

According to Worldwatch Institute projections, China's grain production is likely to fall by at least 20% between 1990 and 2030. Even if China's booming economy resulted in no increases in meat consumption, this 20% drop would mean

that by 2030 China would need to import more than the world's entire 200 million tons of grain exports in 1993 (roughly half from the United States).

However, if the increased demand for meat led to a rise in per capita grain consumption equal to the current level in Taiwan (one-half the current U.S. level), China would have to import more than the entire current grain output of the United States.

The Worldwatch Institute warns that if either of these scenarios is correct, no country or combination of countries has the potential to supply even a small fraction of China's potential food supply deficit. This is not even taking into account the huge food deficits that are projected in other parts of the world by 2030, especially Africa and India.

However, according to a 1997 study by the International Food Policy Institute, China should be able to feed its population and begin exporting grain again by 2020 if the government invests in expanding irrigation and increasing agricultural research. As China's population growth rate declines, the institute projects that the resulting decrease in direct grain consumption will offset the rapid increase

in production of cereal grains to feed meat-producing animals. Recent satellite surveys also show that China has far more potential cropland than previously thought.

Other analysts believe that serious and rapidly growing environmental problems may also limit China's economic growth and its ability to feed its people. So far China has concentrated on rapid industrialization and devoted little attention to sustainable use of its resources and to reducing pollution and environmental degradation.

If China begins acting now to chart a new course, it has a unique opportunity to leapfrog over the traditional Western forms of economic development and show the world how to build an environmentally sustainable economy over the next few decades.

Critical Thinking

If the scenarios about China's growing dependence on food imports are valid, how might this affect **(a)** world food prices, **(b)** your life, and **(c)** the harmful environmental impacts of food production (Figure 7-15)? What ways can you suggest for dealing with this potential problem?

than claimed. The reason is that the yields are based on comparisons between the output per hectare of old and new *monoculture* varieties, rather than between the even higher yields per hectare for *polyculture* cropping systems and the new monoculture varieties that often replace them.

Connections: Will Loss of Genetic Diversity Limit Crop Yields? Some agricultural scientists think that new genetically engineered or crossbred varieties will enable yields of key crops to continue rising. Other scientists question whether this is possible, primarily because of the environmental impacts of current forms of industrialized agriculture (Figure 7-15) and the accelerating loss of biodiversity, which can limit

the genetic raw material needed for future green and gene revolutions.

Scientists can crossbreed varieties of animal and plant life, and genetic engineers can move genes from one organism to another, but they need the genetic materials in the earth's existing species to work with. We are losing much of this genetic diversity as a small number of specially bred monoculture varieties of key crops have replaced thousands of strains of various crops and natural areas have been cleared.

According to the UN Food and Agriculture Organization by the year 2000 two-thirds of all seed planted in developing countries were of uniform strains. Such genetic uniformity increases the vulnerability of food crops to pests, diseases, and harsh weather. Many biol-

Q: What percentage of the original old-growth forests in the United States and Canada has been cut?

ogists argue that this decreased variability, plus growing species extinction, can severely limit the potential of future green and gene revolutions.

Wild varieties of the world's most important plants can be collected and stored in gene or seed banks, agricultural research centers, and botanical gardens. However, space and money severely limit the number of species that can be preserved. Many plants (such as potatoes) cannot be stored successfully as seed in gene banks. Power failures, fires, or unintentional disposal of seeds can also cause irreversible losses.

In addition, because stored seeds don't remain alive indefinitely, periodically they must be planted (germinated) and new seeds collected for storage. Unless this is done, seed banks become seed morgues. Moreover, stored plant species stop evolving, so it may be difficult to reintroduce them into their native habitats, which may have changed in the meantime.

Because of these limitations, ecologists and plant scientists warn that the only effective way to preserve the genetic diversity of most plant and animal species is to protect representative ecosystems throughout the world from agriculture and other forms of development (Chapter 5).

Will People Try New Foods? Some analysts recommend greatly increased cultivation of less widely known plants to supplement or replace such staples as wheat, rice, and corn. One of many possibilities is the winged bean, a protein-rich legume now common only in New Guinea and Southeast Asia. Because of nitrogen-fixing nodules in its roots, this fast-growing plant needs little fertilizer. Indeed, this plant produces so many different edible parts that it has been called a "supermarket on a stalk."

Insects are important food items in many parts of the world. Mopani emperor moth larvae are among several insects eaten in South Africa. Kalahari Desert dwellers eat cockroaches; lightly toasted butterflies are a favorite food in Bali; French-fried ants are sold on the streets of Bogota, Colombia; and Malaysians love deep-fried grasshoppers. Most of these insects are 58–78% protein by weight—three to four times as protein-rich as beef, fish, or eggs. Two basic problems are getting farmers to take the financial risk of cultivating new types of food crops and convincing consumers to try new foods.

Most crops we depend on are tropical annuals. Each year the land is cleared of all vegetation, dug up, and planted with their seeds. Some plant scientists believe we should rely more on polycultures of perennial crops, which are better adapted to regional soil and climate conditions than most annual crops (Solutions, right). Using perennials would also eliminate the need to till soil and replant seeds each year; it would greatly save energy and water and reduce soil erosion and

Growing Perennial Crops on the Kansas Prairie

SOLUTIONS

When you think about farms in Kansas, you probably picture seemingly endless fields of wheat or corn plowed up and planted each year. By 2040 the picture might change, thanks to pioneering work at the nonprofit Land Institute near Salina, Kansas.

The institute, headed by plant geneticist Wes Jackson, is experimenting with an ecological approach to agriculture on the Midwestern prairie. The goal is to grow food crops by planting a mix of *perennial* grasses, legumes, sunflowers, grain crops, and plants that provide natural insecticides in the same field (polyculture). Because these plants are perennials, the soil doesn't have to be plowed up and prepared each year to replant them. The institute's goal is to raise food by mimicking many of the natural conditions of the prairie without losing fertile grassland soil.

Institute researchers believe that perennial polyculture can be blended with modern agriculture to reduce the massive and growing harmful environmental effects of industrialized agriculture (Figure 7-15). Perennial polyculture is especially suitable for marginal land, leaving prime, flat land available for raising annual crops.

By eliminating yearly soil preparation and planting, perennial polyculture requires much less labor than conventional monoculture or diversified organic farms that grow annual crops. It reduces soil erosion because the unplowed soil is not exposed to wind and rain and also reduces pollution caused by chemical fertilizers and pesticides.

If the institute and similar groups doing such earth-sustaining research succeed, within a few decades many people may be eating food made from a mix of perennials such as (1) *eastern gamma grass* (a warm-season grass that is a relative of corn with three times as much protein as corn and twice as much as wheat), (2) *mammoth wildrye* (a cool-season grass that is distantly related to rye, wheat, and barley), (3) *Illinois bundleflower* (a wild nitrogen-producing legume that can enrich the soil and whose seeds may serve as livestock feed), and (4) *Maximillian sunflower* (which produces seeds with as much protein as soybeans).

Critical Thinking

What might be some drawbacks for relying more on perennial food crops? Why will conventional food growing companies be opposed to relying more on perennial crop varieties? How might such opposition be overcome?

A: 85–95% in the United States; 60% in Canada

sediment water pollution. Of course, widespread use of perennials would reduce the profits of agribusinesses selling annual seeds, fertilizers, and pesticides, which explains why they don't favor this approach.

Is Cultivating More Land the Answer? Between 1980 and 1990, the area of the world's cropland expanded by only 2%. Theoretically, the world's cropland could be more than doubled by clearing tropical forests and irrigating arid land (Figure 7-16). However, many analysts believe that this potential for agricultural expansion is often overestimated because much of the land is marginal land, where cultivation is unlikely to be sustainable.

Clearing rain forests to grow crops and graze livestock, for example, can have disastrous ecological consequences, as discussed in Section 5-5. In addition, potential cropland in savanna and other semiarid land in Africa cannot be used for farming or livestock grazing because of the presence of 22 species of the tsetse fly.

Some researchers hope to develop new methods of intensive cultivation in tropical areas. But other scientists argue that it makes more ecological and economic sense to combine various ancient methods of shifting cultivation (Figure 5-8), followed by fallow periods long enough to restore soil fertility with various forms of polyculture.

Much of the world's potentially cultivable land lies in dry areas, especially in Australia and Africa. Large-scale irrigation in these areas would require large, expensive dam projects with a mixture of beneficial and harmful impacts (Figure 4-18) and large inputs of fossil fuel to pump water long distances. Large-scale irrigation could also deplete groundwater supplies by removing water faster than it is replenished. The land would need constant and expensive maintenance to prevent erosion, groundwater contamination, salinization, and waterlogging.

Thus, much of the new cropland that could be developed would be on land that is marginal for raising crops, requiring expensive inputs of fertilizer, water, and energy. Furthermore, these potential increases in cropland would not offset the projected loss of almost one-third of today's cultivated cropland caused by erosion, overgrazing, waterlogging, salinization, mining, and urbanization.

Even if it is financially feasible, such expansion would reduce wildlife habitats and thus the world's biodiversity and ecological integrity. According to the UN Food and Agriculture Organization (FAO), cultivating all potential cropland in developing countries would reduce forests, woodlands, and permanent pasture by 47%.

Can We Harvest More Fish and Shellfish? About 70% of the annual commercial catch of fish and shellfish comes from the ocean using methods shown in Figure 6-5; 99% of this catch is taken from plankton-rich coastal waters. However, this vital coastal zone is being disrupted and polluted at an alarming rate (p. 302). The remainder of the annual catch comes from using aquaculture to raise fish in ponds and underwater cages (20%) and from inland freshwater fishing from lakes and rivers (10%). About one-third of the world fish harvest is not consumed directly by humans and is used primarily as animal feed, fish meal, and oils.

Here is some *good news*. Between 1950 and 1996 the annual world fish catch (marine plus freshwater harvest) increased by about 200% (Figure 7-17, left). And between 1950 and 1988 the per capita seafood catch more than doubled (Figure 7-17, right).

However, mostly because the rate of population growth exceeds the rate of the world's fish catch, the per capita catch fell by 6% between 1989 and 1996 (Figure 7-17, right). The drop would have been more without the almost threefold growth of aquaculture since 1985 (Figure 7-17, left). Because of overfishing, pollution, and population growth, the world's fish catch is not expected to increase significantly; it may even decline as population growth exceeds the growth of the catch.

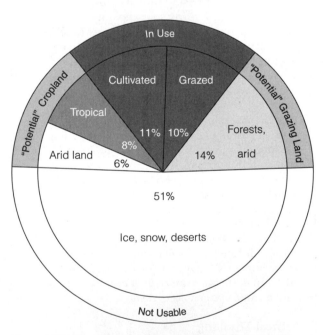

Figure 7-16 Classification of the earth's land. Theoretically, we could double the amount of cropland by clearing tropical forests and irrigating arid lands. However, converting these lands into cropland would destroy valuable forest resources, reduce earth's biodiversity, affect water quality and quantity, and cause other serious environmental problems, usually without being cost-effective.

Q: Between 1978 and 1997, how much money did the Forest Service lose on timber sales?

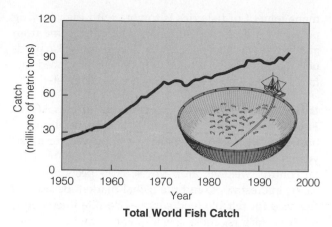

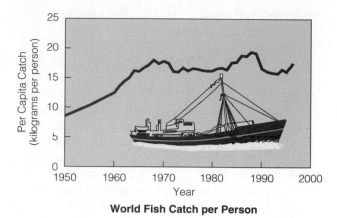

Total World Fish Catch

World Fish Catch per Person

Figure 7-17 World fish catch (marine plus freshwater harvest) (left) and catch per person (right), 1950–96. Worldwide per capita fish catch did not rise much between 1968 and 1989 and has dropped and leveled off since then. Scientists estimate that the sustainable yield of the world's marine fishery is 100 million metric tons (110 million tons)—an amount that is already being exceeded if the unused bycatch is added to the annual fish catch. (Data from UN Food and Agriculture Organization and Worldwatch Institute)

Connections: How Are Overfishing and Habitat Degradation Affecting Fish Harvests? **Overfishing** is the taking of so many fish that too little breeding stock is left to maintain numbers; that is, overfishing is a harvest in excess of the estimated sustainable yield. Prolonged overfishing leads to **commercial extinction**: reduction of a species to the point at which it's no longer profitable to hunt for them. Fishing fleets then move to a new species or a new region, hoping that the overfished species will eventually recover.

According to the UN Food and Agriculture Organization, *11 of the world's 15 major oceanic fishing areas have been fished at or beyond their estimated maximum sustainable yield for commercially valuable species and are in a state of decline.* As a result, 69% of the world's commercial fish stocks are in decline and in need of urgent management.

According to the U.S. National Fish and Wildlife Foundation, 14 major commercial fish species in U.S. waters (accounting for one-fifth of the world's annual catch and half of all U.S. stocks) are so depleted that even if all fishing stopped immediately it would take up to 20 years for stocks to recover.

An even greater threat than overfishing is the possibility of global climate change over the next 50–100 years which can warm ocean waters and enhance the effects of habitat degradation, pollution, and ultraviolet radiation (Sections 9-2 and 9-3). Because fish cannot regulate their internal temperature, they are particularly vulnerable to changes in the temperature of their water environments.

Is Aquaculture the Answer? **Aquaculture**, in which fish and shellfish are raised for food, supplies about 20% of the world's commercial fish harvest.

Aquaculture production increased 3.3-fold between 1984 and 1996; by 2005 it may account for one-third of the world's fish harvest. China is the world leader in aquaculture (producing almost half of the world's output), followed by India and Japan.

There are two basic types of aquaculture. **Fish farming** involves cultivating fish in a controlled environment, often a pond or tank, and harvesting them when they reach the desired size. **Fish ranching** involves holding anadromous species such as salmon (that live part of their lives in fresh water and part in salt water) in captivity for the first few years of their lives (usually in fenced-in areas or floating cages in coastal lagoons and estuaries), releasing them, and then harvesting the adults when they return to spawn.

Species cultivated in developing countries (mostly by inland aquaculture) include carp, tilapia, milkfish, clams, and oysters, all of which feed on phytoplankton and other aquatic plants. In developed countries and some rapidly developing countries in Asia, aquaculture is used mostly to stock lakes and streams with game fish or to raise expensive fish and shellfish such as oysters, catfish, crayfish, rainbow trout, shrimp, and salmon. Aquaculture now produces 90% of all oysters, one-third of all salmon (75% in the United States), and one-quarter of the shrimp and prawns (50% in the United States) sold in the global marketplace.

Aquaculture has several advantages. It is highly efficient and can produce high yields in a small volume of water. Because little fuel is needed, yields and profits are not closely tied to the price of oil (as they are in commercial marine fishing). Crossbreeding and genetic engineering can help increase yields. Some analysts

A: About $7 billion

project that freshwater and saltwater aquaculture production could double during the next 10 years.

There are problems, however. Fish farms are essentially *aquatic feedlots*, requiring large inputs of land, feed, water, and energy and producing large outputs of wastes. Large-scale aquaculture requires considerable capital and scientific knowledge, which are in short supply in developing countries. In addition, scooping out huge ponds for fish and shrimp farming in some developing countries has destroyed ecologically important mangrove forests. Pesticide runoff from nearby croplands can kill fish in aquaculture ponds, and dense populations make fish more vulnerable to bacterial and viral infections.

On average an aquaculture fish pond lasts 5 years before it is too contaminated to use. As a result, shrimp aquaculture farmers buy and flood farms and villages, extract profits of 40–50% return on their investment, then move on, leaving barren, salty land that can't be used to raise shrimp, rice, or anything else. Without adequate pollution control, waste outputs from shrimp farming and other large-scale aquaculture operations can contaminate nearby estuaries, surface water, and groundwater and eliminate some native aquatic species.

Can We Develop a More Sustainable Approach to Fishery Management?

General guidelines for more sustainable management of global fisheries include

- *Shifting the burden of proof to the fishing industry by requiring those who profit from harvesting publicly owned resources in territorial waters and the open ocean to show that their harvests are sustainable before being allowed to operate.*

- *Setting and strictly monitoring and enforcing conservative quotas for fisheries that are well below their estimated maximum sustainable yields.*

- *Establishing and dividing up fishing quotas based on fairness, local needs and conditions, the best available scientific data, and inputs from local communities and fishers.*

- *Greatly strengthening commitment to marine biodiversity protection and integrated coastal management programs that promote both sustainable fishing and the ecological health of marine ecosystems.*

- *Sharply reducing or eliminating fishing subsidies to shrink the size of global fishing fleets, encourage free-market competition, sustain small-scale local fishers, reduce overfishing, and allow economically and biologically depleted stocks to recover.*

It is doubtful that such principles will be put into practice without a public outcry by citizens demanding that policy makers implement such policies. The most important tool consumers have for more sustainable management of fisheries is to vote with their buying power by asking food suppliers where fish came from and how it was harvested and by buying only fish products that have been certified to have been produced sustainably and whose stocks have not been depleted.

How Do Government Agricultural Policies Affect Food Production?

Agriculture is a financially risky business. Whether farmers have a good year or a bad year is determined by factors over which they have little control: weather, crop prices, crop pests and diseases, interest rates, and the global market. Because of the need for reliable food supplies despite fluctuations in these factors, most governments provide various forms of assistance to farmers and consumers.

One approach is to *keep food prices artificially low*. This makes consumers happy but means that farmers may not be able to make a living. Many governments in developing countries keep food prices in cities lower than in the countryside to prevent political unrest. With food prices lower in the cities, more rural people migrate to urban areas, aggravating urban problems and unemployment and increasing the chances of political unrest.

A second approach is to *give farmers subsidies to keep them in business and encourage them to increase food production*. In developed countries government price supports and other subsidies for agriculture total more than $300 billion per year. If government subsidies are too generous and the weather is good, farmers may produce more food than can be sold; food prices and profits then drop because of the surplus. Large amounts of food then become available for export or food aid to developing countries, depressing world food prices; the low prices reduce the financial incentive for farmers in developing countries to increase domestic food production. Moreover, the taxes citizens in developed countries pay to provide agricultural subsidies more than offset the lower food prices they enjoy.

A third policy is to *eliminate most or all price controls and subsidies*, allowing market competition to be the primary factor determining food prices and thus the amount of food produced. Some analysts call for phasing out all government price controls and subsidies over, say, 5–10 years and letting farmers respond to market demand. However, these analysts urge that any phaseout of farm subsidies, which in effect subsidize manufacturers of farm chemicals and machinery, should be coupled with increased aid for the poor and the lower middle class, who would suffer the most from any increase in food prices.

Many environmentalists believe that instead of eliminating all subsidies, we should use them to reward farmers and ranchers who protect the soil, conserve

Freeman, Joe. 1997. "The ABCs of IPM." *Flower & Garden Magazine*, vol. 41, no. 4, 14(2).

water, reforest degraded land, protect and restore wetlands, and conserve wildlife. What do you think?

Another way in which governments and private organizations deal with a lack of food production and hunger is through food aid (Spotlight right).

7-6 PROTECTING FOOD RESOURCES: PESTICIDES AND PEST CONTROL

What Are Pesticides and How Are They Used? A **pest** is any species that competes with us for food, invades lawns and gardens, destroys wood in houses, spreads disease, or is simply a nuisance. In natural ecosystems and many polyculture agroecosystems, natural enemies (predators, parasites, and disease organisms) control the populations of 50–90% of pest species, thus constituting a crucial type of earth capital.

When we simplify natural ecosystems, we upset these natural checks and balances, which keep any one species from taking over for very long. Then we must devise ways to protect our monoculture crops, tree farms, and lawns from insects and other pests that nature once controlled at no charge.

We have done this primarily by developing a variety of **pesticides**: chemicals to kill organisms we consider undesirable. Common types of pesticides include *insecticides* (insect-killers), *herbicides* (weed-killers), *fungicides* (fungus-killers), *nematocides* (roundworm-killers), and *rodenticides* (rat- and mouse-killers).

Worldwide, about 2.3 million metric tons (2.5 million tons) of such pesticides are used yearly—0.45 kilogram (1 pound) for each person on the earth. About 75% of these chemicals are used in developed countries, but use in developing countries is soaring.

About 25% of pesticide use in the United States is for ridding houses, gardens, lawns, parks, playing fields, swimming pools, and golf courses of unwanted pests. According to the EPA, the average lawn in the United States is doused with more than 10 times more synthetic pesticides per hectare than U.S. cropland. Each year, more than 250,000 U.S. residents become ill because of household use of pesticides, and such pesticides are a major source of accidental poisonings and deaths for children under age 5.

Some pesticides, called *broad-spectrum* agents, are toxic to many species; others, called *selective* or *narrow-spectrum* agents, are effective against a narrowly defined group of organisms. Pesticides vary in their *persistence*, the length of time they remain deadly in the environment (Table 7-1). Most organophosphates (except malathion) are highly toxic to humans and other animals, and they account for most human pesticide poisonings and deaths.

SPOTLIGHT — How Useful Is International Food Aid?

Most people view international food aid as a humanitarian effort to prevent people from dying prematurely. However, some analysts contend that giving food to starving people in countries with high population growth rates does more harm than good in the long run. By not helping people grow their own food, they argue, food relief can condemn even greater numbers to premature death from starvation and disease in the future.

Biologist Garrett Hardin has suggested that we use the concept of *lifeboat ethics* to decide which countries get food relief. His basic premise is that there are already too many people in the lifeboat we call earth. Thus, if food relief is given to countries that are not reducing their populations, the effect is to add more people to an already overcrowded lifeboat. Sooner or later the overloaded boat will sink, and most of the passengers will drown.

Large amounts of food relief can also depress local food prices, decrease food production, and stimulate mass migration from farms to already overburdened cities. In addition, food relief discourages local and national governments from investing in the rural agricultural development needed to enable farmers to grow enough food for the population.

Another problem is that much food relief does not reach hunger victims. Transportation networks and storage facilities are often inadequate, so some of the food rots or is devoured by pests before it can reach the hungry. Moreover, officials often steal some of the food and sell it for personal profit; some must also usually be given to officials as bribes for approving the unloading and transporting of the remaining food to the hungry.

Most critics are not against providing aid, but they believe that such aid should help countries control population growth and grow enough food to feed their population by using sustainable agricultural methods (Section 7-7). Temporary food relief, they believe, should be given only when there is a complete breakdown of an area's food supply because of natural disaster.

Critical Thinking

Is sending food to famine victims helpful or harmful in the short run? In the long run? Explain. Are there any conditions you would attach to providing such aid? Explain.

Hint: Enter the search term *integrated pest management* using the Subject Guide.

CHAPTER 7   **219**

Table 7-1 Major Types of Pesticides

Type	Examples	Persistence	Biologically Amplified?
Insecticides			
Chlorinated hydrocarbons	DDT, aldrin, dieldrin, toxaphene, lindane, chlordane, methoxychlor, mirex	High (2–15 years)	Yes (Figure 6-6)
Organophosphates	Malathion, parathion, diazinon, TEPP, DDVP, mevingphos	Low to moderate (1–12 weeks), but some can last several years	No
Carbamates	Aldicarb, carbaryl (Sevin), propoxur, maneb, zineb	Low (days to weeks)	No
Botanicals	Rotenone, pyrethrum, camphor extracted from plants, synthetic pyrethroids (variations of pyrethrum), and rotenoids (variations of rotenone)	Low (days to weeks)	No
Microbotanicals	Various bacteria, fungi, protozoans	Low (days to weeks)	No
Herbicides			
Contact chemicals	Atrazine, simazine, paraquat	Low (days to weeks)	No
Systemic chemicals	2,4-D, 2,4,5-T, Silvex diruon, daminozide (Alar), alachlor (Lasso), glyphosate (Roundup)	Mostly low (days to weeks)	No
Soil sterilants	Trifualin, diphenamid, dalapon, butylate	Low (days)	No
Fungicides			
Various chemicals	Captan, pentachorphenol, zeneb, methyl bromide, carbon bisulfide	Mostly low (days)	No
Fumigants			
Various chemicals	Carbon tetrachloride, ethylene dibromide, methyl bromide	Mostly high	Yes (for most)

What Is the Case for Pesticides? Proponents of pesticides contend that their benefits outweigh their harmful effects.

Pesticides save human lives. Since 1945 DDT and other chlorinated hydrocarbon and organophosphate insecticides have probably prevented the premature deaths of at least 7 million people from insect-transmitted diseases such as malaria (carried by the *Anopheles* mosquito), bubonic plague (rat fleas), typhus (body lice and fleas), and sleeping sickness (tsetse fly).

Pesticides increase food supplies and lower food costs. About 55% of the world's potential human food supply is lost to pests before (35%) or after (20%) harvest. An estimated 37% of the potential U.S. food supply is destroyed by pests before and after harvest; insects cause 13% of these losses, plant pathogens 12%, and weeds 12%. Without pesticides, these losses would be worse, and food prices would rise (by 30–50% in the United States, according to pesticide company officials).

Pesticides increase profits for farmers. Pesticide companies estimate that every $1 spent on pesticides leads to an increase in U.S. crop yields worth approximately $4 (but studies have shown that this benefit drops to about $2 if the harmful effects of pesticides are included).

Pesticides work faster and better than alternatives. Pesticides can control most pests quickly and at a reasonable cost. They have a long shelf life, are easily shipped and applied, and are safe when handled properly. When genetic resistance occurs, farmers can use stronger doses or switch to other pesticides.

Q: What percentage of nonarctic U.S. public rangeland is in unsatisfactory (fair or poor) condition?

The health risks of pesticides are insignificant compared with their benefits. According to Elizabeth Whelan, director of the American Council on Science and Health (ACSH), which presents the position of the pesticide industry, "The reality is that pesticides, when used in the approved regulatory manner, pose no risk to either farm workers or consumers." Pesticide proponents consider media reports describing pesticide health scares to be distorted science and irresponsible reporting. They also point out that about 99.99% of the pesticides we consume in our food are natural chemicals produced by plants.

Proponents point out some beneficial changes in pesticides and their use. For example, *safer and more effective pesticides are being developed.* Industry scientists are developing pesticides, such as botanicals and microbotanicals (Table 7-1), that are safer to users and less damaging to the environment. Genetic engineering also holds promise in developing pest-resistant crop strains. However, total research and development and government approval costs for a new pesticide have risen from $6 million in 1976 to $80–120 million today. As a result, new chemicals are being developed only for crops such as wheat, corn, and soybeans with large markets.

Many new pesticides are used at very low rates per unit area compared to older products. Scientists continue to search for the ideal pest-killing chemical, which would

- *Kill only the target pest*
- *Harm no other species*
- *Disappear or break down into something harmless after doing its job*
- *Not cause genetic resistance in target organisms*
- *Be cheaper than doing nothing*

Unfortunately, no known pesticide meets all these criteria, and most don't even come close.

What Is the Case Against Pesticides? Opponents of widespread use of pesticides believe that their harmful effects outweigh their benefits. The biggest problem is the development of *genetic resistance* to pesticides by pest organisms. Insects breed rapidly, and within 5–10 years (much sooner in tropical areas) they can develop immunity to pesticides through natural selection (Connections, right) and come back stronger than before. Weeds and plant disease organisms also become resistant, but more slowly.

Since 1950 at least 520 species of insects and mites, 273 weed species, 150 plant diseases, and 10 species of rodents (mostly rats) have developed genetic resistance to one or more pesticides. At least 17 species of insect pests are resistant to all major classes of insecticides, and several fungal plant diseases are now immune to most of the widely used fungicides. Because of genetic resistance,

The Pesticide Treadmill

CONNECTIONS

When genetic resistance develops, pesticide sales representatives usually recommend more frequent applications, larger doses, or a switch to new (usually more expensive) chemicals to keep the resistant species under control. This can put farmers on a pesticide treadmill, whereby they pay more and more for a pest control program that often becomes less and less effective.

In 1989 David Pimentel, an expert in insect ecology, evaluated data from more than 300 agricultural scientists and economists and came to the following conclusions:

- Although the use of synthetic pesticides has increased 33-fold since 1942, it is estimated that more of the U.S. food supply is lost to pests today (37%) than in the 1940s (31%). Losses attributed to insects almost doubled (from 7% to 13%) despite a tenfold increase in the use of synthetic insecticides.

- The estimated environmental, health, and social costs of pesticide use in the United States range from $4 billion to $10 billion per year. The International Food Policy Research Institute puts the estimate much higher, at $100–200 billion per year, or $5–10 in damages for every dollar spent on pesticides.

- Alternative pest control practices could halve the use of chemical pesticides on 40 major U.S. crops without reducing crop yields.

- A 50% cut in U.S. pesticide use would cause retail food prices to rise by only about 0.2% but would raise average income for farmers about 9%.

Critical Thinking

What can be done to help farmers get off of the pesticide treadmill?

most widely used insecticides no longer protect people from insect-transmitted diseases in many parts of the world, leading to resurgences of diseases such as malaria.

Another problem is that *broad-spectrum insecticides kill natural predators and parasites that may have been maintaining the population of a pest species at a reasonable level.* With natural enemies out of the way, a rapidly reproducing insect pest species can make a strong comeback only days or weeks after initially being controlled.

U.S. pesticide companies can make and export to other countries pesticides that have been banned or severely restricted—or never even approved—in the United States. But what goes around comes around.

In what environmentalists call a *circle of poison*, residues of some of these banned or unapproved chemicals can return to the United States in or on imported items such as coffee, cocoa, pineapples, and out-of-season melons, tomatoes, and grapes. More than one-fourth of the produce (fruits and vegetables) consumed in the United States is grown overseas. Persistent pesticides such as DDT can also be carried by winds from other countries to the United States.

Environmentalists have urged Congress—without success—to break this deadly circle. Supporters of pesticide exports argue that such sales increase economic growth and provide jobs. They also contend that if the United States didn't export pesticides, other countries would.

In 1998, more than 100 countries began talks to finalize an international treaty curbing trade in potential lethal chemicals such as mercury and its compounds and pesticides such as DDT and organophosphates.

Critical Thinking

Should U.S. companies be allowed to export pesticides to other countries that have been banned, severely restricted, or not registered for use in the United States? Explain.

Wiping out natural predators can also unleash new pests whose populations the predators had previously held in check, causing other unexpected effects. Currently 100 of the 300 most destructive insect pests in the United States were secondary pests that became major pests through the effects of insecticides.

Still another problem is that *pesticides don't stay put.* No more than 2% (and often less than 0.1%) of the insecticides applied to crops by aerial spraying or by ground spraying actually reaches the target pests; less than 5% of herbicides applied to crops reaches the target weeds. Pesticides that miss their target pests end up in the air, surface water, groundwater, bottom sediments, food, and nontarget organisms, including humans and wildlife.

Some pesticides can harm wildlife. During the 1950s and 1960s populations of fish-eating birds such as the osprey, brown pelican, and bald eagle (Figure 6-2) plummeted after exposure to DDT (Figure 6-6). Each year some 20% of U.S. honeybee colonies are wiped out by pesticides and another 15% are damaged, costing farmers at least $200 million per year from reduced pollination of vital crops. Pesticide runoff from cropland, a leading cause of fish kills worldwide, kills 6–14 million fish each year in the United States.

Pesticides can also threaten human health. According to the World Health Organization and the UN Environment Programme, an estimated 25 million agricultural workers in developing countries are seriously poisoned by pesticides each year, resulting in an estimated 220,000 deaths. Health officials believe that the actual number of pesticide-related illnesses and deaths among the world's farm workers is probably greatly underestimated because of poor records, lack of doctors and disease reporting in rural areas, and faulty diagnoses.

Farm workers in developing countries are especially vulnerable to pesticide poisoning because three-quarters of all pesticides are applied by hand. In addition, educational levels are low, warnings are few, pesticide regulations are lax or nonexistent, and use of protective equipment is rare (especially in the hot and humid tropics).

According to a 1987 study by the National Academy of Sciences, exposure to pesticide residues in food causes 4,000–20,000 cases of cancer per year in the United States. Because roughly 50% of those getting cancer die prematurely, this amounts to about 2,000–10,000 premature deaths per year in the United States from exposure to legally allowed pesticide residues in foods. A 1993 study by the National Academy of Sciences concluded that the legal limits for pesticide residues in food may need to be reduced to 1/1,000 of current levels to protect children, who are more vulnerable to such chemicals than are adults.

Representatives from pesticide companies dispute these findings. Indeed, the food industry denies that anyone in the United States has ever been harmed by eating food that has been grown using pesticides for the past 50 years.

Some scientists are becoming increasingly concerned about possible genetic mutations, birth defects, nervous system disorders, and effects on the immune and endocrine systems from long-term exposure to low levels of various pesticides (Section 8-3). Very little research has been conducted on these potentially serious threats to human health.

According to a National Academy of Sciences study, federal laws regulating the use of pesticides in the United States are inadequate and poorly enforced by the EPA, FDA, and USDA. A 1993 study of pesticide safety by the U.S.

Q: How much do 29,000 U.S. ranchers with permits to graze on public lands receive in federal subsidies?

National Academy of Sciences urged the government to do the following things:

- Make human health the primary consideration for setting limits on pesticide levels allowed in food.

- Collect more and better data on exposure to pesticides for different groups, including farm workers, adults, and children.

- Develop new and better test procedures for evaluating the toxicity of pesticides, especially for children.

- Consider cumulative exposures of all pesticides in food and water, especially for children, instead of basing regulations on exposure to a single pesticide.

Between 1972 and 1996, the EPA canceled or severely restricted the use of 55 active pesticide ingredients. The banned chemicals include most chlorinated hydrocarbon insecticides, several carbamates and organophosphates, and the systemic herbicides 2,4,5-T and Silvex (Table 7-1). However, banned or unregistered pesticides may be manufactured in the United States and exported to other countries (Connections, left).

Solutions: What Are Other Ways to Control Pests? A number of *cultivation practices* can help reduce pest damage. The type of crop planted in a field each year can be changed (crop rotation). Rows of hedges or trees can be planted around fields to hinder insect invasions and provide habitats for their natural enemies (with the added benefit of reduced soil erosion). Planting times can be adjusted so that major insect pests either starve or get eaten by their natural predators. Crops can also be grown in areas where their major pests do not exist.

Trap crops can be planted to lure pests away from the main crop. Growers can switch from vulnerable monocultures to intercropping, agroforestry, and polyculture, which use plant diversity to reduce losses to pests. Diseased or infected plants and stalks and other crop residues that harbor pests can be removed from cropfields. Photodegradable plastic can keep weeds from sprouting between crop rows. Vacuum machines can be used to gently remove harmful bugs from plants.

Plants and animals that are *genetically resistant* to certain pest insects, fungi, and diseases can be developed. Genetic engineering is now being used to speed the process.

Biological control using predators, parasites, and pathogens (disease-causing bacteria and viruses) can be encouraged or imported to regulate pest populations. More than 300 biological pest control projects worldwide have been successful, especially in China (Solutions, right). In the United States, biological control has saved farmers an average of $25 for every $1 invested. However, biological agents can't always be mass-produced, and farmers find them slower acting and more difficult to apply than pesticides. Biological agents must

Along Came a Spider

SOLUTIONS

In 1962 biologist Rachel Carson warned against relying on synthetic chemicals to kill insects and other species we deem pests. Some Chinese farmers recently decided it's time to change strategies. Instead of spraying their rice and cotton fields with poisons, they began to build little straw huts around the fields in the fall.

If this sounds crazy, it was crazy like a fox. These farmers were giving aid and comfort to insects' worst enemy, one that has hunted them for millions of years: spiders. The little huts were for hibernating spiders. Protected from the worst of the cold by the huts, far more of the hibernating spiders become active in the spring. Ravenous after their winter fast, they scuttle off into the fields to stalk their insect prey. Even without human help, the world's 30,000 known species of spiders kill far more insects every year than insecticides do.

The idea of encouraging populations of spiders in fields, forests, and even houses scares some people because spiders have bad reputations. A few species of spiders, such as the black widow, the brown recluse, and eastern Australia's Sydney funnel web, are dangerous to people. However, the vast majority of spider species, including the ferocious-looking wolf spider, are harmless to humans. Even the giant tarantula rarely bites people, and its venom is too weak to harm us or other large mammals.

As biologist Thomas Eisner puts it, "Bugs are not going to inherit the earth. They own it now. So we might as well make peace with the landlord." As we seek new ways to coexist with the insect rulers of the planet, we would do well to be sure that spiders are in our corner.

Critical Thinking

Entomologist Willard H. Whitcomb found that one type of harmless banana spider can keep a house clear of cockroaches. Would you deliberately import such spiders into your house to control cockroaches?

also be protected from pesticides sprayed in nearby fields. Once released into the environment, biological control agents can multiply, cause unpredictable harmful ecological effects, and be impossible to recapture. Some may even become pests themselves; others (such as praying mantises) devour beneficial and pest insects alike.

Biopesticides, chemicals produced by plants such as synthetic pyrethroids (Table 7-1), are an increasingly popular method of pest control, and scientists are looking for new plant toxins to synthesize for mass production. Microbes are also being drafted for insect wars, especially by organic farmers. For example, *Bacillus thuringensis (Bt)* toxin is a registered pesticide sold commercially as a dry powder. Each of the thousands of strains of this common soil bacterium will kill a specific pest. The bad news is that genetic resistance is already developing to some *Bt* toxins.

Insect birth control has also been used. Males of some insect pest species (such as the screwworm fly, which can infest and kill livestock) can be raised in the laboratory, sterilized by radiation or chemicals, and then released into an infested area to mate unsuccessfully with fertile wild females. Problems with this approach include high costs, the difficulties in knowing the mating times and behaviors of each target insect, the number of sterile males needed, and the few species for which this strategy works.

In many insect species, a female that is ready to mate releases a minute amount (typically about one-millionth of a gram) of a chemical sex attractant called a *pheromone*. Whether extracted from insects or synthesized in the laboratory, pheromones can lure pests into traps or attract their natural predators into cropfields (usually the more effective approach). These chemicals attract only one species, work in trace amounts, have little chance of causing genetic resistance, and are not harmful to nontarget species. However, it is costly and time-consuming to identify, isolate, and produce the specific sex attractant for each pest or predator.

Each step in the insect life cycle is regulated by the timely natural release of juvenile hormones (JH) and molting hormones (MH) (Figure 7-18). These chemicals, which can be extracted from insects or synthesized in the laboratory, can disrupt an insect's normal life cycle, causing the insect to fail to reach maturity and reproduce (Figure 7-18). Insect hormones have the same advantages as sex attractants, but they take weeks to kill an insect, are often ineffective with large infestations of insects, and sometimes break down before they can act. They must also be applied at exactly the right time in the target insect's life cycle. Moreover, they sometimes affect the target's predators and other nonpest species. Finally, like sex attractants, they are difficult and costly to produce.

Recently some farmers have begun *zapping pests with hot water* by using a machine that sprays boiling water on crops to kill weeds and insects. So far, the system has worked well on cotton, alfalfa, and potato fields and in citrus groves in Florida, where the machine was invented. The cost is roughly equal to that of using chemical pesticides.

Some pests can also be killed by *zapping foods after harvest with gamma radiation.* Such food irradiation extends food shelf life and kills insects, parasitic worms (such as trichinae in pork), and bacteria (such as salmonella, which infect 51,000 Americans and kill 2,000 each year). According to the U.S. Food and Drug Adminis-

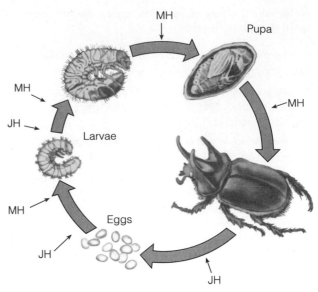

Figure 7-18 For normal insect growth, development, and reproduction to occur, certain juvenile hormones (JH) and molting hormones (MH) must be present at genetically determined stages in the insect's life cycle. If applied at the proper time, synthetic hormones disrupt the life cycles of insect pests and control their populations.

tration and the World Health Organization, over 2,000 studies show that foods exposed to low doses of ionizing radiation are safe for human consumption. However, critics argue that it is too soon to see long-term effects (which might not show up for 30–40 years), and that irradiating food destroys some of its vitamins and other nutrients. Critics also argue that consumers want fresh, wholesome food, not old, possibly less nutritious food made to appear fresh and healthy by irradiation.

Solutions: Is Integrated Pest Management the Wave of the Future? An increasing number of pest control experts and farmers believe that the best way to control crop pests is a carefully designed **integrated pest management (IPM)** program. In this approach, each crop and its pests are evaluated as parts of an ecological system. Then a control program is developed that includes a mix of cultivation and biological and chemical methods applied in proper sequence and with the proper timing.

The overall aim of IPM is not eradication of pest populations, but reduction of crop damage to an economically tolerable level. Fields are carefully monitored; when a damaging level of pests is reached, farmers first use biological and cultivation controls, including vacuuming up harmful bugs. Small amounts of insecticides (mostly botanicals or microbotanicals) are applied only as a last resort, and different chemi-

 Loftus, M., and M. B. Marcus. 1998. "Hold the Chemicals." *U.S. News & World Report,* vol. 124, no. 19, 74(3).

cals are used to slow development of genetic resistance and to avoid killing predators of pest species.

In 1986 the Indonesian government banned the use of 57 of 66 pesticides used on rice, phased out pesticide subsidies over a 2-year period, and used some of the money to help launch a nationwide program to switch to IPM, including a major farmer education program. The results were dramatic: Between 1987 and 1992, pesticide use dropped by 65%, rice production rose by 15%, and more than 250,000 farmers were trained in IPM techniques. By 1993 the program had saved the Indonesian government over $1.2 billion—more than enough to fund its IPM program.

The experiences of countries such as China, Brazil, Indonesia, Australia, and the United States show that a well-designed IPM program can reduce pesticide use and pest control costs by 50–90%. IPM can also reduce preharvest pest-induced crop losses by 50%. It can improve crop yields, reduce inputs of fertilizer and irrigation water, and slow the development of genetic resistance because pests are assaulted less often and with lower doses of pesticides. Thus, IPM is an important form of pollution prevention that reduces risks to wildlife and human health.

However, IPM requires expert knowledge about each pest situation, and it is slower acting than conventional pesticides. Methods developed for a crop in one area might not apply to areas with even slightly different growing conditions. Although long-term costs are typically lower than the costs of using conventional pesticides, initial costs may be higher.

Despite its promise and growth, widespread use of IPM is hindered by government subsidies of conventional chemical pesticides and by opposition from agricultural chemical companies, whose sales would drop sharply. In addition, farmers get most of their information about pest control from pesticide salespeople (and in the United States from USDA county farm agents), few of whom have adequate training in IPM.

In 1996 a study by the National Academy of Sciences recommended that the United States shift from chemically based approaches to ecologically based approaches to pest management. A growing number of scientists urge the USDA to promote IPM in the United States by **(1)** adding a 2% sales tax on pesticides and using the revenue to fund IPM research and education, **(2)** setting up a federally supported IPM demonstration project on at least one farm in every county, **(3)** training USDA field personnel and county farm agents in IPM so that they can help farmers use this alternative, **(4)** providing federal and state subsidies, and perhaps government-backed crop insurance, to farmers who use IPM or other approved alternatives to pesticides, and **(5)** gradually phasing out subsidies to farmers who depend almost entirely on pesticides once effective IPM methods have been developed for major pest species.

Some actions you can take to reduce your use of and exposure to pesticides are listed in Appendix 4.

7-7 SOLUTIONS: SUSTAINABLE AGRICULTURE

What Is Sustainable Agriculture? Many agricultural scientists and experts believe that a key to reducing world hunger, poverty, and the harmful environmental effects of agriculture (Figure 7-15) is to develop systems of **sustainable agriculture**, or **low-input agriculture**, and phase them in over the next three decades.

Low-input farming reduces waste of irrigation water and uses less pesticide and inorganic fertilizer. To maintain and restore soil fertility, farmers rely on good soil conservation practices (Section 7-2) and use manure, compost, and other forms of organic matter. They also emphasize biological and physical methods for controlling pests and use chemical pesticides only as a last resort (and in the smallest amounts possible). A growing number of farmers are discovering that low-input farming is often more profitable than high-input farming because they spend less money in inputs of irrigation water, fertilizer, and pesticides.

Most proponents of more sustainable agriculture are not opposed to high-yield agriculture; indeed, they see it as vital for protecting the earth's biodiversity and ecological integrity by reducing the need to cultivate new and often marginal land. Instead, they believe that current research and economic incentives should be redirected to encourage increases in yield per hectare without depleting or degrading soil, water, and biodiversity.

General guidelines for sustainable agricultural systems suggested by various analysts include the following:

- *Combine traditional high-yield polyculture and modern monoculture methods for growing crops.*

- *Grow more perennial crops (Solutions, p. 215).*

- *Minimize soil erosion, salinization, and waterlogging.*

- *Reduce destruction of forests, grasslands, and wetlands for producing foods by emphasizing increased yields per area of cropland using sustainable methods.*

- *Stabilize aquifers by reducing the rate of water removal to the rate of recharge.*

- *Reduce water waste in irrigation.*

- *Reduce overfishing by implementing the principles listed on p. 218.*

- *Reduce use and waste of fossil fuels, and shift to an energy efficient, solar–hydrogen economy (Chapter 4).*

- *Increase use of organic fertilizers and solar, wind, and biomass energy to grow and process crops.*

- *Emphasize biological pest control and integrated pest management (p. 224).*

- *Protect existing prime cropland from environmental degradation and conversion to urban or industrial uses.*

- *Subsidize sustainable farming and phase out government subsidies for unsustainable farming.*

- *Shift to full-cost pricing that includes the harmful environmental effects of agriculture (Figure 7-15) in food prices.*

- *Reduce food waste.* According to a 1997 USDA study, an estimated 27% of the food produced in the United States (not including crop losses) is wasted by grocery stores, restaurants, and consumers.

- *Greatly increase research on sustainable agriculture.*

- *Set up demonstration projects throughout each country so that farmers can see how sustainable agricultural systems work.*

- *Establish training programs in sustainable agriculture for farmers and government agricultural officials and encourage the creation of college curricula in sustainable agriculture.*

- *Educate the public about the hidden environmental and health costs they are paying for food and the need to gradually incorporate these costs into market prices.*

- *Reduce poverty so that the poor have enough money or land to supply their basic food needs.*

- *Educate people about the right mix of foods to eat for good nutrition and how to prepare foods to reduce losses of essential vitamins and minerals.*

- *Slow population growth to help all of the world's countries reach the more sustainable postindustrial stage of the demographic transition (Figure 3-12).*

- *Integrate agriculture, population, urban and rural, energy, health, climate, water resource, soil resource, land use, pollution, and biodiversity protection policies.*

Can We Make the Transition to Sustainable Agriculture? A growing number agricultural analysts believe that over the next 30 years we must make a transition from unsustainable and environmentally harmful (Figure 7-15) industrialized and conventional subsistence agriculture to more sustainable forms of agriculture.

In developed countries, including the United States, even a partial shift to more environmentally sustainable food production will not be easy. It will be opposed by agribusiness, successful farmers with large investments in unsustainable forms of industrialized agriculture, and specialized farmers unwilling to learn the demanding art of farming sustainably. It might also be resisted by many consumers unwilling or unable to pay higher prices for food, because full-cost accounting would include agriculture's harmful environmental and health costs in the market prices of food.

Despite such difficulties, many environmentalists believe that a new *eco-agricultural revolution* could take place throughout most of the world over the next 30 years. Whether it does occur is primarily a political and ethical issue (Chapter 12). Some actions you can take to help promote sustainable agriculture are listed in Appendix 4.

At some point, either the loss of topsoil from the world's croplands will have to be checked by effective soil conservation practices, or the growth in the world's population will be checked by hunger and malnutrition.

LESTER R. BROWN

CRITICAL THINKING

1. Why should everyone, not just farmers, be concerned with soil conservation?

2. Explain how the Dust Bowl phenomenon of the 1930s in the United States could happen again. How might it affect your life? How would you try to prevent a recurrence?

3. What are the main advantages and disadvantages of commercial inorganic fertilizers? Why should both inorganic and organic fertilizers be used?

4. Summarize the advantages and limitations of each of the following proposals for increasing world food supplies and reducing hunger over the next 30 years: **(a)** cultivating more land by clearing tropical forests and irrigating arid lands, **(b)** catching more fish in the open sea, **(c)** producing more fish and shellfish with aquaculture, and **(d)** increasing the yield per area of cropland.

5. Should all price supports and other government subsidies paid to farmers out of tax revenues be eliminated? Explain. Try to consult one or more farmers in answering this question.

6. Should governments phase in agricultural tax breaks and subsidies to encourage farmers to switch to more sustainable farming? Explain. At the same time, should governments phase in higher taxes and reduce subsidies to discourage farmers from using unsustainable, earth-degrading forms of farming, and then use the resulting revenue to encourage earth-sustaining farming? Explain.

7. Do you believe that because essentially all pesticides eventually fail, their use should be phased out or sharply reduced and that farmers should be given economic incentives for switching to integrated pest management? Explain your position.

8. Should U.S. companies be allowed to make and export pesticides that have been banned, severely restricted, or not registered for use in the United States? Explain.

8 RISK, TOXICOLOGY, AND HUMAN HEALTH

For the first time in the history of the world, every human being is now subjected to dangerous chemicals from the moment of conception until death.

RACHEL CARSON

8-1   TYPES OF HAZARDS

What Is Risk?   Risk is the possibility of suffering harm from a *hazard* that can cause injury, disease, economic loss, or environmental damage. Risk is expressed in terms of **probability**: a mathematical statement about how likely it is that some event or effect will occur.

The probability of a risk is expressed as a fraction ranging from 0 (absolute certainty that there is no risk, which can never be shown) to 1.0 (absolute certainty that there is a risk). For example, the lifetime cancer risk from exposure to a particular chemical may have a probability of 0.000001, or 1 in 1 million. This means that one of every 1 million people exposed to the chemical at a specified average daily dose will develop cancer over a typical lifetime (usually considered to be 70 years).

Risk assessment involves using data, hypotheses, and models to estimate the probability of harm to human health, society, or the environment that may result from exposure to specific hazards.

What Are the Major Types of Hazards?   The various kinds of hazards we face can be categorized as follows:

- *Cultural hazards* such as unsafe working conditions, smoking (Spotlight, p. 228), poor diet, drugs, drinking, driving, criminal assault, unsafe sex, and poverty
- *Chemical hazards* from harmful chemicals in the air (Chapter 9), water (Chapter 10), soil, and food
- *Physical hazards* such as ionizing radiation, noise, fire, tornadoes, hurricanes, earthquakes, volcanic eruptions, and floods
- *Biological hazards* from pathogens (bacteria, viruses, and parasites), pollen and other allergens, and animals such as bees and poisonous snakes.

8-2   TOXICOLOGY

What Determines Whether a Chemical Is Harmful? Dose and Response   Toxicity is a measure of how harmful a substance is. Whether a chemical (or other agent such as ionizing radiation) is harmful depends on several factors. One is the **dose**, the amount of a potentially harmful substance a person has ingested, inhaled, or absorbed through the skin. Whether a chemical is harmful depends on the size of the dose over a certain period of time, how often an exposure occurs, who is exposed (adult or child, for example), and how well the body's detoxification systems (liver, lungs, and kidneys) work.

The type and amount of health damage that result from exposure to a chemical or other agent are called the **response**. An *acute effect* is an immediate or rapid harmful reaction to an exposure; it can range from dizziness or a rash to death. A *chronic effect* is a permanent or long-lasting consequence (kidney or liver damage, for example) of exposure to a harmful substance. Some individuals are more susceptible to specific toxins than others because of genetic differences, allergic responses, weakened immune systems, acute sensitivity, or age (children are usually more susceptible than adults).

The detection of trace amounts of a chemical in air, water, or food does not necessarily mean that it is there at a level harmful to most people. In some cases, all we may be doing is finding trace levels we could not detect before. Indeed, practically any synthetic or natural chemical, even water, can be harmful if ingested in a large enough quantity. The critical question is how much exposure to a particular toxic chemical causes a harmful response.

Some people have the mistaken idea that all natural chemicals are safe and all synthetic chemicals are harmful. In fact, many synthetic chemicals are quite safe if used as intended, and many natural chemicals are deadly.

What Is a Poison?   Legally, a **poison** is a chemical that has an LD_{50} of 50 milligrams or less per kilogram of body weight. An LD_{50} is the **median lethal dose,** or the amount of a chemical received in one dose that kills exactly 50% of the animals (usually rats and mice) in a test population (usually 60–200 animals) within a 14-day period. *LD* stands for *lethal dose,* and the

Hint: Enter the search terms *pollution, health* using Key Words.

227

What is roughly the diameter of a 30-caliber bullet, can be bought almost anywhere, is highly addictive, and kills about 8,200 people every day? It's a cigarette.

Cigarette smoking is the single most preventable cause of death and suffering among adults. The World Health Organization (WHO) estimates that each year tobacco contributes to the premature deaths of at least 3 million people from heart disease, lung cancer, other cancers, bronchitis, emphysema, and stroke. The annual death toll from smoking-related diseases is projected by WHO to reach 10 million by 2020 and 12 million (primarily in developing countries) by 2050—an average of almost 33,000 preventable deaths per day.

In 1993 smoking killed about 419,000 Americans (up from 390,000 in 1985), an average of 1,150 deaths per day (Figure 8-1). This death toll is equivalent to three fully loaded jumbo jets crashing every day with no survivors. A 1998 report estimated that more than 500,000 people in Europe and 750,000 people in China die from smoking-related disease every year.

The overwhelming consensus in the scientific community is that the nicotine (and probably the acetaldehyde) in inhaled tobacco smoke is highly addictive. Only 1 in 10 people who try to quit smoking succeed—about the same relapse rate as for recovering alcoholics and those addicted to heroin or crack cocaine. A British government study showed that adolescents who smoke more than one cigarette have an 85% chance of becoming smokers. About 80% of all smokers say they wish they had never started smoking and have tried to quit. Recent evidence suggests that some people may be genetically prone to becoming addicted to nicotine.

Government agencies and independent economists estimate that the country's 48 million smokers cost the United States up to $100 billion a year in medical bills, increased insurance costs, disability, lost earnings and productivity because of illness, and property damage from smoking-caused fires. This is an average of $4.20 per pack of cigarettes sold. Some put this figure much lower because when smokers die prematurely they don't draw benefits from social security and private pensions and cut down nursing home bills—certainly not a good argument for the benefits of smoking.

Many health experts urge that a $2–4 federal tax be added to the price of a pack of cigarettes. Such a tax would mean that the users of cigarettes (and other tobacco products), not the rest of society, would pay a much greater share of the health, economic, and social costs associated with their smoking—a *user-pays* approach.

Other suggestions for reducing the death toll and health effects of smoking in the United States include (1) banning all cigarette advertising (as has been done in France), (2) forbidding the sale of cigarettes and other tobacco products to anyone under 21 (with strict penalties for violators), (3) banning all cigarette vending machines, (4) classifying nicotine as an addictive and dangerous drug (and placing its use in tobacco or other products under the jurisdiction of the Food and Drug Administration), (5) eliminating all federal subsidies and tax breaks to U.S. tobacco farmers and tobacco companies, and (6) using cigarette tax income to finance a massive antitobacco advertising and education program.

Critical Thinking

Explain why you agree or disagree with imposing a heavy tax on cigarettes and implementing the other six suggestions given above for reducing the death toll and health effects of smoking.

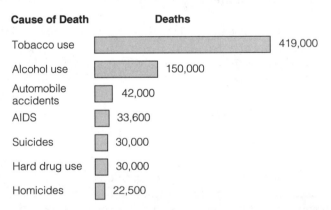

Figure 8-1
Deaths in the United States from tobacco use and other causes in 1993. Smoking is by far the nation's leading cause of preventable death, causing more premature deaths each year than all the other categories in this figure combined. Cardiovascular disease causes about 180,000 of the 419,000 smoking-related deaths per year, followed by 120,000 from lung cancer, and 30,000–60,000 from inhaling secondhand smoke. (Data from National Center for Health Statistics)

Q: What percentage of all U.S. land area is protected as wilderness?

Table 8-1 Toxicity Ratings and Average Lethal Doses for Humans

Toxicity Ratings	LD_{50} (milligrams per kilogram of body weight)*	Average Lethal Dose†	Examples
Super toxic	Less than 0.01	Less than 1 drop	Nerve gases, botulism toxin, mushroom toxins, dioxin (TCDD)
Extremely toxic	Less than 5	Less than 7 drops	Potassium cyanide, heroin, atropine, parathion, nicotine
Very toxic	5–50	7 drops to 1 teaspoon	Mercury salts, morphine, codeine
Toxic	50–500	1 teaspoon to 1 ounce	Lead salts, DDT, sodium hydroxide, fluoride, sulfuric acid, caffeine, carbon tetrachloride
Moderately toxic	500–5,000	1 ounce to 1 pint	Methyl (wood) alcohol, ether, phenobarbitol, amphetamine, kerosene, aspirin
Slightly toxic	5,000–15,000	1 pint to 1 quart	Ethyl alcohol, Lysol, soaps
Essentially nontoxic	15,000 or greater	More than 1 quart	Water, glycerin, table sugar

*Dosage that kills 50% of individuals exposed.

†Amounts of substances that are liquids at room temperature when given to a 70.4-kilogram (155-pound) human.

subscript *50* refers to the percentage of test organisms for which the dose was lethal.

Chemicals vary widely in their toxicity (Table 8-1). Some poisons can cause serious harm or death after a single acute exposure at extremely low doses, whereas others cause such harm only at such huge doses that it is nearly impossible to get enough into the body. Most chemicals fall between these two extremes.

How Do Scientists Determine Toxicity?

Three methods are used to determine the level at which a substance poses a health threat; each has certain limitations. One is *case reports* (usually made by physicians) about people suffering some adverse health effect or death after exposure to a chemical. Such information often involves accidental poisonings, drug overdoses, homicides, or suicide attempts.

A second method relies on *laboratory investigations* (usually on test animals) to determine toxicity, residence time, what parts of the body are affected, and (sometimes) how the harm takes place. The third method is *epidemiology*, which involves studies of populations of humans exposed to certain chemicals or diseases.

Case reports are usually the least valuable source for determining toxicity because the actual dose and the exposed person's health status are often not known. However, such reports can provide clues about environmental hazards and suggest the need for laboratory investigations.

Acute toxicity and chronic toxicity are usually determined by tests on live laboratory animals (especially mice and rats, which are small and prolific and can be housed inexpensively in large numbers) and on bacteria. Such tests are also made on cell and tissue cultures and chicken egg membranes.

Acute toxicity tests are run to develop a **dose–response curve**, which shows the effects of various doses of a toxic agent on a group of test organisms (Figure 8-2). Such tests are *controlled experiments* in which the effects of the chemical on a *test group* are compared with the responses of a *control group* of organisms not exposed to the chemical. Care is taken to ensure that organisms in each group are as identical in age, health status, and genetic makeup as possible and that they are exposed to the same environmental conditions.

Fairly high dose levels are used in order to reduce the number of test animals needed, obtain results quickly, and lower costs. Otherwise, tests would have to be run on millions of laboratory animals for many years, and manufacturers couldn't afford to test most chemicals. For the same reasons, the results of high-dose exposures are usually extrapolated to low-dose levels using mathematical models. Then the extrapolated low-dose results on the test organisms are extrapolated to humans to estimate LD_{50} values for acute toxicity (Table 8-1).

According to the *linear dose–response model*, any dose of a toxic chemical or ionizing radiation has a certain risk of causing harm (Figure 8-2, left). With the *threshold*

A: About 4% (only 1.8% in the lower 48 states)

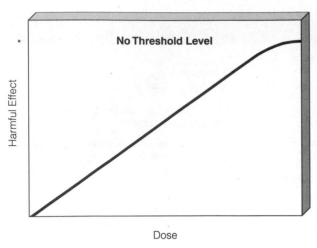

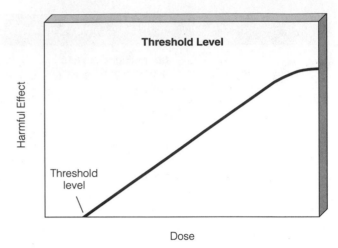

Figure 8-2 Two hypothetical dose–response curves. The curve on the left represents harmful effects that occur with increasing doses of a chemical or ionizing radiation; no dose is considered to be safe. The curve on the right shows the response of exposure to a chemical or ionizing radiation in which harmful effects appear only when the dose exceeds a certain threshold level. There is considerable uncertainty and controversy over which of these models applies to various harmful agents because of the difficulty in estimating the response to very low doses. Nonlinear threshold and nonthreshold dose–response curves, in which the graphs are curved instead of being straight lines, have also been observed.

dose–response model (Figure 8-2, right) there is a threshold dose below which no detectable harmful effects occur, presumably because the body can repair the damage caused by low doses of some substances. It is extremely difficult to establish which of these models applies at low doses. To err on the side of safety, the linear or non-threshold dose–response model is often assumed.

Some scientists challenge the validity of extrapolating data from test animals to humans because human physiology and metabolism are often different from those of the test animals. Others counter that such tests and models work fairly well (especially for revealing cancer risks) when the correct experimental animal is chosen or when a chemical is toxic to several different species of test animals.

Animal tests take 2–5 years and cost $200,000 to $2 million per substance tested. They are opposed by animal rights groups that want to ban all use of test animals (or to ensure that experimental animals are treated in the most humane manner possible). Scientists are looking for substitute methods, but they point out that some animal testing is needed because the known alternatives cannot adequately mimic the complex biochemical interactions of a live animal.

Another approach to testing for toxicity and identifying the agents causing diseases is **epidemiology**: the study of the patterns of disease or toxicity to find out why some people get sick and others do not. Typically, the health of people exposed to a particular toxic agent or disease organism (the experimental group) is compared with the health of another group of statistically similar people not exposed to these conditions (the control group).

Epidemiology has limitations. For many toxic agents, too few people have been exposed to sufficiently high levels to allow detection of statistically significant differences. Because people are exposed to many different toxic agents and disease-causing factors throughout their lives, it is usually impossible to conclusively link an observed epidemiological effect with exposure to a particular toxic agent. All an epidemiological study can do is to establish strong, moderate, weak, or no statistical associations between a hazard and a health problem. Because epidemiology can be used to evaluate only the hazards to which people have already been exposed, it is rarely useful for predicting the effects of new technologies, substances, or diseases.

According to risk assessment expert Joseph V. Rodricks, "Toxicologists know a great deal about a few chemicals, a little about many, and next to nothing about most." There are three reasons for this lack of information. *First*, under existing laws, most chemicals are considered innocent until proven guilty. No one is required to investigate whether they are harmful. *Second*, there are not enough money, personnel, facilities, and test animals to provide such information for more than a small fraction of the many chemicals we encounter in our daily lives.

Third, even if we could make a reasonable estimate of the biggest risks associated with particular technologies or chemicals (a very difficult and expensive thing to do), we know little about their possible interactions with

Q: What percentage of the medicines sold worldwide have active ingredients extracted from wild species (mostly plants)?

other technologies and chemicals or about the effects of such interactions on human health and ecosystems. For example, we would have to carry out almost 125,000 different experiments to study the possible two-chemical combinations among the 500 most widely used industrial chemicals. If each experiment cost only $100,000, the 125,000 experiments would cost $12.5 billion. To study all the different three-chemical interactions among the top 500 chemicals would require 20.7 million experiments—a physical and financial impossibility.

Thus, all methods for estimating toxicity levels and risks have serious limitations, but they are all we have. To take this uncertainty into account and minimize harm, standards for allowed exposure to toxic substances and radiation are typically set at levels 100 or even 1,000 times lower than the estimated harmful levels.

The difficulty and expense of getting information about the harmful effects of chemicals is one reason an increasing number of environmentalists and health officials are pushing for much greater emphasis on *pollution prevention*. This strategy greatly reduces the need for statistically uncertain and controversial toxicity studies and exposure standards. It also reduces the risk posed by potentially hazardous chemicals and products and their possible but poorly understood multiple interactions.

Despite their many limitations, carefully conducted and evaluated toxicity studies are important sources of information used to help us understand dose and response effects and to estimate and set exposure standards. But citizens, lawmakers, and regulatory officials must recognize the huge uncertainties and guesswork involved in all such studies.

8-3   CHEMICAL HAZARDS

What Are Toxic and Hazardous Chemicals? **Toxic chemicals** are generally defined as substances that are fatal to over 50% of test animals (LD_{50}) at given concentrations. **Hazardous chemicals** cause harm by **(1)** being flammable or explosive, **(2)** irritating or damaging the skin or lungs (strong acidic or alkaline substances such as oven cleaners), **(3)** interfering with or preventing oxygen uptake and distribution (asphyxiants such as carbon monoxide and hydrogen sulfide), or **(4)** inducing allergic reactions of the immune system (allergens).

What Are Mutagens, Teratogens, and Carcinogens? **Mutagens** are agents, such as chemicals and radiation, that cause random *mutations*, or changes in the DNA molecules found in cells. Mutations in a sperm or egg cell may be passed on to future generations and cause diseases such as manic depression, cystic fibrosis,

hemophilia, sickle-cell anemia, Down's syndrome, and some types of cancer. Those in other cells are not inherited but may cause harmful effects such as tumors. Although some mutations are harmful, most are of no consequence, probably because all organisms have biochemical repair mechanisms that can find and correct mistakes or changes in the DNA code.

Teratogens are chemicals, radiation, or viruses that cause birth defects while the human embryo is growing and developing during pregnancy, especially during the first 3 months. Chemicals known to cause birth defects in laboratory animals include PCBs, thalidomide, steroid hormones, and heavy metals such as arsenic, cadmium, lead, and mercury.

Carcinogens are chemicals, radiation, or viruses that cause or promote the growth of a malignant (cancerous) tumor, in which certain cells multiply uncontrollably. Many cancerous tumors spread by **metastasis** when malignant cells break off from tumors and travel in body fluids to other parts of the body. There, they start new tumors, making treatment much more difficult.

Because there are more than 100 types of cancer (depending on the types of cells involved), there are many different causes. These include genetic predisposition, viral infections, and exposure to various mutagens and carcinogens.

According to the World Health Organization (WHO), environmental and lifestyle factors play a key role in causing or promoting up to 80% of all cancers. Major sources of carcinogens are cigarette smoke (35–40% of cancers), diet (20–30%), occupational exposure (5–15%), and environmental pollutants (1–10%). About 10–20% of cancers are believed to be caused by inherited genetic factors or certain viruses.

Typically, 10–40 years may elapse between the initial exposure to a carcinogen and the appearance of detectable symptoms. Partly because of this time lag, many healthy teenagers and young adults have trouble believing that their smoking (Spotlight, p. 228), drinking, eating, and other lifestyle habits today could lead to some form of cancer before they reach age 50.

How Can Chemicals Harm the Immune, Nervous, and Endocrine Systems? Since the 1970s a growing body of research on wildlife and laboratory animals and epidemiological studies of humans indicates that long-term (often low-level) exposure to various toxic chemicals in the environment can cause damage by disrupting the body's immune, nervous, and endocrine systems.

The *immune system* consists of numerous specialized cells and tissues that protect the body from disease and harmful substances by forming antibodies to invading agents and rendering them harmless. Synthetic

Are Hormone Disrupters a Health Threat?

Over the last few years experts from a number of disciplines have been piecing together field studies on wildlife, studies on laboratory animals, and epidemiological studies of human populations indicating that a variety of human-made chemicals can act as *hormone disrupters*.

Numerous studies indicate that extremely low levels of hormone disrupters, like trace levels of natural hormones, can affect developing embryos and human fetuses. There is also growing concern about pollutants that can act as *thyroid disrupters* and cause growth, weight, brain, and behavioral disorders.

Research suggests that various cells can sometimes mistake human-made *hormone mimics* for natural hormones; the resulting disruption in the growth and development processes regulated by hormones occurs because certain biochemical pathways, especially those in the reproductive system, are turned off, slowed down, or speeded up. Others, called *hormone blockers*, disrupt the endocrine system by preventing natural hormones from carrying out their biochemical roles.

Most natural hormones are broken down or excreted. However, many of the synthetic hormone impostors are stable, fat-soluble compounds whose concentrations can be biomagnified as they move through food chains and webs (Figure 6-6). Thus, they can pose a special threat to humans and other carnivores dining at the top of food webs.

Exposure to PCBs, for example, has reduced penis size in some test animals and in 118 boys born to women who were exposed to a PCB spill in Taiwan in 1979. In 1973, estrogen mimics called PBBs accidentally got into cattle feed in Michigan, and from there into beef. Pregnant women who ate the beef (and whose breast milk had high levels of PBBs) had sons with undersized penises and malformed testicles.

Much more research is needed to verify such findings and to determine whether low levels of most hormone-disrupting chemicals in the environment pose a threat to the human population. If exposure to small amounts of hormone disrupters is shown to be harmful to humans and some forms of wildlife, the only reasonable choice may be to prevent such chemicals from reaching the environment—a shift from pollution control to pollution prevention.

Critical Thinking

Do you consider the possible threat from hormone disrupters a problem that could affect you? Any child you might have? Give three important ways for dealing with this problem.

chemicals, viruses such as HIV, and ionizing radiation that weaken the human immune system leave the body open to attacks by allergens and infectious bacteria, viruses, and protozoans. Recent studies of laboratory animals and wildlife as well as epidemiological studies of humans (especially in developing countries) have linked suppression of the immune system to several widely used pesticides.

The human *nervous system* (brain, spinal cord, and peripheral nerves) is also being threatened by synthetic chemicals in the environment. Many poisons are *neurotoxins*, which attack nerve cells (neurons). Examples are **(1)** chlorinated hydrocarbons (DDT, PCBs, dioxins);**(2)** organophosphate pesticides; **(3)** formaldehyde; **(4)** various compounds of arsenic, mercury, lead, and cadmium; and **(5)** widely used industrial solvents such as trichloroethylene (TCE), toluene, and xylene.

The *endocrine system* is a complex set of organs and tissues whose actions are coordinated by chemical messengers called *hormones*. These hormones (which are secreted in extremely low levels into the bloodstream) control sexual reproduction, growth, development, and behavior in humans and other animals (Connections, left).

Mostly because of a lack of money and the scientific complexity involved, the U.S. National Academy of Sciences estimates that only about 10% of the 72,000 chemicals in commercial use have been thoroughly screened for toxicity, and only 2% have been adequately tested to determine whether they are carcinogens, teratogens, or mutagens. Hardly any of the chemicals in commercial use have been screened for damage to the nervous, endocrine, and immune systems. Currently, about 99.5% of the commercially used chemicals in the United States are entirely unregulated by federal and state governments.

8-4 BIOLOGICAL HAZARDS: DISEASE IN DEVELOPED AND DEVELOPING COUNTRIES

What Are the Major Types of Disease? A transmissible disease is caused by a living organism (such as a bacterium, virus, protozoa, or parasite) and can be spread from one person to another. The infectious agents, called *pathogens*, are spread by air, water, food, body fluids, and some insects (such as mosquitoes, flies, and ticks), animals (such as rodents and monkeys), and other nonhuman carriers called *vectors*. Table 8-2 gives information on the world's eight deadliest infectious diseases.

Typically, a *bacterium* is a one-celled microorganism capable of replicating itself by simple division. A *virus* is a microscopic, noncellular infectious agent. Its DNA contains instructions for making more viruses, but it has no apparatus to do so. In order to replicate, a

Q: What percentage of the world's estimated plant species have been evaluated for their medical uses?

Table 8-2 The World's Eight Most Deadly Infectious Diseases

Disease	Cause	Estimated New Cases per Year	Estimated Deaths per Year
Acute respiratory infections*	Bacteria, viruses	1 billion	4.7 million
Diarrheal diseases†	Bacteria, viruses, parasites	1.8 billion	3.1 million
Tuberculosis	Bacteria	9 million	3.1 million
Malaria	Parasitic protozoa	110 million	2.5–2.7 million
AIDS	Virus (HIV)	600,000 (AIDS) 4 million (HIV)	2.3 million
Measles	Viruses	200 million	1 million
Hepatitis B	Virus	200 million	1 million
Tetanus	Bacteria	1 million	500,000

Source: World Health Organization (WHO), *The World Health Report 1997* (Geneva, Switzerland: WHO, 1998)
* Includes pneumonia, influenza, and whooping cough.
† Includes amoebic dysentery, cryptosporidiosis, and gastrochloritis.

virus must invade a host cell and take over the cell's DNA to create a factory for producing more viruses.

In developing countries, infectious diseases accounted for about 44% of the deaths, compared to only 5% in developed countries in 1997. In 1997, infectious diseases were the world's leading cause of death, resulting in 52.2 million deaths. According to WHO and UNICEF, every year in developing countries at least 11 million children under the age of 5 die of mostly preventable infectious diseases—an average of at least 30,000 premature deaths per day. About 80% of all illnesses in developing countries are caused by waterborne infectious diseases (such as diarrhea, hepatitis, typhoid fever, and cholera), mainly from unsafe drinking water and inadequate sanitation systems (Spotlight, p. 234). Table 8-2 shows that the biggest killers by far are infectious diseases—such as respiratory infections, diarrheal diseases, and tuberculosis—that have been around for a long time.

One of the world's most underreported stories in the 1990s has been the rapid spread of tuberculosis (TB), a highly infectious bacterial disease that currently kills about 3.1 million people per year (Table 8-2). The current TB epidemic is so severe that in 1993 WHO declared a global state of emergency. By contrast, the highly publicized ebola viruses have killed about 650 people over the past 20 years—an average of 33 people per year.

Major reasons for the recent increase in TB are **(1)** poor TB screening and control programs (especially in developing countries, where about 95% of the new cases occur), **(2)** development of strains of the tuberculosis bacterium that are genetically resistant to virtually all effective antibiotics (typically leading to mortality rates of over 50%), **(3)** population growth and increased urbanization and crowding (which increase contacts among people), **(4)** poverty, and **(5)** the spread of acquired immune deficiency syndrome (AIDS), which greatly weakens the immune system and allows TB bacterium to multiply.

Treatment with a combination of four inexpensive drugs can cure 90% of those with active TB. However, the patients must take the drugs *every* day for 6 to 8 months. Because the symptoms disappear after a few weeks, many patients think they are cured and stop taking the drugs. This allows the disease to recur in a hard-to-treat form, it spreads to other people, and drug-resistant strains of TB bacteria develop.

Diseases such as cardiovascular (heart and blood vessel) disorders, most cancers, diabetes, bronchitis, emphysema, and malnutrition typically have multiple (and often unknown) causes. They also tend to develop slowly and progressively over time. Because they are not caused by living organisms and do not spread from one person to another, they are classified as **nontransmissible diseases**.

A: About 2%

Are We Losing the War Against Infectious Bacteria?

There is growing and alarming evidence that we may be losing our war against infectious bacterial diseases because bacteria are among the earth's ultimate survivors. When a colony of bacteria is dosed with an antibiotic such as penicillin, most die, but a few have mutant genes that make them immune to the drug. Through natural selection (Section 2-8), a single mutant can pass such traits on to most of its offspring, which can amount to 16,777,216 in only 24 hours!

Each time this strain of bacterium is exposed to penicillin or some other antibiotic, a larger proportion of its offspring are genetically resistant to the drug. The rapid multiplication of resistant bacteria in a victim is made easier because the antibiotics also wipe out their bacterial competitors.

Even worse, bacteria can become genetically resistant to antibiotics they have never been exposed to. When a resistant and a nonresistant bacterium touch one another (say in a hospital bedsheet or in a human stomach), they can exchange a loop of DNA called a plasmid, thereby transferring genetic resistance from one organism to another.

The incredible genetic adaptability of bacteria is one reason the world now faces a potentially serious rise in the incidence of some infectious bacterial diseases once controlled by antibiotics. Other factors also play a key role, including (1) spread of bacteria (some beneficial and some harmful) around the globe by human travel and the trade of goods; (2) overuse of antibiotics by doctors, often at the insistence of their patients; (3) failure of many patients to take all of their prescribed antibiotics, which promotes bacterial resistance; (4) availability of antibiotics without prescriptions in many countries; and (5) widespread use of antibiotics in the livestock and dairy industries.

The result of these factors acting together is that every major disease-causing bacterium now has strains that resist at least one of the roughly 160 antibiotics we use to treat bacterial infections. In 1998, officials at the Centers for Disease Control and Prevention estimated that about 2 million patients (most with a weakened immunity system) develop a hospital-acquired infection in the United States each year and about 90,000 of these patients die.

Critical Thinking

What role, if any, have you played in the increase in genetic resistance of bacteria to widely used antibiotics? List three ways to reduce this threat.

How Rapidly Are Viral Diseases Spreading?

Viral diseases include *influenza* or *flu* (transmitted by the bodily fluids or airborne emissions of an infected person), *ebola* (transmitted by the blood or other body fluids of an infected person), and *AIDS*. Like bacteria, viruses can genetically adapt rapidly to different conditions.

Flu viruses move through the air and are highly contagious. In 1918–19 a flu epidemic infected almost half the world's population and killed 20–30 million people. Although health officials worry about the emergence of new viral diseases (such as those caused by ebola viruses), they recognize that the greatest viral health threat to humans is the emergence of new, very virulent strains of influenza.

There is a growing threat from the spread of *AIDS*, which is caused by the human immunodeficiency virus (HIV). The virus itself isn't deadly, but it kills immune cells and leaves the body defenseless against all sorts of other infectious bacteria and viruses. The HIV virus can be transmitted during unprotected sexual activity, from one intravenous drug user to another through shared needles, from an infected mother to an infant during birth, and by exposure to infected blood.

According to WHO, in 1997 at least 31 million people (21 million of them in sub-Saharan Africa) were infected with HIV. An average of 16,000 people are newly infected with the virus each day, 80% of them in Asia and Africa.

Within about 7–10 years, 95% of those with HIV develop AIDS. This long incubation period means that infected people often spread the virus for several years before they learn that they are infected. There is no cure for AIDS yet, although drugs may help some infected people live longer. By the end of 1997, 12.5 million of the 31 million people infected with HIV had acquired full-blown AIDS, with a record number (2.3 million) dying from AIDS in 1997.

Once a viral infection starts, it is much harder to fight than infections by bacteria and protozoans. Only a few antiviral drugs exist because most drugs that kill a virus also harm the cells of its host. Treating viral infections with antibiotics is useless and merely increases genetic resistance in disease-causing bacteria. Medicine's only effective weapon against viruses are preventive vaccines that provide antibodies to ward off viral infections. Immunization with vaccines has helped tame viral diseases such as smallpox, polio, rabies, influenza, and measles.

Case Study: Malaria, a Protozoal Disease

About 45% (or 2.6 billion people) of the world's population live in tropical and subtropical regions in which malaria is present (Figure 8-3). Currently, an estimated 300–500 million people are infected with malaria parasites worldwide, and at least 110 million new cases

Q: How many of earth's species are believed to become extinct each year because of human activities?

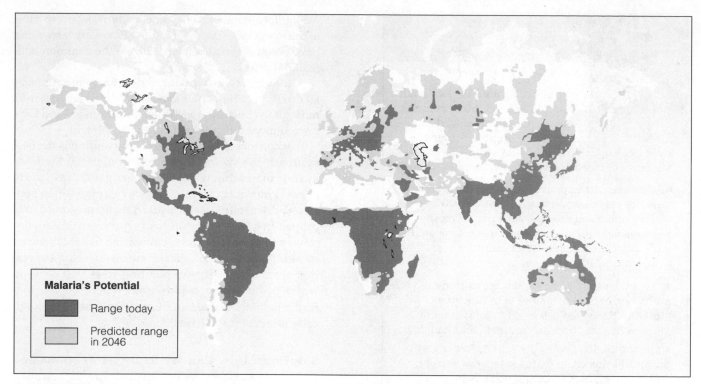

Malaria's Potential

▨ Range today

▨ Predicted range in 2046

Figure 8-3 Worldwide distribution of malaria. About 45% of the world's current population lives in areas in which malaria is present, with the disease killing 2.7 million people a year. If the world becomes warmer, as projected by current climate models, by 2046 malaria could affect 60% of the world's population. (Data from the World Health Organization)

occur each year (97 million of them in Africa). Malaria's symptoms come and go and include fever and chills, anemia, an enlarged spleen, severe abdominal pain and headaches, extreme weakness, and greater susceptibility to other diseases. The disease kills 2.7 million people each year, more than half of them children under age 5.

Malaria is caused by four species of protozoa of the genus *Plasmodium*. Most cases of the disease are transmitted when an uninfected female of any one of 60 species of *Anopheles* mosquito bites an infected person, ingests blood that contains the parasite, and later bites an uninfected person (Figure 8-4). When this happens, *Plasmodium* parasites move out of the mosquito and into the human's bloodstream, multiply in the liver, and then enter blood cells to continue multiplying. Malaria can also be transmitted by blood transfusions or by sharing needles. This cycle repeats itself until immunity develops, treatment is given, or the victim dies.

During the 1950s and 1960s, the spread of malaria was sharply curtailed by draining swamplands and marshes, spraying breeding areas with insecticides, and using drugs to kill the parasites in the bloodstream. Since 1970, however, malaria has come roaring back. Most species of the malaria-carrying *Anopheles*

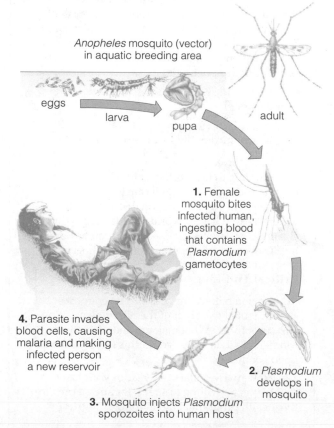

Anopheles mosquito (vector) in aquatic breeding area

eggs
larva
pupa
adult

1. Female mosquito bites infected human, ingesting blood that contains *Plasmodium* gametocytes

2. *Plasmodium* develops in mosquito

3. Mosquito injects *Plasmodium* sporozoites into human host

4. Parasite invades blood cells, causing malaria and making infected person a new reservoir

Figure 8-4 The life cycle of malaria.

Improving Health Care in Developing Countries

SOLUTIONS

With adequate funding, the health of people in developing countries (and the poor in developed countries) can be improved dramatically, quickly, and cheaply by providing the following forms of mostly preventive health care:

- Better nutrition, prenatal care, and birth assistance for pregnant women. At least 600,000 women in developing countries die each year of mostly preventable pregnancy-related causes, compared with about 6,000 in developed countries.

- Better nutrition for children.

- Greatly improved postnatal care (including promotion of breast-feeding) to reduce infant mortality. Breast-fed babies get natural immunity to many diseases from antibodies in breast milk.

- Immunization against the world's five largest preventable infectious diseases: tetanus, measles, diphtheria, typhoid fever, and polio. Between 1971 and 1992, the percentage of children in developing countries immunized against these diseases rose from 10% to 80%, saving about 9 million lives a year

- Oral rehydration therapy for victims of diarrheal diseases, which cause about one-fourth of all deaths of children under the age of 5. A simple solution of boiled water, salt, and sugar or rice, at a cost of only a few cents per person, can prevent death from dehydration.

- Careful and selective use of antibiotics for infections.

- Clean drinking water and sanitation facilities for the one-third of the world's population that lacks them.

According to the World Health Organization, extending such primary health care to all the world's people would cost an additional $10 billion per year, a mere 4% of what the world spends every year on cigarettes or devotes every 4 days to military spending. The cost of this program is about $1 per child.

Critical Thinking

1. Do you believe that developed countries should foot at least half of the bill implementing such proposals? What major economic and environmental advantages would this have for developed countries?

2. How many dollars per year of your taxes would you be willing to spend for such a preventive health program in developing countries?

mosquitoes have become genetically resistant to most of the insecticides used. Worse, the *Plasmodium* parasites have become genetically resistant to the common antimalarial drugs.

Researchers are working to develop new antimalarial drugs, vaccines, and biological controls for *Anopheles* mosquitoes. Such approaches are underfunded and have proved more difficult than originally thought.

According to health experts, prevention is the best approach to slowing the spread of malaria. Methods include increasing water flow in irrigation systems to prevent mosquito larvae from developing (an expensive and wasteful use of water), using mosquito nets dipped in a nontoxic insecticide (permethrin) in windows and doors of homes, cultivating fish that feed on mosquito larvae (biological control), clearing vegetation around houses, and planting trees that soak up water in low-lying marsh areas where mosquitoes thrive (a method that can degrade or destroy ecologically important wetlands).

Solutions: How Can We Reduce Infectious and Other Diseases? According to health scientists and public health officials, disease in developing and developed countries could be greatly reduced by

- Greatly increasing research on tropical diseases. WHO estimates that only 3% of the money spent worldwide each year for such research is devoted to malaria and other tropical diseases, even though more people suffer and die worldwide from these diseases than from all others combined.

- Mounting a global campaign to reduce overcrowding, unsafe drinking water, poor sanitation, malnutrition, and poverty.

- Not using powerful antibiotics to treat minor infections or undiagnosed symptoms.

- Educating the public about the dangers of overuse of antibiotics and the need to take all of the antibiotics in any prescription.

- Not selling antibiotics without a prescription (allowed in many countries).

- Sharply reducing the use of antibiotics in livestock.

- Insisting that doctors, nurses, and orderlies strictly maintain hygienic standards at all times.

- Putting much more money into the development of vaccines to prevent infections by bacteria and viruses responsible for most disease and death.

- Emphasizing preventive health care, especially in developing countries (Solutions, left).

Slovic, Paul. 1997. "Public Perception of Risk." *Journal of Environmental Health*, vol. 59, no. 9, 22(3).

8-5 RISK ANALYSIS

How Can We Estimate Risks? Risk analysis involves identifying hazards and evaluating their associated risks (*risk assessment*), ranking risks (*comparative risk analysis*), determining options and making decisions about reducing or eliminating risks (*risk management*), and informing decision makers and the public about risks (*risk communication*).

Risk assessment involves determining the types of hazards involved, estimating the probability that each hazard will occur, and estimating how many people are likely to be exposed to it and how many may suffer serious harm. Statistical probabilities based on past experience, animal testing and other tests, and epidemiological studies are used to estimate risks from older technologies and products (Section 8-2). For new technologies and products, much more uncertain statistical probabilities, based on models rather than actual experience, must be calculated.

The left side of Figure 8-5 is an example of *comparative risk analysis*, summarizing the greatest ecological and health risks identified by a panel of scientists acting as advisers to the U.S. Environmental Protection Agency. Note the considerable difference between the comparison of relative risk by scientists (Figure 8-5, left) and the general public (Figure 8-5, right). These differences result largely from failure of professional risk evaluators to communicate the nature of risks and their relative importance to the public, teachers, and members of the media. Much of our risk education is based on misleading media reports on the latest risk scare of the week (based mainly on frontier science) that do not put such risks in perspective.

The key question is whether the estimated short- and long-term risks of using a particular technology or product outweigh the estimated short- and long-term

---

Scientists
(Not in rank order
in each category)

Citizens
(In rank order)

High-Risk Health Problems
- Indoor air pollution
- Outdoor air pollution
- Worker exposure to industrial or farm chemicals
- Pollutants in drinking water
- Pesticide residues on food
- Toxic chemicals in consumer products

High-Risk Ecological Problems
- Global climate change
- Stratospheric ozone depletion
- Wildlife habitat alteration and destruction
- Species extinction and loss of biodiversity

Medium-Risk Ecological Problems
- Acid deposition
- Pesticides
- Airborne toxic chemicals
- Toxic chemicals, nutrients, and sediment in surface waters

Low-Risk Ecological Problems
- Oil spills
- Groundwater pollution
- Radioactive isotopes
- Acid runoff to surface waters
- Thermal pollution

High-Risk Problems
- Hazardous waste sites
- Industrial water pollution
- Occupational exposure to chemicals
- Oil spills
- Stratospheric ozone depletion
- Nuclear power-plant accidents
- Industrial accidents releasing pollutants
- Radioactive wastes
- Air pollution from factories
- Leaking underground tanks

Medium-Risk Problems
- Coastal water contamination
- Solid waste and litter
- Pesticide risks to farmworkers
- Water pollution from sewage plants

Low-Risk Problems
- Air pollution from vehicles
- Pesticide residues in foods
- Global climate change
- Drinking water contamination

Figure 8-5 Comparative risk analysis of the most serious ecological and health problems according to scientists acting as advisers to the U.S. Environmental Protection Agency (left). Risks in each of these categories are not listed in rank order. The right side of this figure represents polls showing how U.S. citizens rank the ecological and health risks they perceive as being the most serious. Why do you think there is such a great difference between the ranking by risk experts and by the general public? (Data from Science Advisory Board, *Reducing Risks*, Washington, D.C.: Environmental Protection Agency, 1990)

Hint: Enter the search terms *risk assessment, public perception* using Key Words.

benefits of other alternatives. One method for making such evaluations is **risk–benefit analysis**, which involves estimating such benefits and the risks involved.

What Are the Greatest Risks People Face? The greatest risks most people face today are rarely dramatic enough to make the daily news. In terms of reduced life span from malnutrition, exposure to disease-causing organisms and dangerous chemicals, and lack of basic health care, *the greatest risk by far is poverty* (Figure 8-6).

After the health risks associated with poverty, the greatest risks of premature death are mostly the result of voluntary—and thus correctable—choices people make about their lifestyles (Figures 8-5 and 8-6). By far

the best ways to reduce one's risk of premature death and serious health risks are to not smoke (Spotlight, p. 228), avoid excess sunlight (which ages skin and causes skin cancer), not drink alcohol or drink only in moderation (no more than two drinks per day), reduce consumption of foods containing cholesterol and saturated fats, eat a variety of fruits and vegetables, exercise regularly, lose excess weight, and (for those who can afford a car) drive as safely as possible in a vehicle with the best available safety equipment.

How Can We Estimate Risks for Technological Systems? The more complex a technological system and the more people needed to design and run it, the more difficult it is to estimate the risks. The overall relia-

Figure 8-6 Comparison of risks people face, expressed in terms of shorter average life span. After poverty, the greatest risks people face are mostly from voluntary choices they make about their lifestyles. (Data from Bernard L. Cohen)

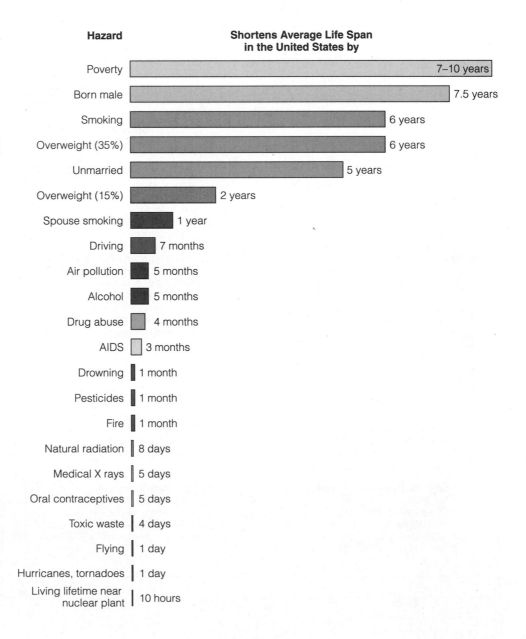

Hazard	Shortens Average Life Span in the United States by
Poverty	7–10 years
Born male	7.5 years
Smoking	6 years
Overweight (35%)	6 years
Unmarried	5 years
Overweight (15%)	2 years
Spouse smoking	1 year
Driving	7 months
Air pollution	5 months
Alcohol	5 months
Drug abuse	4 months
AIDS	3 months
Drowning	1 month
Pesticides	1 month
Fire	1 month
Natural radiation	8 days
Medical X rays	5 days
Oral contraceptives	5 days
Toxic waste	4 days
Flying	1 day
Hurricanes, tornadoes	1 day
Living lifetime near nuclear plant	10 hours

Q: What percentage of earth's land surface has been set aside to protect wildlife?

bility of any technological system (expressed as a percentage) is the product of two factors (multiplied by 100):

$$\text{System reliability (\%)} = \text{Technology reliability} \times \text{Human reliability} \times 100$$

With careful design, quality control, maintenance, and monitoring, a highly complex system such as a nuclear power plant or a space shuttle can achieve a high degree of technology reliability. However, human reliability is almost always much lower than technology reliability and is virtually impossible to predict; to err is human. Suppose, for example, that the technology reliability of a nuclear power plant is 95% (0.95) and that human reliability is 75% (0.75). Then the overall system reliability is only 71% ($0.95 \times 0.75 \times 100 = 71\%$). Even if we could make the technology 100% reliable (1.0), the overall system reliability would still be only 75% ($1.0 \times 0.75 \times 100 = 75\%$). The crucial dependence of even the most carefully designed systems on unpredictable human reliability helps explain essentially "impossible" tragedies such as the Chernobyl nuclear power-plant accident (Figure 4-32) and the explosion of the space shuttle *Challenger*.

One way to make a system more foolproof or fail-safe is to move more of the potentially fallible elements from the human side to the technical side. However, chance events such as a lightning bolt can knock out an automatic control system. No machine or computer program can completely replace human judgment in ensuring that a complex system operates properly and safely. Of course, the parts in any automated control system are manufactured, assembled, tested, certified, and maintained by fallible human beings. Computer software programs used to monitor and control complex systems can also contain human errors, or they can be deliberately modified by computer viruses to malfunction. The pros and cons of genetic engineering reveal the difficulty in evaluating a new technology (Pro/Con, p. 240).

What Are the Limitations of Risk Assessment?
Risk assessment is a young science that has many built-in uncertainties and limitations. It depends on toxicology assessments that have scientific and economic limitations (Section 8-2). Each additional step in risk assessment also has uncertainties and economic limitations.

Here are some of the key questions involved in risk assessment and risk–benefit analysis:

- How reliable are risk assessment data and models?

- Who profits from allowing certain levels of harmful chemicals into the environment, and who suffers? Who decides this?

- Should estimates emphasize short-term risks, or should more weight be put on long-term risks? Who should make this decision?

- Should the primary goal of risk analysis be to determine how much risk is acceptable (the current approach) or to figure out how to do the least damage (a prevention approach)?

- Who should do a particular risk–benefit analysis or risk assessment, and who should review the results? A government agency? Independent scientists? The general public?

- Should cumulative effects of various risks be considered, or should risks be considered separately, as is usually done? Suppose a pesticide is found to have an annual risk of killing one person in a million from cancer, the current EPA limit. Cumulatively, however, effects from 40 such pesticides might kill 40, or even 400, of every 1 million people. Is this acceptable?

- Should risk levels be higher for workers (as is almost always the case) than for the general public? What say should workers and their families have in this decision? According to government estimates, the exposure of workers to toxic chemicals in the United States causes 50,000–70,000 deaths, at least half from cancer, and 350,000 new cases of illness per year. The situation is much worse in developing countries. Is this a necessary cost of doing business?

Some see risk analysis as a useful and much-needed tool. Others see it as a way to justify premeditated murder in the name of profit. According to the National Academy of Sciences, exposure to toxic chemicals is responsible for 2–4% of the 521,000 cancer deaths in the United States; this amounts to 10,400–20,800 premature cancer deaths per year. According to hazardous waste expert Peter Montague, "The explicit aim of risk assessment is to convince people that some number of citizens *must* be killed each year to maintain a national lifestyle based on necessities like Saran Wrap, throwaway cameras, and lawns without dandelions." Risk evaluator Mary O'Brien says, "Risk assessment is industry's attempt to make the intolerable appear tolerable."

Critics understand that not all dangerous chemicals can be banned, but they argue that the emphasis should shift from determining acceptable risk levels to trying to reduce the risks as much as possible by pollution prevention. Less toxic or nontoxic chemicals can often be substituted in manufacturing processes or in products, toxic chemicals can be recycled or reused instead of being buried or burned, and industrial processes can be redesigned to eliminate or sharply reduce the use and release of toxic chemicals into the en-

Genetic engineers have learned how to splice genes and recombine sequences of existing DNA molecules in organisms to produce DNA with new genetic characteristics (recombinant DNA). Thus, they can transfer traits from one species to another without waiting for new genetic combinations to evolve through the process of natural selection.

This rapidly developing technology excites many scientists and investors, who see it as a way to produce, patent, and sell high-yield crops and livestock with more protein and greater resistance to diseases, pests, frost, and drought. They also hope to create bacteria that can destroy oil spills, degrade toxic wastes, concentrate metals found in low-grade ores, and serve as biological factories for new vaccines, drugs, and therapeutic hormones. Gene therapy, its proponents argue, could eliminate certain genetic diseases.

When the U.S. Supreme Court ruled in 1980 that genetically engineered organisms could be patented, investors began pouring billions of dollars into the fledgling biotech industry. Genetic engineering has produced a drug that reduces heart attack damage and agents that fight diabetes, hemophilia, and some forms of cancer; it has also been used to diagnose AIDS and cancer.

Genetically altered bacteria have been used to manufacture more effective vaccines and hormones. In agriculture, gene transfer has been used to develop strawberries and tomatoes that resist frost because they contain "antifreeze" genes from fish, tomatoes that stay fresh and tasty longer, potatoes that resist disease because they contain chicken genes, and smaller cows that produce more milk.

Some people worry that biotechnology may run amok. They argue that it should be kept under strict control because we don't understand well enough how nature works to allow unregulated genetic alteration of humans and other species. They also point to serious and widespread ecological problems that have resulted from the accidental or deliberate introduction of nonnative organisms into biological communities (Section 6-3).

To such critics the resulting *genetic pollution* may eventually prove to be more harmful than chemical pollution. Unlike chemicals, genetically engineered organisms might also mutate, change their form and behavior, migrate to other areas, and alter the genetic traits of existing wild species; unlike defective cars and other products, they couldn't be recalled. Proponents argue that the risk of biotech-caused ecological catastrophes is small.

Critics are also concerned that one of the most serious effects of widespread use of biotechnology is reduction of the world's vital genetic diversity and thus biodiversity. Already the world's 20 major food crops have become 70% less genetically diverse because a wide range of wild strains have been replaced with only a few varieties developed by crossbreeding; using bioengineered crop strains could hasten this loss of vital biodiversity.

Development of all new varieties by crossbreeding or genetic engineering is based on mixing certain traits found in wild varieties. Thus, reducing the natural genetic diversity of wild plants and animals undermines the genetic engineers' ability to produce new combinations in the laboratory and undermines the future success of the biotech industry.

In 1989, a committee of prominent ecologists appointed by the Ecological Society of America warned that the ecological effects of new combinations of genetic traits from different species would be difficult to predict. Their report called for a case-by-case review of any proposed environmental releases, as well as carefully regulated, small-scale field tests before any bioengineered organism is put into commercial use—practices that are not being followed adequately.

This controversy illustrates the difficulty of balancing the actual and potential benefits of a technology with its actual and potential risks of harm. Proponents of this new technology may be right in their belief that the benefits far exceed the potential and mostly unknown risks. But what if they are wrong?

Critical Thinking

1. What government controls, if any, do you believe should be applied to the development and use of genetic engineering? How would you enforce any restrictions?

2. Use the library or the Internet to find out what controls now exist on biotechnology in the United States (or the country where you live) and how well such controls are enforced.

vironment—solutions a small but growing number of manufacturing industries are beginning to use.

These critics also accuse industries of favoring risk analysis because so little is known about health risks from pollutants and because the data that do exist are controversial. The result is that risk assessment and risk–benefit analysis (Spotlight, right) can be crafted to support almost any conclusion. The huge

SPOTLIGHT

When we add new chemicals or technologies we are mostly flying blind about their possible harmful effects. For example, a recent study has documented the significant uncertainties involved in even simple risk assessment. Eleven European governments established 11 different teams of their best scientists and engineers (including those from private companies) to assess the hazards and risks from a small plant storing only one hazardous chemical (ammonia). The 11 teams, consisting of world-class experts analyzing this very simple system, disagreed with one another on fundamental points and varied in their assessments of the hazards by a factor of 25,000!

By analogy, the current built-in uncertainty in risk–benefit analysis is analogous to a radar device that can detect a car speeding at 160 kilometers (100 miles) per hour, but can tell us only that the car is traveling somewhere between 0.16 kph (0.1 mph) and 160,000 kph (100,000 mph). Such inherent uncertainty in risk–benefit analysis explains

why regulators setting human exposure levels for toxic substances usually divide the best results by 100 to 1,000, to provide the public with a margin of safety.

Despite the inevitable uncertainties involved, risk assessment and risk–benefit analysis are useful ways to **(1)** organize available information, **(2)** identify significant hazards, **(3)** focus on areas that need more research, **(4)** help regulators decide how money for reducing risks should be allocated, and **(5)** stimulate people to make more informed decisions about health and environmental goals and priorities.

However, at best risk assessments and risk–benefit analyses yield only a range of probabilities and uncertainties based on different assumptions, not the precise numbers that decision makers and the general public wish they had. Environmentalists and health officials believe that all such analyses should contain the wide range of assumptions, probabilities, and uncertainties involved and be open to full public review.

Critics of risk assessment and risk–benefit analysis argue that

the main decision-making tools we should rely on are *looking at the available alternatives to a particular chemical or technology and having a full public discussion of the major advantages (benefits) and disadvantages (costs) of those alternatives.*

The goal is not to find out how much risk is acceptable but to find out the least-damaging reasonable alternatives by asking, "Which alternative will bring sufficient benefits and minimize damage to humans and to the earth?" They argue that if the alternatives are examined fairly, emphasis will shift from trying to determine "acceptable" risk levels to trying to reduce the risks as much as possible by *pollution prevention.*

Critical Thinking

Do you believe that **(a)** risk assessment, risk–benefit analysis, and cost–benefit analysis should be used as the *primary* tool for establishing any federal health, safety, or environmental regulation? or that **(b)** the emphasis should be placed on fair evaluation of alternatives, with the goal of finding the least harmful (and most affordable) alternative? Explain.

uncertainties in risk assessment and risk–benefit analysis also allow industries to delay regulatory decisions for decades by challenging data in the courts.

How Should Risks Be Managed? Once an assessment of risk is made, decisions must be made about what to do about the risk. **Risk management** includes the administrative, political, and economic actions taken to decide whether and how to reduce a particular societal risk to a certain level—and at what cost.

Risk management involves deciding **(1)** which of the vast number of risks facing society should be evaluated and managed and in what order or priority with the limited funds available, **(2)** how reliable the risk–benefit analysis or risk assessment performed for each risk is, **(3)** how much risk is acceptable, **(4)** how much

money it will take to reduce each risk to an acceptable level, **(5)** how much each risk will be reduced if available funds are limited (as is almost always the case), **(6)** and how the risk management plan will be communicated to the public, monitored, and enforced. Each step in this process involves making value judgments and weighing trade-offs to find a reasonable compromise among conflicting political, economic, health, and environmental interests.

How Well Do We Perceive Risks? Most of us do poorly in assessing the risks from the hazards that surround us (Figures 8-5 and 8-6), and we tend to be full of contradictions. Many people deny or shrug off the high-risk chances of dying (or injury) from voluntary activities they enjoy, such as motorcycling (1 death in 50 participating), smoking (1 in 300 participants by age 65 for a pack-a-day smoker), hang-gliding (1 in 1,250), and driving (1 in 2,500 without a seat belt and 1 death in 5,000 with a seat belt).

Yet some of these same people may be terrified about the possibility of dying from a commercial airplane crash (1 in 4.6 million), train crash (1 in 20 million), snakebite (1 in 36 million), shark attack (1 in 300 million), exposure to asbestos in schools (1 in 11 million), or exposure to trichloroethylene (TCE) in drinking water at the trace levels allowed by the EPA (1 in 2 billion).

Being bombarded with news about people killed or harmed by various hazards distorts our sense of risk. However, *the most important news each year is that about 99.1% of the people on the earth didn't die.** But that's not what we see on TV, hear, or read about every day. Despite the greatly increased use of synthetic chemicals in food production and processing, the general health and average life expectancy of people in the United States (and most developed countries) have increased during the past 50 years.

Our perceptions of risk and our responses to perceived risks often have little to do with how risky the experts say something is (Figures 8-5 and 8-6). The public generally sees a technology or a product as being riskier than experts do when

■ *It is new or complex rather than familiar.* Examples might include genetic engineering or nuclear power, as opposed to large dams or coal-fired power plants.

■ *It is perceived as being mostly involuntary.* Examples include nuclear power plants or food additives, as opposed to driving or smoking.

■ *It is viewed as unnecessary rather than as beneficial or necessary.* Examples might include using chlorofluorocarbon (CFC) and hydrocarbon propellants in aerosol spray cans or using food additives that increase sales appeal, as opposed to automobiles or aspirin.

■ *Its use involves a large, well-publicized death toll from a single catastrophic accident rather than the same or an even larger death toll spread out over a longer time.* Examples might include a severe nuclear power plant accident, an industrial explosion, or a plane crash, as opposed to coal-burning power plants, automobiles, or smoking.

■ *Its use involves unfair distribution of the risks.* Citizens are outraged when government officials decide to put a hazardous-waste landfill or incinerator in or near their neighborhood, even when the decision is based on risk–benefit analysis. This is usually seen as politics, not science. Residents will not be satisfied by estimates that the lifetime risks of cancer death from the facility are no greater than, say, 1 in 100,000. Living near the facility means that they, not the 99,000 or more people living farther away, have a much higher risk of dying from cancer by having this risk involuntarily imposed on them.

■ *The people affected are not involved in the decision-making process from start to finish.*

■ *Its use does not involve a sincere search for and evaluation of alternatives.* People who believe that their lives and the lives of their children are being threatened want to know what the alternatives are and which alternative provides the least harm to them and the earth.

Better education and communication about the nature of risks will help bring the public's perceptions of various risks closer to those of professional risk evaluators. However, such education will not eliminate the emotional, cultural, and ethical factors that decision makers must take into account in determining the acceptability of a particular risk and in evaluating the possible alternatives.

Not all waste and pollution can be eliminated. . . . What is absolutely crucial, however, is to recognize that pollution prevention should be the first choice and the option against which all other options are judged. The burden of proof imposed on individuals, companies, and institutions should be to show that pollution prevention options have been thoroughly examined, evaluated, and used before lesser options are chosen.

JOEL HIRSCHORN

CRITICAL THINKING

1. Should we have zero pollution levels for all hazardous chemicals? Explain.

2. Evaluate the following statements:
   a. Because almost any chemical can cause some harm in a large enough dose, we shouldn't get so worked up about exposure to toxic chemicals.
   b. We shouldn't worry so much about exposure to toxic chemicals because through genetic adaptation we can develop immunity to such chemicals.
   c. We shouldn't worry so much about exposure to toxic chemicals because we can use genetic engineering to reduce or eliminate such problems.

---

*The world's crude death rate in 1998 was 9 per 1,000 people. With 5.9 billion people, this amounts to about 53.1 million deaths (9/1,000 × 5.9 billion = 53.1 million). In other words, these deaths made up 0.9% of the world's population in 1998 (53.1 million/5.9 billion × 100 = 0.9%).

Q: Worldwide, how much topsoil is eroded each year?

3. Explain why you agree or disagree with the proposals for reducing the death toll and other harmful effects from smoking given on p. 228.

4. Do you agree that poverty is the greatest human risk in terms of sickness and premature death? Why do you think little is being done to reduce poverty compared to huge expenditures to reduce other, much less serious risks?

5. Do you believe that health and safety standards in the workplace should be strengthened and enforced more vigorously, even if this causes a loss of jobs when companies transfer operations to countries with weaker standards? Explain.

6. What controls, if any, do you believe should be applied to the development and use of genetic engineering? How would you enforce any restrictions?

7. Do you believe that we should shift the emphasis from risk assessment and risk–benefit analysis to pollution prevention and risk reduction? Explain. What beneficial and harmful effects might such a shift have on your life? On the life of any child you might have?

8. How would you answer each of the questions raised about risk assessment on p. 239? Explain each of your answers.

9. What are the five major risks that you as an individual face from your lifestyle, where you live, and what you do for a living? Which of these risks are voluntary and which are involuntary? List the five most important things you can do to reduce these risks. Which of these things do you actually plan to do?

*In our every deliberation, we must consider the impact
of our decisions on the next seven generations.*

IROQUOIS CONFEDERATION, EIGHTEENTH CENTURY

9-1 ATMOSPHERE, WEATHER, AND CLIMATE

**What Is the Troposphere? Weather
Breeder** We live at the bottom of a sea of air called
the **atmosphere**. This thin envelope of life-sustaining
gases surrounding the earth is divided into several
spherical layers characterized by abrupt changes in
temperature, the result of differences in the absorption
of incoming solar energy (Figure 9-1).

About 75% of the mass of the earth's air is found
in the atmosphere's innermost layer, the **troposphere**,
which extends about 17 kilometers (11 miles) above sea
level at the equator and about 8 kilometers (5 miles)
over the poles. If the earth were an apple, this lower
layer (containing the air we breathe) would be no
thicker than the apple's skin. This thin but turbulent
layer of rising and falling air currents and winds is the
planet's weather breeder.

Throughout the earth's long history the composition of the troposphere has varied considerably. Today,
about 99% of the volume of clean, dry air in the troposphere consists of two gases: nitrogen (78%) and oxygen
(21%). The remainder has slightly less than 1% argon
(Ar), 0.036% carbon dioxide (CO_2), and trace amounts
of several other gases. Air in the troposphere also holds
water vapor in amounts varying from 0.01% by volume
at the frigid poles to 5% in the humid tropics.

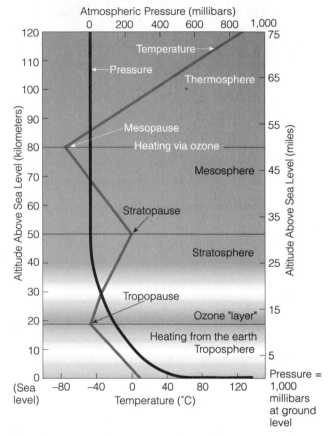

Figure 9-1 The earth's current atmosphere consists of several
layers. Most ultraviolet radiation from the sun is absorbed by
ozone (O_3) in the stratosphere, which is found primarily in the
ozone layer, between 17 and 26 kilometers (11–16 miles)
above sea level.

**What Is the Stratosphere? Earth's Global Sun-
screen** The atmosphere's second layer is the **stratosphere**, which extends from about 17 to 48 kilometers
(11–30 miles) above the earth's surface (Figure 9-1).Un-
like air in the troposphere, air in the stratosphere is calm,
with little vertical mixing. Although the stratosphere
contains less matter than the troposphere, its composition
is similar, with two notable exceptions: Its volume of
water vapor is about 1/1,000 as much and its volume of
ozone (O_3) is about 1,000 times greater.

Stratospheric ozone is produced when some of the
oxygen molecules there interact with UV radiation emitted by the sun. In the lower stratosphere oxygen (O_2) is
continuously converted to ozone (O_3) and back to oxygen by a sequence of reactions initiated by ultraviolet radiation from the sun. The result is a thin veil of renewable
ozone at very low concentrations. These ozone molecules absorb about 99% of the harmful incoming ultraviolet radiation from the sun and prevent it from
reaching the earth's surface. Normally, the average levels of ozone in this layer don't change much because the
rate of ozone destruction is equal to its rate of formation.

This "global sunscreen" of ozone **(1)** allows humans
and other forms of life to exist on land; **(2)** helps protect

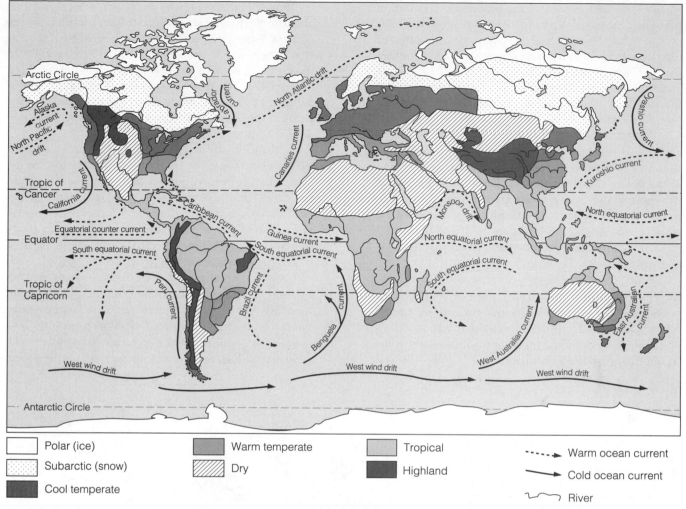

Figure 9-2 Generalized map of global climate zones, showing the major contributing ocean currents and drifts. Large variations in climate are dictated mainly by temperature (with its seasonal variations) and by the quantity and distribution of precipitation.

Legend:
- Polar (ice)
- Subarctic (snow)
- Cool temperate
- Warm temperate
- Dry
- Tropical
- Highland
- -----▶ Warm ocean current
- ——▶ Cold ocean current
- ⌐⌐⌐ River

humans from sunburn, skin and eye cancer, cataracts, and damage to the immune system; and **(3)** prevents much of the oxygen in the troposphere from being converted to ozone, a harmful air pollutant. The trace amounts of ozone that do form in the troposphere are a component of urban smog and they damage plants, the respiratory systems of humans and other animals, and materials such as rubber.

Thus, our good health, and that of many other species, depends on having enough ozone in the stratosphere and as little ozone as possible in the troposphere. There is considerable evidence that some human activities are increasing the amount of harmful ozone in the tropospheric air we breathe and decreasing the amount of beneficial ozone in the stratosphere.

How Does Weather Differ from Climate? At every moment at any spot on the earth, the troposphere (the inner layer of the atmosphere containing most of the earth's air) has a particular set of physical properties. Ex-

amples are temperature, pressure, humidity, precipitation, sunshine, cloud cover, and wind direction and speed. These short-term properties of the troposphere at a given place and time are what we call **weather**.

Climate is the average long-term weather of an area; it is a region's general pattern of atmospheric or weather conditions, including seasonal variations and weather extremes (such as hurricanes or prolonged drought or rain) averaged over a long period (at least 30 years).

What Factors Influence Climate? The two main factors determining an area's climate are *temperature*, with its seasonal variations, and the amount and distribution of *precipitation*. Figure 9-2 is a generalized map of the earth's major climate zones. The temperature and precipitation patterns that lead to different climates shown in Figure 9-2 are caused primarily by the way air circulates over the earth's surface.

Several factors determine global air circulation patterns. One is *long-term variations in the amount of solar energy striking the earth*. Such variation occurs because of occasional changes in solar output, slight planetary shifts in which the earth's axis wobbles (22,000-year cycle) and tilts (44,000-year cycle) as it revolves around the sun, and minute changes in the shape of its orbit around the sun (100,000-year cycle).

A second factor is the *uneven heating of the earth's surface*. Air is heated much more at the equator (where the sun's rays strike directly throughout the year) than at the poles (where sunlight strikes at an angle and is thus spread out over a much greater area). These differences help explain why tropical regions near the equator are hot, polar regions are cold, and temperate regions in between generally have intermediate average temperatures (Figure 9-2).

Third, *seasonal changes occur because the earth's axis* (an imaginary line connecting the north and south poles) *is tilted*; as a result, various regions are tipped toward or away from the sun as the earth makes its annual revolution. This creates opposite seasons in the northern and southern hemispheres.

Fourth, *the earth rotates on its axis*, which prevents air currents from moving due north and south from the equator. Forces in the atmosphere created by this rotation deflect winds (moving air masses) to the right in the northern hemisphere and to the left in the southern hemisphere. The result is six huge convection cells of swirling air masses—three north and three south of the equator—that convey or move heat and water from one area to another (Figure 9-3).

Finally, climate and global air circulation are affected by the *properties of air and water*. When heated by the sun, ocean water evaporates and removes heat from the oceans to the atmosphere, especially near the hot equator. This moist, hot air expands, becomes less dense (weighs less per unit of volume), and rises in fairly narrow vortices. These upward spirals create an area of low pressure at the earth's surface.

As this moisture-laden air rises, it cools and releases moisture as condensation (because cold air can hold less water vapor than warm air). When water vapor condenses it releases heat, which radiates into space. The resulting cooler, drier air becomes denser, sinks (subsides), and creates an area of high pressure.

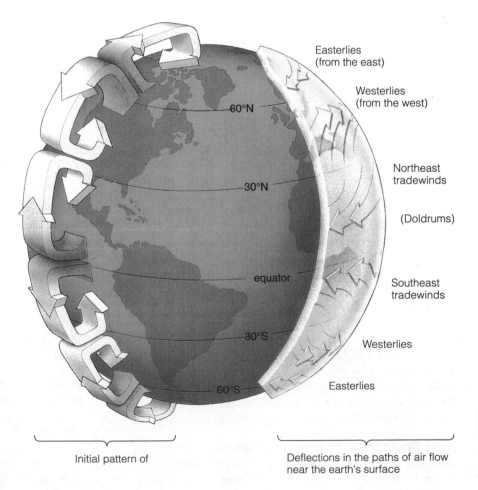

Figure 9-3 Formation of prevailing surface winds, which disrupt the general flow of air from the equator to the poles and back to the equator. As the earth rotates, its surface turns faster beneath air masses at the equator and slower beneath those at the poles. This deflects air masses moving north and south to the west or east, creating six huge convection cells in which air swirls upward and then descends toward the earth's surface at different latitudes. The direction of air movement in these cells sets up belts of prevailing winds that distribute air and moisture over the earth's surface. These winds affect the general types of climate found in different areas and drive the circulation of ocean currents. (Used by permission from Cecie Starr and Ralph Taggart, *Biology: The Unity and Diversity of Life*, 7th ed., Belmont, Calif.: Wadsworth, 1995)

Easterlies (from the east)

Westerlies (from the west)

Northeast tradewinds

(Doldrums)

Southeast tradewinds

Westerlies

Easterlies

60°N

30°N

equator

30°S

60°S

Initial pattern of

Deflections in the paths of air flow near the earth's surface

Kaiser, Jocelyn. 1998. "New Network Aims to Take the World's CO_2 Pulse." *Science*, vol. 281, no. 5376, 506(2).

As this air mass flows across the earth's surface, it picks up heat and moisture and begins to rise again. The resulting small and giant convection cells circulate air, heat, and moisture both vertically and from place to place in the troposphere, leading to different climates and patterns of vegetation (Figures 9-4 and 2-28).

Ocean currents also affect climate. The factors just listed, plus differences in water density, cause warm and cold ocean currents (Figure 9-2). These currents, like air currents, redistribute heat received from the sun and thus influence climate and vegetation, especially near coastal areas. For example, without the warm Gulf Stream, which transports 25 times more water than all the world's rivers, the climate of northwestern Europe would be subarctic. Currents also help mix ocean waters and distribute nutrients and dissolved oxygen needed by aquatic organisms.

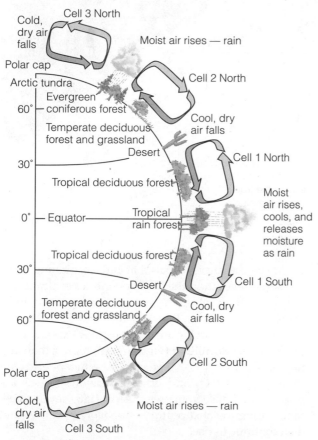

Figure 9-4 Model of global air circulation and biomes. Heat and moisture are distributed over the earth's surface by vertical convection currents that form six large convection cells (called Hadley cells) at different latitudes. The direction of air flow and the ascent and descent of air masses in these convection cells determine the earth's general climatic zones. The uneven distribution of heat and moisture over the planet's surface leads to the forests, grasslands, and deserts that make up the earth's biomes.

Another factor affecting climate is the *chemical makeup of the atmosphere*. Small amounts of carbon dioxide and water vapor, as well as trace amounts of ozone, methane, nitrous oxide, chlorofluorocarbons, and other gases in the troposphere, play a key role in determining the earth's average temperatures and thus its climates.

Together, these gases, known as **greenhouse gases**, act somewhat like the glass panes of a greenhouse: They allow light, infrared radiation, and some ultraviolet radiation from the sun (Figure 2-5) to pass through the troposphere. The earth's surface then absorbs much of this solar energy and degrades it to longer-wave, infrared radiation (that is, heat), which then rises into the troposphere (Figure 2-10). Some of this heat escapes into space; some is absorbed by molecules of greenhouse gases, warming the air; and some radiates back toward the earth's surface. This natural trapping of heat in the troposphere is called the **greenhouse effect** (Figure 9-5).

9-2 GLOBAL WARMING: HOW SERIOUS IS THE THREAT

What Is the Scientific Consensus About the Greenhouse Effect? The greenhouse effect, first proposed by Swedish chemist Svante Arrhenius in 1896, has been confirmed by numerous laboratory experiments and atmospheric measurements. It is one of the most widely accepted theories in the atmospheric sciences. Indeed, without its current greenhouse gases (especially water vapor), the earth would be a cold and lifeless planet with an average surface temperature of −18°C (0°F) instead of its current 15°C (59°F).

In 1990 and 1995 the Intergovernmental Panel on Climate Change (IPCC, a network of about 2,500 of the world's leading climate experts from 70 nations) published several reports evaluating the best available evidence concerning the greenhouse effect, past changes in global temperatures, and climate models projecting future changes in global temperatures and climate.

According to this scientific consensus, the amount of heat trapped in the troposphere depends mainly on the concentrations of heat-trapping or *greenhouse gases* and the length of time they stay in the atmosphere. The major greenhouse gases are water vapor (H_2O), carbon dioxide (CO_2), ozone (O_3), methane (CH_4), nitrous oxide (N_2O), and chlorofluorocarbons (CFCs). Recently scientists have identified another greenhouse gas—perfluorocarbons (PFCs, such as CF_4)—emitted mostly from aluminum production. They remain in the atmosphere from 2,000 to 50,000 years.

The primary heat-trapping gas in the atmosphere is water vapor. However, because its concentration in the atmosphere is fairly high (1–5%), inputs of water

Hint: Enter the search term *carbon dioxide monitoring* using the Subject Guide.

CHAPTER 9 **247**

Figure 9-5 The greenhouse effect. Without the atmospheric warming provided by this natural effect, the earth would be a cold and mostly lifeless planet. According to the widely accepted greenhouse theory, when concentrations of greenhouse gases in the atmosphere rise, the average temperature of the troposphere also rises. (Used by permission from Cecie Starr and Ralph Taggart, *Biology: The Unity and Diversity of Life*, 7th ed., Belmont, Calif.: Wadsworth, 1995)

Inside the figure:

1. Sunlight penetrating the atmosphere warms the earth's surface.

2. The earth's surface radiates heat (infrared wavelengths) to the atmosphere, and some escapes into space. Greenhouse gases and water vapor absorb some infrared wavelengths and reradiate a portion of them toward the earth.

3. When greenhouse gases build up in the atmosphere, more heat is trapped near the earth's surface. Ocean surface temperatures rise, more water vapor enters the atmosphere, and the earth's surface temperature increases.

vapor from human activities have little effect on this chemical's greenhouse effects. By contrast, the concentration of carbon dioxide in the atmosphere is so small (0.036%) that fairly large input of CO_2 from human activities can significantly affect the amount of heat trapped in the atmosphere.

Measured atmospheric levels of certain greenhouse gases—CO_2, CFCs, methane, and nitrous oxide—have risen substantially in recent decades (Figure 9-6) and are projected to enhance the earth's natural greenhouse effect, a phenomenon called *global warming*. Most of the increased levels of these greenhouse gases since 1958 have been caused by human activities: burning fossil fuels, agriculture, deforestation, and use of CFCs. Although molecules of CFCs, methane, and nitrous oxide trap much more heat per molecule than CO_2, the much larger input of CO_2 makes it the most important greenhouse gas produced by human activities.

Ice core analysis reveals that at the beginning of the industrial revolution the atmospheric concentration of CO_2 was about 280 parts per million (referred to as the *preindustrial level*). Between 1860 and 1997 the concentration of CO_2 in the atmosphere grew exponentially to 364 parts per million (Figure 9-6a), higher than any time in the past 160,000 years. The atmospheric concentrations of CO_2 and other greenhouse gases are projected to double from preindustrial (1860) levels sometime during the next century—probably by 2050—and then continue to rise.

What Is the Scientific Consensus About the Earth's Past Temperatures? Analysis of the content of ancient ice in Antarctic glaciers and other data show that the earth's average surface temperature has fluctuated considerably over geologic time. These data show that during the past 800,000 years several

Q: What major food-producing country is doing the most to reduce soil erosion?

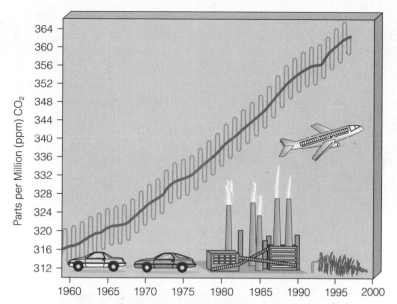

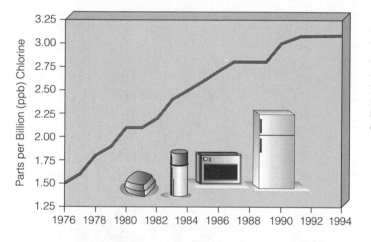

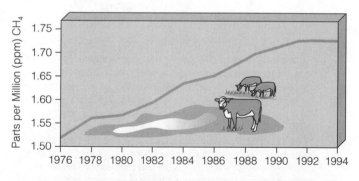

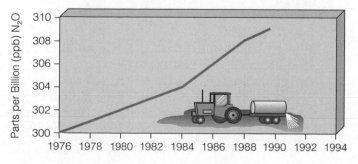

Figure 9-6 Increases in average concentrations of major greenhouse gases in the troposphere, mostly as a result of human activities. (Data from Electric Power Research Institute. Adapted and updated by permission from Cecie Starr and Ralph Taggart, *Biology: The Unity and Diversity of Life*, 6th ed., Belmont, Calif.: Wadsworth, 1992)

a. Carbon dioxide (CO2) is responsible for 50–60% of the global warming from greenhouse gases produced by human activities since preindustrial times. The main sources are fossil-fuel burning (75%) and land clearing and burning (25%). Most of the CO_2 comes from burning coal, but an increasing fraction is released from motor vehicle exhaust. CO_2 remains in the atmosphere for 50–200 years. The annual rise and fall of CO_2 levels shown in the graph result from less photosynthesis during winter and more during summer.

b. Chlorofluorocarbons (CFCs) contribute to global warming in the troposphere and deplete ozone in the stratosphere. The main sources are leaking air conditioners and refrigerators, evaporation of industrial solvents, production of plastic foams, and aerosol propellants. CFCs take 10–20 years to reach the stratosphere and generally trap 1,500–7,000 times as much heat per molecule as CO_2 while they are in the troposphere. This heating effect in the troposphere may be partially offset by the cooling caused when CFCs deplete ozone during their 65- to 135-year stay in the stratosphere. Their use is being phased out.

c. Methane (CH4) is produced when anaerobic bacteria break down dead organic matter in moist places that lack oxygen. These areas include swamps and other natural wetlands, rice paddies, and landfills, and the intestinal tracts of cattle, sheep, and termites. Production and use of oil and natural gas (especially from leaks in natural gas pipelines) and incomplete burning of organic materials (including biomass burning in the tropics) also are significant sources. CH_4 stays in the troposphere for 9–15 years. Each CH_4 molecule traps about 25 times as much heat as a CO_2 molecule. Methane levels have stopped growing since 1991, possibly because of slightly better control of massive leaks in Russia's natural gas system.

d. Nitrous oxide (N2O) can trap heat in the troposphere and can also deplete ozone in the stratosphere. It is released from nylon production, burning of biomass and nitrogen-rich fuels (especially coal), and the breakdown of nitrogen fertilizers in soil, livestock wastes, and nitrate-contaminated groundwater. Its life span in the troposphere is about 120 years, and it traps about 230 times as much heat per molecule as CO_2.

A: United States

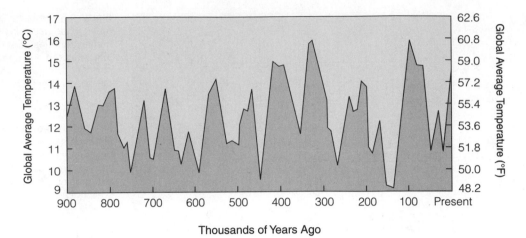

Figure 9-7 During the past 900,000 years the earth has experienced cycles of ice ages, each lasting about 100,000 years and followed by warmer interglacial periods lasting 10,000–12,500 years. The warm interglacial period during the past 10,000 years has been a major factor in the development of agriculture, human civilizations, and population growth.

Thousands of Years Ago

ice ages have covered much of the planet with thick ice. Each glacial period lasted about 100,000 years and was followed by a warmer interglacial period of 10,000–12,500 years (Figure 9-7).

For the past 10,000 years we have enjoyed the warmth of the latest interglacial period. This climatic stability has prevented drastic changes in the nature of soils and vegetation patterns throughout the world, allowing large increases in food production and thus in population. However, even small temperature changes during this period have led to large migrations of peoples in response to changed agricultural and grazing conditions.

Analysis of gases in bubbles trapped in ancient ice show that over the past 160,000 years tropospheric water vapor levels (the dominant greenhouse gas) have remained fairly constant. During most of this period levels of CO_2 have fluctuated between 190 and 290 parts per million. Estimated changes in the levels of tropospheric CO_2 correlate fairly closely with estimated variations in the earth's mean surface temperature during the past 160,000 years (Figure 9-8).

Since 1860 (when measurements began), mean global temperature (after correcting for urban heat island effects; see Figure 3-17) has risen 0.3–0.6°C (0.5–1.1°F), with most of this rise occurring since 1946 (Figure 9-9). Since 1860, the 12 warmest years occurred between 1979 and 1998, with 1990, 1995, 1997, and 1998 being the four hottest years.

Many uncertainties remain. Some or even all of the roughly 0.5°C rise since 1860 could result from normal fluctuations in the mean global temperature. On balance, however, the IPCC concluded in its 1995 report that "the observed increase over the last century is unlikely to be entirely due to natural causes" and "the balance of evidence suggests that there is a discernible human influence on global climate."

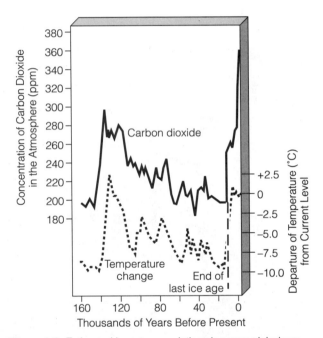

Figure 9-8 Estimated long-term variations in mean global surface temperature and average tropospheric carbon dioxide levels over the past 160,000 years. These CO_2 levels were obtained by inserting metal tubes deep into Antarctic glaciers, removing the ice, and analyzing bubbles of ancient air trapped in ice at various depths throughout the past. Such analyses reveal that since the last great ice age ended about 10,000 years ago, we have enjoyed a warm interglacial period. The rough correlation between tropospheric CO_2 levels and temperature shown in these estimates based on ice core data suggests a connection.

What Is the Scientific Consensus About Future Global Warming and Its Effects? To project the effects of increases in greenhouse gases on average global temperature and changes in the earth's climate, scientists develop *mathematical models* of such systems and run them on supercomputers. How well the results

Q: What percentage of the world's irrigated cropland suffers reduced yields because of soil salinization?

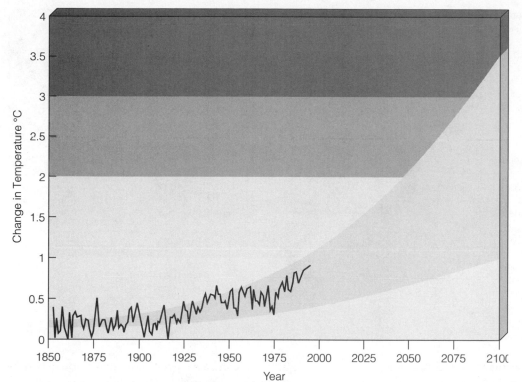

Figure 9-9 Recorded changes in the earth's mean surface temperature between 1860 and 1997 (dark line). The curved shaded region shows global warming projected by various computer models of the earth's climate systems. Note that the climate model projections roughly match the 0.3–0.6°C (0.5–1.1°F) recorded temperature increase between 1860 and 1997. Current models indicate that the average global temperature will rise by 1–3.5°C (1.8–6.3°F) sometime during the next century. However, this projected warming could be *overestimated* or *underestimated* by a factor of two. (Data from U.S. National Academy of Sciences and National Center for Atmospheric Research)

correspond to the real world depends on the design and assumptions of the model, the accuracy of the data used, factors in the earth's climate system that amplify or dampen changes in average global temperatures, and the effects of totally unexpected or unpredictable events (chaos).

Current models generally agree that global temperature will increase, but generally disagree concerning the effects of such changes on individual regions. According to the latest climate models, the IPCC projects that the earth's mean surface temperature should rise 1–3.5°C (1.8–6.3°F) between 1990 and 2100 (Figure 9-9); the most likely rise in temperature before 2100 would be about 2°C (3.6°F). This may not seem like much, but even at the lowest projected increase of 1.0°C, the earth would be warmer than it has been for 10,000 years. Current models project that climate change, once begun, will continue for hundreds of years.

According to the models, the northern hemisphere should warm more and faster than the southern hemisphere and there should be more pronounced warming at the earth's poles. Measurements reveal that the surface temperatures at nine stations north of the Arctic circle have risen by about 5.5°C (9.9°F) since 1968. Since 1947 the average summer temperature at Antarctica has risen by almost 2°C (3.6°F) and five of the nine massive, floating ice shelves surrounding the continent have broken up since 1950 (three between 1994 and 1998).

Other possible signs of global warming include **(1)** increased retreat of some glaciers on the tops of mountains in the Alps, Andes, Himalayas, and Northern Cascades of Washington during the last 30 years, **(2)** northward migrations of some warm-climate fish and trees, **(3)** spread of some tropical diseases away from the equator, and **(4)** bleaching of coral reefs in tropical areas with warmer water.

With a warmer climate global sea levels will rise, mainly because water expands slightly when heated. Between 1900 and 1990, global sea levels have risen by 9–18 centimeters (3.5–7 inches). Climate models estimate that two-thirds of this rise is the result of the expansion of the warmer water. Current climate models of global warming project that global sea levels will rise by 15–95 centimeters (6–37 inches) between 1990 and 2100, with a best estimate of 48 centimeters (19 inches).

IPCC scientists warn that warming or cooling by more than 1°C (1.8°F) over a few decades (instead of over many centuries, as has been the pattern during the last 10,000 years) will cause serious disruptions of the current structure and functioning of earth's ecosystems and of human economic and social systems.

Will the Earth Really Get Warmer? There is much controversy over whether we are already experiencing global warming, how warm temperatures might be in the future, and the effects of such temperature

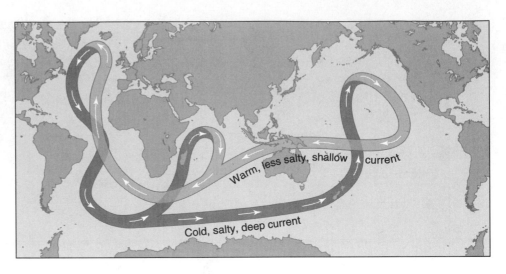

Figure 9-10 This loop of ocean water stores carbon dioxide in the deep sea and brings warmth to Europe. It occurs when ocean water in the North Atlantic is dense enough (because of its salt content and cold temperature) to sink to the ocean bottom and well up in the warmer Pacific (helping cool that part of the world). Then a shallower return current aided by winds brings warmer and less salty—and thus less dense—water to the Atlantic, which can then cool and sink to begin the cycle again. If this heat conveyor belt or loop stalls out because of a drop in density of ocean water in the North Atlantic (possibly from changes caused by global warming), massive climate changes over much of the earth's surface could occur within only a few decades. Models indicate that this oceanic heat conveyor belt would return, but only after hundreds or thousands of years.

increases. One problem is that many of the past measurements and estimates of the earth's average temperature are imprecise, and we have only about 100 years of accurate data. With such limited data, it is difficult to separate out the normal short-term ups and downs of global temperatures (called *climate noise*) from an overall rise in average global temperature.

What Factors Might Amplify or Dampen Global Warming? Scientists have also identified a number of factors that might amplify or dampen a rise in average atmospheric temperature. These factors influence both how fast temperatures might climb and what the effects might be on various areas. Let's look more closely at the possible effects of such factors.

One factor is *changes in the amount of solar energy reaching the earth*. Solar output varies by about 0.1% over the 11-year and 22-year sunspot cycles and over 80-year and other, much longer cycles. These up-and-down changes in solar output can temporarily warm or cool the earth and thus affect the projections of climate models. In 1998, two Danish scientists published an untested hypothesis suggesting that virtually all of rises and falls of the earth's temperatures are the result of cosmic rays produced by solar cycles on cloud cover. Two 1992 studies concluded that the projected warming power of greenhouse gases should outweigh the climatic influence of the sun over at least the next 50 years. Suppose that increases in atmospheric temperatures during this century were caused mostly by a slightly brighter sun. If this is a valid hypothesis, then analysts point out that a

brighter sun can cause even more greenhouse warming as human activities continue adding greenhouse gases to the troposphere.

Another problem is understanding the *effects of oceans on climate*. The world's oceans might amplify global warming by releasing more CO_2 into the atmosphere or might dampen it by absorbing more heat. We know that the oceans currently help moderate tropospheric temperature by removing about 29% of the excess CO_2 we pump into the atmosphere, but we don't know whether they can absorb more. If the oceans warm up enough, some of the dissolved CO_2 will bubble out into the atmosphere (just as in a glass of carbonated ginger ale left out in the sun), amplifying and accelerating global warming.

Global warming could be dampened if the oceans absorbed more heat, but this depends on how long the heat takes to reach deeper layers. Recent measurements indicate that deep vertical mixing in the ocean occurs extremely slowly (taking hundreds of years) in most places because water density increases with depth, inhibiting mixing of different layers.

There is also concern that deep ocean currents could be disrupted. At present, these currents (driven largely by differences in water density and winds) act like a gigantic conveyor belt, transferring heat from one place to another and storing carbon dioxide in the deep sea (Figure 9-10). There is concern that global warming could halt this thermal conveyor belt by reducing the density and salinity of water in the North Atlantic so that the water would not sink. If this loop stalls out, evidence from past climate changes indicates that this

Q: What percentage of the world's irrigated cropland suffers from waterlogging?

could trigger atmosphere temperature changes of more than 5°C (9°F) over periods as short as 40 years.

Changes in the atmosphere's *water vapor content and the amount and types of cloud cover* also affect climate. Warmer temperatures would increase evaporation and the water-holding capacity of the air and create more clouds. Significant increases in water vapor, a potent greenhouse gas, could enhance warming.

However, it is difficult to predict the net effect of additional clouds on climate. They could have a warming effect by trapping heat or a cooling effect by reflecting sunlight back into space. The net result of these two opposing effects depends on whether it is day or night and on the type (thin or thick) and altitude of clouds. Scientists don't know which of these factors might predominate or how cloud types and heights might vary in different parts of the world as a result of global warming.

Climate is also influenced by *changes in polar ice*. The light-colored Greenland and Antarctic ice sheets act like enormous mirrors, reflecting sunlight back into space. If warmer temperatures melted some of this ice and exposed darker ground or ocean, more sunlight would be absorbed and warming would be accelerated. Then more ice would melt, further accelerating the rise in atmospheric temperature.

On the other hand, the early stages of global warming might increase the amount of the earth's water stored as ice. Warmer air would carry more water vapor, which could drop more snow on some polar glaciers, especially the gigantic Antarctic ice sheet. If snow accumulated faster than ice was lost, the ice sheet would grow, reflect more sunlight, and help cool the atmosphere, perhaps leading to a new ice age within a thousand years.

Climate can also be affected by *air pollution*. Projected global warming might be partially offset by *aerosols* (tiny droplets and solid particles) of various air pollutants released or formed in the atmosphere from volcanic eruptions and human activities. This occurs because during daytime they reflect some of the incoming sunlight back into space. However, nights would be warmer because the clouds would still be there and prevent some of the heat stored in the earth's surface (land and water) during the day from being radiated back into space. These pollutants may explain why most recent warming in the northern hemisphere occurs at night.

However, these interactions are complex. Pollutants in the lower troposphere can either warm or cool the air and the surface below them, depending on the reflectivity of the underlying surface. We also know little about the effects of aerosols on the properties of clouds. These contradictory and patchy effects and uncertainties, plus improved air pollution control, make it unlikely that air pollutants will counteract projected global warming very much in the next half century.

Aerosols also fall back to the earth or are washed out of the atmosphere within weeks or months, whereas CO_2 and other greenhouse gases remain in the atmosphere for decades to several hundred years.

We could maintain or increase levels of aerosol air pollutants to offset possible global warming. However, because these air pollutants already kill hundreds of thousands of people a year and damage vegetation (including food crops), they are being reduced.

Another uncertainty is the *effect of increased CO_2 on photosynthesis*. Some studies suggest that more CO_2 in the atmosphere is likely to increase the rate of photosynthesis in areas with adequate amounts of water and other soil nutrients. This would remove more CO_2 from the atmosphere and help slow global warming. However, other studies suggest that much of any increased plant growth could also be offset by plant-eating insects that breed more rapidly and year-round in warmer temperatures. Weeds may also grow more rapidly at the expense of food crops.

Another factor that can affect CO_2 levels is *forest turnover*: how fast trees grow and die in a forest. A 1994 study indicated that the turnover in tropical forests worldwide is accelerating. Besides reducing the biodiversity of tree species in tropical forests, this change could enhance global warming by reducing removal of CO_2 from the atmosphere because less dense, faster-growing trees require less CO_2 for growth.

Global warming could be accelerated by *increased release of methane, a potent greenhouse gas, from wetlands*. A 1994 study indicated that increased uptake of CO_2 by wetland plants could boost emissions of methane by providing more organic matter for methane-producing anaerobic bacteria to decompose. Some scientists also speculate that in a warmer world huge amounts of methane now tied up in arctic tundra soils and in muds on the bottom of the Arctic Ocean might be released if the blanket of permafrost in tundra soils melts and the oceans warm considerably. Conversely, some scientists believe that bacteria in tundra soils would rapidly oxidize the escaping methane to CO_2, a less potent but still important greenhouse gas.

There is also concern over *how rapidly the earth's climate might change*. If moderate change takes place gradually over several hundred years, people in areas with unfavorable climate changes may be able to adapt to the new conditions. However, if the projected global temperature change takes place over several decades or all within the next century, we may not be able to switch food-growing regions and relocate the large portion of the world's population living near coastal areas fast enough. The result would be large numbers of premature deaths from lack of food, as well as social and economic chaos. Such rapid changes would also reduce the earth's biodiversity because many species couldn't move or adapt.

Recent data from analysis of ice cores suggest that the earth's climate has shifted often and more drastically and quickly than previously thought. This new evidence indicates that average temperatures during the warm interglacial period that began about 125,000 years ago (Figure 9-7) varied as much as 10°C (18°F) in only a decade or two and that such warming and cooling periods each lasted 1,000 years or more.

If these findings are correct and also apply to the current interglacial period, fairly small rises in greenhouse gas concentrations could trigger rapid up-and-down shifts in average global temperatures. Such rapid shifts would be disastrous for humans and many other forms of life on the earth.

As a result of the factors discussed in this section, climate scientists estimate that their projections about *global warming and rises in average sea levels during the next 50–100 years could be half the current projections (the best-case scenario) or double them (the worst-case scenario)*. In any event, possible climate change and its effects are likely to be erratic and mostly unpredictable.

9-3 SOME POSSIBLE EFFECTS OF A WARMER WORLD

Why Should We Worry if the Earth's Temperature Rises a Few Degrees? So what's the big deal? Why should we worry about a possible rise of only a few degrees in the earth's average temperature? We often have that much change between May and July, or even between yesterday and today.

This is a common critical thinking trap that many people fall into. The key point is that we are not talking about normal swings in *local weather*, but about a projected *global* change in *climate*.

A warmer global climate could have a number of possible effects. One is *changes in food production*, which could increase in some areas and drop in others. Archeological evidence and computer models indicate that climate belts would shift northward by 100–150 kilometers (60–90 miles) or upward 150 meters (500 feet) in altitude for each 1°C (1.8°F) rise in global temperature.

Whether such poleward shifts would actually lead to increased crop productivity in new crop-growing areas depends mainly on two factors: the fertility of the soil in such regions and the availability of enormous amounts of money to build a new agricultural infrastructure (for irrigation and food storage and distribution). In parts of Asia, food production in more northern areas could increase because of favorable soils. However, in North America the northward expansion of crop-growing regions from the Midwestern United States into Canada would be limited by the thinner and less fertile soils there.

Current climate models project 10–70% declines in the global yield of key food crops and a loss in current cropland area of 10–50%, especially in most poor countries. Other studies project a 1–8% drop in the global production of wheat, rice, and other grains by 2060 because of projected warming. Currently, we can't predict where changes in crop-growing capacity might occur or how long such changes might last. However, we do know that drops in global crop yields of only 10% would cause large increases in hunger and starvation (especially in poor countries) and cause economic and social chaos.

Global warming would also *reduce water supplies* in some areas. Lakes, streams, and aquifers in some areas that have provided water to ecosystems, croplands, and cities for centuries could shrink or dry up altogether. This would force entire populations to migrate to areas with adequate water supplies—if they could. So far we can't say with much certainty where this might happen.

Global warming will also *change the makeup and location of many of the world's forests*. Forests in temperate and subarctic regions would move toward the poles or to higher altitudes, leaving more grassland and shrubland in their wake.

However, tree species move slowly through the growth of new trees along forest edges—typically about 0.9 kilometer (0.5 mile) per year or 9 kilometers (5 miles) per decade. According to the 1995 report of the IPCC, midlatitude climate zones are projected to shift northward by 550 kilometers (340 miles) over the next century. At that rate some tree species such as beech (Figure 9-11) might not be able to migrate fast enough and would die out. According to the IPCC, over the next century "entire forest types might disappear, including half of the world's dry tropical forests." Such forest diebacks would release carbon stored in their biomass and in surrounding soils and accelerate global warming.

Oregon State University scientists project that drying from global warming could cause massive *wildfires* in up to 90% of North American forests. If widespread fires occurred, large numbers of homes and large areas of wildlife habitats would be destroyed. In addition, huge amounts of CO_2 would be injected into the atmosphere, which would accelerate global warming.

Climate change would lead to *reductions in biodiversity* in many areas. Large-scale forest diebacks would cause mass extinction of plant and animal species that couldn't migrate to new areas. Fish would die as temperatures soared in streams and lakes and as lowered water levels concentrated pesticides. Any shifts in regional climate would threaten many parks, wildlife reserves, wilderness areas, wetlands, and

Q: How many plants feed most of the world's people?

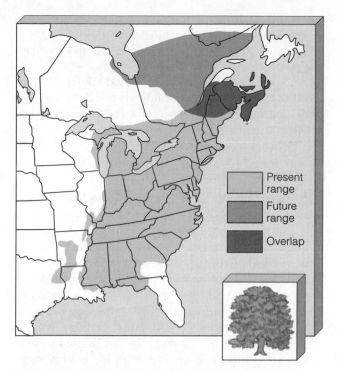

Figure 9-11 Possible effects of global warming on the geographic range of beech trees. According to one projection, if CO$_2$ emissions doubled between 1990 and 2050, beech trees (now common throughout the eastern United States) would survive only in a greatly reduced range in northern Maine and southeastern Canada. This is only one of a number of tree species whose geographic ranges would be drastically changed by global warming. (Data from Margaret B. Davis and Catherine Zabinski, University of Minnesota)

coral reefs, wiping out many current efforts to stem the loss of biodiversity.

Projected global warming may sharply reduce populations of some species, especially those with specialized niches. For example, Kirtland's warbler, an endangered bird that nests exclusively under young jack pines in northern Michigan, would probably become extinct. The reason is that if northern Michigan ends up with less rain and more heat, jack pines there will probably die off in the next 30–90 years.

In a warmer world water in the world's oceans would expand and lead to a *rise in sea level*. Even the best estimate of a 48-centimeter (19-inch) rise in sea level projected to occur by 2100 would flood coastal regions—where about one-third of the world's people and economic infrastructure are concentrated—as well as lowlands and deltas where crops are grown. Most or all of low-lying islands such as the Cook and Marshall Islands in the Pacific, the Maldives, and some Caribbean island nations would be covered with water and disappear. It would also destroy most coral reefs, move bar-

rier islands farther inland, accelerate coastal erosion, contaminate coastal aquifers with salt water, reduce already declining global fish catches, and flood tanks storing oil and other hazardous chemicals in coastal areas.

If warming at the poles caused ice sheets and glaciers to melt even partially, the global sea level would rise far more. One comedian jokes that he plans to buy land in Kansas because it will probably become valuable beachfront property; another boasts that she isn't worried because she lives in a houseboat—the "Noah strategy."

In a warmer world, *weather extremes* are expected to increase in number and severity. Prolonged heat waves and droughts could become the norm in many areas, taking a huge toll on many humans and ecosystems. As the upper layers of seawater warm, the intensity of damaging hurricanes, typhoons, tornadoes, and violent storms is expected to increase.

Alarmed by an unprecedented series of hurricanes, floods, droughts, and wildfires in recent years, a growing number of the world's insurance companies are sharply raising their premiums for coverage of damages from such events and eliminating coverage in high-risk areas. Executives of some insurance companies are also pressuring government leaders to get more serious about slowing possible global warming and are spurring improvements in energy efficiency and increased use of renewable energy through their procurement and investment policies.

Global warming also poses *threats to human health*. According to the 1995 IPCC report, global warming would bring more heat waves. This would double or triple heat-related deaths among the elderly and people with heart disease; it would also increase suffering from respiratory ailments such as asthma and bronchitis.

A warmer world would also disrupt supplies of food and fresh water, displacing millions of people and altering disease patterns in unpredictable ways. The spread of warmer and wetter tropical climates from the equator would bring malaria (Figure 8-3), encephalitis, yellow fever, dengue fever, and other insectborne diseases to formerly temperate zones. Scientists also project that higher ocean temperatures could trigger algal blooms, leading to cholera epidemics.

Atmospheric warming also affects the respiratory tract by increasing air pollution in winter and by increasing exposure to dusts, pollens, and smog in summer. Sea-level rise could spread infectious disease by flooding coastal sewage and sanitation systems.

Climate change would lead to *a growing number of environmental refugees*. According to environmental expert Norman Myers, by 2050 global warming could produce as many as 50–150 million environmental refugees (compared to 7 million war refugees in Europe after World War II). Most of these refugees would illegally

A: 15 (especially wheat, rice, corn, and potato)

migrate to other countries, causing much social disorder and political instability. Thus, projected global warming has serious implications for the foreign, military, and economic security policies of nations.

9-4 SOLUTIONS: DEALING WITH POSSIBLE GLOBAL WARMING

Should We Do More Research or Act Now? There are three schools of thought concerning global warming. A *very small* group of scientists (many of them not experts in climate research) contend that global warming is not a threat; some popular press commentators and writers even claim that it is a hoax. Widespread reporting of this *no-problem* minority view in the media has clouded the issue, cooled public support for action, and slowed international negotiations to deal with this threat.

A second group of scientists and economists believe we should wait until we have more information about the global climate system, possible global warming, and its effects before we take any action. Proponents of this *waiting strategy* question whether we should spend hundreds of billions of dollars phasing out fossil fuels and replacing deforestation with reforestation (and in the process risk disrupting national and global economies) to help ward off something that might not happen. They call for more research before making such far-reaching decisions.

A third group of scientists and economists point out that greatly increased spending on research about the possibility and effects of global warming will not provide the certainty decision makers want because the global climate system is so complex. These scientists urge us to adopt a *precautionary strategy*. They believe that when dealing with risky and far-reaching environmental problems such as possible global warming, the safest course is to take informed preventive action *before* there is overwhelming scientific knowledge to justify acting.

Economists at a 1997 meeting of the American Economics Association, led by Nobel laureates Kenneth Arrow and Robert Solow, declared, "As economists, we believe that global climate change carries with it significant environmental, economic, social, and geopolitical risks and that preventive steps are justified." In 1997, CEOs of several major oil companies, including British Petroleum, Royal Dutch Shell, and Sun Oil, joined most insurance company executives and expressed the view that there was enough evidence that human activities were contributing to global warming to begin taking precautionary action.

Those who favor doing nothing or waiting before acting point out that there is a 50% chance that we are *overestimating* the impact of rising greenhouse gases. However, those urging action point out that there is also a 50% chance that we are *underestimating* such effects.

Some analysts say that we should take the actions needed to slow global warming even if there were no threat because of their important environmental and economic benefits (Solutions, right). This so called *no-regrets strategy* is an important part of the precautionary strategy.

How Can We Slow Possible Global Warming?
According to the 1995 IPCC report, stabilizing CO_2 levels at the current level would require reducing current global CO_2 emissions by 66–83%. This is a highly unlikely and politically charged change.

Figure 9-12 presents a variety of solutions analysts have suggested to slow possible global warming; none of these solutions is being vigorously pursued. The quickest, cheapest, and most effective way to reduce emissions of CO_2 and other air pollutants over the next two to three decades is to use energy more efficiently (Section 4-2 and Solutions, right).

Some analysts call for increased use of nuclear power (Section 4-7) because it produces only about one-sixth as much CO_2 per unit of electricity as coal. Other analysts argue that the danger of large-scale releases of highly radioactive materials from nuclear power-plant accidents and the very high cost of nuclear power make it a much less desirable option than improving energy efficiency and relying more on renewable energy resources (Chapter 4).

Using natural gas (Section 4-6) could help make the 40- to 50-year transition to an age of energy efficiency and renewable energy. When burned, natural

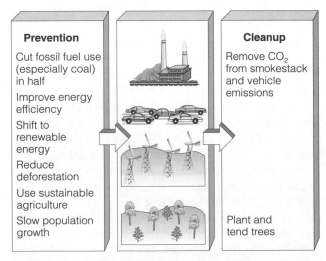

Figure 9-12 Solutions: methods for slowing possible global warming.

Q: What percentage of the earth's land area is suitable for cultivation?

gas emits only half as much CO_2 per unit of energy as coal, and it emits far smaller amounts of most other air pollutants. Because burning natural gas still emits CO_2, however, this approach would merely buy some time to increase energy efficiency and switch to renewable energy resources.

One method for reducing CO_2 emissions would be to phase out government subsidies for fossil fuels over a decade and gradually phase in *carbon taxes* on fossil fuels (especially coal and gasoline) based on their emissions of CO_2 and other air pollutants. To be politically feasible, analysts warn that these consumption tax increases should be matched by declines in taxes on income, labor, or capital. In 1997 more than 2,500 economists, including eight Nobel laureates, signed a statement **(1)** stating that sound economic analysis shows that greenhouse emissions can be cut without harming American living standards and **(2)** calling for carbon taxes as part of an international system of tradable permits for greenhouse gas emissions.

It has also been suggested that we remove CO_2 from the exhaust gases of fossil fuel–burning vehicles, furnaces, and industrial boilers. However, available methods can remove only about 30% of the CO_2, and using them would double the cost of electricity.

Reducing deforestation (Sections 5-3 and 5-5) and switching to more sustainable agriculture (Section 7-7) would reduce CO_2 emissions and help preserve biodiversity and ecological integrity. According to most analysts, slowing population growth is also crucial. If we cut per capita greenhouse gas emissions in half but world population doubles, we're back where we started.

Some call for a massive global reforestation program as a strategy for slowing global warming. However, studies suggest that such a program (requiring each person in the world to plant and tend an average of 1,000 trees every year) would offset only about 3 years of our current CO_2 emissions from burning fossil fuels. However, a global program for planting and tending trees would help restore deforested and degraded land and reduce soil erosion and loss of biodiversity.

Some scientists have suggested various technofixes for dealing with possible global warming, including **(1)** adding iron to the oceans to stimulate the growth of marine algae (which could remove more CO_2 through photosynthesis), **(2)** unfurling gigantic foil-surfaced sun mirrors in space to reduce solar input, and **(3)** injecting sunlight-reflecting sulfate particulates into the stratosphere to cool the earth's surface.

Many of these costly schemes might not work, and most would probably produce unpredictable short- and long-term harmful environmental effects. Moreover, once started, those that work could never be stopped without a renewed rise in CO_2 levels. Instead

Energy Efficiency to the Rescue

SOLUTIONS

According to energy expert Amory Lovins, *the major remedies for slowing possible global warming are things we should be doing already even if there were no threat of global warming.* He argues that if we waste less energy, reduce air pollution by cutting down on our use of fossil fuels and switching to renewable forms of energy, and harvest trees sustainably, we and other forms of life—in this and in future generations—would be better off even if these actions had nothing to do with global climate.

According to Lovins, improving energy efficiency (Section 4-2) would be the fastest, cheapest, and surest way to slash emissions of CO_2 and most other air pollutants within two decades using existing technology. He estimates that increased energy efficiency would also save the world up to $1 trillion per year in reduced energy costs—as much as the annual global military budget.

Using energy more efficiently would also reduce pollution, help protect biodiversity, and deter arguments among governments about how CO_2 reductions should be divided up and enforced. This approach would also make the world's supplies of fossil fuel last longer, reduce international tensions over who gets the dwindling oil supplies, and allow more time to phase in renewable energy.

According to a 1990 government report, controlling emissions of greenhouse gases in the United States will cost about $10 billion per year over the next century, for a total of $1 trillion. Cutting annual U.S. oil imports by 20% by wasting less energy would cover the annual $10 billion projected costs of reducing greenhouse gas emissions.

To Lovins and most environmentalists, greatly improving worldwide energy efficiency *now* is a money-saving, life-saving, biodiversity-saving, win–win, no-regrets proposition that we should not refuse, even if climate change were not an issue. Atmospheric scientist Fred Singer, a strong critic of global warming projections (as well as of ozone depletion), agrees.

Critical Thinking

1. Do you agree that improving energy efficiency **(a)** is an important way to reduce the input of CO_2 into the atmosphere and **(b)** should be done regardless of its impact on the threat of global warming?

2. Why do you think there has been little emphasis on improving energy efficiency? Explain. (See Section 4-2.)

of spending huge sums of money on such schemes, many scientists believe it would be much more effective and cheaper to improve energy efficiency and shift to renewable forms of energy that don't produce carbon dioxide (Chapter 4).

What Has Been Done to Reduce Greenhouse Gas Emissions? At the 1992 Earth Summit in Rio de Janeiro, Brazil, 106 nations approved a Convention on Climate Change, in which developed countries committed themselves to reducing their emissions of CO_2 and other greenhouse gases to 1990 levels by the year 2000. However, the convention did not *require* countries to reach this goal and most will not achieve this goal.

In December 1997 more than 2,200 delegates from 161 nations met in Kyoto, Japan, to negotiate a new treaty to help slow global warming. The resulting treaty would **(1)** require 38 developed countries to cut greenhouse emissions by an average of 5.2% below 1990 levels between 2008 and 2012, **(2)** not require developing countries to make any cuts in their greenhouse gas emissions because of an earlier treaty unless they choose to do so, and **(3)** allow emissions trading in which a country that beats its target goal for reducing greenhouse gas emissions can sell its excess reductions to countries that failed to meet their reduction goals.

Some analysts praise the agreement as a small but important step in dealing with the problem of potential global warming. However, there is strong opposition in the U.S. Senate for ratifying the treaty because developing countries are not required to meet any emission reduction goals. In addition, climate scientists estimate that it would take at least a 60% reduction in global emissions of greenhouse gases below 1990 levels to slow projected global warming to an acceptable level—compared to the only 5.2% reduction goal set by the Kyoto treaty.

How Can We Prepare for Possible Global Warming? It seems clear that many (perhaps most) of the things climate experts have recommended either will not be done or will be done too slowly. As a result, some analysts suggest that we should also begin to prepare for the effects of long-term global warming. Figure 9-13 shows some ways to do this.

Implementing the key measures for slowing or responding to climate change listed in Figures 9-12 and 9-13 will cost a lot of money. However, studies indicate that in the long run the savings would greatly exceed the costs. Some actions you can take to reduce the threat of global warming are given in Appendix 4.

9-5 OZONE DEPLETION: IS IT A SERIOUS THREAT?

What Is the Threat from Ozone Depletion? Evolution of photosynthetic, oxygen-producing bacteria has produced a stratospheric global sunscreen: the ozone layer (Figure 9-1). Its presence for the past 450 million years has allowed life to develop and expand on land and in the surface layers of aquatic systems.

A handful of scientists dismiss the threat of ozone depletion from human-produced chemicals. Based on measurements and models, however, the overwhelming consensus of researchers in this field is that ozone depletion by certain chlorine- and bromine-containing chemicals emitted into the atmosphere by human activities is a serious long-term threat to human health, animal life, and the sunlight-driven primary producers (mostly plants) that support the earth's food chains and webs.

Measurements of the concentrations of ozone in the stratosphere at numerous sites around the world show that during the 1980s normal ozone levels dropped 5–15% in winter above the temperate and tropical zones of both hemispheres—three times the losses measured in the 1970s. Globally, the earth lost an average of about 4% of its stratospheric ozone between 1979 and 1994. According to a 1995 report by prominent atmospheric scientists, average global ozone levels are projected to drop 7–13% during the 1990s.

What Causes Ozone Depletion? From Dream Chemicals to Nightmare Chemicals This situation started when Thomas Midgley, Jr., a General Motors chemist, discovered the first chlorofluorocarbon (CFC) in 1930, and chemists then made similar compounds to create a family of highly useful CFCs. The two most widely used are CFC-11 (trichlorofluoromethane, CCl_3F) and CFC-12 (dichlorofloromethane, CCl_2F_2), known by their trade name as Freons.

These chemically stable, odorless, nonflammable, nontoxic, and noncorrosive compounds seemed to be dream chemicals. Cheap to make, they became popular as coolants in air conditioners and refrigerators (replacing toxic sulfur dioxide and ammonia), propellants in aerosol spray cans, cleaners for electronic parts such as computer chips, sterilants for hospital instruments, fumigants for granaries and ship cargo holds, and bubbles in plastic foam used for insulation and packaging.

But CFCs were too good to be true. In 1974 calculations by chemists Sherwood Rowland and Mario Molina (building on earlier work by Paul Crutzen) indicated that CFCs were creating a global chemical time bomb by lowering the average concentration of ozone

Kellner, Tomas. 1998. "Cool operators." *The Sciences*, vol. 38, no. 5, 19(5).

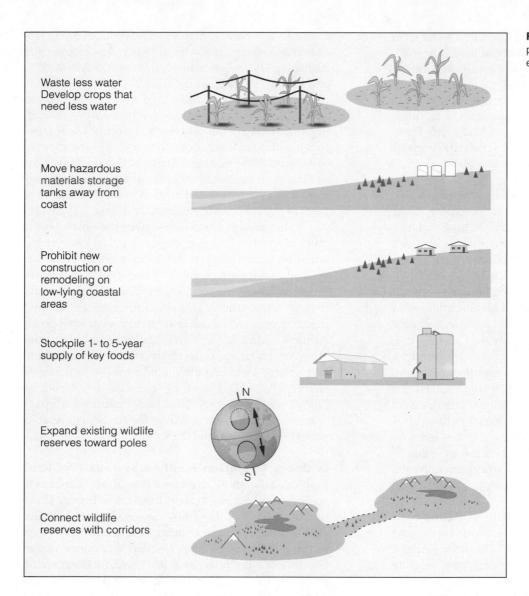

Figure 9-13 Solutions: ways to prepare for the possible long-term effects of global warming.

Waste less water
Develop crops that need less water

Move hazardous materials storage tanks away from coast

Prohibit new construction or remodeling on low-lying coastal areas

Stockpile 1- to 5-year supply of key foods

Expand existing wildlife reserves toward poles

Connect wildlife reserves with corridors

in the stratosphere. They shocked both the scientific community and the $28-billion-per-year CFC industry by calling for an immediate ban of CFCs in spray cans (for which substitutes were readily available).

Here's what Rowland and Molina found: Spray cans, discarded or leaky refrigeration and air conditioning equipment, and the production and burning of plastic foam products release CFCs into the atmosphere. Because these molecules are insoluble in water and are chemically unreactive, they are not removed from the troposphere. As a result—mostly through convection, random drift, and the turbulent mixing of air in the troposphere—they rise slowly into the stratosphere, taking 10–20 years to make the journey.

In the stratosphere, under the influence of high-energy UV radiation, these molecules break down and release highly reactive chlorine atoms, which speed up the breakdown of highly reactive ozone (O_3) into O_2 and O. This causes ozone to be destroyed faster than it is formed.

Each CFC molecule can last in the stratosphere for 65–110 years (depending on its type). During that time each chlorine atom released from these molecules can convert up to 100,000 molecules of O_3 to O_2. If Rowland and Molina's calculations and later models and atmospheric measurements of CFCs in the stratosphere are correct (as almost all scientists in this field believe), these dream molecules have turned into a nightmare of global ozone destroyers.

Although Rowland and Molina warned us of this problem in 1974, it took 15 years of interaction between the scientific and political communities before countries agreed to begin slowly phasing out CFCs.* The CFC industry (led by the Du Pont Company), a powerful, well-funded adversary with a lot of profits and jobs at stake, attacked Rowland and Molina. However, they held their ground, expanded their research, and explained the meaning of their calculations to other scientists, elected officials, and the media. In 1995, Rowland and Molina (along with Paul Crutzen) received the Nobel prize in chemistry for their work.

What Other Chemicals Deplete Stratospheric Ozone? CFCs are not the only ozone-eaters; other chemicals can release highly reactive chlorine and bromine atoms if they reach the stratosphere and are exposed to intense UV radiation. Collectively, all ozone-depleting compounds are called *ODCs*.

One group consists of long-lived bromine-containing compounds such as *halons* and *HBFCs*, both used in fire extinguishers. Another is *methyl bromide*, a widely used fumigant. Another group consists of chlorine-containing compounds such as *carbon tetrachloride*, a cheap, highly toxic solvent, and toxic *methyl chloroform*, or 1,1,1-trichloroethane, used as a cleaning solvent for clothes and metals and as a propellant in more than 160 consumer products, such as correction fluid, dry-cleaning sprays, spray adhesives, and other aerosols. Another source of ozone depletion is the emission of hydrogen chloride (HCl) into the stratosphere by the U.S. space shuttle vehicles.

Why Is There Seasonal Thinning of Ozone over the Poles? Sometimes the news about ozone loss has taken scientists by surprise. The first surprise came in 1984, when researchers analyzing satellite data discovered that 40–50% of the ozone in the upper stratosphere over Antarctica was being destroyed during the Antarctic spring and early summer (September–December), when sunlight returned after the dark Antarctic winter. Since then, this seasonal Antarctic *ozone thinning* (incorrectly called an ozone hole) has expanded in most years, with a typical seasonal loss of about 50% (but as high as 100% in some spots). In 1992 and 1993, seasonal ozone thinning above Antarctica covered an area three times the size of the continental United States.

Measurements have indicated that CFCs are the primary culprits. After weeks-long ozone depletion when sunlight returns to the Antarctic, huge masses of ozone-depleted air above Antarctica flow northward and linger for a few more weeks over parts of

Australia and New Zealand and over the southern tips of South America and Africa. This raises biologically damaging UV radiation levels in these areas by 3–10%, and in some years by as much as 20%.

In 1988 scientists discovered that similar but less severe ozone thinning occurs over the north pole during the arctic spring and early summer (February–June), with a seasonal ozone loss of 10–38% (compared to a typical 50% loss above Antarctica). When this mass of air above the Arctic breaks up each spring, large masses of ozone-depleted air flow south to linger over parts of Europe, North America, and Asia.

So far, seasonal ozone loss over the north pole is much lower than that over the south pole. However, the situation is changing. In 1992 atmospheric scientists warned that if rising levels of greenhouse gases warm the troposphere, the resulting insulating blanket could cool the stratosphere. This could increase the size and duration of seasonal ozone depletion over both poles, perhaps leading to large and regular arctic ozone thinning. As a result, ozone levels over parts of the northern hemisphere (including the United States) would decline sharply. According to a 1998 model developed by NASA scientists at the Goddard Institute for Space Studies, ozone destruction over the Arctic will be at its worst between 2010 and 2019.

Is Ozone Depletion Really a Serious Problem? Political talk show commentator Rush Limbaugh, zoologist and former head of the Atomic Energy Commission Dixy Lee Ray (now deceased), physicist and climate scientist S. Fred Singer, and articles and books in the popular press have claimed that ozone depletion by CFCs is a hoax, or at least a vastly exaggerated problem. The Pro/Con box on the right summarizes some of their charges and the answers by prominent research scientists studying ozone depletion.

Why Should We Be Worried About Ozone Depletion? Life in the Ultraviolet Zone Why should we care about ozone loss? From a human standpoint the answer is that with less ozone in the stratosphere, more biologically damaging UV radiation will reach the earth's surface and give humans worse sunburns, more cataracts (a clouding of the lens that reduces vision and can cause blindness if not corrected), and more skin cancers (Connections, p. 262).

According to UN Environment Programme estimates, the additional UV radiation reaching the earth's surface resulting from an annual 10% loss of global ozone (already a likely possibility within a few years) could lead to 300,000 additional cases of squamous cell cancer and basal cell cancer worldwide each year, 4,500–9,000 additional cases of potentially fatal malignant melanoma each year, and 1.5 million new cases of

---

*For a fascinating account of how corporate stalling, politics, economics, and science can interact, see Sharon Roan's *Ozone Crisis: The 15-Year Evolution of a Sudden Global Emergency* (New York: Wiley, 1989).

Q: How many people on average does one U.S. farmer feed?

Is Ozone Depletion a Hoax?

PRO/CON

Charge: There is no ozone hole, and the whole idea is merely a scientific theory.

Response: Technically, it is not correct to call this phenomenon an ozone *hole*. Instead, it is a thinning or partial depletion of normal ozone levels in the stratosphere. A scientific theory represents a widely accepted idea or principle that has a very high degree of certainty because it has been supported by a great deal of evidence. No theory in science can be proven absolutely, but there is overwhelming scientific evidence of ozone thinning in the stratosphere.

Charge: There is no scientific proof that an ozone hole exists.

Response: Ground-based and satellite measurements clearly show seasonal ozone thinning above Antarctica and the Arctic, and other measurements reveal a lower overall thinning everywhere except over the tropics.

Charge: CFC molecules can't reach the stratosphere because they are heavier than air, and no measurements have detected them in the stratosphere.

Response: The atmosphere is like a turbulent fluid in which gases, both lighter and heavier than air, are mixed thoroughly by the churning movements of large air masses. Since 1975 measuring instruments on balloons and satellites have clearly shown that CFCs are transported high into the stratosphere.

Charge: Sodium chloride (NaCl) from the evaporation of sea spray contributes more chlorine to the stratosphere than do CFCs.

Response: Particles of NaCl from sea spray (unlike CFCs) are soluble in water and are washed out of the lower atmosphere. Measurements have detected no sodium in the lower stratosphere.

Charge: Volcanic eruptions and biomass burning have added much more chlorine (Cl) to the stratosphere than have human-caused CFC emissions.

Response: Most of the water-soluble HCl injected into the troposphere from occasional large-scale volcanic eruptions is washed out by rain before it reaches the stratosphere. Even Fred Singer (who is highly skeptical about some aspects of ozone depletion models) stated in 1993 that "CFCs make the major contribution to stratospheric chlorine." However, recent research indicates that aerosol particles of sulfate salts produced from sulfur dioxide emitted by volcanic eruptions can destroy stratospheric ozone for several years after an eruption. These volcanic aerosols plus the human-related inputs of ozone-destroying chemicals can seriously deplete ozone in the stratosphere.

Charge: Seasonal ozone thinning over Antarctica is a natural phenomenon because it appeared long before CFCs were in wide use.

Response: In the late 1950s, British scientist Gordon Dobson discovered that the Antarctic polar vortex leads to some seasonal ozone loss through natural causes. Measurements indicate that between 1956 and 1976 the natural pattern of slight ozone loss above Antarctica did not change significantly. Since 1976, however, seasonal losses of ozone above Antarctica have increased dramatically and have been linked by measurements and models to rising levels of CFCs in the stratosphere. A 1994 review of Dobson's 1958 data concluded that in 1958 there was no credible evidence for a significant ozone thinning (such as that observed since 1976) over Antarctica.

Charge: The best models of ozone production and destruction estimate that global ozone levels should be 10% less than we actually observe.

Response: Scientists recognize that their models need to be improved. However, they point out that their models of ozone depletion by chlorine- and bromine-containing chemicals produced by human activities still account for about 90% of the observed ozone depletion.

Charge: The expected increase in UV-B radiation due to stratospheric ozone loss has not been detected in urban areas in the United States and most other developed countries.

Response: Increased UV levels have been detected in urban areas in countries in the southern hemisphere such as Australia, New Zealand, South Africa, Argentina, and Chile (as well as in Toronto, Canada) that are exposed to ozone-depleted air drifting away from Antarctica after the seasonal polar vortex breaks up each year. Some ozone depletion experts hypothesize that significant ground-level increases in UV levels have not been observed in many urban areas (especially in industrialized countries in the northern hemisphere) because ozone-laden smog over most cities may be filtering out some of the UV radiation. They argue that polluting our way out of increased UV levels in such cities is unacceptable: It will continue to cause premature death and widespread health problems for humans and damage trees and crops.

Critical Thinking

Do you believe that ozone depletion in the stratosphere is a serious problem or one that has been greatly exaggerated? Explain.

A: 140 (105 at home and 35 abroad)

The Cancer You Are Most Likely to Get

Considerable research indicates that years of exposure to UV ionizing radiation in sunlight is the primary cause of *squamous-cell* and *basal-cell skin cancers*, which together make up 95% of all skin cancers. Typically there is a 15- to 40-year lag between excessive exposure to UV radiation and development of these cancers.

Caucasian children and adolescents who get only a single severe sunburn double their chances of getting these two types of cancers. Some 90–95% of these types of skin cancer can be cured if detected early enough, although their removal may leave disfiguring scars.* In 1997, about 900,000 Americans developed such skin cancers. These cancers kill only 1–2% of their victims, but this still amounts to about 2,300 deaths in the United States each year.

A third type of skin cancer, *malignant melanoma*, can spread rapidly (within a few months) to other organs. It kills about one-fourth of its victims (most under age 40) within 5 years, despite surgery, chemotherapy, and radiation treatments. Each year it kills about 100,000 people (including 7,000 Americans in 1997), mostly

*I have had six basal-cell cancers on my face and neck because of "catching too many rays" in my younger years. I wish I had known then what I know now.

Caucasians. It can often be cured if detected early enough, but recent studies show that some melanoma survivors have a recurrence more than 15 years later.

The lag time between first substantial exposure to UV radiation (apparently mostly UV-A) and the occurrence of melanoma is 15–25 years.

Evidence indicates that people, especially Caucasians, who get three or more blistering sunburns before age 20 subsequently are five times more likely to develop malignant melanoma than those who have never had severe sunburns. About 10% of those who get malignant melanoma have an inherited gene that makes them especially susceptible to the disease.

To protect yourself, the safest course is to stay out of the sun (especially between 10 A.M. and 3 P.M., when UV levels are highest) and avoid tanning parlors. When you are in the sun, wear tightly woven protective clothing, a wide-brimmed hat, and sunglasses that protect against UV radiation (ordinary sunglasses may actually harm your eyes by dilating your pupils so that more UV radiation strikes the retina). Because UV rays can penetrate clouds, overcast does not protect you; neither does shade because UV rays can reflect off sand, snow, water, or patio floors. People who take antibiotics and women who take birth control pills are more susceptible to UV damage.

Use a sunscreen that offers protection against both UV-A and UV-B and has a protection factor of 15 or more (25 if you have light skin). Apply to all exposed skin and reapply it after swimming or excessive perspiration. Most people don't realize that the protection factors for sunscreens are based on using one full ounce of the product—far more than most people apply. Children who use a sunscreen with a protection factor of 15 every time they are in the sun from birth to age 18 decrease their chance of getting skin cancer by 80%; babies under a year old should not be exposed to the sun at all.

Become familiar with your moles and examine your skin at least once a month. The warning signs of skin cancer are a change in the size, shape, or color of a mole or wart (the major sign of malignant melanoma, which must be treated quickly); sudden appearance of dark spots on the skin; or a sore that keeps oozing, bleeding, and crusting over but does not heal. Be alert for precancerous growths (reddish-brown spots with a scaly crust). If you observe any of these signs, consult a doctor immediately.

Critical Thinking

What precautions, if any, do you take to reduce your chances of getting skin cancer from exposure to sunlight? Explain why you do or don't take such precautions.

---

cataracts (which account for over half of the world's 25–35 million cases of blindness) each year.

Assuming that we phase out *all* ozone-destroying chemicals over the next three decades, the EPA estimates that projected ozone thinning during the 1980s and 1990s will lead to 12 million new cases of skin cancer and 200,000 additional skin cancer deaths in the United States alone over the next 50 years.

Other effects from increased UV exposure are (1) suppression of the human immune system, which makes the body more susceptible to infectious diseases and some forms of cancer, (2) an increase in eye-burning, highly damaging acid deposition and ozone in smog in the troposphere (Section 9-7), (3) lower yields of key crops such as corn, rice, soybeans, cotton, beans, peas, sorghum, and wheat, with

Q: How many units of energy are required to put one unit of food energy on the table in the United States?

"I MISS THE OZONE LAYER...."

1988, Los Angeles Times Syndicate
Reprinted with permission.

Ray Turner and His Refrigerator

Ray Turner, an aerospace manager at Hughes Aircraft in California, made an important low-tech, ozone-saving discovery by using his head—and his refrigerator. His concern for the environment led him to look for a cheap and simple substitute for the CFCs used as cleaning agents for removing films of oxidation in the manufacture of most electronic circuit boards at his plant and elsewhere.

He started his search by looking in his refrigerator for a better circuit board cleaner. He decided to put drops of various substances on a corroded penny to see whether any of them would remove the film of oxidation. Then he used his soldering gun to see whether solder would stick to the cleaned surface of the penny, indicating that the film had been cleaned off.

First, he tried vinegar. No luck. Then he tried some ground-up lemon peel, also a failure. Next he tried a drop of lemon juice and watched as the solder took hold. The rest, as they say, is history.

Today, Hughes Aircraft uses inexpensive CFC-free, citrus-based solvents to clean circuit boards. This new cleaning technique has reduced circuit board defects by about 75% at Hughes. And Turner got a hefty bonus. Now other companies, such as AT&T, clean computer boards and chips using acidic chemicals extracted from cantaloupes, peaches, and plums. Maybe you can find a solution to an environmental problem in your refrigerator, grocery or drugstore, or backyard.

estimated losses totaling $2.5 billion per year in the United States alone before the middle of the 21st century, and **(4)** reduction in the productivity of surface-dwelling phytoplankton, which could upset aquatic food webs, decrease yields of seafood eaten by humans, and possibly accelerate global warming by decreasing the oceanic uptake of CO_2 by phytoplankton.

In a worst-case ozone depletion scenario, most people would have to avoid the sun altogether (see cartoon). Even cattle could graze only at dusk, and farmers and other outdoor workers might need to limit their exposure to the sun to minutes. Fortunately, we are acting to prevent such a scenario.

Some critics who believe that the threat of ozone depletion has been exaggerated argue that we should not worry about a 10%, 20%, or even 50% increase in UV radiation reaching the earth's surface. According to Fred Singer, moving just 97 kilometers (60 miles) closer to the equator involves potential exposure to 10% more UV radiation.

Many scientists contend that this argument ignores two crucial points. First, the southward movement of the U.S. population (and people in many other countries) is one reason skin-cancer rates have risen. Second, we are also talking about the likely possibility of rapid (within decades), widespread, long-lasting, and essentially unpredictable ecological disruptions in species adapted to existing levels of background UV radiation.

Humans can quickly make cultural adaptations to increased UV-B radiation by staying out of the sun, protecting their skin with clothing, and applying sunscreens. However, plants and other animals that help support us and other forms of life can't make such changes except through the long process of biological evolution.

9-6   SOLUTIONS: PROTECTING THE OZONE LAYER

How Can We Protect the Ozone Layer? The scientific consensus of researchers in this field is that we should immediately stop producing all ozone-depleting chemicals. Even with immediate action, the models indicate that it will take 50–60 years for the ozone layer to return to 1975 levels and another 100–200 years for full recovery to pre-1950 levels.

Substitutes are already available for most uses of CFCs, and others are being developed (Individuals Matter, p. above). One substitute for CFCs is *hydrochlorofluorocarbons* or *HCFCs* (such as $CHClF_2$). If used in massive quantities, however, HCFCs would still cause ozone depletion and act as potent greenhouse gases.

A: About 10 (a loss of 9 units of energy)

Another substitute for CFCs is *hydrofluorocarbons* or *HFCs* (such as CF_3, containing fluorine but no chlorine or bromine). Recent research indicates that HFCs have a negligible effect on ozone depletion. However, they may need to be restricted or phased out because they are powerful greenhouse gases that remain in the atmosphere much longer than CFCs or CO_2.

To a growing number of scientists, hydrocarbons (HCs) such as propane and butane are a better way to reduce ozone depletion while doing little to enhance global warming. Developing countries can use HC refrigerator technology to leapfrog ahead of industrialized countries without having to invest in costly HFC and HCFC technologies that will have to be phased out within a few decades. This approach is also less costly because HCs cannot be patented and can be manufactured locally, reducing the need to import expensive HFCs and HCFCs.

Can Technofixes Save Us? What about a quick fix from technology, so that we can keep on using CFCs? Physicist Alfred Wong has proposed that each year we launch a fleet of 20–30 football-field–long, radio-controlled blimps into the stratosphere above Antarctica. Hanging from each blimp would be a huge curtain of electrical wires that would inject negatively charged electrons into the stratosphere when exposed to high voltages (produced by electricity from huge panels of solar cells). Based on 4 years of laboratory experiments, Wong believes that ozone-destroying chlorine atoms (Cl) in the stratosphere would each pick up an electron and be converted to chloride ions ($Cl + e \longrightarrow Cl^-$) that would not react with ozone. A second suspended sheet of positively charged wires could be used to attract the negatively charged ions and remove them from the stratosphere. Wong estimates that it would cost about $400 million a year to remove 10–30% of the chlorine atoms formed in the stratosphere each year.

However, atmospheric chemist Ralph Ciecerone believes that this plan won't work because other chemical species in the stratosphere snatch electrons more readily than does chlorine. This scheme could also have unpredictable side effects on atmospheric chemistry.

Others have suggested using tens of thousands of lasers to blast CFCs out of the atmosphere before they can reach the stratosphere. However, the energy required to do this would be enormous and expensive, and decades of research would be needed to perfect the types of lasers needed. Moreover, we can't predict the possible effects of such powerful laser blasts on climate, birds, or planes.

What Is Being Done to Reduce Ozone Depletion? Some Hopeful Progress In 1987, 36 nations meeting in Montreal developed a treaty, commonly known as the *Montreal Protocol*, to cut emissions of CFCs (but not other ozone depleters) into the atmosphere by about 35% be-

tween 1989 and 2000. After hearing more bad news about ozone depletion, representatives of 93 countries met in London in 1990 and in Copenhagen in 1992 and adopted a protocol accelerating the phaseout of key ozone-depleting chemicals, with some phaseout schedules accelerated in 1995 and 1997.

The agreements reached so far are important examples of global cooperation in response to serious threats to global environmental security. Because of these agreements, CFC production fell by 76% between 1988 (its peak production year) and 1995. Global production of halons, carbon tetrachloride, and methyl chloroform has also dropped sharply, but that of methyl bromide and HCFCs continues to rise.

However, there is growing concern that the requirements of the Copenhagen agreement might not be met. By 1995 there was a political and economic backlash in the United States against this treaty. This was caused by widely publicized (but scientifically refuted; see Pro/Con on p. 261) attacks on the overwhelming scientific consensus that ozone depletion is a very serious problem.

The effectiveness of the treaty is also being undermined by a rapidly growing black market in CFCs (apparently being smuggled into the United States and other developed countries from Russia, China, and Mexico). This black market is being stimulated by dwindling supplies of CFCs (phased out by 1996), the high prices of some CFC substitutes, and the costly conversion of some older air conditioners and refrigerators to use replacement chemicals. There have also been signs that some countries are cheating and not living up to the requirements of the ozone treaty. A cheating rate of only 10% can keep stratospheric levels of CFCs from declining as projected.

Even if the ozone treaty is only partially implemented, it has set an important precedent for global cooperation and action when faced with potential global disaster. However, international cooperation in dealing with projected global warming is much more difficult because the evidence for global warming is less clear-cut. Moreover, lowering our inputs of greenhouse gases by greatly reducing use of fossil fuels and greatly slowing deforestation is economically and politically difficult to do.

9-7 TYPES AND SOURCES OF OUTDOOR AND INDOOR AIR POLLUTION

What Are the Major Types and Sources of Air Pollution? Air pollution is the presence of one or more chemicals in the atmosphere in quantities and duration that cause harm to humans, other forms of life, and materials. As clean air in the troposphere moves across the earth's surface, it collects the products of natural events (dust storms and volcanic eruptions) and human activities (emissions from cars and smokestacks). These potential pollutants, called **primary pollutants**, are mixed vertically and horizontally and are dispersed and diluted by the churning air in the troposphere. While in the troposphere, some of these primary pollutants may react with one another or with the basic components of air to form new pollutants, called **secondary pollutants** (Figure 9-14).

Long-lived primary and secondary pollutants can travel great distances before they return to the earth's surface as solid particles, droplets, or chemicals dissolved in precipitation. Table 9-1 lists the major classes of pollutants commonly found in outdoor air.

Pollutants are also found indoors from infiltration of polluted outside air and from various chemicals used or produced inside buildings. Risk analysis experts rate indoor and outdoor air pollution as high-risk human health problems (Figure 8-5, left).

In developed countries, most pollutants enter the atmosphere from the burning of fossil fuels in both power plants and factories (*stationary sources*) and in motor vehicles (*mobile sources*). Motor vehicles produce more air pollution than any other human activity. In car-clogged cities such as Los Angeles, California; São Paulo, Brazil; Bangkok, Thailand; Rome, Italy; and Mexico City, Mexico (Case Study, p. 89), motor vehicles are responsible for 80–88% of the air pollution. According to the World Health Organization, more than 1.1 billion people—one of every five—live in urban areas where the air is unhealthy to breathe.

Because they contain large concentrations of cars and factories, cities normally have higher air pollution levels than rural areas. However, prevailing winds can spread long-lived primary and secondary air pollutants from emissions in urban and industrial areas to the countryside and to other downwind urban areas.

What Is Photochemical Smog? Brown-Air Smog

Air pollution known as **photochemical smog** is a mixture of primary and secondary pollutants formed under the influence of sunlight (Figure 9-15). The resulting mixture of more than 100 chemicals is dominated by ozone, a highly reactive gas that harms most living organisms.

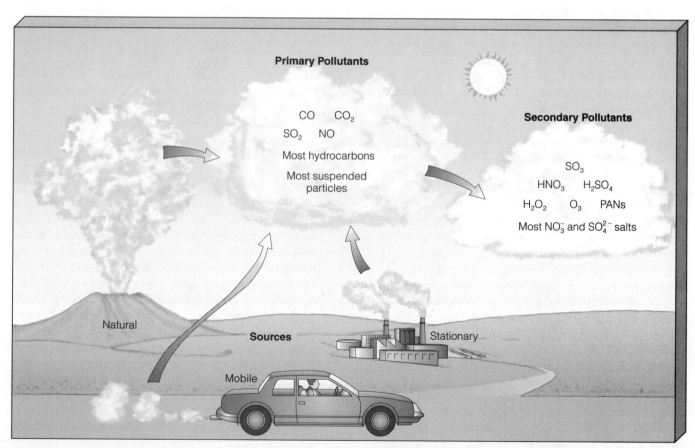

Figure 9-14 Sources and types of air pollutants. Human inputs of air pollutants may come from mobile sources (cars) and stationary sources (industrial and power plants). Some primary air pollutants may react with one another or with other chemicals in the air to form secondary air pollutants.

Hint: Enter the search term *indoor air quality* using the Subject Guide.

Table 9-1 Major Classes of Outdoor Air Pollutants

Class	Examples
Carbon oxides	Carbon monoxide (CO), carbon dioxide (CO_2)
Sulfur oxides	Sulfur dioxide (SO_2), sulfur trioxide (SO_3)
Nitrogen oxides	Nitric oxide (NO), nitrogen dioxide (NO_2), nitrous oxide (N_2O) (NO and NO_2 are often lumped together and labeled as NO_x)
Volatile organic compounds	Methane (CH_4), propane (C_3H_8), benzene (C_6H_6), chlorofluorocarbons (CFCs)
Suspended particles	Solid particles (dust, soot, asbestos, lead, nitrate and sulfate salts), liquid droplets (sulfuric acid, PCBs, dioxins, pesticides)
Photochemical oxidants	Ozone (O_3), peroxyacyl nitrates (PANs), hydrogen peroxide (H_2O_2), aldehydes
Radioactive substances	Radon-222, iodine-131, strontium-90, plutonium-239
Toxic compounds	Trace amounts of at least 600 toxic substances (many of them volatile organic compounds), 60 of them known carcinogens

Virtually all modern cities have photochemical smog. However, it is much more common in cities with sunny, warm, dry climates and lots of motor vehicles such as Los Angeles, California; Denver, Colorado; and Salt Lake City, Utah, in the United States, as well as Sydney, Australia; Mexico City, Mexico; and São Paulo and Buenos Aires, Brazil.

What Is Industrial Smog? Gray-Air Smog Thirty years ago cities such as London, England, and Chicago and Pittsburgh in the United States burned large amounts of coal and heavy oil (which contain sulfur impurities) in power plants and factories and for space heating. During winter, people in such cities were exposed to **industrial smog** consisting mostly of sulfur dioxide, suspended droplets of sulfuric acid (formed from some of the sulfur dioxide), and a variety of suspended solid particles and droplets (called aerosols).

Urban industrial smog is rarely a problem today in most developed countries because coal and heavy oil are burned only in large boilers with reasonably good pollution control or with tall smokestacks. However, industrial smog is a problem in industrialized urban areas of China, India, Ukraine, and some eastern European countries, where large quantities of coal are burned with inadequate pollution controls.

What Factors Influence the Formation of Photochemical and Industrial Smog? The frequency and severity of smog in an area depend on several things: the local climate and topography, the population density, the amount of industry, and the fuels used in industry, heating, and transportation. In areas with high average annual precipitation, rain and snow help cleanse the air of pollutants. Winds help sweep pollutants away and bring in fresh air, but they may also transfer some pollutants to downwind areas.

Hills and mountains tend to reduce the flow of air in valleys below them and allow pollutant levels to build up at ground level. Buildings in cities generally slow wind speed, thereby reducing dilution and removal of pollutants.

During the day the sun warms the air near the earth's surface. Normally this heated air expands and rises, carrying low-lying pollutants higher into the troposphere (Figure 9-16, left). Colder, denser air from surrounding high-pressure areas then sinks into the low-pressure area created when the hot air rises. This continual mixing of the air helps keep pollutants from reaching dangerous concentrations near the ground.

Sometimes, however, a layer of dense, cool air beneath can be trapped beneath a layer of less dense, warm air in an urban basin or valley, causing a phenomenon known as a **temperature inversion**, or a **thermal inversion** (Figure 9-16, right). The changing temperature (temperature gradient) in the warm air above the pool of cool air prevents ascending air currents (that would disperse and dilute pollutants) from developing. These inversions usually last for only a few hours, but when a high-pressure air mass stalls over an area, they can last

Q: What percentage of the world's of the world's grain production is consumed by livestock?

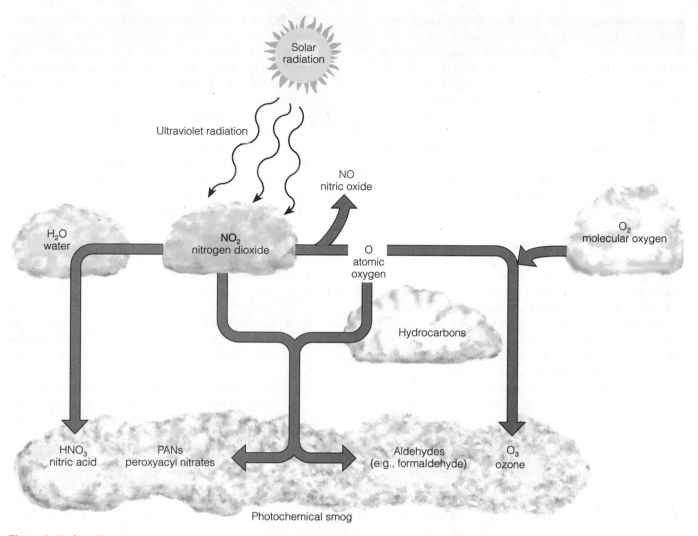

Figure 9-15 Simplified scheme of the formation of photochemical smog. The severity of smog is generally associated with atmospheric concentrations of ozone at ground level.

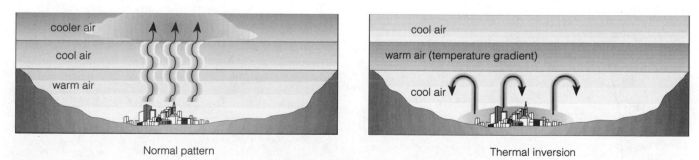

Normal pattern

Thermal inversion

Figure 9-16 Thermal inversion. The change in temperature (temperature gradient) in the warm air (right) prevents ascending air currents from rising and dispersing and diluting pollutants in the cool pool of near the ground. Because of their topography, Los Angeles in the United States and Mexico City in Mexico have frequent thermal inversions, many of them prolonged during the summer.

A: About 38% (70% in the United States)

for several days, allowing air pollutants at ground level to build up to harmful and even lethal concentrations.

The first U.S. air pollution disaster occurred in 1948, when fog laden with sulfur dioxide and suspended particulate matter stagnated for 5 days over the town of Donora, in Pennsylvania's Monongahela Valley south of Pittsburgh. About 6,000 of the town's 14,000 inhabitants fell ill, and 20 of them died. This killer fog resulted from a combination of mountainous terrain surrounding the valley and weather conditions that trapped and concentrated deadly pollutants emitted by the community's steel mill, zinc smelter, and sulfuric acid plant.

A city with several million people and motor vehicles in an area with a sunny climate, light winds, mountains on three sides, and the ocean on the other has ideal conditions for photochemical smog worsened by frequent thermal inversions. This describes California's Los Angeles basin, which has 14 million people, 23 million motor vehicles, thousands of factories, and thermal inversions at least half of the year. Despite having the world's toughest air-pollution control program, Los Angeles is the air pollution capital of the United States. Other cities with frequent thermal inversions are Denver, Colorado, in the United States, Mexico City in Mexico, Rio de Janeiro and São Paulo in Brazil, and Beijing and Shenyang in China.

What Is Acid Deposition? Operators of coal-burning power plants, industrial plants, and ore smelters in developed countries have found a way to reduce local air pollution and meet government air pollution standards without having to add expensive air-pollution control devices. They use tall smokestacks to emit sulfur dioxide, suspended particles, and nitrogen oxides above the inversion layer. This "dilution solution" reduces local air pollution. However, it increases pollution downwind because what goes up must come down—another example of connections or unintended consequences.

As these primary pollutants are transported as much as 1,000 kilometers (600 miles) by prevailing winds, they form secondary pollutants such as nitric acid vapor, droplets of sulfuric acid, and particles of acid-forming sulfate and nitrate salts. These chemicals descend to the earth's surface in two forms: *wet* (as acidic rain, snow, fog, and cloud vapor) and *dry* (as acidic particles). The resulting mixture is called **acid deposition** (Figure 9-17). Although this form of pollution is commonly called acid rain, *acid deposition* is a better term because the acidity can reach the earth's surface not only in rain but also as gases and solid particles.

Acidity of substances in water is commonly expressed in terms of pH. Solutions with pH values less than 7 are acidic, and those with pH values greater than 7

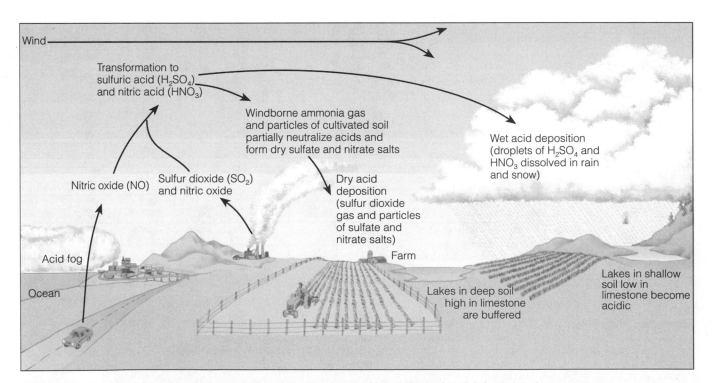

Figure 9-17 Acid deposition, which consists of rain, snow, dust, or gas with a pH lower than 5.6, is commonly called acid rain. Soils and lakes vary in their ability to buffer or remove excess acidity.

Q: How many people are undernourished or malnourished?

are alkaline or basic (Figure 7-5). Natural precipitation is slightly acidic, with a pH of 5.0–5.6. However, because of acid deposition, typical rain in the eastern United States is now about 10 times more acidic, with a pH of 4.3. In some areas it is 100 times more acidic, with a pH of 3— as acidic as vinegar. Some cities and mountaintops are bathed in a fog as acidic as lemon juice, with a pH of 2.3—about 1,000 times the acidity of normal precipitation.

What Areas Are Most Affected by Acid Deposition? Acid deposition occurs on a regional rather than a global basis because the acidic components remain in the atmosphere only for a few days. However, acid deposition is a serious regional problem (Figure 9-18) in many areas downwind from coal-burning power plants, smelters, factories, and large urban areas. How seriously vegetation and aquatic life in nearby lakes are affected by an area receiving acid deposition depends mostly on whether its soils are acidic or basic.

In some areas, soils are basic enough to neutralize or buffer some inputs of acids. The ecosystems most harmed by acid deposition are those containing thin, acidic soils without such natural buffering (Figure 9-18) and those where the buffering capacity of soils has been depleted because of decades of exposure to acid deposition.

Many of the acid-producing chemicals generated by power plants, factories, smelters, and cars in one country may be exported to others by prevailing winds. For example, more than three-fourths of the acid deposition in Norway, Switzerland, Austria, Sweden, the Netherlands, and Finland is blown to those countries from industrialized areas of western Europe (especially the United Kingdom and Germany) and eastern Europe.

Chemical detective work indicates that more than half the acid deposition in southeastern Canada and the eastern United States originates from coal- and oil-burning power plants and factories in the states of Ohio, Indiana, Pennsylvania, Illinois, Missouri, West Virginia, and Tennessee. In areas near and downwind from large urban areas, emissions of NO and NO_2 (mostly from cars) leading to the formation of nitric acid may be the main culprit.

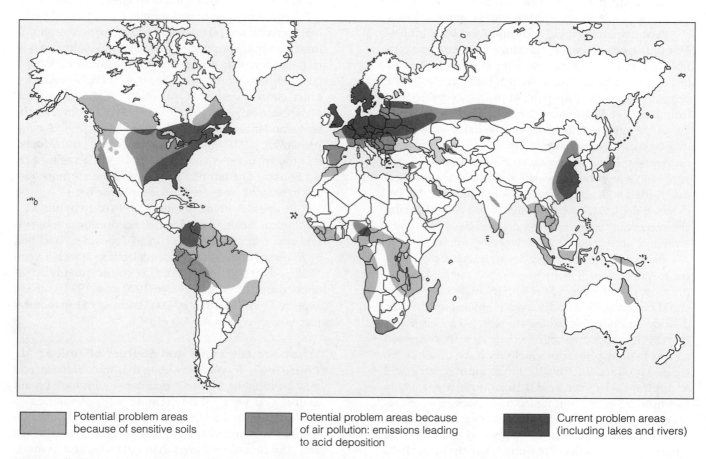

| | Potential problem areas because of sensitive soils | | Potential problem areas because of air pollution: emissions leading to acid deposition | | Current problem areas (including lakes and rivers) |

Figure 9-18 Regions where acid deposition is now a problem and regions with the potential to develop this problem, either because of increased air pollution (mostly from power plants, industrial plants, and ore smelters) or because of soils that cannot neutralize inputs of acidic compounds. (Data from World Resource Institute and U.S. Environmental Protection Agency)

A: About 840 million in 1995 (down from 940 million in 1970)

Within a few decades, NO_x and SO_2 emissions from developing countries are expected to outstrip those from developed countries, leading to greatly increased damage from acid deposition over a much wider area, especially where soils are sensitive to acidification.

What Are the Effects of Acid Deposition? Risk analysis experts rate acid deposition as a medium-risk ecological problem and a high risk to human health (Figure 8-5, left). Acid deposition has many harmful ecological effects, especially when the pH falls below 5.1 for terrestrial systems and below 5.5 for aquatic systems. It also contributes to human respiratory diseases such as bronchitis and asthma (which can cause premature death), and it damages statues, buildings, metals, and car finishes.

Acid deposition can damage tree foliage directly, but the most serious effect is weakening trees so they become more susceptible to other types of damage. The areas hardest hit by acid deposition are mountaintop forests, which tend to have thin soils without much buffering capacity. Trees on mountaintops, especially conifers such as red spruce that keep their leaves year-round, are bathed almost continuously in very acidic fog and clouds.

A combination of acid deposition and other air pollutants (especially ozone) can make trees more susceptible to stresses such as cold temperatures, diseases, insects, drought, and fungi (which thrive under acidic conditions), and to a drop in net primary productivity from loss of soil plant nutrients.

Although the final cause of tree damage or death may be mosses, insect attacks, diseases, and lack of plant nutrients, the underlying cause is often years of exposure to an atmospheric cocktail of air pollutants and soil overloaded with acids. Drops in forest productivity because of depletion of soil nutrients and acid-buffering chemicals can reduce biodiversity and have significant economic implications for timber companies.

Because it can take decades to hundreds of years for soil to replenish nutrients leached out by acid deposition, losses in plant productivity in damaged areas could continue for decades even if emissions of sulfur dioxide and nitrogen oxides are reduced by air-pollution control programs. Until recently scientists expected some of the lost forest productivity to be offset by increased productivity from the larger input of nitric acid and nitrate salts from acid deposition in areas where this nutrient is the limiting factor. However, recent research revealed that much of the nitrate raining down on forest areas in parts of Germany and Norway damaged by air pollution is not being taken up by the trees or by nitrogen-using microbes in the soil.

Acid deposition can also release aluminum ions attached to soil minerals. Once released from soil particles, these water-soluble ions can damage tree roots. When washed into lakes, aluminum ions can also kill many kinds of fish by stimulating excessive mucus formation, which asphyxiates the fish by clogging their gills.

Excess acidity can contaminate fish in some lakes with highly toxic methylmercury. Increased acidity of lakes apparently converts moderately toxic inorganic mercury compounds in lake-bottom sediments into highly toxic methylmercury, which is more soluble in the fatty tissue of animals and can be biomagnified to higher concentrations in aquatic food chains and webs. However, acid runoff into lakes and streams is rated by risk analysis experts as a low-risk ecological problem (Figure 8-5, left).

How Serious Is Acid Deposition in the United States? A large-scale, government-sponsored research study on the *ecological effects* of acid deposition in the United States in the 1980s concluded that the problem was serious but not yet at a crisis stage. However, numerous health studies have shown that the effects from exposure to the chemical components of acid deposition are a *serious health problem* (Figure 8-5, left and Section 9-8) and also damages materials.

Representatives of coal companies and of industries that burn coal and oil claim that adding expensive air pollution control equipment or burning low-sulfur coal or oil costs more than the resulting health and environmental benefits are worth. According to the EPA, however, the actual cleanup costs of SO_2 in 1994 were about one-tenth of the estimate given by industry when they opposed the new standards set by the Clean Air Act of 1990. A comprehensive 1997 study by Resources for the Future found that the environmental and public health benefits of reductions in SO_2 from 1995 to 2030 will generate more than $12 in benefits for every $1 in compliance costs.

Progress is being made. A 1993 study by the U.S. Geological Survey found that the concentration of sulfate ions, a key component of acid deposition, declined at 26 out of 33 U.S. rainwater collection sites between 1980 and 1991. In addition, U.S. sulfur dioxide emissions dropped 30% between 1970 and 1993 and are expected to fall further by 2000 because of the requirements of the Clean Air Act of 1990.

What Are the Types and Sources of Indoor Air Pollution? If you are reading this book indoors, you may be inhaling more air pollutants with each breath than if you were outside (Figure 9-19). According to EPA studies, in the United States levels of 11 common pollutants are generally 2–5 times higher inside homes and commercial buildings than outdoors, and as much as 70 times higher in some cases. A 1993 study found that pollution levels inside cars in traffic-clogged U.S. urban areas can be up to 18 times higher than those outside the vehicles.

Q: How many people die each year from hunger-related causes?

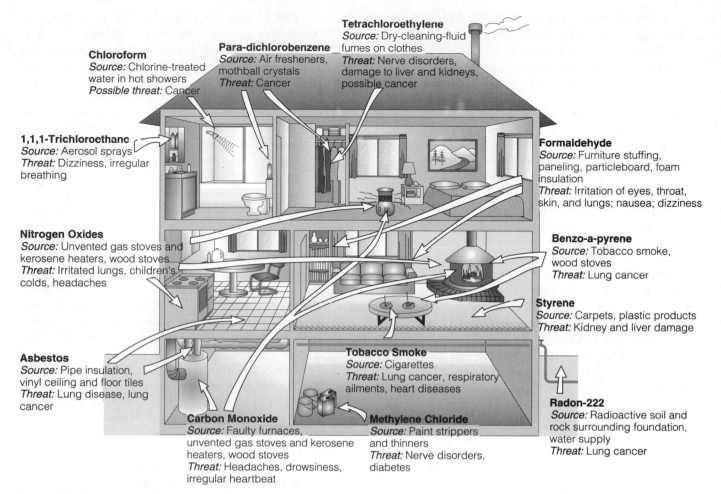

Chloroform
Source: Chlorine-treated water in hot showers
Possible threat: Cancer

Para-dichlorobenzene
Source: Air fresheners, mothball crystals
Threat: Cancer

Tetrachloroethylene
Source: Dry-cleaning-fluid fumes on clothes
Threat: Nerve disorders, damage to liver and kidneys, possible cancer

1,1,1-Trichloroethane
Source: Aerosol sprays
Threat: Dizziness, irregular breathing

Formaldehyde
Source: Furniture stuffing, paneling, particleboard, foam insulation
Threat: Irritation of eyes, throat, skin, and lungs; nausea; dizziness

Nitrogen Oxides
Source: Unvented gas stoves and kerosene heaters, wood stoves
Threat: Irritated lungs, children's colds, headaches

Benzo-a-pyrene
Source: Tobacco smoke, wood stoves
Threat: Lung cancer

Styrene
Source: Carpets, plastic products
Threat: Kidney and liver damage

Asbestos
Source: Pipe insulation, vinyl ceiling and floor tiles
Threat: Lung disease, lung cancer

Tobacco Smoke
Source: Cigarettes
Threat: Lung cancer, respiratory ailments, heart diseases

Radon-222
Source: Radioactive soil and rock surrounding foundation, water supply
Threat: Lung cancer

Carbon Monoxide
Source: Faulty furnaces, unvented gas stoves and kerosene heaters, wood stoves
Threat: Headaches, drowsiness, irregular heartbeat

Methylene Chloride
Source: Paint strippers and thinners
Threat: Nerve disorders, diabetes

Figure 9-19 Some important indoor air pollutants. (Data from U.S. Environmental Protection Agency)

The health risks from exposure to such chemicals are magnified because people spend 70–98% of their time indoors. In 1990 the EPA placed indoor air pollution at the top of the list of 18 sources of cancer risk, and it is rated by risk analysis scientists as a high-risk health problem for humans (Figure 8-5, left). At greatest risk are smokers, infants and children under age 5, the old, the sick, pregnant women, people with respiratory or heart problems, and factory workers.

Danish and EPA studies have linked pollutants found in buildings to dizziness, headaches, coughing, sneezing, nausea, burning eyes, chronic fatigue, and flulike symptoms, known as the *sick building syndrome*. New buildings are more commonly "sick" than old ones because of reduced air exchange (to save energy) and chemicals released from new carpeting and furniture. According to the EPA, at least 17% of the 4 million commercial buildings in the United States are considered "sick" (including EPA's headquarters). Indoor air pollution in the United States costs an estimated $100 billion per year in absenteeism, reduced productivity, and health costs.

According to the EPA and public health officials, cigarette smoke (Spotlight, p. 228), formaldehyde, asbestos, and radioactive radon-222 gas are the four most dangerous indoor air pollutants. A number of research studies on laboratory animals have also identified tiny fibers of *fiberglass* as a widespread and potentially potent carcinogen in indoor air.

The chemical that causes most people difficulty is *formaldehyde*, an extremely irritating gas. As many as 20 million Americans suffer from chronic breathing problems, dizziness, rash, headaches, sore throat, sinus and eye irritation, and nausea caused by daily exposure to low levels of formaldehyde emitted (outgassed) from common building materials (such as plywood, particleboard, and paneling), furniture, drapes, upholstery, and adhesives in carpeting and wallpaper (Figure 9-19). The EPA estimates that as many as 1 out of every 5,000 people who live in manufactured homes for more than 10 years will develop cancer from formaldehyde exposure.

In developing countries, the burning of wood, dung, and crop residues in open fires or in unvented or poorly vented stoves for cooking and heating exposes inhabitants, especially women and young children, to very high levels of particulate air pollution. Partly as a result, respiratory illnesses are a major cause of death and illness among the poor in most developing countries.

Case Study: What Should Be Done About Asbestos? There is intense controversy over what to do about possible exposure to tiny fibers of asbestos, a name given to several different fibrous forms of silicate minerals widely used since the 1940s for fireproofing and thermal insulation. Unless completely sealed within a product, asbestos easily crumbles into a dust of fibers tiny enough to become suspended in the air and inhaled deep into the lungs, where they remain for many years.

Prolonged exposure to asbestos fibers can cause *asbestosis* (a chronic, sometimes fatal disease that eventually makes breathing nearly impossible and was recognized as a hazard among asbestos workers as early as 1924), *lung cancer*, and *mesothelioma* (an inoperable cancer of the chest cavity lining). Epidemiological studies have shown that smokers exposed to asbestos fibers have a much greater chance of dying from lung cancer than do nonsmokers exposed to such fibers.

Most of these diseases occur in people exposed for years to high levels of asbestos fibers. This group includes asbestos miners, insulators, pipefitters, shipyard employees, and workers in asbestos-producing factories. By the year 2000 it is estimated that 300,000 American workers will have died prematurely because of exposure to asbestos fibers. After being swamped with health claims from workers, most U.S. asbestos manufacturing companies have either declared bankruptcy or have moved their operations to other countries (such as Mexico and Brazil) with weaker environmental laws and lax enforcement.

In recent years the focus has shifted from asbestos workers to concern over possible health effects of inhalation of low levels of asbestos fibers by the public in buildings. In the United States between 1900 and 1984, asbestos was sprayed on ceilings and walls of schools and other public and private buildings for fireproofing, soundproofing, insulation of heaters and pipes, and wall and ceiling decoration. The EPA banned those uses in 1984.

In 1989 the EPA ordered a ban on almost all remaining uses of asbestos (such as brake linings, roofing shingles, and water pipes) in the United States by 1997. Representatives of the asbestos industry in the United States and Canada (which now produces most of the asbestos used in the United States) challenged the ban in court, contending that with proper precautions asbestos products can be used safely and that the costs of the ban outweigh the benefits. In 1991 a federal appeals court overturned the 1989 EPA ban.

Critics contend that the health benefits of asbestos removal from many schools, homes, and other buildings are not worth the costs unless measurements (not just visual inspection) indicate that the buildings have high levels of airborne asbestos fibers.* They call for sealing, wrapping, and other forms of containment instead of removal of most asbestos, and they point out that improper or unnecessary removal can release more asbestos fibers than sealing off asbestos that is not crumbling.

After much controversy and huge expenditures of money on asbestos removal, there is now general agreement that the degree of risk from low-level exposure to asbestos fibers is unclear and asbestos should not be removed from buildings where it has not been damaged or disturbed. Instead it should be sealed or wrapped, with removal only as a last, carefully conducted resort.

In 1998 chemists developed a foam that lets building owners treat asbestos-containing fireproofing material without removing it. The foam initiates chemical reactions that bind the minerals in the asbestos together to form a hard material that is nontoxic and still acts as a fireproofing material.

Critics of environmentalists charge that much of the government-required removal of asbestos from schools and other public buildings was unnecessary and wasted billions of dollars—an example of environmental and regulatory overkill.

Case Study: Is Your Home Contaminated with Radon Gas? Radon-222 is a colorless, odorless, tasteless, naturally occurring radioactive gas produced by the radioactive decay of uranium-238. Small amounts of uranium-238 are found in most soil and rock, but this isotope is much more concentrated in underground deposits of minerals such as uranium, phosphate, granite, and shale.

When radon gas from such deposits seeps upward through the soil and is released outdoors, it disperses quickly in the atmosphere and decays to harmless levels. However, when the gas is drawn into buildings through cracks, drains, and hollow concrete blocks (Figure 9-20), or seeps into groundwater in underground wells over such deposits, it can build up to high levels.

Radon-222 gas quickly decays into solid particles of other radioactive elements that, if inhaled, expose lung tissue to a large amount of radiation from alpha particles. Several epidemiological studies indicate that

---

*If you plan to buy or live in a house built before 1980, you may want to have its air tested for asbestos fibers. To get a free list of certified asbestos laboratories that charge $25–50 to test a sample, send a self-addressed stamped envelope to NIST/NVLAP, Building 411, Room A124, Gaithersburg, MD 20899, or call the EPA's Toxic Substances Control Hotline at 202-554-1404.

Tennesen, Michael. 1997. "On a Clear Day." *National Parks*, vol. 71, no. 11–12, 26(4).

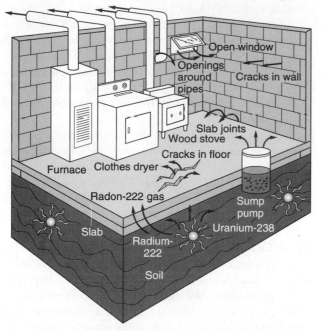

Figure 9-20 Sources and paths of entry for indoor radon-222 gas. (Data from U.S. Environmental Protection Agency)

prolonged exposure (defined as 75% of one's time spent in the same home for a lifetime of 70 years) to low levels of radon or radon acting together with smoking is responsible for 6,000–36,000 of the 130,000 lung cancer deaths each year in the United States (13,600 deaths is the best estimate).

If this estimate is accurate, indoor radon exposure is second only to smoking as the leading cause of lung cancer deaths. However, the results of several epidemiological studies in various countries are mixed and provide no clear evidence to support or refute a connection between lung cancer deaths and inhaled radon in homes.

According to the EPA, prolonged exposure to average radon levels above 4 picocuries* per liter of air in a closed house is considered unsafe. Other researchers cite evidence suggesting that radon becomes dangerous only if indoor levels exceed 20 picocuries per liter—the level accepted in Canada, Sweden, and Norway. Such controversy over acceptable radon levels demonstrates the uncertainties and problems inherent in risk assessment and risk management (Section 8-5).

EPA indoor radon surveys suggest that 4–5 million U.S. homes may have annual radon levels above 4 picocuries per liter of air and that 50,000–100,000 homes

may have levels above 20 picocuries per liter. If the 4 picocuries per liter standard is adopted (as proposed by the EPA), the cost of testing and correcting the problem could run about $50 billion, with a 15–20% reduction in radon-related deaths. Some researchers argue that it makes more sense to spend perhaps only $500 million to find and fix homes and buildings with radon levels above 20 picocuries per liter until more reliable data are available on the threat from exposure to lower levels of radon.

Because radon "hot spots" can occur almost anywhere, it's impossible to know which buildings have unsafe levels of radon without conducting tests. In 1988 the EPA and the U.S. Surgeon General's Office recommended that everyone living in a detached house, a town house, a mobile home, or on the first three floors of an apartment building test for radon.* Ideally, radon levels should be continuously monitored in the main living areas (not basements or crawl spaces) for 2 months to a year. By 1998 only about 6% of U.S. households had conducted radon tests (most lasting only 2 to 7 days and costing $20–100 per home).

If testing reveals an unacceptable level, you can consult the free EPA publication *Radon Reduction Methods* for ways to reduce radon levels and health risks. According to the EPA, radon control could add $350–500 to the cost of a new home, and correcting a radon problem in an existing house could run $800–2,500.

9-8 EFFECTS OF AIR POLLUTION ON LIVING ORGANISMS AND MATERIALS

How Is Human Health Harmed by Air Pollutants? Your respiratory system has a number of mechanisms that help protect you from air pollution. Hairs in your nose filter out large particles. Sticky mucus in the lining of your upper respiratory tract captures smaller (but not the smallest) particles and dissolves some gaseous pollutants. Sneezing and coughing expel contaminated air and mucus when your respiratory system is irritated by pollutants. The cells of your upper respiratory tract are also lined with hundreds of thousands of tiny, mucus-coated hairlike structures called *cilia* that continually wave back and forth, transporting mucus and the pollutants they trap to your throat (where they are then either swallowed or expelled).

Years of smoking and exposure to air pollutants can overload or break down these natural defenses, causing

---

*A picocurie is a trillionth of a curie, which is the amount of radioactivity emitted by a gram of radium.

*For information, see "Radon Detectors: How to Find Out if Your House Has a Radon Problem," *Consumer Reports*, July 1987.

or contributing to respiratory diseases. Examples are (1) *lung cancer*, (2) *asthma* (typically an allergic reaction causing sudden episodes of muscle spasms in the bronchial walls, resulting in acute shortness of breath), (3) *chronic bronchitis* (persistent inflammation and damage to the cells lining the bronchi and bronchioles, causing mucus buildup, painful coughing, and shortness of breath), and (4) *emphysema* (irreversible damage to air sacs or alveoli leading to abnormal dilation of air spaces, loss of lung elasticity, and acute shortness of breath. Elderly people, infants, pregnant women, and people with heart disease, asthma, or other respiratory diseases are especially vulnerable to air pollution.

It is difficult to estimate by risk analysis how many people die prematurely from respiratory or cardiac problems caused or aggravated by air pollution because people are exposed to so many different pollutants over their lifetimes. In the United States, estimates of annual deaths related to outdoor air pollution range from 65,000 to 200,000. If indoor air pollution is included, estimated annual deaths from air pollution in the United States range from 150,000 to 350,000 people—equivalent to 1–2 fully loaded 400-passenger jumbo jets crashing *each day* with no survivors.

Millions more become ill and lose work time. According to the EPA and the American Lung Association, air pollution in the United States costs at least $150 billion annually in health care and lost work productivity, with $100 billion of that caused by indoor air pollution.

According to a 1997 study by the World Health Organization and the World Bank, worldwide at least 2.7 million people (most of them in Asia) die prematurely each year from the effects of indoor (2.2 million deaths) and outdoor (0.5 million deaths) air pollution.

How Are Plants Damaged by Air Pollutants?

Some gaseous pollutants (especially ozone) damage leaves of crop plants and trees directly when they enter leaf pores. Spruce, fir, and other conifers, especially at high elevations, are most vulnerable to air pollution because of their long life spans and the year-round exposure of their needles to a mixture of air pollutants.

However, the effects may not become visible for several decades, when large numbers of trees suddenly begin dying off because of depletion of soil nutrients and increased susceptibility to pests, diseases, fungi, and drought. This phenomenon, known as *Waldsterben* (forest death), has turned whole forests of spruce, fir, and beech into stump-studded meadows and mountainsides. It is estimated that air pollution has been a key factor in reducing the overall productivity of European forests by about 16% and in causing damage valued at roughly $30 billion per year.

Forest diebacks in the United States have occurred, mostly on high-elevation slopes that face moving air masses and are dominated by red spruce. The most seriously affected areas are in the Appalachian Mountains. Ozone and other air pollutants are believed to be responsible for this forest degradation and loss of biodiversity.

Air pollution, mostly by ozone, also threatens some crops—especially corn, wheat, and soybeans, the three most important U.S. crops—and is reducing U.S. food production by 5–10%. In the United States, estimates of economic losses to agriculture as a result of air pollution are $1.9–5.4 billion per year.

How Can Air Pollutants Damage Aquatic Life?

High acidity (low pH) can severely harm the aquatic life in freshwater lakes, both where the surrounding soils have little acid-neutralizing capacity and in the northern hemisphere, where there is significant winter snowfall. Much of the damage to aquatic life in such areas is a result of *acid shock* caused by the sudden runoff of large amounts of highly acidic water and aluminum ions into lakes and streams, when snow melts in the spring or after unusually heavy rains. The aluminum ions leached from the soil and lake sediment by this sudden input of acid can kill fish and inhibit their reproduction.

As the acidity of a lake increases and its food chain is disrupted, there is a decline in net primary productivity. This can turn a moderately eutrophic lake (Figure 2-36, top) into a clear blue oligotrophic lake (Figure 2-36, bottom).

Because of excess acidity, at least 16,000 lakes in Norway and Sweden contain no fish, and 52,000 more lakes have lost most of their acid-neutralizing capacity. In Canada some 14,000 acidified lakes are almost fish graveyards, and 150,000 more are in peril.

In the United States, about 9,000 lakes are threatened with excess acidity, one-third of them seriously. Most of them are concentrated in the Northeast and the upper Midwest—especially Minnesota, Wisconsin, and the upper Great Lakes—where 80% of the lakes and streams are threatened by excess acidity. Over 10% of some 200 lakes in New York's Adirondack Mountains are too acidic to support fish.

What Are the Harmful Effects of Air Pollutants on Materials?
Each year, air pollutants cause billions of dollars in damage to various materials we use (Table 9-2). The fallout of soot and grit on buildings, cars, and clothing requires costly cleaning. Air pollutants break down exterior paint

Koontz, Michael. 1998. "Clean Air and Transportation: The Facts May Surprise You." *Public Roads*, vol. 62, no. 1, 42(5).

Table 9-2 Harmful Effects of Air Pollution on Materials

Material	Effects	Principal Air Pollutants
Stone and concrete	Surface erosion, discoloration, soiling	Sulfur dioxide, sulfuric acid, nitric acid, particulate matter
Metals	Corrosion, tarnishing, loss of strength	Sulfur dioxide, sulfuric acid, nitric acid, particulate matter, hydrogen sulfide
Ceramics and glass	Surface erosion	Hydrogen fluoride, particulate matter
Paints	Surface erosion, discoloration, soiling	Sulfur dioxide, hydrogen sulfide, ozone, particulate matter
Paper	Embrittlement, discoloration	Sulfur dioxide
Rubber	Cracking, loss of strength	Ozone
Leather	Surface deterioration, loss of strength	Sulfur dioxide
Textile	Deterioration, fading, soiling	Sulfur dioxide, nitrogen dioxide, ozone, particulate matter

on cars and houses and they deteriorate roofing materials. Irreplaceable marble statues, historic buildings, and stained glass windows around the world have been pitted, gouged, and discolored by air pollutants. Damage to buildings in the United States from acid deposition alone is estimated at $5 billion per year.

9-9 SOLUTIONS: PREVENTING AND REDUCING AIR POLLUTION

How Have Laws Been Used to Reduce Air Pollution in the United States? The U.S. Congress passed Clean Air Acts in 1970, 1977, and 1990, providing federal air pollution regulations that are enforced by each state. These laws required the EPA to establish *national ambient air quality standards (NAAQS)* for seven outdoor pollutants: suspended particulate matter, sulfur oxides, carbon monoxide, nitrogen oxides, ozone, volatile organic compounds, and lead. Each standard specifies the maximum allowable level, averaged over a specific period, for a certain pollutant in outdoor (ambient) air.

The Clean Air Act has worked. Between 1970 and 1995, levels of major air pollutants decreased nationally almost 30%, despite significant increases in population size and a doubling of economic growth. According to EPA data, between 1986 and 1995 lead levels in U.S. air decreased by 78% (98% since 1970), carbon monoxide 37%, sulfur dioxide 37%, suspended particulate matter 10 micrometers or less in diameter 17%, nitrogen dioxide 14%, and ground-level ozone 6%.

Between 1990 and 1995 ozone levels in U.S. urban areas fell by 50%. Result: 50 million people breathe cleaner air. Nitrogen dioxide levels have not dropped much since 1980 because of a combination of inadequate automobile emission standards and more vehicles traveling longer average distances.

Without the 1970 standards for emissions of pollutants, air pollution levels would be much higher today. Even so, in 1997 the EPA estimates that 80 million Americans still live in areas that exceed at least one air pollution standard.

A 1996 study by the EPA found that the benefits of the Clean Air Act greatly exceed costs. Between 1970 and 1990, the United States spent about $436 billion (in 1990 dollars) to comply with clean air regulations. Total human health and ecological benefits during the same 20-year period were estimated at $2.7–14.6 trillion (in 1990 dollars)—6–33 times higher than the costs.

How Could U.S. Air Pollution Laws Be Improved? The Clean Air Act of 1990 was an important step in the right direction, but most environmentalists point to the following major deficiencies in this law:

- *Continuing to rely almost entirely on pollution cleanup rather than pollution prevention.* In the United States, the air pollutant with the largest drop (98% between 1970 and 1995) in its atmospheric level was lead, which was virtually banned in gasoline. This shows the effectiveness of the pollution prevention approach.

- *Failing to sharply increase the fuel efficiency standards for cars and light trucks,* which environmentalists believe would reduce oil imports and air pollution more quickly and effectively than any other method and would save consumers enormous amounts of money.

- *Not requiring stricter emission standards for fine particulates.* In 1997 the EPA proposed stricter standards for fine particles that would prevent an estimated 20,000 premature deaths, 500,000 asthma attacks, and 9,000 hospital admissions per year and save $1 billion worth of crop losses. The EPA estimates the cost of implementing the standards at $7 billion per year, with the resulting health and other benefits estimated at $120 billion per year. Industries opposing these regulations put the cost of implementing the regulations at $200 billion per year.*

- *Giving municipal trash incinerators 30-year permits,* which locks the nation into hazardous air pollution emissions and toxic waste from incinerators well into the 21st century. This also undermines pollution prevention, recycling, and reuse (Chapter 11).

- *Doing too little to reduce emissions of carbon dioxide and other greenhouse gases.*

Should We Use the Marketplace to Reduce Pollution? To help reduce SO_2 emissions, the Clean Air Act of 1990 allows an *emissions trading policy*, which enables the 110 most polluting power plants in 21 states (primarily in the Midwest and East) to buy and sell SO_2 pollution rights.

Each year a power plant is given a certain number of pollution credits or rights that allow it to emit a certain amount of SO_2. A utility that emits less SO_2 than its limit receives more pollution credits. It can use these credits to avoid reductions in SO_2 emissions from some of its other facilities, bank them for future plant expansions, or sell them to other utilities, private citizens, or environmental groups.

Proponents of this system argue that it allows the marketplace to determine the cheapest, most efficient way to get the job done instead of having the government dictate how to control pollution. If this market-based approach works for reducing SO_2 emissions, it could be applied to other air and water pollutants.

Some environmentalists see this market approach as an improvement over the current regulatory approach, as long as it achieves net reduction in SO_2 pollution. This would be done by limiting the total number of credits and gradually lowering the annual num-

ber of credits (to encourage pollution prevention and the development of better pollution control technologies), something that is not required by the 1990 Clean Air Act. Without such reductions, critics contend that the system of tradable pollution rights is essentially an economic shell game, with no continuing progress in reducing overall SO_2 emissions.

Some environmentalists also contend that marketing pollution rights allows utilities with older, dirtier power plants to buy their way out and keep on emitting unacceptable levels of SO_2. They also warn that this approach creates incentives to cheat. Because air quality regulation is based largely on self-reporting of emissions and because pollution monitoring is incomplete and imprecise, sellers of permits will benefit by understating their reductions (to get more permits) and permit buyers will benefit by underreporting emissions (to reduce their permit purchases).

Between 1994 and 1997, the emissions trading system helped reduce SO_2 emissions in the United States by 30% at less than one-tenth the cost projected by industry because this market-based system has motivated companies to reduce emissions in efficient ways. In 1997, the EPA proposed a voluntary emissions trading program involving smog-forming nitrogen oxides (NO_x) for 22 eastern states and the District of Columbia. Emissions trading programs may also be implemented for particulates emissions.

How Can We Reduce Outdoor Air Pollution? Figure 9-21 summarizes ways to reduce emissions of sulfur oxides, nitrogen oxides, and particulate matter from stationary sources (such as electric power plants and industrial plants that burn coal). Until recently,

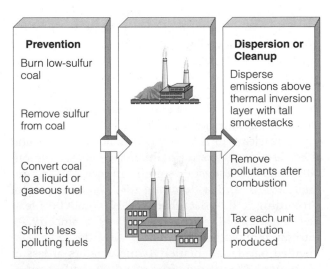

Figure 9-21 Solutions: methods for reducing emissions of sulfur oxides, nitrogen oxides, and particulate matter from stationary sources such as coal-burning electric power plants and industrial plants.

---

*In 1990 industry lobbyists predicted that new air pollution standards to reduce acid deposition would cost $1,500 for each one-ton reduction in sulfur dioxide emissions, compared to a $500 per ton estimate by the EPA. Because the regulations spurred companies to innovate and look at alternatives, the real cost turned out to be less than $100 per ton.

Q: What percentage of the food produced in the United States is wasted?

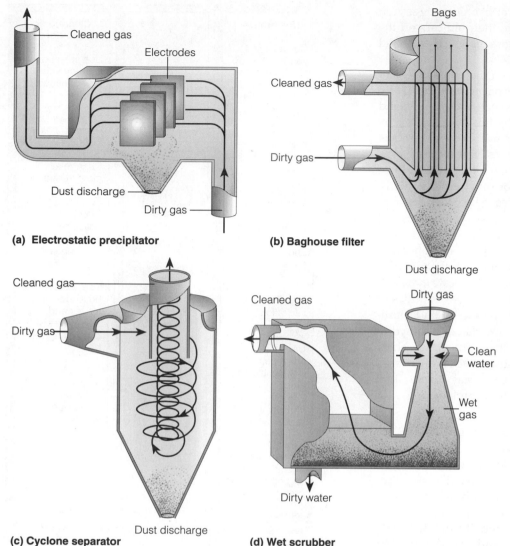

(a) **Electrostatic precipitator**

(b) **Baghouse filter**

(c) **Cyclone separator**

(d) **Wet scrubber**

Figure 9-22 Solutions: four commonly used methods for removing particulates from the exhaust gases of electric power and industrial plants. Of these, only baghouse filters remove many of the more hazardous fine particles. All these methods produce hazardous materials that must be disposed of safely, and except for cyclone separators, all of them are expensive. The wet scrubber can also reduce sulfur dioxide emissions.

emphasis has been primarily on dispersing and diluting the pollutants by using tall smokestacks or adding equipment that removes some of the particulate pollutants after they are produced (Figure 9-22). Under the sulfur-reduction requirements of the Clean Air Act of 1990, more utilities are switching to low-sulfur coal to reduce SO_2 emissions. Environmentalists call for taxes on air pollutant emissions and greater emphasis on prevention methods.

Figure 9-23 lists ways to reduce emissions from motor vehicles, the primary culprits in producing photochemical smog. Use of alternative vehicle fuels to reduce air pollution was evaluated in Table 4-1 (p. 118).

An important way to make significant reductions in air pollution is to get older, high-polluting vehicles off the road. According to EPA estimates, 10% of the vehicles on the road emit 50–60% of the pollutants. A problem is that many old cars are owned by people

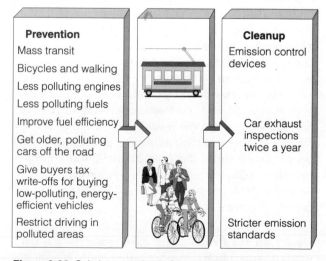

Figure 9-23 Solutions: methods for reducing emissions from motor vehicles.

who can't afford to buy a newer car. One suggestion would be to pay people to take their old cars off the road, which would result in huge savings in health and air pollution control costs.

California's South Coast Air Quality Management District Council developed a drastic program to produce an 80% reduction in ozone, photochemical smog, and other major air pollutants in the Los Angeles area by 2009. This plan would **(1)** sharply reduce use of gasoline-burning engines over two decades by converting cars, trucks, buses, chain saws, outboard motors, and lawn mowers to run on electricity or on alternative fuels; **(2)** outlaw drive-through facilities to keep vehicles from idling in lines; **(3)** substantially raise parking fees and assess high fees for families owning more than one car; **(4)** require gas stations to use a hydrocarbon-vapor recovery system on gas pumps and to sell alternative fuels (Table 4-1); **(5)** strictly control or relocate industrial plants and businesses that release large quantities of hydrocarbons and other pollutants; and **(6)** find substitutes for or ban consumer products that release hydrocarbons, including aerosol propellants, paints, household cleaners, and barbecue starter fluids. Such measures are a glimpse of what many other cities may have to do if people, cars, and industries continue proliferating.

How Can We Reduce Indoor Air Pollution? In the United States indoor air pollution poses a greater threat to health for many people than outdoor air pollution. Yet the EPA spends about $500 million per year fighting outdoor air pollution and only about $13 million a year on indoor air pollution.

To reduce indoor air pollution, it's not necessary to impose indoor air quality standards and monitor the more than 100 million homes and buildings in the United States. Instead, air pollution experts suggest that indoor air pollution reduction can be achieved by several means (Figure 9-24). Another possibility for cleaner indoor air in high-rise buildings is rooftop greenhouses through which building air can be circulated. Some actions you can take to reduce your exposure to indoor air pollutants are listed in Appendix 4.

In developing countries, indoor air pollution from open fires and leaky and inefficient wood- or charcoal-burning stoves (and the resulting high levels of respiratory illnesses) could be reduced if governments gave people simple stoves that burn biofuels more efficiently (which would also reduce deforestation) and that are vented outside, or provided them with simple solar cookers (Figure 4-16d).

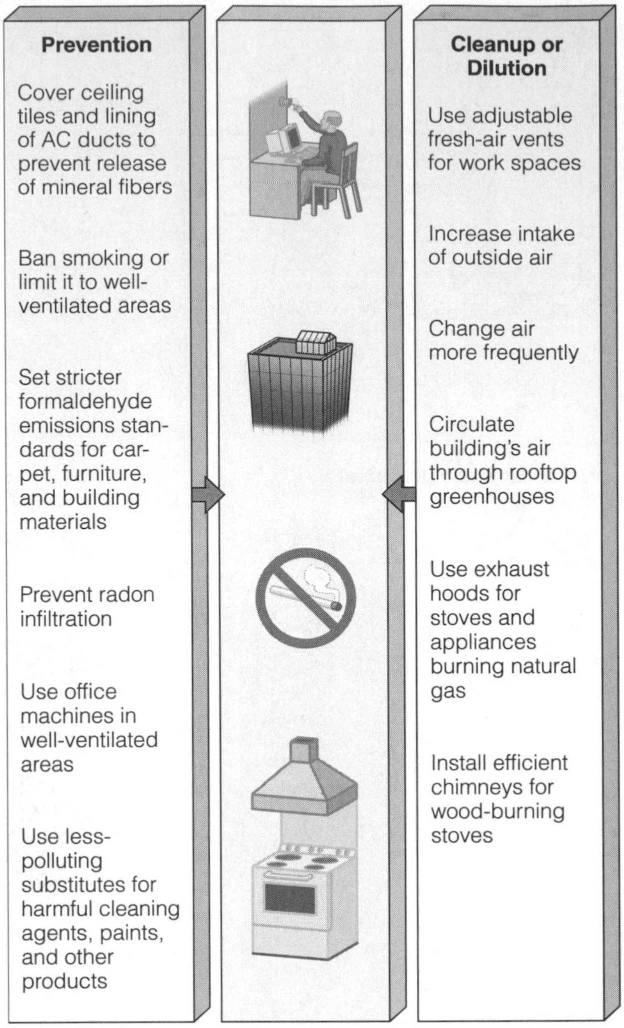

How Can We Protect the Atmosphere? An Integrated Approach As population and consumption rise, we can generate new air pollution faster than we

Figure 9-24 Solutions: ways to prevent and reduce indoor air pollution.

can clean up the old, even in developed countries with strict air-pollution control laws. As a result, environmentalists believe that protecting the atmosphere, and thus the health of people and many other organisms, will require a global approach that integrates many different strategies. Suggestions for doing this over the next 40–50 years include the following:

- *Putting more emphasis on pollution prevention*
- *Improving energy efficiency*
- *Reducing use of fossil fuels (especially coal and oil)*
- *Increasing use of renewable energy*
- *Slowing population growth*
- *Integrating air pollution, water pollution, energy, land-use, population, economic, and trade policies*

Q: What percentage of the USDA's annual research budget is spent on development of sustainable agriculture?

- *Phasing in full-cost pricing, mostly by taxing the production of air pollutants*

- *Distributing cheap and efficient cookstoves and solar cookstoves in developing countries*

- *Transferring the latest energy-efficiency, renewable-energy, pollution prevention, and pollution control technologies to developing countries*

Some people have wondered whether there is intelligent life in other parts of the universe; others wonder whether there is intelligent life on the earth. To them, if we can seriously deal with the interconnected global problems of loss of biodiversity (Chapters 5, 6, and 7), possible climate change, and depletion of stratospheric ozone from human activities, then the answer is a hopeful *yes*. According to most scientists, this means recognizing that *prevention* is the best (and in the long run, the least costly) way to deal with global environmental problems. Otherwise, they believe the answer is a tragic *no*.

The atmosphere is the key symbol of global interdependence. If we can't solve some of our problems in the face of threats to this global commons, then I can't be very optimistic about the future of the world.

MARGARET MEAD

CRITICAL THINKING

1. Do you believe that possible global warming from an enhanced greenhouse effect caused at least partly by human activities is a serious problem or one that has been greatly exaggerated? Explain.

2. What consumption patterns and other features of your lifestyle directly add greenhouse gases to the atmosphere? Which, if any, of those things would you be willing to give up to slow projected global warming and reduce other forms of air pollution?

3. Explain why you agree or disagree with each of the proposals listed in Figure 9-12 for slowing down emissions of greenhouse gases into the atmosphere, and those given in Figure 9-13 for preparing for the effects of global warming. What effects would carrying out these proposals have on your lifestyle and on that of your descendants? What might be the effects of *not* carrying out these actions?

4. Do you believe that ozone depletion in the stratosphere is a serious problem or one that has been greatly exaggerated? Explain.

5. What topographical and climate factors increase or help decrease air pollution in your community?

6. Should all tall smokestacks be banned? Explain.

7. Have dormitories and other buildings on your campus been tested for radon? If so, what were the results? What has been done about areas with unacceptable levels? If this testing has not been done, talk with school officials about having it done.

8. Should annual government-held auctions of marketable permits be used as a primary way of controlling and reducing air pollution? Explain. What conditions, if any, would you put on this approach?

9. Evaluate your exposure to some or all of the indoor air pollutants in Figure 9-19 in your home and in any indoor area where you work. Come up with a plan for reducing your exposure to these pollutants.

A: About 1%

10 WATER

Our liquid planet glows like a soft blue sapphire in the hard-edged darkness of space. There is nothing else like it in the solar system. It is because of water.

JOHN TODD

10-1   WATER'S IMPORTANCE AND UNIQUE PROPERTIES

Why Is Water So Important?   We live on the water planet, with a precious film of water—most of it salt water—covering about 71% of the earth's surface (Figure 2-33). The world's oceans help regulate the planet's climate, dilute and degrade some of our wastes, and are a major habitat for many of the planet's living creatures. The earth's organisms are made up mostly of water; a tree is about 60% water by weight, and most animals (including humans) are about 50–65% water.

Each of us needs only a dozen or so cupfuls of water per day to survive, but huge amounts of water are needed to supply us with food, shelter, and our other needs and wants. Water also plays a key role in sculpting the earth's surface, moderating climate, and diluting pollutants. In fact, without water the earth would have no oceans, no life as we know it, and no people.

What Are Some Important Properties of Water?   Water is a remarkable substance with a unique combination of properties:

- *There are strong forces of attraction (called hydrogen bonds) between molecules of water.* These attractive forces between water molecules are the major factor determining water's unique properties.

- *Water exists as a liquid over a wide temperature range because of the strong forces of attraction between water molecules.* Its high boiling point of 100°C (212°F) and low freezing point of 0°C (32°F) mean that water remains a liquid in most climates on the earth.

- *Liquid water changes temperature very slowly because it can store a large amount of heat without a large change in temperature.* This high heat capacity helps protect living organisms from the shock of abrupt temperature changes; it also moderates the earth's climate and makes water an excellent coolant for car engines,

power plants, and heat-producing industrial processes.

- *It takes a lot of heat to evaporate liquid water because of the strong forces of attraction between its molecules.* Water's ability to absorb large amounts of heat as it changes into water vapor—and to release this heat as the vapor condenses back to liquid water—is a primary factor in distributing heat throughout the world (Figure 9-4). This property makes evaporation of water an effective cooling process, which is why you feel cooler when perspiration or bathwater evaporates from your skin.

- *Liquid water can dissolve a variety of compounds.* This enables it to carry dissolved nutrients into the tissues of living organisms, to flush waste products out of those tissues, to serve as an all-purpose cleanser, and to help remove and dilute the water-soluble wastes of civilization. Water's superiority as a solvent also means that it is easily polluted by water-soluble wastes.

- *The strong attractive forces between the molecules of liquid water cause its surface to contract (high surface tension) and also to adhere to and coat a solid (high wetting ability).* Together these properties allow water to rise through a plant from the roots to the leaves (capillary action).

- *Unlike most liquids, water expands when it freezes and becomes ice.* This means that ice has a lower density (mass per unit of volume) than liquid water. Thus, ice floats on water, and as air temperatures fall below freezing, bodies of water freeze from the top down instead of from the bottom up. Without this property, lakes and streams in cold climates would freeze solid, and most of their current forms of aquatic life could not exist. Because water expands upon freezing, it can also break pipes, crack engine blocks (which is why we use antifreeze), and break up streets and fracture rocks (thus forming soil).

Water—the lifeblood of the ecosphere—is truly a wondrous substance; it connects us to one another, to other forms of life, and to the entire planet. Despite its importance, water is one of our most poorly managed resources. We waste it and pollute it, and we also charge too little for making it available, thus encouraging still greater waste and pollution of this potentially renewable resource, for which there is no substitute.

Q: What percentage of the world's potential food supply is lost to pests?

10-2 SUPPLY, RENEWAL, AND USE OF WATER RESOURCES

How Much Fresh Water Is Available? Only a tiny fraction of the planet's abundant water is available to us as fresh water (Figure 10-1). About 97% by volume is found in the oceans and is too salty for drinking, irrigation, or industry (except as a coolant).

The remaining 3% is fresh water. About 2.997% of all water is locked up in ice caps or glaciers or is buried so deep that it costs too much to extract. Only about 0.003% of the earth's total volume of water is easily available to us as soil moisture, usable groundwater, water vapor, and lakes and streams. If the world's water supply were only 100 liters (26 gallons), our usable supply of fresh water would be only about 0.003 liter (one-half teaspoon).

Fortunately, the available fresh water amounts to a generous supply that is continuously collected, purified, recycled, and distributed in the solar-powered *hydrologic cycle* (Figure 2-26) as long as we don't overload it with slowly degradable and nondegradable wastes or withdraw it from underground supplies faster than it is replenished. Unfortunately, we are doing both.

Differences in average annual precipitation divide the world's countries and people into water haves and have-nots. For example, Canada, with only 0.5% of the world's population, has 20% of the world's fresh water supply. By contrast, China, with 21% of the world's people, has only 7% of the world's fresh water supply.

As population, irrigation, and industrialization increase, water shortages in already water-short regions will intensify and wars over water may erupt. Projected global warming also might cause changes in rainfall patterns and disrupt water supplies in unpredictable ways.

What Is Surface Water? The fresh water we use first arrives as the result of precipitation (Figure 10-2). Precipitation that does not infiltrate the ground or return to the atmosphere by evaporation (including transpiration) is called **surface runoff** that flows into streams, lakes, wetlands, and reservoirs.

A **watershed**, also called a **drainage basin**, is a region from which water drains into a stream, stream system, lake, reservoir, or other water body.

What Is Groundwater? Some precipitation infiltrates the ground and percolates downward through voids (pores, fractures, crevices, and other spaces) in soil and rock (Figure 10-3). The water in these voids is called **groundwater**.

Close to the surface, the voids have little moisture in them. However, below some depth, in what is called

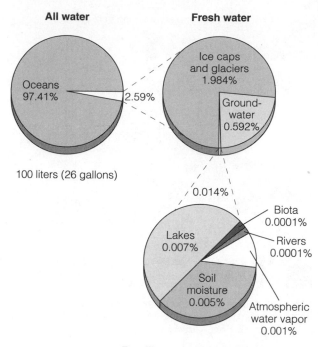

Figure 10-1 The planet's water budget. Only a tiny fraction by volume of the world's water supply is fresh water that is available for human use.

the **zone of saturation**, they are filled with water except for an occasional air bubble. The upper surface of the zone of saturation, at the boundary with the unsaturated zone, is the **water table**. The water table falls in dry weather and rises in wet weather.

Porous, water-saturated layers of sand, gravel, or bedrock through which groundwater flows are called **aquifers** (Figure 10-3). Any area of land through which water passes downward or laterally into an aquifer is called a **recharge area**. Aquifers are replenished naturally by precipitation that percolates downward through soil and rock in what is called **natural recharge**, but some are recharged from the side by *lateral recharge*.

Groundwater moves from the recharge area through an aquifer and out to a discharge area (well, spring, lake, geyser, stream, or ocean) as part of the hydrologic cycle. Groundwater normally moves from points of high elevation and pressure to points of lower elevation and pressure. This movement is quite slow, typically only a meter or so (about 3 feet) per year and rarely more than 0.3 meter (1 foot) per day.

Some aquifers get very little (if any) recharge. Often found fairly deep underground and formed tens of thousands of years ago, they are (on a human time scale) nonrenewable resources. Withdrawals from such aquifers amount to *water mining* that, if kept up, will deplete these ancient deposits of liquid earth capital.

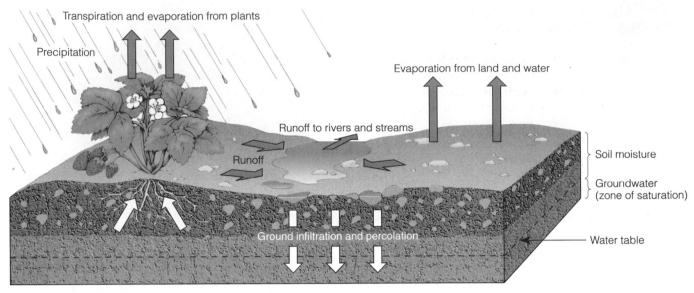

Figure 10-2 Main routes and destinations of local precipitation: surface runoff into surface waters, ground infiltration, and evaporation and transpiration into the atmosphere.

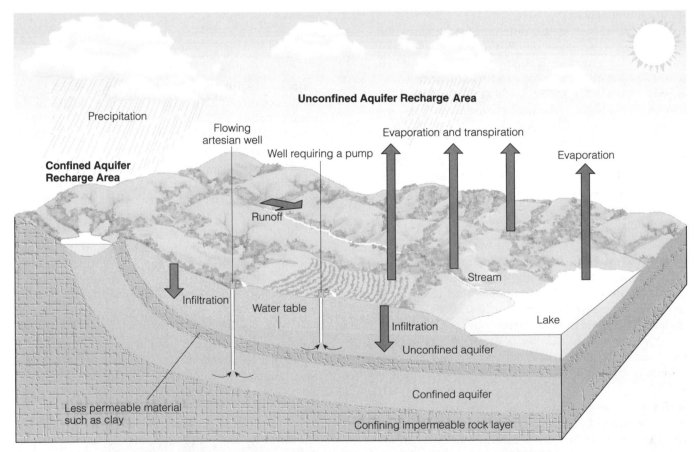

Figure 10-3 The groundwater system. An *unconfined aquifer* is an aquifer with a water table. A *confined aquifer* is bounded above and below by less permeable beds of rock. Groundwater in this type of aquifer is confined under pressure.

How Do We Use the World's Fresh Water? Since 1950, the global rate of water withdrawal from surface and groundwater sources has increased almost fivefold and per capita use has tripled. According to a 1996 study, humans currently use about 54% of the global surface runoff that is realistically available from the hydrologic cycle. Because of increased population growth and economic development, global withdrawal rates of surface water are projected to at least double in the next two decades and exceed the available surface runoff in a growing number of areas.

Uses of withdrawn water vary from one region to another and from one country to another (Figure 10-4). Averaged globally, about 65% of all water withdrawn each year from rivers, lakes, and aquifers is used to irrigate 16% of the world's cropland. Some 60–80% of this water either evaporates or seeps into the ground before reaching crops. In the western United States, irrigation accounts for about 85% of all water use.

Worldwide, about 25% of the water withdrawn is used for energy production (oil and gas production and power-plant cooling) and industrial processing, cleaning, and waste removal. Agricultural and manufactured products require large amounts of water, much of which could be used more efficiently and reused. It takes about 380,000 liters (100,000 gallons) to make an automobile, 3,800 liters (1,000 gallons) to produce 454 grams (1 pound) of aluminum, 3,000 liters (800 gallons) to produce 454 grams (1 pound) of grain-fed beef in a feedlot (where large numbers of cattle are confined to a fairly small area), and 100 liters (26 gallons) to produce 1 kilogram (2.2 pounds) of paper.

Domestic and municipal use accounts for about 10% of worldwide water withdrawals and about 13–16% of withdrawals in developed countries. As population, urbanization, and industrialization grow, the volume of wastewater needing treatment will increase enormously.

10-3 WATER RESOURCE PROBLEMS: TOO LITTLE WATER AND TOO MUCH WATER

What Causes Freshwater Shortages? According to water expert Malin Falkenmark, there are four causes of water scarcity: **(1)** a *dry climate* (Figure 9-2), **(2)** *drought* (a period in which precipitation is much lower and evaporation is higher than normal), **(3)** *desiccation* (drying of the soil because of such activities as deforestation and overgrazing by livestock), and **(4)** *water stress* (low per capita availability of water because of increasing numbers of people relying on relatively fixed levels of runoff).

Since the 1970s water scarcity intensified by prolonged drought (mostly in areas with dry climates, where 40% of the world's people live) has killed more than 24,000 people per year and created many environmental refugees. In water-short areas, many women and children must walk long distances each day, carrying heavy jars or cans, to get a meager supply of sometimes contaminated water. Millions of poor people in developing countries have no choice but to try to survive on drought-prone land. If global warming occurs as projected, severe droughts may become more common in some areas of the world.

According to a 1995 World Bank study, 30 countries containing 42% of the world's population (2.4 billion people) now experience chronic water shortages that threaten their agriculture and industry and the health of their people. By 2025 at least 3 billion people in 90 countries are expected to face severe water stress. In most of these countries the problem is not a shortage of water but the wasteful and unsustainable use of normally available supplies.

In the United States, many major urban centers (especially those in the West and Midwest) are located in areas that don't have enough water or are projected to have water shortages by 2000 (Figure 10-5). Experts project that current shortages and conflicts over water supplies will get much worse as more industries and people migrate west and compete with farmers for scarce water. These shortages could worsen even more if climate warms as a result of an enhanced greenhouse effect.

Some areas have lots of water, but the largest rivers (which carry most of the runoff) are far from agricultural and population centers where the water is needed. For example, South America has the largest annual water runoff of any continent, but 60% of the runoff flows through the Amazon River in remote areas where few people live.

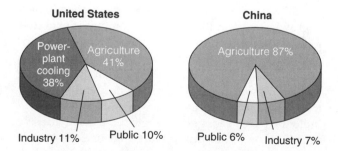

Figure 10-4 Use of water in the United States and China. The United States has the world's highest per capita use of water, amounting to an average of 6,000 liters (1,600 gallons) per person every day. About half of the water used in the United States is unnecessarily wasted. (Data from Worldwatch Institute and World Resources Institute)

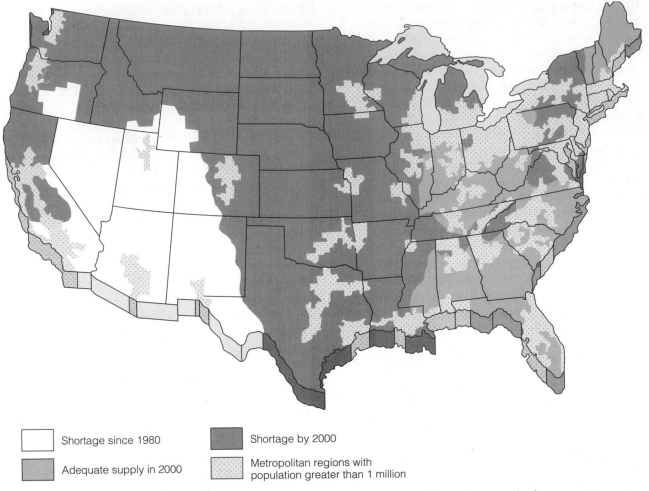

Shortage since 1980

Shortage by 2000

Adequate supply in 2000

Metropolitan regions with
population greater than 1 million

Figure 10-5 Water deficit regions in the continental United States and their proximity to metropolitan areas having populations greater than 1 million. (Data from U.S. Water Resources Council and U.S. Geological Survey)

Case Study: Water Conflicts in the Middle East
A number of analysts believe that *access to water resources, a key foreign policy and environmental security issue for water-short countries in the 1990s, will become even more important early in the next century.* Almost 150 of the world's 214 major river systems (57 of them in Africa) are shared by 2 countries, and another 50 are shared by 3 to 10 nations. Some 40% of the world's population already clashes over water, especially in the Middle East.

The next war in the Middle East may well be fought over water, not oil. Most water in this dry region comes from three shared river basins: the Jordan, the Tigris–Euphrates, and the Nile (Figure 10-6). Water is already in short supply in much of this arid region, where the human population is projected to double within only 25 years.

Ethiopia, which controls the headwaters of 80% of the Nile's flow, has plans to divert more of this water; so does Sudan. This could greatly reduce the amount of water available to desperately water-short Egypt, whose terrain is desert except for a thin strip of irrigated cropland running down its middle along the Nile and its delta.

Between 1996 and 2025, Egypt's population (which is growing by 1 million every 9 months) is expected to increase from 66 million to 96 million, greatly increasing the demand for already scarce water. Egypt's options are to **(1)** go to war with Sudan and Ethiopia to obtain more water, **(2)** slash population growth, **(3)** improve irrigation efficiency, **(4)** spend $2 billion to build the world's longest concrete canal and a massive pumping station to pump water out of Lake Nasser (the reservoir created from the Nile by the Aswan High Dam) and create a lush new valley of irrigated farmland in the middle of the desert, **(5)** import more grain to reduce the need for irrigation water, **(6)** work out water-sharing agreements with other countries, or **(7)** suffer the harsh human and economic consequences.

Q: What is the most serious drawback to using chemicals to control pests (especially insects)?

There is also fierce competition for water among Jordan, Syria, and Israel, which get most of their water from the Jordan River basin (Figure 10-6). Israel irrigates two-thirds of its croplands and uses water more efficiently than any other country in the world. However, within the next few years its supply is projected to fall by up to 30% short of demand because of increased immigration.

Jordan, which gets about 75% of its water from the Jordan River system, must double its supply over the next 20 years just to keep up with projected population growth. In 1990, King Hussein declared that water was the only issue that could cause him to go to war against Israel.

Some 90 million people currently live in the water-short basins of the Tigris and Euphrates rivers (Figure 10-6), and by the year 2020 the population there is projected to almost double to 170 million. Turkey, located at the headwaters of these two rivers, has abundant water, and it plans to build 22 dams along the upper Tigris and Euphrates to generate huge quantities of electricity and irrigate a large area of land. These dams will reduce the flow of water to downstream Syria and Iraq.

Clearly, distribution of water resources will be a key issue in any future peace talks in this region. Resolving these problems will require a combination of regional cooperation in allocating water supplies, slowed population growth, improved efficiency in the use of water resources, and eliminating water subsidies and raising the price of water to encourage conservation and improve irrigation efficiency.

What Are the Causes and Effects of Flooding?
Some countries have enough annual precipitation but get most of it at one time of the year. In India, for example, 90% of the annual precipitation falls in the monsoon season between June and September. This prolonged downpour causes floods, waterlogs soils, leaches soil nutrients, and washes away topsoil and crops.

Natural flooding by streams, the most common type of flooding, is caused primarily by heavy rain or rapid melting of snow; this causes water in the stream to overflow its normal channel and to cover the adjacent area, called a **floodplain** (Figure 10-7, left). Floodplains, which include highly productive wetlands, provide natural flood and erosion control, help maintain high water quality, and contribute to the recharging of groundwater.

People have settled on floodplains since the beginnings of agriculture. The soil is fertile and ample water is available for irrigation. Communities can use the water for transportation of people or goods, and floodplains (being flat) are suitable for cropland, buildings, highways, and railroads. People may decide that all of these benefits outweigh the risk of flooding (if they are even aware of the risk).

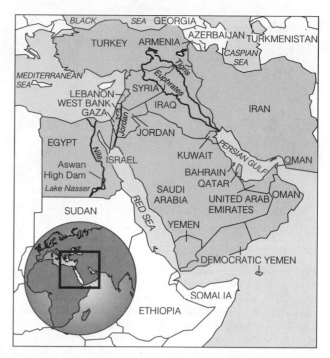

Figure 10-6 The Middle East, whose countries have some of the highest population growth rates in the world. Because of the dry climate, food production depends heavily on irrigation. Existing conflicts among countries in this region over access to water may soon overshadow both long-standing religious and ethnic clashes and attempts to take over valuable oil supplies.

Prolonged rains anywhere can cause streams and lakes to overflow and flood the surrounding floodplain land, but low-lying river basins such as the Ganges River basin in India and Bangladesh are especially vulnerable. Hurricanes and typhoons can also flood low-lying coastal areas.

Floods are a natural phenomenon and have several benefits. They provide the world's most productive farmland because they are regularly covered with nutrient-rich silt left after flood waters recede. Floods also recharge groundwater under plains and refill wetlands, which help keep rivers flowing during droughts and provide important breeding and feeding grounds for fish and waterfowl.

Each year, flooding (Figure 10-7, right) kills thousands of people and causes tens of billions of dollars in property damage. Floods, like droughts, are usually considered natural disasters, but since the 1960s human activities have contributed to the sharp increase in flood deaths and damages. Indeed, floods account for about 39% of all deaths from natural disaster—more than any other type.

One way humans increase the likelihood of flooding and the resulting damage is by removing water-absorbing vegetation, especially on hillsides (Figure 10-8); another is by living on floodplains. In many developing

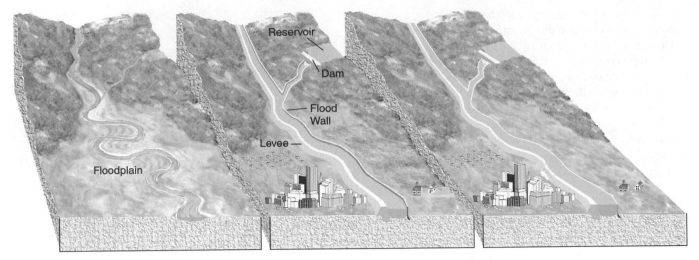

Figure 10-7 Land in a natural floodplain (left) is often flooded after prolonged rains. When the floodwaters recede, alluvial deposits of silt are left behind, creating a nutrient-rich soil. To reduce the threat of flooding (and thus allow people to live in floodplains), rivers have been dammed to create reservoirs, to store and release water as needed. They have also been narrowed, straightened, and equipped with protective levees and walls (middle). However, these alterations can give a false sense of security to floodplain dwellers, who actually live in high-risk areas. In the long run, such measures can greatly increase flood damage. Although dams, levees, and walls prevent flooding in most years, they can be overwhelmed by prolonged rains (right), as happened in the Midwestern United States during the summer of 1993.

countries, the poor have little choice but to try to survive in flood-prone areas. In developed countries, however, people deliberately settle on floodplains and then expect dams, levees, and other devices to protect them from flood waters—solutions that don't work when heavier-than-normal rains come.

Urbanization increases flooding (even with only moderate rainfall) by replacing vegetation and soil with highways, parking lots, and buildings, which leads to rapid runoff of rainwater. If sea levels rise during the next century, as projected, many low-lying coastal cities, wetlands, and croplands will be under water.

Ways humans can reduce the risk include **(1)** straightening and deepening streams (channelization), **(2)** building levees and dams (Figure 10-7, middle), **(3)** restoring wetlands to take advantage of the natural flood control provided by floodplains, and **(4)** instituting floodplain management to get people out of flood-prone areas. This last approach, viewed by many water resource experts as the best approach, is based on thousands of years of experience that can be summed up in one idea: *Sooner or later the river (or the ocean) always wins.*

Case Study: Living Dangerously in Bangladesh
Bangladesh is one of the world's most densely populated countries, with 123 million people packed into an area roughly the size of Wisconsin. Its population is projected to reach 166 million by 2025. Bangladesh is also one of the world's poorest countries, with an average per capita GNP of about $260, or 71¢ per day.

More than 80% of the country consists of floodplains and shifting islands of silt formed by a delta at the mouths of three major rivers. Runoff from annual monsoon rains in the Himalaya Mountains of India, Nepal, Bhutan, and China flows down the rivers through Bangladesh into the Bay of Bengal.

The people of Bangladesh are used to moderate annual flooding during the summer monsoon season, and they depend on the floodwaters to grow rice, their primary source of food. The annual deposit of eroded Himalayan soil in the delta basin also helps maintain soil fertility.

In the past, great floods occurred every 50 years or so, but during the 1970s and 1980s they came about every 4 years. Bangladesh's flood problems begin in the Himalayan watershed, where a combination of rapid population growth, deforestation, overgrazing, and unsustainable farming on steep, easily erodible mountain slopes has greatly diminished the ability of the soil to absorb water. Instead of being absorbed and released slowly, water from the monsoon rains runs off the denuded Himalayan foothills, carrying vital topsoil with it (Figure 10-8). This runoff, combined with heavier-than-normal monsoon rains, has increased the severity of

McManus, Reed. 1998. "Down Come the Dams." *Sierra*, vol. 83, no. 3, 16(2).

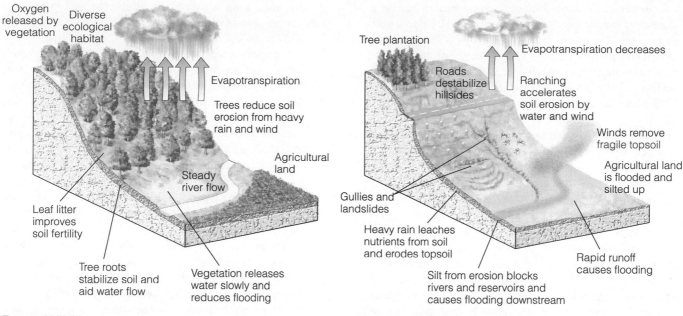

Oxygen released by vegetation
Diverse ecological habitat
Evapotranspiration
Trees reduce soil erosion from heavy rain and wind
Agricultural land
Steady river flow
Leaf litter improves soil fertility
Tree roots stabilize soil and aid water flow
Vegetation releases water slowly and reduces flooding

Forested hillside

Tree plantation
Roads destabilize hillsides
Evapotranspiration decreases
Ranching accelerates soil erosion by water and wind
Winds remove fragile topsoil
Agricultural land is flooded and silted up
Gullies and landslides
Heavy rain leaches nutrients from soil and erodes topsoil
Silt from erosion blocks rivers and reservoirs and causes flooding downstream
Rapid runoff causes flooding

After deforestation

Figure 10-8 A hillside before and after deforestation. Once a hillside has been deforested—for timber and fuelwood, grazing livestock, or unsustainable farming—water from precipitation rushes down the denuded slopes, eroding precious topsoil and flooding downstream areas. A 3,000-year-old Chinese proverb says, "To protect your rivers, protect your mountains."

flooding along Himalayan rivers and in Bangladesh—another example of connections.

In 1988 a disastrous flood covered two-thirds of Bangladesh's land area for 3 weeks and leveled 2 million homes after the heaviest monsoon rains in 70 years. At least 2,000 people drowned, and 30 million people—one of every four—were left homeless. More than a quarter of the country's crops were destroyed, costing at least $1.5 billion and causing thousands of people to die of starvation.

In their struggle to survive, the poor in Bangladesh have cleared many of the country's coastal mangrove forests for fuelwood and for cultivation of crops. This has led to more severe flooding because these coastal wetlands shelter the low-lying coastal areas from storm surges and cyclones.

Sixteen devastating cyclones have slammed into Bangladesh since 1961; in 1970 as many as 1 million people drowned in one storm. Another surge killed an estimated 130,000 people in 1991. Damages and deaths from cyclones in areas still protected by mangrove forests are much lower than in areas where the forests have been cleared.

The severity of this problem can be reduced if Bangladesh, Bhutan, China, India, and Nepal cooperate in reforestation and flood control measures and reduce their population growth.

10-4 SOLUTIONS: SUPPLYING MORE WATER

How Can Water Supplies Be Increased? There are five ways to increase supply of fresh water in a particular area: (1) build dams and reservoirs to store runoff, (2) bring in surface water from another area, (3) withdraw groundwater, (4) convert salt water to fresh water (desalination), and (5) improve the efficiency of water use.

In developed countries, people tend to live where the climate is favorable and then bring in water from another watershed. In developing countries, most people (especially the rural poor) must settle where the water is and try to capture and use as much precipitation as they can.

What Are the Pros and Cons of Large Dams and Reservoirs? Large dams and reservoirs have benefits and drawbacks (Figure 4-18). Large reservoirs created by damming streams can capture and store water from rain and melting snow. This water can then be released as desired to produce hydroelectric power at the dam site, irrigate land below the dam, control flooding of land below the reservoir, and provide water carried to towns and cities by aqueducts. Reservoirs also provide recreational activities such as swimming, fishing, and boating.

Hint: Enter the search terms *dams, economic aspects* using the Subject Guide.

Between 25% and 50% of the total runoff on every continent is now captured and controlled by dams and reservoirs, and many more large projects are planned. When completed, China's Three Gorges project on the mountainous upper reaches of the Yangtze River will be the world's largest hydroelectric dam and reservoir. This superdam will supply power to industries and to 150 million Chinese. The project will also help China reduce its dependence on coal and help hold back the Yangtze River's floodwaters (which have claimed at least 500,000 lives in this century).

However, the project will also flood large areas of some of the best farmland in the region and 800 existing factories; it will also displace 1.9 million people, including 100 towns and 2 cities (each with about 100,000 people), and flood some of China's most scenic land. Critics charge that the dam will transform the Yangtze into a sewer for industrial wastes and the immense pressure from the huge volume of water in the reservoir could trigger landslides and earthquakes. Furthermore, as the reservoir fills up with sediment, the dam could overflow and expose half a million people to severe flooding (especially if the reservoir is kept filled at a high level, as planned, to provide maximum hydroelectric power).

Many analysts see a series of small dams as a better, cheaper, and less destructive alternative to large dams. Chinese officials claim that both big and small dams are needed.

What Are the Pros and Cons of Watershed Transfers? The California Experience Tunnels, aqueducts, and underground pipes can transfer stream runoff collected by dams and reservoirs from water-rich watersheds to water-poor areas. Although such transfers have benefits, they also create environmental problems. Indeed, most of the world's large-scale water transfers illustrate the important ecological principle that *you can't do just one thing.*

One of the world's largest watershed transfer projects is the California Water Project. In California, the basic water problem is that 75% of the population lives south of Sacramento but 75% of the rain falls north of it. The California Water Project uses a maze of giant dams, pumps, and aqueducts to transport water from water-rich northern California to heavily populated areas and to arid and semiarid agricultural regions, mostly in southern California (Figure 10-9).

For decades, northern and southern Californians have been feuding over how the state's water should be allocated under this project. Southern Californians say they need more water from the north to support Los Angeles, San Diego, and other growing urban areas and to grow more crops. Although agriculture uses 82% of the water withdrawn in California, it accounts for only 2.5% of the state's economic output.

Irrigation for just two crops, alfalfa (to feed cattle) and cotton, uses as much water as the residential needs of all 32 million Californians.

Opponents in the north say that sending more water south would degrade the Sacramento River, threaten fisheries, and reduce the flushing action that helps clean San Francisco Bay of pollutants. They also argue that much of the water already sent south is unnecessarily wasted and that making irrigation just 10% more efficient would provide enough water for domestic and industrial uses in southern California.

If water supplies in California were to drop sharply because of projected global warming, the amount of water delivered by the huge distribution system would plummet. Most irrigated agriculture in California would have to be abandoned and much of the population of southern California might have to move to areas with more water. The 6-year drought that California experienced between 1986 and 1992 was perhaps just a small taste of the future.

Pumping out more groundwater is not the answer. Throughout much of California, groundwater is already being withdrawn faster than it is replenished. Most analysts see improving irrigation efficiency and allowing farmers to sell their legal rights to withdraw certain amounts of water from the river as much quicker and cheaper solutions.

Case Study: The James Bay Watershed Transfer Project Another major watershed transfer project is the James Bay project, a $60-billion, 50-year scheme to harness the wild rivers that flow into Quebec's James and Hudson Bays to produce electric power for Cana-

Figure 10-9 The California Water Project and the Central Arizona Project involve large-scale transfers of water from one watershed to another. Arrows show the general direction of water flow.

Q: How many premature cancer deaths in the United States are caused by exposure to pesticide residues in foods?

Figure 10-10 If completed, the James Bay project in northern Quebec will alter or reverse the flow of 19 major rivers and flood an area the size of Washington State to produce hydropower for consumers in Quebec and the United States, especially in New York State. Phase I of this 50-year project is completed; phase II was postponed indefinitely in 1994 because of a surplus of electricity and opposition by environmentalists and the indigenous Cree, whose ancestral hunting grounds would have been flooded.

Figure 10-11 Once the world's fourth-largest fresh water lake, the Aral Sea has been shrinking and getting saltier since 1960 because most of the water from the rivers that replenish it has been diverted to grow cotton and food crops. As the lake shrinks, it leaves behind a salty desert, economic ruin, increasing health problems, and severe ecological disruption.

dian and U.S. consumers (Figure 10-10). If completed, this megaproject would **(1)** construct 600 dams and dikes that will reverse or alter the flow of 19 giant rivers covering a watershed three times the size of New York State, **(2)** flood an area of boreal forest and tundra equal in area to Washington State or Germany, and **(3)** displace thousands of indigenous Cree and Inuit, who for 5,000 years have lived off James Bay by subsistence hunting, fishing, and trapping.

After 20 years and $16 billion, phase I has been completed. The second and much larger phase was postponed indefinitely in 1994 because of an excess of power generated, opposition by the Cree (whose ancestral hunting grounds would have been flooded) and Canadian and U.S. environmentalists, and New York State's cancellation of two contacts to buy electricity produced by phase II.

Case Study: The Aral Sea Watershed Transfer Disaster The shrinking of the Aral Sea (Figure 10-11) is a result of a large-scale water transfer project in an area of the former Soviet Union with the driest climate in Central Asia. Since 1960, enormous amounts of irrigation water have been diverted from the inland Aral Sea

and its two feeder rivers to irrigate cotton, vegetable, fruit, and rice crops to supply much of the region's needs. The irrigation canal, the world's longest, stretches over 1,300 kilometers (800 miles).

This water diversion (coupled with droughts) has caused a regional ecological, economic, and health disaster, described by one former Soviet official as "ten times worse than the 1986 Chernobyl nuclear power-plant accident." The sea's salinity has tripled, its surface area has shrunk by 54%, and its volume has decreased by almost 75%; the two supply rivers are now mere trickles. About 30,000 square kilometers (11,600 square miles) of former lake bottom has become a human-made salt desert. The process continues, and within a few decades the once enormous Aral Sea may be reduced to a few small brine lakes.

Of the 24 native fish species, 20 have become extinct, devastating the area's fishing industry, which once provided work for more than 60,000 people. Roughly half of the area's bird and mammal species have also disappeared.

Winds pick up the salty dust that encrusts the lake's now-exposed bed and blow it onto fields as far away as 300 kilometers (190 miles) away. As the salt spreads, it kills crops, trees, and wildlife and destroys pastureland. This phenomenon has added a new term to our list of environmental ills: *salt rain*.

These changes have also affected the area's already semiarid climate. The once-huge Aral Sea acted as a thermal buffer, moderating the heat of summer and the extreme cold of winter. Now there is less rain, summers are hotter, winters are colder, and the growing season is shorter. Cotton and crop yields have dropped dramatically.

To raise yields, farmers have increased inputs of herbicides, insecticides, and fertilizers on some crops. Many of these chemicals have percolated downward and accumulated to dangerous levels in the groundwater, from which most of the region's drinking water comes. The low river flows have also concentrated salts, pesticides, and other toxic chemicals, making surface water supplies hazardous to drink.

During the mid-1980s, the area near the shrunken Aral Sea had the highest levels of infant mortality and maternal mortality in the former Soviet Union. Between 1980 and 1993, kidney and liver diseases (especially cancers) increased 30- to 40-fold, arthritic diseases 60-fold, chronic bronchitis 30-fold, typhoid fever 30-fold, and hepatitis 7-fold.

Ways to deal with the ecological, economic, and health problems caused by this disaster include (1) charging farmers more for irrigation water to reduce waste and encourage a shift to less water-intensive crops, (2) decreasing irrigation water quotas, (3) introducing water-saving technologies, at a cost of at least $50 million, (4) developing a regional integrated water management plan, (5) planting protective forest belts, (6) using underground water to supplement irrigation water and to lower the water table to reduce waterlogging and salinization, (7) improving health services, and (8) slowing the area's rapid population growth.

Even with help from foreign countries, the United Nations, and agencies such as the World Bank, the money needed to save the Aral Sea and restore the ecological and economic services may not be available to this extremely poor area. What has happened to the Aral Sea basin is a stark reminder that everything in nature is connected and that preventing an ecological problem is much cheaper than trying to deal with its harmful consequences.

What Are the Pros and Cons of Withdrawing Groundwater? In the United States, about half of the drinking water (96% in rural areas and 20% in urban areas) and 40% of irrigation water is pumped from aquifers.

Overuse of groundwater can cause or intensify several problems: *aquifer depletion, aquifer subsidence* (sinking of land when groundwater is withdrawn), and *intrusion of salt water into aquifers* (Figure 10-12).

Groundwater can also become contaminated from industrial and agricultural activities, septic tanks, and other sources. Because groundwater is the source of about 40% of the stream flow in the United States, groundwater depletion also robs streams of water.

Currently, *about one-fourth of the groundwater withdrawn in the United States is not replenished.* The most serious overdrafts are occurring in parts of the huge Ogallala Aquifer, extending from southern South Dakota to central Texas (Case Study, p. 292) and in parts of the arid southwestern United States (Figure 10-12), especially California's Central Valley, which is the country's vegetable basket. Aquifer depletion is also a problem in northern China, northern Africa, southern Europe, the Middle East, and parts of Mexico, Thailand, and India.

When fresh water from an aquifer near a coast is withdrawn faster than it is recharged, salt water intrudes into the aquifer (Figure 10-13).Such intrusion threatens to irreversibly contaminate the drinking water of many towns and cities along the Atlantic and Gulf coasts (Figure 10-12) and in the coastal areas of Israel, Syria, and the Arabian Gulf states.

Ways to slow groundwater depletion include controlling population growth, not planting water-thirsty crops in dry areas, developing crop strains that require less water and are more resistant to heat stress, and wasting less irrigation water.

How Useful Is Desalination? The removal of dissolved salts from ocean water or from brackish (slightly salty) groundwater—called **desalination**—is another way to increase fresh water supplies. Distillation and reverse osmosis are the two most widely used methods. *Distillation* involves heating salt water until it evaporates and condenses as fresh water, leaving salts behind in solid form. In *reverse osmosis*, salt water is pumped at high pressure through a thin membrane whose pores allow water molecules, but not dissolved salts, to pass through.

About 7,500 desalination plants in 120 countries provide about 0.1% of the fresh water used by humans. However, desalination has a downside. Because it uses vast amounts of electricity, water produced in this way costs three to five times more than water from conventional sources. Distributing the water from coastal desalination plants costs even more in terms of the energy needed to pump the water uphill and inland.

Desalination also produces large quantities of brine containing high levels of salt and other minerals. Dumping the concentrated brine into the ocean near the plants might seem to be the logical solution, but this would increase the local salt concentration and threaten food resources in estuary waters. If these wastes were

Q: What do most scientists believe is the best way to control pests?

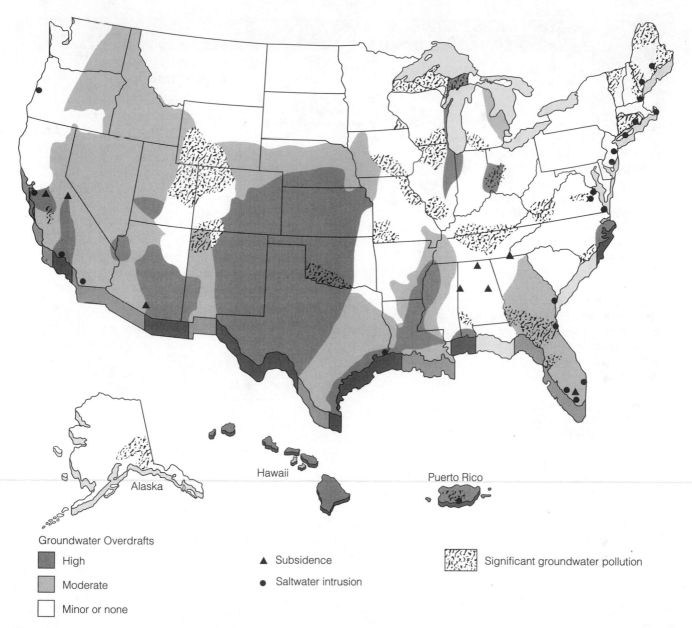

Groundwater Overdrafts

■ High

■ Moderate

□ Minor or none

▲ Subsidence

● Saltwater intrusion

▦ Significant groundwater pollution

Figure 10-12 Areas of greatest aquifer depletion, subsidence, saltwater intrusion, and groundwater contamination in the United States. (Data from U.S. Water Resources Council and U.S. Geological Survey)

dumped on the land, they could contaminate groundwater and surface water.

Desalination can provide fresh water for coastal cities in arid countries (such as sparsely populated Saudi Arabia), where the cost of getting fresh water by any method is high. However, desalinated water will probably never be cheap enough to irrigate conventional crops or to meet much of the world's demand for fresh water unless affordable, efficient solar-powered distillation plants can be developed and someone can figure out what to do with the resulting mountains of salt. Instead of spending less money to import more wheat, Saudi Arabia uses desalinated water to cultivate wheat in the desert at an estimated cost seven times the world price per bushel.

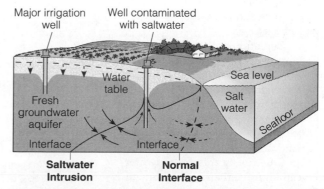

Figure 10-13 Saltwater intrusion along a coastal region. When the water table is lowered, the normal interface (dotted line) between fresh and saline groundwater moves inland (solid line).

Mining Groundwater: The Shrinking Ogallala Aquifer

Vast amounts of water pumped from the Ogallala, the world's largest known aquifer (Figure 10-14), have helped transform vast areas of arid high plains prairie land into one of the largest and most productive agricultural regions in the United States. More than 90% of the water withdrawn from this aquifer irrigates crops, amounting to about 30% of all U.S. groundwater used for irrigation.

Although this aquifer is gigantic, it is essentially a nonrenewable aquifer (stored during the retreat of the last ice age about 15,000–30,000 years ago) with an extremely slow recharge rate. In parts of the southern states where the aquifer is thinner (Figure 10-14), water is being pumped out of the aquifer 8–10 times faster than the natural recharge rate.

Water experts project that at the current rate of withdrawal, one-fourth of the aquifer's original supply will be depleted by 2020, and much sooner in areas where it is shallow. It will take thousands of years to replenish the aquifer.

Government farm policies encourage depletion of the aquifer by subsidizing water-thirsty crops that require irrigation and by providing crop-disaster payments. Federal groundwater depletion allowances for farmers also encourage groundwater mining. The greater the use of groundwater, the greater this tax break.

Depletion of this essentially nonrenewable water resource can be delayed. Farmers can use more efficient forms of irrigation (see Figure 10-15); in some areas they may have to switch to crops that require less water, or to dryland farming.

Cities using this groundwater can improve their water management by wasting less water and using less for lawns and golf courses. Individuals enjoying the benefits of this aquifer can pitch in by installing water-saving toilets and showerheads and converting their lawns to plants that can survive in an arid climate with little watering.

Critical Thinking

Do you believe that farmers and cattle raisers using water withdrawn from the Ogallala should receive federal subsidies to raise livestock and grow crops that require large amounts of irrigation water? Explain. What are the alternatives?

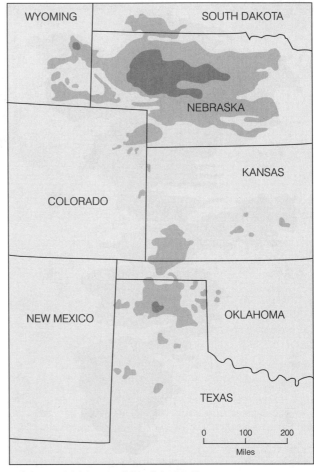

Saturated thickness of Ogallala Aquifer:

☐	Less than 61 meters (200 ft.)
▨	61–183 meters (200–600 ft.)
■	More than 61 meters (600 ft.) (as much as 370 meters or 1,200 ft. in places)

Figure 10-14 The Ogallala, the world's largest known aquifer. If the water in this aquifer were above ground, it could cover all 50 states with 0.5 meter (1.5 feet) of water. This fossil aquifer, which is renewed very slowly, is being depleted (especially at its thin southern end in parts of Texas, New Mexico, Oklahoma, and Kansas) to grow crops, raise cattle, and provide urban dwellers and industries with water. (Data from U.S. Geological Survey)

Q: What percentage of the USDA's research and education budget is spent on integrated pest management?

Can Cloud Seeding and Towing Icebergs Improve Water Supplies? For years several countries, particularly the United States, have been experimenting with seeding clouds with tiny particles of chemicals (such as silver iodide) to form water condensation nuclei and thus produce more rain over dry regions and more snow over mountains. However, cloud seeding is not useful in very dry areas, where it is most needed, because rain clouds are rarely available there. Furthermore, widespread cloud seeding would introduce large amounts of the cloud-seeding chemicals into soil and water systems, possibly harming people, wildlife, and agricultural productivity.

Another obstacle to cloud seeding is legal disputes over the ownership of water in clouds. During the 1977 drought in the western United States, the attorney general of Idaho accused officials in neighboring Washington of "cloud rustling" and threatened to file suit in federal court.

There also have been proposals to tow massive icebergs to arid coastal areas (such as Saudi Arabia and southern California) and then to pump the fresh water from the melting bergs ashore. However, the technology for doing this is not available, and the costs may be too high, especially for water-short developing countries.

Solutions: Why Is Reducing Water Waste So Important? Increasing the water supply in some areas is important, but soaring population, food needs, and industrialization (and unpredictable shifts in water supplies) will eventually outstrip this approach. It makes much more sense, economically and environmentally, to use water more efficiently.

Mohamed El-Ashry of the World Resources Institute estimates that *65–70% of the water people use throughout the world is wasted through evaporation, leaks, and other losses*. The United States, the world's largest user of water, does slightly better but still loses about 50% of the water it withdraws. El-Ashry believes that it is economically and technically feasible to reduce water waste to 15%, thereby meeting most of the world's water needs for the foreseeable future.

Conserving water would have many other benefits, including reducing the burden on wastewater plants and septic systems, decreasing pollution of surface water and groundwater, reducing the number of expensive dams and water transfer projects that destroy wildlife habitats and displace people, slowing depletion of groundwater aquifers, and saving energy and money needed to treat and distribute water.

Why Is So Much Water Wasted? A prime cause of water waste in the United States (and in most countries) is artificially low water prices resulting from government subsidies. Cheap water is the only reason farmers in Arizona and southern California can grow water-thirsty crops such as alfalfa in the middle of the desert. It also enables affluent people in desert areas to keep their lawns and golf courses green.

Water subsidies are paid for by all taxpayers through higher taxes. Because the harmful environmental and groundwater depletion costs caused by these perverse subsidies don't show up on monthly water bills, consumers have little incentive to use less water or to adopt water-conserving devices and processes. Some analysts believe that sharply raising the price of federally subsidized water would encourage investments in improving water efficiency, and many of the West's water supply problems could be eased. However, so far western members of the U.S. Congress have prevented such increases.

Another reason for the water waste in the United States (and many other countries) is that the responsibility for water resource management in a particular watershed may be divided among many state and local governments rather than being handled by one authority. The Chicago metropolitan area, for example, has 349 water supply systems divided among some 2,000 local units of government over a six-county area.

In sharp contrast is the regional approach to water management used in England and Wales. The British Water Act of 1973 replaced more than 1,600 agencies with 10 regional water authorities based on natural watershed boundaries. Each water authority owns, finances, and manages all water supply and waste treatment facilities in its region. The responsibilities of each authority include water pollution control, water-based recreation, land drainage and flood control, inland navigation, and inland fisheries. Each water authority is managed by a group of elected local officials and a smaller number of officials appointed by the national government.

Solutions: How Can We Waste Less Water in Irrigation? Worldwide, about 65% of the water that is diverted from rivers or pumped out of aquifers is used for irrigating about 16% of the world's cropland to produce almost 40% of the world's food. Because only about 40% of this water reaches crops, more efficient use of even a small amount of irrigation water would free large amounts of water for other uses.

Most irrigation systems distribute water from a groundwater well or a surface canal by downslope or gravity flow through unlined ditches in cropfields so that the water can be absorbed by crops (Figure 10-15, left). However, this flood irrigation method delivers far more water than needed for crop growth, and because

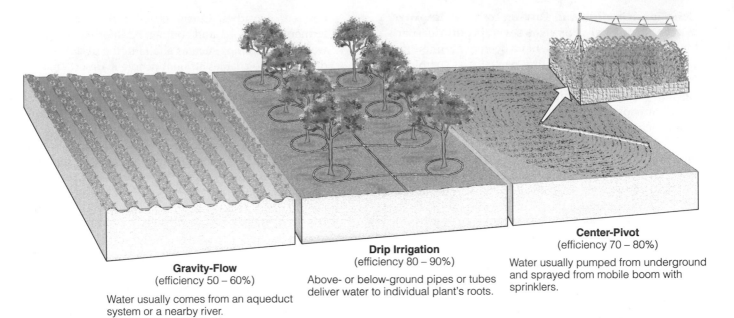

Figure 10-15 Major irrigation systems.

Gravity-Flow
(efficiency 50 – 60%)

Water usually comes from an aqueduct system or a nearby river.

Drip Irrigation
(efficiency 80 – 90%)

Above- or below-ground pipes or tubes deliver water to individual plant's roots.

Center-Pivot
(efficiency 70 – 80%)

Water usually pumped from underground and sprayed from mobile boom with sprinklers.

of evaporation, deep percolation (seepage), and runoff, only 50–60% of the water reaches crops.

Seepage can be reduced by placing plastic, concrete, or tile liners in the large irrigation canals that distribute water to unlined irrigation ditches. Lasers can be used to make sure that fields are level so that water flowing into unlined irrigation ditches is distributed more evenly. Small check dams of earth and stone can capture runoff from hillsides and channel it to fields. Holding ponds can store rainfall or capture irrigation water for recycling to crops.

Many U.S. farmers served by the dwindling Ogallala Aquifer now use center-pivot sprinkler systems (Figure 10-15, right), with which 70–80% of the water reaches crops. Some of these farmers are switching to low-energy precision-application (LEPA) sprinklers. These systems bring 75–85% (some claim 85–95%) of the water to crops by spraying it closer to the ground and in larger droplets than the center-pivot system; they also reduce energy use and costs by 20–30%. However, because of the high initial costs, such sprinklers are used on only about 1% of the world's irrigated cropland.

In the 1960s, highly efficient trickle or drip irrigation systems were developed in arid Israel. A network of perforated piping, installed at or below the ground surface, releases a trickle of water close to plant roots (Figure 10-15, center), minimizing evaporation and seepage and bringing 80–90% of the water to crops. Drip systems tend to be too expensive for most poor farmers and for use on low-value row crops. However, they are economically feasible for high-profit fruit, vegetable, and orchard crops and for home gardens. Researchers are testing drip irrigation systems that cost one-third to one-tenth as much as current drip systems.

Irrigation efficiency can also be improved by computer-controlled systems that monitor soil moisture and provide water only when necessary. Farmers can also switch to more water-efficient, drought-resistant, and salt-tolerant crop varieties and use nighttime irrigation to reduce loss to evaporation. In addition, farmers can use organic farming techniques, which produce higher crop yields per hectare and require only one-fourth of the water and commercial fertilizer of conventional farming.

Since 1950 Israel has used many of these techniques to slash irrigation water waste by about 84% while irrigating 44% more land. The government also gradually removed most government subsidies to raise the price of irrigation water to one of the highest rates in the world. Israel also imports most of its wheat and meat, concentrating on growing fruits, vegetables, and flowers that require less water. However, in other countries where government-funded water supply schemes and other subsidies provide artificially cheap irrigation water, farmers have little incentive to invest in water-saving techniques.

As fresh water becomes scarce and cities consume water once used for irrigation, carefully treated urban wastewater (which is rich in nitrate and phosphate plant nutrients) could be used for irrigation. Israel now

Q: What environmental and lifestyle factor causes the most death and suffering?

treats and reuses 55% of its municipal sewage water, mostly for irrigating nonedible crops such as cotton and flax, and plans to reuse 80% of this flow within the next few years.

Solutions: How Can We Waste Less Water in Industry, Homes, and Businesses? Manufacturing processes can use recycled water or be redesigned to save water. Japan and Israel lead the world in conserving and recycling water in industry. One paper mill in Hadera, Israel, uses one-tenth as much water as most other paper mills. Manufacturing aluminum from recycled scrap rather than from virgin ores can reduce water needs by 97%.

Nearly half of the water supplied by municipal water systems in the United States is used to flush toilets and water lawns, and another 15% is lost through leaky pipes. In a typical U.S. home, flushing toilets, washing hands, and bathing account for about 78% of the water used. In the arid western United States and in dry Australia, lawn and garden watering can take 80% of a household's daily usage, and much of this water is unnecessarily wasted.

Green lawns in an arid or semiarid area can be replaced with vegetation adapted to a dry climate—a form of landscaping called *xeriscaping*, from the Greek word *xeros*, meaning "dry." A xeriscaped yard typically uses 30–80% less water than a conventional one. Drip irrigation can water gardens and other vegetation around homes.

More than half the water supply in Cairo, Egypt; Lima, Peru; Mexico City, Mexico; and Jakarta, Indonesia, disappears before it can be used, mostly from leaks. In Cairo people often have to wade ankle deep across streets because of leaky water pipes. Leaky pipes, water mains, toilets, bathtubs, and faucets waste 20–35% of water withdrawn from public supplies in the United States and the United Kingdom.

Many cities offer no incentive to reduce leaks and waste. About one-fifth of all U.S. public water systems don't have water meters and charge a single low rate (often less than $100 a year for an average family) for virtually unlimited use of high-quality water. Many apartment dwellers have little incentive to conserve water because their water use is included in their rent.

In Boulder, Colorado, the introduction of water meters reduced water use by more than one-third. Tucson, Arizona, a desert city with ordinances that require conserving and reusing water, now consumes half as much water per capita as Las Vegas, a desert city where water conservation is still voluntary (Spotlight, right).

On January 1, 1994, the Comprehensive Energy Act of 1992 required that all new toilets sold in the United States use no more than 6 liters (1.6 gallons) per flush.

Similar laws have been passed in Mexico and in Ontario, Canada. A low-flow showerhead costing about $20 saves about $34–56 per year in water heating costs—a one-time investment that can give you a 50% return on your money with no financial risk. Audits conducted in Brown University's environmental studies program showed that the school could save $44,000 a year by using low-flow showerheads in dormitories.

Two decades of experience with droughts in northern California has shown that water demand can be cut by more than 50% for homes, 60% for parks, and 20% for businesses without economic hardships. Savings in water use—called *negaliters* or *negagallons*—can also save money. For instance, a 1994 New York City study showed that providing a $240 rebate to customers who convert to water-saving toilets and showerheads would cost only one-third of the $8–12 billion price tag for developing new water supplies for the city's residents.

Used water (gray water) from bathtubs, showers, bathroom sinks, and clothes washers can be collected, stored, treated, and reused for irrigation and other purposes. California has become the first state to legalize reuse of gray water to irrigate landscapes. An estimated 50–75% of the water used by a typical house could be reused as gray water.

In some parts of the United States, people can lease systems that purify and completely recycle wastewater from houses, apartments, or office buildings. Such a system can be installed in a small outside shed and serviced for a monthly fee roughly equal to that charged by most city water-and-sewer systems. In Tokyo, all the water used in Mitsubishi's 60-story office building is purified for reuse by an automated recycling system. Some actions you can take to waste less water are listed in Appendix 4.

 10-5 WATER POLLUTION PROBLEMS

What Are the Major Water Pollutants?
Water pollution is any chemical, biological, or physical change in water quality that has a harmful effect on living organisms or that makes water unsuitable for desired uses. There are several classes of water pollutants. One is *disease-causing agents (pathogens)*, which include bacteria, viruses, protozoa, and parasitic worms that enter water from domestic sewage and untreated human and animal wastes. According to a 1995 World Bank study, contaminated water (lack of clean drinking water and lack of sanitation) causes about 80% of the diseases in developing countries and kills about 10 million people annually—an average of 27,000 premature deaths per day, more than half of them children under age 5.

A good indicator of the quality of water for drinking or swimming is the number of colonies of *coliform bacteria* present in a 100-milliliter (0.1-quart) sample of water. The World Health Organization recommends a coliform bacteria count of 0 colonies per 100 milliliters for drinking water, and the EPA recommends a maximum level for swimming water of 200 colonies per 100 milliliters.

A second category of water pollutants is *oxygen-demanding wastes*, organic wastes that can be decomposed by aerobic (oxygen-requiring) bacteria. Large populations of bacteria decomposing these wastes can degrade water quality by depleting water of dissolved oxygen, causing fish and other forms of oxygen-consuming aquatic life to die. The quantity of oxygen-demanding wastes in water

can be determined by measuring the **biological oxygen demand (BOD)**: the amount of dissolved oxygen needed by aerobic decomposers to break down the organic materials in a certain volume of water over a 5-day incubation period at 20°C (68°F).

A third class of water pollutants is *water-soluble inorganic chemicals*, which include acids, salts, and compounds of toxic metals such as mercury and lead. High levels of these chemicals can make water unfit to drink, harm fish and other aquatic life, lower crop yields, and accelerate corrosion of metals exposed to such water.

Inorganic plant nutrients are another class of water pollutants. They are water-soluble nitrates and phosphates that can cause excessive growth of algae and other aquatic plants, which then die and decay, depleting water of dissolved oxygen and killing fish. Drinking water with excessive levels of nitrates lowers the oxygen-carrying capacity of the blood; this can kill unborn children and infants, especially those under 1 year old.

Water can also be polluted by a variety of *organic chemicals*, which include oil, gasoline, plastics, pesticides, cleaning solvents, detergents, and many other chemicals. They threaten human health and harm fish and other aquatic life.

By far the biggest class of water pollutants by weight is *sediment, or suspended matter*: insoluble particles of soil and other solids that become suspended in water, mostly when soil is eroded from the land. Sediment clouds water and reduces photosynthesis; it also disrupts aquatic food webs and carries pesticides, bacteria, and other harmful substances. Sediment that settles out destroys feeding and spawning grounds of fish. It also clogs and fills lakes, artificial reservoirs, stream channels, and harbors.

Water can also be polluted by *water-soluble radioactive isotopes*, some of which are concentrated or biologically magnified in various tissues and organs as they pass through food chains and webs. Radiation emitted by such isotopes can cause birth defects, cancer, and genetic damage.

Heat absorbed by water used to cool industrial and power plants can lower water quality. The resulting rise in water temperature, called *thermal pollution*, lowers dissolved oxygen levels and makes aquatic organisms more vulnerable to disease, parasites, and toxic chemicals.

Another form of water pollution, *genetic pollution*, occurs when aquatic systems are disrupted by the deliberate or accidental introduction of *nonnative species*. Some of these species can crowd out native species, reduce biodiversity, and cause economic losses.

What Are Point and Nonpoint Sources of Water Pollution? **Point sources** discharge pollutants at specific locations through pipes, ditches, or sewers into bodies of surface water. Examples include factories, sewage treatment plants (which remove some but not all pollutants), active and abandoned underground mines, offshore oil wells, and oil tankers. Because point sources are at specific places, they are fairly easy to identify, monitor, and regulate. In developed countries many industrial discharges are strictly controlled, whereas in most developing countries such discharges are largely uncontrolled.

Nonpoint sources are sources that cannot be traced to any single site of discharge. They are usually large land areas or airsheds that pollute water by runoff, subsurface flow, or deposition from the atmosphere. Examples include acid deposition (Figure 9-17), runoff of chemicals into surface water (including stormwater), and seepage into the ground from croplands, livestock feedlots, logged forests, streets, lawns, and parking lots.

In the United States, nonpoint pollution from agriculture—mostly in the form of sediment, inorganic fertilizers, manure, salts dissolved in irrigation water, and pesticides—is responsible for an estimated 64% of the total mass of pollutants entering streams and 57% of those entering lakes. Little progress has been made in controlling nonpoint water pollution because of the difficulty and expense of identifying and controlling discharges from so many diffuse sources.

Is the Water Safe to Drink? Many rivers in eastern Europe, Latin America, and Asia are used as sources of drinking water but are severely polluted, as are some rivers in developed countries. Aquifers used as sources of drinking water in many developed countries and developing countries are becoming contaminated with pesticides, fertilizers, and hazardous organic chemicals. In China, 41 large cities get their drinking water from polluted groundwater. In Russia, half of all tap water is unfit to drink, and a third of the aquifers are too contaminated for drinking purposes.

According to the World Health Organization, 1.2 billion people—over one-fifth of humanity—don't have a safe supply of drinking water, and 1.8 billion people lack adequate sanitation facilities. At least 5 million people, most of them children under age 5, die every year from waterborne diseases that could be prevented by clean drinking water and better sanitation.

In 1980, the UN recommended spending $300 billion to supply all the world's people with clean drinking water and adequate sanitation by 1990. The $30-billion annual cost of this program is about what the world spends every 10 days for military purposes. Only about $1.5 billion per year was actually spent. Researchers are trying to find cheap and simple ways to purify drinking water in developing nations (Individuals Matter, right).

Using UV Light and Horseradish to Purify Water

INDIVIDUALS MATTER

Ashok J. Gadgil, a physicist at California's Lawrence Berkeley National Laboratory, and his colleagues have recently developed a simple device that uses ultraviolet light to kill disease-causing organisms in drinking water. When water from a well or hand pump (in a village or household in a developing country) is passed through this table-top system, UV radiation from a mercury vapor lamp zaps germs in the water.

This $300 device weighs only 7 kilograms (15 pounds) and can disinfect 57 liters (15 gallons) of water per minute at a very low cost. It draws only 40 watts of power, supplied by solar cells, and can run unsupervised in remote areas of developing countries.

Recently, Pennsylvania State soil biochemists have discovered that chopped horseradish mixed with hydrogen peroxide (H_2O_2) helps rid contaminated water of organic pollutants called *phenols*. The horseradish contains an enzyme that speeds up the breakdown of the phenols.

What Are the Pollution Problems of Streams? Flowing streams, including large ones called *rivers*, can recover rapidly from degradable, oxygen-demanding wastes and excess heat by a combination of dilution and bacterial decay. This natural recovery process works as long as streams are not overloaded with these pollutants and as long as their flow is not reduced by drought, damming, or diversion for agriculture and industry. However, these natural dilution and biodegradation processes do not eliminate slowly degradable and nondegradable pollutants.

The breakdown of degradable wastes by bacteria depletes dissolved oxygen, which reduces or eliminates populations of organisms with high oxygen requirements until the stream is cleansed of wastes. The depth and width of the resulting *oxygen sag curve* (Figure 10-16) (and thus the time and distance required for a stream to recover) depend on the stream's volume, flow rate, temperature, and pH level and the volume of incoming degradable wastes. Similar oxygen sag curves can be plotted when heated water from industrial and power plants is discharged into streams (Figure 4-30).

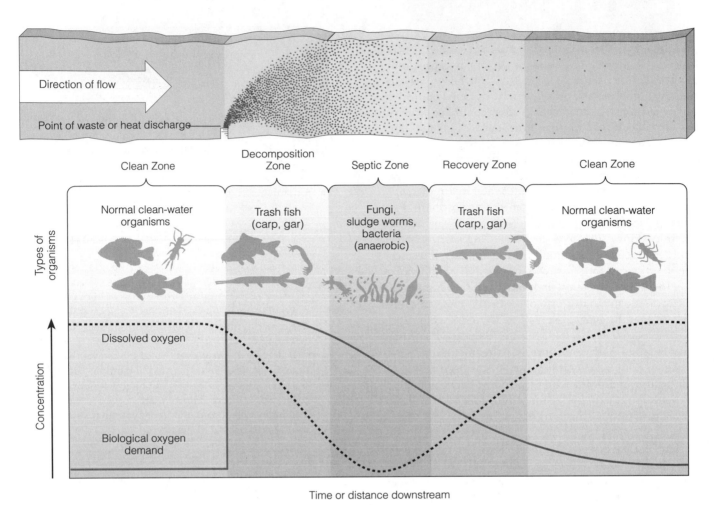

Figure 10-16 Dilution and decay of degradable, oxygen-demanding wastes and heat, showing the oxygen sag curve (dotted line) and the curve of oxygen demand (solid line). Depending on flow rates and the amount of pollutants, streams recover from oxygen-demanding wastes and heat if they are given enough time and are not overloaded.

Requiring cities to withdraw their drinking water downstream rather than upstream (as is done now) would dramatically improve water quality because each city would be forced to clean up its own waste outputs rather than passing them downstream. However, upstream users, who already have the use of fairly clean water without high cleanup costs, fight this pollution prevention approach.

Here is some good news. Water pollution control laws enacted in the 1970s have greatly increased the number and quality of wastewater treatment plants in the United States and many other developed countries; laws have also required industries to reduce or eliminate point-source discharges into surface waters. These efforts have enabled the United States to hold the line against increased pollution of most of its streams by disease-causing agents and oxygen-demanding wastes. This is an impressive accomplish-ment considering the rise in economic activity and population since the laws were passed.

One success story is the cleanup of Ohio's Cuyahoga River, which was so polluted that in 1959 and again in 1969 it caught fire and burned for several days as it flowed through Cleveland, Ohio. The highly publicized image of this burning river prompted city, state, and federal elected officials to enact laws limiting the discharge of industrial wastes into the river and sewage systems and to appropriate funds to upgrade sewage treatment facilities. Today the river has made a comeback and is widely used by boaters and anglers.

Despite progress in improving stream quality in most developed countries, large fish kills and contamination of drinking water still occur. Most of these disasters are caused by accidental or deliberate releases of toxic inorganic and organic chemicals by industries, malfunctioning sewage treatment plants, and nonpoint runoff of

Q: What is the projected death toll from smoking by 2020?

pesticides from cropland and nutrients (eroded soil, fertilizer, and animal waste) from cropland, urban development, and animal feedlots (Individuals Matter, below).

Available data indicate that stream pollution from huge discharges of sewage and industrial wastes is a serious and growing problem in most developing countries, where waste treatment is practically nonexistent. Numerous streams in the former Soviet Union and in eastern European countries are severely polluted. Currently, more than two-thirds of India's water resources are polluted with industrial wastes and sewage. Of the 78 streams monitored in China, 54 are seriously polluted

INDIVIDUALS MATTER

Tracking Down the Cell From Hell

Dr. JoAnn M. Burkholder is an aquatic ecologist at North Carolina State University. She knows what it is to be crippled by a dangerous new type of microbe and to experience political heat when you go public with your research to alert people to a potentially serious health threat.

In 1986 she investigated why a colleague's laboratory research fish were dying mysteriously and discovered that the culprit was a new microbe so tiny that dozens could fit on the head of a pin. She named it *Pfiesteria piscida* (pronounced fee-STEER-e-ah pis-kuh-SEED-uh)—Latin for "fish killer"—and some biologists call it the cell from hell.

She discovered that this tiny microscopic organism can assume at least 24 guises in its lifetime as it moves from river bottoms to midwaters to feed on fish. Without suitable prey, the microbe can masquerade as a plant or lie dormant on the bottom sediment of rivers and bays for years.

However, under certain conditions these benign microbes change from algae eaters into rapidly moving dinoflagellates. These rapidly moving organisms then release a water-soluble neurotoxin that stuns the fish or other prey and usually kills them within 10 minutes to several hours.

In 1993, Dr. Burkholder and her chief research aid experienced nausea, burning eyes and cramps, weakness, slow-healing sores, difficulty breathing, and severe loss of memory and mental powers from breathing toxic fumes released in tanks of fish dying from *Pfiesteria* attacks. They eventually recovered, but still cannot exercise strenuously without severe shortness of breath and the onset of respiratory illness. Since then more than 100 researchers, fishermen, and waterskiers in North Carolina, Virginia, and Maryland have experienced one or more of these symptoms when exposed to water or air contaminated by *Pfiesteria* toxins.

She developed evidence through lab and field research that connected outbreaks or blooms of *Pfiesteria* with excessive levels of nitrogen (as nitrates) and phosphorus (as phosphates) in rivers and estuaries. High levels of such nutrients are found in runoff from fertilized croplands, industrial development, and feedlots (especially those used to raise hogs and chickens) into rivers flowing into coastal estuaries.

In 1991 she went public with her findings and urged North Carolina state legislators to put curbs on hog farming (which other researchers claim is also polluting groundwater) and enact much tougher laws to reduce the flow of nutrients and other pollutants into the state's rivers. Hog farmers, developers, farming interests, fishing industry officials (worried about whether it is safe to eat fish and shellfish from affected rivers and estuaries), tourist-industry officials (alarmed about a negative image on the state's huge coastal recreational industry), and some state officials reacted negatively to her political activism. Some challenged her character and competence and accused her of using the results of preliminary research to push for questionable policies. She also received some anonymous death threats.

But she did not back down and continued criticizing state health officials and legislators for not taking her concerns about public health seriously enough. Eventually, under the glare of state and national publicity, state officials softened their public criticism.* The state now supports research on the problem and in 1997 the North Carolina legislature passed a bill putting a 2-year moratorium on new hog farms, reducing nutrient levels allowed in wastewater discharges, and requiring better management of land draining into rivers.

Her initial concerns have been vindicated as *Pfiesteria* have been implicated in fish kills in rivers flowing into the Chesapeake Bay in Maryland, the Rappahanock River in Virginia, and other rivers in Alabama, Delaware, and Florida. With the proper nutrient environment and other conditions, the organism can live in waters as far south as the Gulf of Mexico and thrive throughout much of the world.

By 1997, a number of federal and state environmental, health, and agricultural agencies had set up a coordinated research effort to learn more about what triggers outbreaks of the organism and how they affect human health and other organisms.

*For a popularized description of her research and political battle to alert the public and elected officials to the dangers posed by this microbe see Rodney Barker's, *And the Waters Turned to Blood: The Ultimate Biological Threat.* New York: Simon & Schuster, 1997).

with untreated sewage and industrial wastes, and 20% of China's rivers are too polluted to use for irrigation. In Latin America and Africa, most streams passing through urban or industrial areas are severely polluted.

What Are the Pollution Problems of Lakes? In lakes, reservoirs, and ponds, dilution is often less effective than in streams. Lakes and reservoirs often contain stratified layers (Figure 2-35) that undergo little vertical mixing, and ponds contain small volumes of water. Stratification also reduces levels of dissolved oxygen, especially in the bottom layer. In addition, lakes, reservoirs, and ponds have little flow, further reducing dilution and replenishment of dissolved oxygen. The flushing and changing of water in lakes and large artificial reservoirs can take from 1 to 100 years, compared with several days to several weeks for streams.

Thus, lakes are more vulnerable than streams to contamination by plant nutrients, oil, pesticides, and toxic substances that can destroy both bottom life and fish and birds that feed on contaminated aquatic organisms. Atmospheric fallout and runoff of acids is a serious problem in lakes subject to acid deposition (Figures 9-17 and 9-18).

Lakes receive inputs of nutrients and silt from the surrounding land basin as a result of natural erosion and runoff. This natural nutrient enrichment of lakes is called **eutrophication**. Over time, some of these lakes become more eutrophic (Figure 2-36, top), but others don't because of differences in the surrounding water basin. Near urban or agricultural areas, human activities can greatly accelerate the input of nutrients to a lake, which results in a process known as **cultural eutrophication**. Such a change is caused mostly by nitrate- and phosphate-containing effluents from sewage treatment plants, runoff of fertilizers and animal wastes, and accelerated erosion of nutrient-rich topsoil (Figure 10-17).

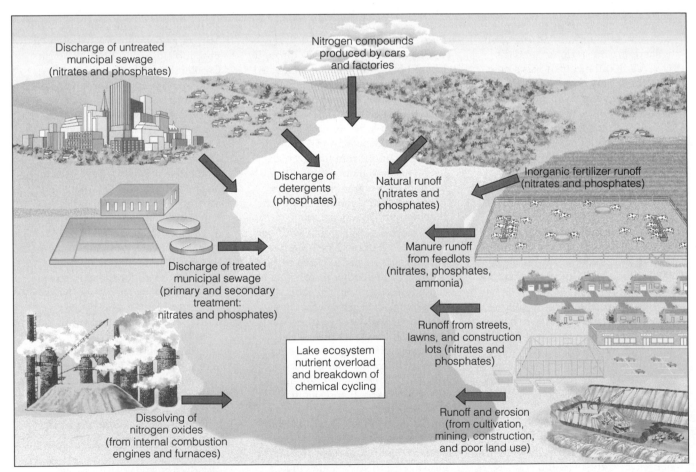

Figure 10-17 Principal sources of nutrient overload causing cultural eutrophication in lakes. The amount of nutrients from each source varies according to the types and amounts of human activities occurring in each airshed and watershed. Levels of dissolved oxygen drop when enlarged populations of algae and plants (stimulated by increased nutrient input) die and are decomposed by aerobic bacteria. Lowered oxygen levels can kill fish and other aquatic life and reduce the aesthetic and recreational values of the lake.

Q: How many Americans die because of exposure to other people's smoke (passive smoke)?

During hot weather or drought, this nutrient overload produces dense growths of organisms such as algae, cyanobacteria, water hyacinths, and duckweed. Dissolved oxygen in both the surface layer of water near the shore and in the bottom layer is depleted when large masses of algae die, fall to the bottom, and are decomposed by aerobic bacteria. This oxygen depletion can kill fish and other aquatic animals. If excess nutrients continue to flow into a lake, anaerobic bacteria take over and produce gaseous decomposition products such as smelly, highly toxic hydrogen sulfide and flammable methane.

About one-third of the 100,000 medium to large lakes and about 85% of the large lakes near major population centers in the United States suffer from some degree of cultural eutrophication. One-fourth of China's lakes are classified as eutrophic.

Ways to *prevent* or reduce cultural eutrophication include advanced waste treatment, bans or limits on phosphates in household detergents and other cleaning agents, and soil conservation and land-use control to reduce nutrient runoff.

Major *cleanup methods* are dredging bottom sediments to remove excess nutrient buildup, removing excess weeds, controlling undesirable plant growth with herbicides and algicides, and pumping air through lakes and reservoirs to avoid oxygen depletion (an expensive and energy-intensive method).

As usual, pollution prevention is more effective and usually cheaper in the long run than pollution control. If excessive inputs of limiting plant nutrients stop, a lake can usually return to its previous state.

Case Study: Chemical and Genetic Pollution in the Great Lakes

The five interconnected Great Lakes contain at least 95% of the surface fresh water in the United States and 20% of the world's fresh surface water. The Great Lakes basin is home for about 38 million people, about 30% of the Canadian population and 14% of the U.S. population.

Despite their enormous size, these lakes are vulnerable to pollution from point and nonpoint sources because less than 1% of the water entering the Great Lakes flows out to the St. Lawrence River each year. In addition to land runoff, these lakes receive large quantities of acids, pesticides, and other toxic chemicals by deposition from the atmosphere, often blown in from hundreds or thousands of kilometers away.

By the 1960s many areas of the Great Lakes were suffering from severe cultural eutrophication, huge fish kills, and contamination from bacteria and other wastes. The impact on Lake Erie was particularly intense because it is the shallowest of the Great Lakes. Many bathing beaches had to be closed, and by 1970 the lake had lost nearly all its native fish.

Since 1972, a $20-billion pollution-control program, carried out jointly by Canada and the United States, has significantly decreased levels of phosphates, coliform bacteria, and many toxic industrial chemicals in the Great Lakes. Algae blooms have also decreased, dissolved oxygen levels and sport and commercial fishing have increased, and most swimming beaches have reopened. These improvements were brought about mainly by new or upgraded sewage treatment plants, better treatment of industrial wastes, and banning of phosphate detergents, household cleaners, and water conditioners. Even so, less than 3% of the lakes' shoreline is clean enough for swimming or for supplying drinking water.

Levels of several toxic chlorinated hydrocarbon pollutants such as DDT and PCBs in Great Lakes water have dropped to their lowest levels in two decades. Despite this progress, contamination from toxic wastes flowing into the lakes (especially Lakes Erie and Ontario) from land runoff, streams, and atmospheric deposition (which accounts for an estimated 50% of the input of toxic compounds) is still a serious problem.

Toxic chemicals such as PCBs have built up in food chains and webs, contaminated many types of sport fish, and depleted populations of birds, river otters, and other animals feeding on contaminated fish. There is growing evidence and concern about possible effects of exposure to small amounts of many of these substances on the hormone systems of wildlife and humans (Connections, p. 232). A survey by Wisconsin biologists revealed that one fish in four taken from the Great Lakes is unsafe for human consumption.

In 1991 the U.S. government passed a law requiring accelerated cleanup of the lakes, especially of toxic hot spots, and an immediate reduction in air pollutant emissions in the region. However, meeting these goals may be delayed by a lack of federal and state funds.

Some environmentalists call for a ban on the use of chlorine as a bleach in the pulp and paper industry around the Great Lakes, a ban on all new incinerators in the area, and an immediate ban on discharge into the lakes of 70 toxic chemicals that threaten human health and wildlife. Understandably, officials of these industries strongly oppose such bans.

Great lakes fisheries also face threats from genetic pollution. In 1986, larvae of a nonnative species, the *zebra mussel*, arrived in ballast water discharged from a European ship near Detroit. With no known natural enemies, these tiny mussels have run amok; they deplete the food supply for other lake species, clog irrigation pipes, shut down water intake systems for power plants and city water supplies, foul beaches, and grow in huge masses on boat hulls, piers, and other surfaces.

Zebra mussels cost the Great Lakes basin at least $500 million per year, and the annual costs could reach $5 billion within a few years. The zebra mussel is

A: At least 3,000 and as many as 60,000 per year

expected to spread unchecked and dramatically alter most freshwater communities in parts of the United States and southern Canada within a few years, with damage costing tens of billions of dollars.

However, zebra mussels may be good news for a number of aquatic plants. By consuming algae and other microorganisms, the mussels increase water clarity. Clearer waters permit deeper penetration of sunlight and more photosynthesis, allowing some native plants to thrive and return the plant composition of Lake Erie (and presumably other lakes) closer to what it was 100 years ago. Because the plants provide food and increase dissolved oxygen, their comeback may benefit certain aquatic animals (including the mussels).

There is more bad news, however. In 1991 a larger and potentially more destructive species, the *quagga mussel*, invaded the Great Lakes, probably discharged in the ballast water of a Russian freighter. It can survive at greater depths and tolerate more extreme temperatures than the zebra mussel. There is concern that it may eventually colonize areas such as the Chesapeake Bay and waterways in parts of Florida.

How Much Pollution Can the Oceans Tolerate?

The oceans are the ultimate sink for much of the waste matter we produce, as summarized in the African proverb, "Water may flow in a thousand channels, but it all returns to the sea."

Oceans can dilute, disperse, and degrade large amounts of raw sewage, sewage sludge, oil, and some types of industrial waste, especially in deep-water areas. Marine life has also proved to be much more resilient than some scientists had expected, leading some to suggest that it is generally safer to dump sewage sludge and most other hazardous wastes into the deep ocean than to bury them on land or burn them in incinerators.

Other scientists dispute this idea, pointing out that we know less about the deep ocean than we do about outer space. They add that dumping waste in the ocean would delay urgently needed pollution prevention and promote further degradation of this vital part of the earth's life-support system.

Coastal areas, especially wetlands and estuaries, coral reefs, and mangrove swamps, bear the brunt of our enormous input of wastes into the ocean. This is not surprising, for half the world's population lives on or within 100 kilometers (160 miles) of the coast, and coastal populations are growing at a more rapid rate than the global population.

In most coastal developing countries (and in some coastal developed countries), municipal sewage and industrial wastes are often dumped into the sea without treatment. In the United States about 35% of all municipal sewage ends up virtually untreated in marine waters. Most U.S. harbors and bays are badly polluted from municipal sewage, industrial wastes, and oil.

California's Santa Monica Bay is the filming site for the widely watched television show *Baywatch*, which gives an illusion of a clean California beach lifestyle. What viewers don't know is that the bay is so polluted that the actors get extra pay each time they enter the water and are chemically cleaned afterward.

Runoff of sewage and agricultural wastes into coastal waters and acid deposition from the atmosphere (Figure 9-17) introduce large quantities of nitrogen and phosphorus, which can cause explosive growth or blooms of harmful algae or other toxic organisms. The organisms responsible for red and green tides can release waterborne and airborne toxins that damage fisheries, kill some fish-eating birds, reduce tourism, and poison seafood. In recent years there has been an increase in the frequency and magnitude of harmful blooms of such organisms in coastal waters of the United States and in other areas.

Case Study: The Chesapeake Bay The Chesapeake Bay, the largest estuary in the United States, is in trouble because of human activities. Between 1940 and 1995, the number of people living in the Chesapeake Bay area grew from 3.7 million to 15 million, and within a few years its population may reach 18 million.

The estuary receives wastes from point and nonpoint sources scattered throughout a huge drainage basin that includes 9 large rivers and 141 smaller streams and creeks in parts of six states (Figure 10-18). The bay has become a huge pollution sink because it is quite shallow and because only 1% of the waste entering it is flushed into the Atlantic Ocean.

Levels of phosphates and nitrates have risen sharply in many parts of the bay, causing algae blooms and oxygen depletion (Figure 10-18). Studies have shown that point sources, primarily sewage treatment plants, contribute about 60% by weight of the phosphates. Nonpoint sources—mostly runoff from urban, suburban, and agricultural land and deposition from the atmosphere—are the origins of about 60% by weight of the nitrates.

Air pollutants account for nearly 35% of the nitrogen entering the estuary. In addition, large quantities of pesticides run off cropland and urban lawns, and industries discharge large amounts of toxic wastes, often in violation of their discharge permits. Commercial harvests of oysters, crabs, and several important fish have fallen sharply since 1960 because of a combination of overfishing, pollution, and disease.

In the 1980s the Chesapeake Bay Program—the country's most ambitious attempt at the integrated coastal management—was implemented. Results have been impressive. Since 1983, more than $700 million in state and federal funds has been spent on a Chesapeake Bay cleanup program that will ultimately cost several

Q: How many U.S. workers die prematurely from exposure to toxic substances?

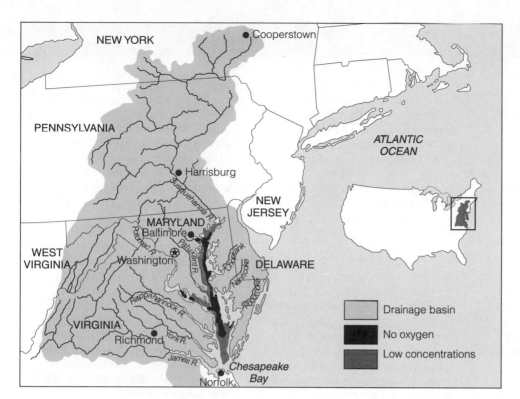

Figure 10-18 Chesapeake Bay, the largest estuary in the United States, is severely degraded as a result of water pollution from point and non-point sources in six states and from deposition of air pollutants.

billion dollars. Between 1985 and 1992, phosphorus levels declined 16% and nitrogen levels dropped 7%—a significant achievement given the increasing population in the watershed and the fact that more than a third of the nitrogen inputs come from the atmosphere.

Reaching the goals of a 40% reduction in nutrient levels and a significant improvement in habitat water quality throughout the bay will be especially difficult because the area's population is expected to grow by 25% between 1995 and 2020. Moreover, the bay will soon be invaded by zebra and quagga mussels. So far, however, the Chesapeake Bay Program shows what can be done when diverse parties work together to achieve goals that benefit both wildlife and people.

What Pollutants Are Dumped into the Ocean?
Dumping of industrial waste off U.S. coasts has stopped, although it still occurs in a number of other developed countries and some developing countries. However, barges and ships still legally dump large quantities of **dredge spoils** (materials, often laden with toxic metals, scraped from the bottoms of harbors and rivers to maintain shipping channels) at 110 sites off the Atlantic, Pacific, and Gulf coasts.

In addition, many countries dump into the ocean large quantities of **sewage sludge**: a gooey mixture of toxic chemicals, infectious agents, and settled solids removed from wastewater at sewage treatment plants. Since 1992 this practice has been banned in the United States.

Fifty countries with at least 80% of the world's merchant fleet have agreed not to dump sewage and garbage at sea, but this agreement is difficult to enforce and is often violated. Most ship owners save money by dumping wastes at sea and risk only small fines if they are caught. Each year as many as 2 million seabirds and more than 100,000 marine mammals (including whales, seals, dolphins, and sea lions) die when they ingest or become entangled in fishing nets, ropes, and other debris dumped into the sea and discarded on beaches.

Under the London Dumping Convention of 1972, 100 countries agreed not to dump highly toxic pollutants and high-level radioactive wastes in the open sea beyond the boundaries of their national jurisdiction. Since 1983 these same nations have observed a moratorium on the dumping of low-level radioactive wastes at sea, which in 1994 became a permanent ban. However, France, Great Britain, Russia, China, and Belgium may legally exempt themselves from this ban. In 1992 it was learned that for decades the former Soviet Union had been dumping large quantities of high- and low-level radioactive wastes into the Arctic Ocean and its tributaries.

What Are the Effects of Oil on Ocean Ecosystems? *Crude petroleum* (oil as it comes out of the ground) and *refined petroleum* (fuel oil, gasoline, and other processed petroleum products; Figure 4-24) are accidentally or deliberately released into the environment from a number of sources.

A: 50,000–70,000 per year

The *Exxon Valdez* Oil Spill

Crude oil from Alaska's North Slope fields near Prudhoe Bay is carried by pipeline to the port of Valdez and then shipped by tanker to the West Coast. On March 24, 1989, the *Exxon Valdez*, a tanker more than three football fields long, went off course in a 16-kilometer-wide (11-mile-wide) channel in Prince William Sound near Valdez, Alaska. It hit submerged rocks, creating the worst oil spill ever in U.S. waters.

The rapidly spreading oil slick coated more than 1,600 kilometers (1,000 miles) of shoreline, almost the length of the shoreline between New Jersey and South Carolina. The full loss of wildlife will never be known because most of the dead animals sank and decomposed without being counted.

Exxon spent $2.2 billion directly on the cleanup, but some aspects of the cleanup effort did more harm than good. For example, the use of high-pressure jets of hot water to clean beaches killed coastal plants and animals that had survived the spill. As a result, a year after the spill the oiled sites had recovered more rapidly than the washed sites.

In 1994 a jury awarded members of the fishing industry, land-owners, and other Alaska residents $5 billion in damages and penalties. However, Exxon has appealed this decision and it may be tied up in the courts for decades.

This roughly $8.5-billion accident might have been prevented if Exxon had spent only $22.5 million to fit the tanker with a double hull (which it still has not done). In the early 1970s interior secretary Rogers Morton told Congress that all oil tankers using Alaskan waters would have double hulls, but under pressure from oil companies the requirement was dropped.

Today, virtually all merchant ships have double hulls, but only 15% of oil tankers have such hulls. Legislation passed since the spill requires all new tankers to have double hulls and all existing large single-hulled oil tankers to be phased out between 1995 and 2015. However, the oil industry is working to weaken these and other requirements enacted since the spill as the public memory of the accident fades. Oil-company officials also hope to return the repaired but still single-hulled *Exxon Valdez* tanker (under a new name) to operation in Prince William Sound.

This spill highlighted the importance of pollution prevention and the advantages of shifting to improving energy efficiency and renewable energy (Chapter 4) to reduce dependence on oil. Even with the best technology and a fast response by well-trained people, scientists estimate that no more than 11–15% of the oil from a major spill can be recovered.

Critical Thinking

Explain how people who drive fuel inefficient cars and those who oppose increased fuel-efficiency standards share part of the blame for the *Exxon Valdez* oil spill.

Although tanker accidents (Case Study, above) and blowouts at offshore drilling rigs (when oil escapes under high pressure from a borehole in the ocean floor) get most of the publicity. However, more oil is released during normal operation of offshore wells, from washing tankers and releasing the oily water, and from pipeline and storage tank leaks. A 1993 Friends of the Earth study estimated that each year U.S. oil companies unnecessarily spill, leak, or waste an amount of oil equal to that shipped by 1,000 *Exxon Valdez* tankers—more oil than Australia uses. Oil pollution from shipping in the Mediterranean Sea is equivalent to 17 *Exxon Valdez* tankers emptying their tanks per year.

Natural oil seeps also release large amounts of oil into the ocean at some sites, but most ocean oil pollution comes from activities on land. Almost half (some experts estimate 90%) of the oil reaching the oceans is waste oil dumped, spilled, or leaked onto the land or into sewers by cities, individuals, and industries. Each year, a volume of oil equal to 20 times the amount spilled by the *Exxon Valdez* is improperly disposed of by about 50 million U.S. motorists changing their own motor oil. Worldwide, about 10% of the oil that reaches the ocean comes from the atmosphere, mostly from smoke emitted by oil fires.

The effects of oil on ocean ecosystems depend on a number of factors: type of oil (crude or refined), amount released, distance of release from shore, time of year, weather conditions, average water temperature, and ocean currents. Research shows that most (but not all) forms of marine life recover from exposure to large amounts of crude oil within 3 years. However, recovery from exposure to refined oil, especially in estuaries, may take 10 years or longer. The effects of spills in cold waters and in shallow enclosed gulfs and bays generally last longer.

Oil slicks that wash onto beaches can have a serious economic impact on coastal residents, who lose income from fishing and tourist activities. Oil-polluted beaches washed by strong waves or currents become clean after about a year, but beaches in sheltered areas remain contaminated for several years. Estuaries and salt marshes suffer the most and longest-lasting damage. Despite their localized harmful effects, oil spills

Q: What percentage of the 72,000 chemicals in commercial use have been thoroughly screened for toxicity?

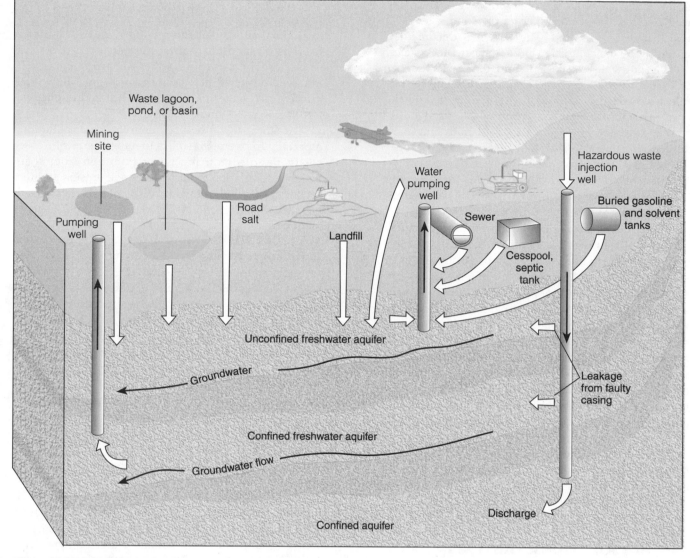

Figure 10-19 Principal sources of groundwater contamination in the United States.

are rated by experts as a low-risk ecological problem (Figure 8-5, left).

Why Is Groundwater Pollution Such a Serious Problem? Highly visible oil spills get a lot of media attention, but a much greater threat to human health is the out-of-sight pollution of groundwater (Figure 10-19), a prime source of water for drinking and irrigation. This vital form of earth capital is easy to deplete and pollute because much of it is renewed slowly. Although experts rate groundwater pollution as a low-risk ecological problem, they consider pollutants in drinking water (much of it from groundwater) a high-risk health problem (Figure 8-5, left). Laws protecting groundwater are weak in the United States and nonexistent in most countries.

When groundwater becomes contaminated, it cannot cleanse itself of degradable wastes, as surface water can if it is not overloaded. Because groundwater flows are slow and not turbulent, contaminants are not effectively diluted and dispersed. Groundwater also has much smaller populations of decomposing bacteria than surface-water systems, and its cold temperature slows down decomposition reactions. Thus, it can take hundreds to thousands of years for contaminated groundwater to cleanse itself of degradable wastes; nondegradable wastes are there permanently on a human time scale.

Crude estimates indicate that up to 25% of the usable groundwater in the United States is contaminated (and in some areas as much as 75%). In New Jersey, for example, every major aquifer is contaminated. In California, pesticides contaminate the drinking water of more than 1 million people. In Florida, where 92% of the residents rely on groundwater for drinking, over 1,000 wells have been closed. The EPA has documented groundwater contamination by 74 pesticides in 38 states.

Groundwater can be contaminated from a number of sources, including underground storage tanks, landfills, abandoned hazardous-waste dumps, deep wells used to dispose of liquid hazardous wastes, and industrial waste storage lagoons located above or near aquifers (Figure 10-19). An EPA survey found that one-third of 26,000 industrial waste ponds and lagoons have no liners to prevent toxic liquid wastes from seeping into aquifers, and one-third of those sites are within 1.6 kilometers (1 mile) of a drinking water well.

The EPA estimates that at least 1 million underground tanks storing gasoline, diesel fuel, and toxic solvents are leaking their contents into groundwater. A slow gasoline leak of just 4 liters (1 gallon) per day can seriously contaminate the water supply for 50,000 people. Such slow leaks usually remain undetected until someone discovers that a well is contaminated.

Determining the extent of a leak can cost $25,000–250,000. Cleanup costs range from $10,000 for a small spill to $250,000 or more if the chemical reaches an aquifer, and complete cleanup is rarely possible. Replacing a leaking tank adds an additional $10,000–60,000. Legal fees and damages to injured parties can run into the millions.

Current regulations should reduce leakage from new tanks but would do little about the millions of older tanks that are toxic time bombs. Some analysts call for aboveground storage of hazardous liquids so that leaks can be easily detected and rectified.

10-6 SOLUTIONS: PREVENTING AND CONTROLLING WATER POLLUTION

What Can We Do About Water Pollution from Nonpoint Sources? The leading nonpoint source of water pollution is agriculture. Farmers can sharply reduce fertilizer runoff into surface waters and leaching into aquifers by using only moderate amounts of fertilizer and by using none at all on steeply sloped land. They can use slow-release fertilizers and alternate their plantings between row crops and soybeans or other nitrogen-fixing plants to reduce the need for fertilizer. Farmers can also plant buffer zones of permanent vegetation between cultivated fields and nearby surface water.

Applying pesticides only when needed can reduce pesticide runoff and leaching. Farmers can also reduce the need for pesticides by using biological control or integrated pest management (Section 7-7). Nonfarm uses of inorganic fertilizers and pesticides—on golf courses, lawns, and public lands, for example—can also be sharply reduced and replaced with organic methods.

Livestock growers can control runoff and infiltration of manure from feedlots and barnyards by managing animal density, planting buffers, and not locating feedlots on land near surface water when that land slopes steeply toward the water. Diverting the runoff into well-designed detention basins would allow this nutrient-rich water to be pumped out and applied as fertilizer to cropland or forestland.

Another way to reduce nonpoint water pollution, especially from eroded soil, is to reforest critical watersheds. Besides reducing water pollution from sediments, reforestation would reduce soil erosion and the severity of flooding; it would also help slow projected global warming (Section 9-2) and loss of wildlife habitat.

What Can We Do About Water Pollution from Point Sources? The Legal Approach In many developing countries and in some developed countries, sewage and waterborne industrial wastes are discharged without treatment into the nearest waterway or into wastewater lagoons. In Latin America, less than 2% of urban sewage is treated. Only 15% of the urban wastewater in China receives treatment, and in India treatment facilities protect water quality for less than a third of the urban population.

In developed countries, most wastes from point sources are purified to varying degrees. The Federal Water Pollution Control Act of 1972 (renamed the Clean Water Act when it was amended in 1977) and the 1987 Water Quality Act form the basis of U.S. efforts to control pollution of the country's surface waters. The main goals of the Clean Water Act were to make all U.S. surface waters safe for fishing and swimming by 1983 and to restore and maintain the chemical, physical, and biological integrity of the nation's waters. Progress has been made, but these goals have not been met.

Here is some good news. The Clean Water Act of 1972 has led to significant improvements in U.S. water quality between 1972 and 1992. The percentage of U.S. rivers and lakes tested that have become fishable and swimmable increased from 36% to 62%.

Here is some bad news. Despite this significant progress, a 1994 report by the EPA revealed a number of problems. Antiquated sewage systems in 1,100 cities still dump poorly treated sewage into streams, lakes, and coastal waters. About 44% of lakes, 37% of rivers, and 32% of estuaries in the United States are still unsafe for fishing, swimming, and other recreational uses. Fish caught in more than 1,400 different waterways are unsafe to eat because of high levels of pesticides and other toxic substances.

What Can We Do About Water Pollution from Point Sources? The Technological Approach In rural and suburban areas with suitable soils, sewage from each house is usually discharged into a **septic tank** (Figure 10-20). About 25% of all homes in the United States are served by septic tanks, which must be cleaned out every 3–5 years by a reputable contractor so that they won't contribute to groundwater pollution.

In urban areas, most waterborne wastes from homes, businesses, factories, and storm runoff flow through a network of sewer pipes to wastewater treatment plants. Some cities have separate lines for stormwater runoff, but in 1,200 U.S. cities the lines for these two systems are combined because it is cheaper. When rains cause combined sewer systems to overflow, they discharge untreated sewage directly into surface waters.

When sewage reaches a treatment plant, it can undergo up to three levels of purification, depending on the type of plant and the degree of purity desired. **Primary sewage treatment** is a mechanical process that uses screens to filter out debris such as sticks, stones, and rags; suspended solids settle out as sludge in a settling tank (Figure 10-21). Improved primary treatment uses chemically treated polymers to remove suspended solids more thoroughly.

Secondary sewage treatment is a biological process in which aerobic bacteria are used to remove up to 90% of biodegradable, oxygen-demanding organic wastes (Figure 10-21). Some plants use *trickling filters*, in which aerobic bacteria degrade sewage as it seeps through a bed of crushed stones covered with bacteria and protozoa. Others use an *activated sludge process*, in which the sewage is pumped into a large tank and mixed for several hours with bacteria-rich sludge and air bubbles to facilitate degradation by microorganisms. The water then goes to a sedimentation tank, where most of the suspended solids and microorganisms settle out as sludge. The sludge produced from primary or secondary treatment is broken down in an anaerobic digester and either incinerated, dumped into the ocean or a landfill, or applied to land as fertilizer.

Even after secondary treatment, wastewater still contains about 3–5% by weight of the original oxygen-demanding wastes, 3% of the suspended solids, 50% of the nitrogen (mostly as nitrates), 70% of the phosphorus (mostly as phosphates), and 30% of most toxic metal compounds and synthetic organic chemicals. Virtually no long-lived radioactive isotopes or persistent organic substances such as pesticides are removed.

As a result of the Clean Water Act, most U.S. cities have secondary sewage treatment plants. In 1989, however, the EPA found that more than 66% of sewage treatment plants have either water quality or public health problems, and studies by the General Accounting Office have shown that most industries have violated regulations. Moreover, 500 cities have failed to meet federal standards for sewage treatment plants, and 34 East Coast cities simply screen out large floating objects from their sewage before discharging it into coastal waters.

Advanced sewage treatment is a series of specialized chemical and physical processes that remove specific pollutants left in the water after primary and secondary treatment. Types of advanced treatment vary according to the specific contaminants to be removed. Advanced treatment is rarely used because such plants typically cost twice as much to build and four times as much to operate as secondary plants.

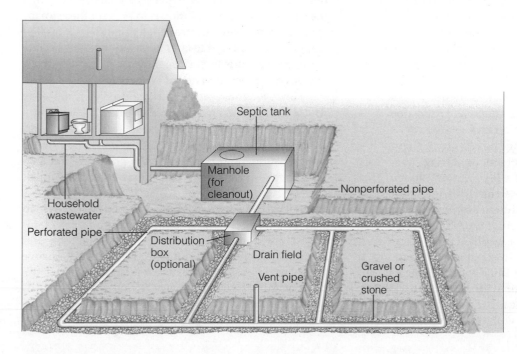

Figure 10-20 Septic tank system used for disposal of domestic sewage and wastewater in rural and suburban areas. This system traps greases and large solids and discharges the remaining wastes over a large drainage field. As these wastes percolate downward, the soil filters out some potential pollutants and soil bacteria decompose biodegradable materials. To be effective, septic tank systems must be properly installed in soils with adequate drainage, not placed too close together or too near well sites, and pumped out when the settling tank becomes full.

Hint: Enter the search term *Clean Water Act of 1977* using the Subject Guide.

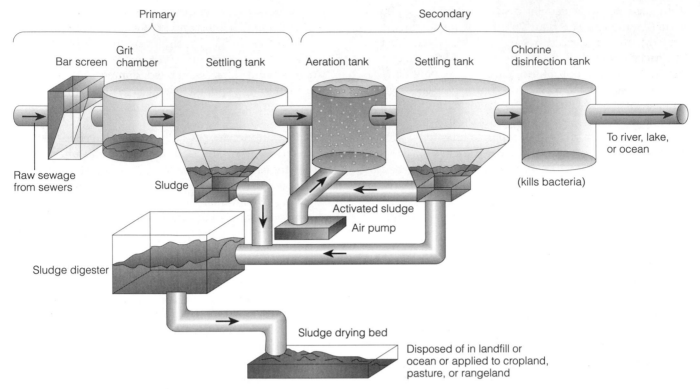

Figure 10-21 Primary and secondary sewage treatment.

Before water is discharged after primary, secondary, or advanced treatment, it is bleached (to remove water coloration) and disinfected (to kill disease-carrying bacteria and some but not all viruses). The usual method for doing this is *chlorination*. However, chlorine can react with organic materials in water to form small amounts of chlorinated hydrocarbons, some of which cause cancers in test animals. According to some preliminary research in 1992, chlorinated drinking water may cause 7–10% of all cancers in the United States. There is also growing evidence that some chlorinated hydrocarbons may damage the human nervous, immune, and endocrine systems (Section 8-3). Other disinfectants such as ozone and ultraviolet light are used in some places, but they cost more than chlorination and are not as long-lasting.

Sewage treatment produces a toxic, gooey sludge that must be disposed of or recycled as fertilizer. About 54% by weight of all municipal sludge produced in the United States is applied to farmland, forests, highway medians, and degraded land as fertilizer, and 9% is composted. The rest is dumped in conventional landfills (where it can contaminate groundwater) or incinerated (which can pollute the air with traces of toxic chemicals, and the resulting toxic ash is usually buried in landfills that EPA experts say will eventually leak).

Before it is applied to land, sewage sludge can be heated to kill harmful bacteria, as is done in Switzerland and parts of Germany; it can also be treated to remove toxic metals and organic chemicals before application, but such treatment can be expensive. (Scientists say that the best and cheapest solution is to prevent these toxins from reaching sewage treatment plants.) Untreated sludge can be applied to land not used for crops or livestock, such as forests, surface-mined land, golf courses, lawns, cemeteries, and highway medians.

However, most sewage sludge used as fertilizer in the United States is not adequately treated to kill harmful bacteria and remove toxic metals and organic chemicals, which may end up in food products or leach into groundwater. A growing number of health problems and lawsuits have resulted from use of sludge to fertilize crops.

Solutions: How Can We Use Nature's Ways to Purify Sewage? Some communities and individuals are seeking better ways to purify contaminated water by working with nature. A low-tech, low-cost alternative to expensive waste treatment plants is to create an artificial wetland, as the residents of Arcata, California, did.

In this coastal town of 17,000, some 63 hectares (155 acres) of wetlands has been created between the town and the adjacent Humboldt Bay. The marshes, developed on land that was once a dump, act as an inexpensive natural waste treatment plant. The project was completed in 1974 for less than half the estimated cost of a conventional treatment plant.

Here's how it works: First, sewage is held in sedimentation tanks, where the solids settle out as sludge that is removed and processed for use as fertilizer. The liquid is pumped into oxidation ponds, where remaining wastes are broken down by bacteria. After a month or so, the water is released into the artificial marshes, where it is further filtered and cleansed by plants and bacteria. Although the water is clean enough to be discharged directly into the bay, state law requires that it first be chlorinated. So the town chlorinates the water and then dechlorinates it before sending it into the bay, where oyster beds thrive.

The marshes and lagoons are an Audubon Society bird sanctuary and provide habitats for thousands of otters, seabirds, and marine animals; the treatment center is a city park and attracts more than 150,000 visitors a year. The town even celebrates its natural sewage treatment system with an annual "Flush with Pride" festival. Over 150 cities and towns in the United States now use natural and artificial wetlands for treating sewage.

Is it possible to use natural processes for treating wastewater if there isn't a wetland available or enough land on which to create one? According to ecologist John Todd, it is: Just set up a greenhouse lagoon and use sunshine the way nature does. The process begins when sewage flows into a greenhouse containing rows of large tanks full of aquatic plants such as water hyacinths, cattails, and bulrushes. In these tanks, algae and microorganisms decompose wastes into nutrients absorbed by the plants.

The decomposition is speeded up by sunlight streaming into the greenhouse. Then the water passes through an artificial marsh of sand, gravel, and bulrush plants, which filters out algae and organic waste. Next the water flows into aquarium tanks, where snails and zooplankton consume microorganisms and are in turn consumed by crayfish, tilapia, and other fish that can be eaten or sold as bait. After 10 days, the now-clear water flows into a second artificial marsh for final filtering and cleansing.

When working properly, such solar–aquatic treatment systems have produced water fit for drinking. The chief byproducts of such systems are plants, trees, snails, and fish that can be sold as compost, ornamental plants, or baitfish. Todd's living purification systems now operate in 13 states around the United States and in seven other countries.

How Is Drinking Water Purified? Treatment of water for drinking by city dwellers is much like wastewater treatment. Areas that depend on surface water usually store it in a reservoir for several days to improve clarity and taste by allowing the dissolved oxygen content to increase and suspended matter to settle out. The water is then pumped to a purification plant, where it is treated to meet government drinking water standards. Usually the water is run through sand filters and activated charcoal before it is disinfected. In areas with very pure sources of groundwater, little treatment is necessary.

How Is Drinking Water Quality Protected?

About 54 countries, most of them in North America and Europe, have safe drinking water standards. The U.S. Safe Drinking Water Act of 1974 requires the EPA to establish national drinking water standards, called *maximum contaminant levels*, for any pollutants that *may* have adverse effects on human health. This act has helped improve drinking water in much of the United States, but attempts to weaken this law continue. At least 700 potential pollutants have been found in municipal drinking water supplies. Maximum contaminant levels have not been set for potentially dangerous water pollutants such as certain synthetic organic compounds, radioactive materials, toxic metals, and pathogens.

Privately owned wells are not required to meet federal drinking water standards, primarily because of the costs of testing each well regularly (at least $1,000) and ideological opposition to mandatory testing and compliance by some homeowners.

According to a 1994 study by the Natural Resources Defense Council (NRDC), in 1992 the drinking water of 50 million people—one American in five—violated one or more EPA pollutant standards. In most cases people were not notified (as required by the 1974 law) when their drinking water was contaminated. This study also estimated that contaminated drinking water is responsible for 7 million illnesses and 1,200 deaths per year in the United States.

Such dangers were revealed when residents of Milwaukee, Wisconsin, were told on April 7, 1993, that water from their taps was unsafe to drink. Before this health crisis was over, 104 people had died and 400,000 people had become sick as a result of exposure to *Cryptosporidium*, a parasitic organism. In May 1994, another outbreak of this parasite killed 19 and sickened more than 100 people in Las Vegas, Nevada. A survey in 1992 showed that nearly 40% of treated drinking water supplies in the United States contained either *Cryptosporidium* or *Giardia* (a protozoan that infects the small intestine, causing nausea and diarrhea). Mostly as a result of the Milwaukee crisis and pressure from environmentalists and health officials, the EPA began requiring municipal drinking water suppliers to begin testing for cryptosporidium in 1996.

A: About 0.5%

According to the NRDC, U.S. drinking water supplies could be made safer at a cost of only about $30 a year per household. Currently, more money is spent on military bands each year than on the EPA's enforcement of the Safe Drinking Water Act. One might expect that such information would lead to public pressure to upgrade drinking water in the United States. Instead, Congress is being pressured by industry to weaken the Safe Drinking Water Act.

Is Bottled Water the Answer? The United States has one of the world's best drinking water supply systems. Yet contaminated wells and concern about possible contamination of public drinking water supplies have stimulated many U.S. citizens to drink bottled water at costs about 1,500 times more than that of tap water, or to add water purification devices to their home systems. Studies indicate that many of these consumers are being ripped off and in some cases may end up drinking water that is dirtier than they can get from their taps.

To be safe, consumers purchasing bottled water should determine whether the bottler belongs to the International Bottled Water Association (IBWA) and adheres to its testing requirements.* The IBWA requires its members to test for 181 contaminants, and annually it sends an inspector to bottling plants to check all pertinent records and ensure that the plant is run cleanly. Some companies pay $2,500 annually to obtain more stringent certification by the National Sanitation Foundation, an independent agency that tests for 200 chemical and biological contaminants.

Before buying expensive home water purifiers, health officials suggest that consumers have their water tested by local health authorities or private labs† to identify what contaminants, if any, must be removed; then they should buy a unit that does the required job. Independent experts contend that unless tests show otherwise, for most urban and suburban Americans served by large municipal drinking water systems, home water treatment systems aren't worth the expense and maintenance hassles.

Buyers should be suspicious of door-to-door salespeople, telephone appeals, scare tactics, and companies offering free water tests (which often are neither accurate nor carried out by certified labs). They should carefully check out companies selling such equipment and demand a copy of purifying claims by EPA-certified laboratories. Buyers should also be wary of claims that a treatment device has been *approved* by the EPA. Although the EPA does *register* such devices, it neither tests nor approves them.

How Can We Protect Coastal Waters? The key to protecting oceans is to reduce the flow of pollution from the land and from streams emptying into the ocean. Such efforts must also be integrated with efforts to prevent and control air pollution because an estimated 33% of all pollutants entering the ocean worldwide comes from air emissions from land-based sources.

Some ways various analysts have suggested to prevent and reduce excessive pollution of coastal waters are as follows:

Prevention

- *Encourage or require separate sewage and storm runoff lines in coastal urban areas.*
- *Discourage ocean dumping of sludge and hazardous dredged materials.*
- *Protect sensitive and ecologically valuable coastal areas from development, oil drilling, and oil shipping.*
- *Use ecological land-use planning to control and regulate coastal development.*
- *Require double hulls for all oil tankers by 2002.*
- *Recycle used oil.*
- *Reduce genetic pollution from nonnative aquatic species by using heat to kill organisms in ballast water or developing filters to trap the organisms when ballast water is taken on or discharged from a ship.*

Cleanup

- *Improve oil-spill cleanup capabilities.*
- *Require at least secondary treatment of coastal sewage, or use wetlands, solar aquatic, or other environmentally acceptable methods.*

How Can We Protect Groundwater? Pumping polluted groundwater to the surface, cleaning it up, and returning it to the aquifer is usually too expensive (for a single aquifer, $5 million or more). Recent attempts to pump and treat contaminated aquifers indicate that it may take 50–1,000 years of continuous pumping before all the contamination is forced to the surface and drinking-water quality is achieved. Thus, *preventing contamination by various means is considered the only effective way to protect groundwater resources.* Ways to do this include

- *Monitoring aquifers near landfills and underground tanks*
- *Requiring leak detection systems for existing and new underground tanks used to store hazardous liquids*

---

*Check for the IBWA seal of approval on the bottle or contact International Bottled Water Association (113 North Henry Street, Alexandria, VA 22314; phone: 703-683-5213) for a member list.

†Contact the state health or other appropriate department for help in locating laboratories that are certified to do tests. You can also call EPA's Safe Drinking Water Hotline from 8:30 A.M. to 4:30 P.M., EST at 800-426-4791 or 202-382-5533.

Q: Worldwide, how many children under age 5 die each year in developing countries from mostly preventable infectious diseases?

- *Requiring liability insurance for old and new underground tanks used to store hazardous liquids*

- *Banning or more strictly regulating disposal of hazardous wastes in deep injection wells and landfills*

- *Storing hazardous liquids above ground in tanks with systems for detecting and containing any leaks*

Solutions: How Can We Use Water Resources More Sustainably? According to water resource expert Sandra Postel, none of the technological solutions for supplying more water—dams, watershed transfers, tapping groundwater, and using water more efficiently—deal with the three underlying forces that can lead humans to use such a potentially renewable resource in unsustainable ways: **(1)** depletion or degradation of a shared resource, which shrinks the resource "pie" to be shared (the tragedy of the commons); **(2)** population growth, which forces the resource pie to be divided into smaller slices; and **(3)** unequal distribution or access (primarily because of poverty), which means that some countries and people get larger slices than others.

Sustainable water use is based on the common-sense principle stated in an old Inca proverb: "The frog does not drink up the pond in which it lives."

Sustainable use of potentially renewable groundwater means that the rate of extraction should not exceed the rate of recharge. Determining what constitutes sustainable use of shared rivers is much more complex because available water from a river varies with the time of year and with conditions such as drought and higher-than-normal precipitation. Sustainable water use also requires an integrated plan governing water use, sewage treatment, and water pollution among all users of a water basin (as is done in Great Britain).

Many analysts believe that a key to reducing water waste and providing more equitable access to water resources is for governments to phase out subsidies that reduce the price of water for those benefiting from such subsidies (often large farmers and industries).

According to environmentalists, a sustainable approach to dealing with water pollution requires that we shift our emphasis from pollution cleanup to pollution prevention. This involves **(1)** *source reduction* to reduce the toxicity or volume of pollutants (for example, replacing organic solvent-based inks and paints with water-based materials), **(2)** *reuse* of wastewater instead of discharging it (for example, reusing treated wastewater for irrigation), and **(3)** *recycling* pollutants (for example, cleaning up and recycling contaminated solvents for reuse) instead of discharging them.

To make such a shift, we need to accept that the environment—air, water, soil, life—is an interconnected whole. Without an integrated approach to all forms of pollution, environmentalists argue that we will continue to shift environmental problems from one part of the environment to another. Some actions you can take to help reduce water pollution are listed in Appendix 4.

It is not until the well runs dry that we know the worth of water.

BENJAMIN FRANKLIN

CRITICAL THINKING

1. How do human activities increase the harmful effects of prolonged drought? How can these effects be reduced?

2. How do human activities contribute to flooding and flood damage? How can these effects be reduced?

3. Do you agree or disagree with the statement that flooding is a natural part of a river's complex ecological cycle and is not a pathological state that needs to be fixed by building dams and reservoirs? Explain. How would adopting such an approach affect the large numbers of people living in floodplains along the world's river systems?

4. Should prices of water for all uses be raised sharply to include more of its environmental costs and to encourage water conservation? Explain. What harmful and beneficial effects might this have on business and jobs, on your lifestyle and the lifestyles of any children or grandchildren you might have, on the poor, and on the environment?

5. List five major ways you can conserve water on a personal level (see Appendix 4). Which, if any, of these practices do you now use or intend to use?

6. Why is dilution not always the solution to water pollution? Give examples and conditions for which this solution is, and is not, applicable.

7. How can a stream cleanse itself of oxygen-demanding wastes? Under what conditions will this natural cleansing system fail?

8. Should all dumping of wastes in the ocean be banned? Explain. If so, where would you put the wastes instead? What exceptions would you permit, and why?

9. Should the injection of hazardous wastes into deep underground wells be banned? Explain. What would you do with these wastes?

Solid wastes are only raw materials we're too stupid to use.

ARTHUR C. CLARKE

11-1 WASTING RESOURCES: THE HIGH-WASTE APPROACH

What Is Solid Waste, and How Much Is Produced? The United States, with only 4.6% of the world's population, produces about 33% of the world's **solid waste**: any unwanted or discarded material that is not a liquid or a gas. Although garbage produced directly by households and businesses is a significant problem, *about 98.5% of the solid waste in the United States comes from mining, oil and natural gas production, agriculture, and industrial activities used to produce goods and services for consumers* (Figure 11-1).

The remaining 1.5% of solid waste produced in the United States is **municipal solid waste** (MSW) from homes and businesses in or near urban areas. The amount of municipal solid waste, often called *garbage*, produced in the United States in 1996 was enough to fill a bumper-to-bumper convoy of garbage trucks encircling the globe almost eight times. This amounted to an average of 680 kilograms (1,500 pounds) per person in the United States—two to three times that in most other developed countries and many times that in developing countries.

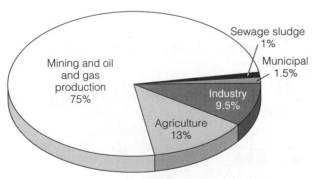

Figure 11-1 Sources of the estimated 10 billion metric tons (11 billion tons) of solid waste produced each year in the United States. Mining and industrial activities produce 65 times as much solid waste as household activities. (Data from U.S. Environmental Protection Agency and U.S. Bureau of Mines)

About 27% of the resources in MSW produced in the United States in 1996 was recycled or composted (up from 7% in 1960); the other 73% was hauled away and either dumped in landfills (58%) or burned in incinerators and waste-to-energy plants (15%) at a cost of about $40 billion (projected to rise to $75 billion by 2005).

What Does It Mean to Live in a High-Waste Society? U.S. consumers throw away astounding amounts of solid waste, including the following quantities:

- Enough aluminum to rebuild the country's entire commercial airline fleet every 3 months

- Enough tires each year to encircle the planet almost three times

- About 18 billion disposable diapers per year, which if linked end-to-end would reach to the moon and back seven times

- About 2 billion disposable razors, 10 million computers, and 8 million television sets each year

- About 2.5 million nonreturnable plastic bottles each hour

- Some 14 billion catalogs (an average of 54 per American) and 38 billion pieces of junk mail each year

This is only part of the 1.5% of all solid waste labeled "municipal" in Figure 11-1.

What Is Hazardous Waste, and How Much Is Produced? In the United States, **hazardous waste** is legally defined as any discarded solid or liquid material that **(1)** contains one or more of 39 toxic, carcinogenic, mutagenic, or teratogenic compounds at levels that exceed established limits (including many solvents, pesticides, and paint strippers), **(2)** catches fire easily (gasoline, paints, and solvents), **(3)** is reactive or unstable enough to explode or release toxic fumes (acids, bases, ammonia, chlorine bleach), or **(4)** is capable of corroding metal containers such as tanks, drums, and barrels (industrial cleaning agents and oven and drain cleaners).

This narrow official definition of hazardous wastes (mandated by Congress) does *not* include the following materials: **(1)** radioactive wastes (Section 4-7), **(2)** hazardous and toxic materials discarded by households

(Table 11-1), **(3)** mining wastes, **(4)** oil- and gas-drilling wastes (routinely discharged into surface waters or dumped into unlined pits and landfills), **(5)** liquid waste containing organic hydrocarbon compounds (80% of all liquid hazardous waste), **(6)** cement kiln dust produced when liquid hazardous wastes are burned in a cement kiln (a practice classified as recycling by the EPA but considered dangerous *sham recycling* by environmentalists), and **(7)** wastes from the thousands of small businesses and factories that generate less than 100 kilograms (220 pounds) of hazardous waste per month.

Designating these excluded categories as hazardous waste would shift efforts from waste management and pollution control to waste reduction and pollution prevention. Industry representatives say that having to manage these wastes would bankrupt them or force them to move their operations to other countries with less strict waste regulations.

The EPA estimates that at least 5.5 billion metric tons (6 billion tons) of hazardous waste are produced each year in the United States—an average of 21 metric tons (23 tons) per person. However, because only about 6% of the total is *legally* defined as hazardous waste, *94% of the country's hazardous waste is not regulated by hazardous-waste laws.* In most other countries, especially developing countries, even less of the hazardous waste is regulated.

11-2 PRODUCING LESS WASTE AND POLLUTION: REDUCING THROUGHPUT

What Are Our Options? There are two ways to deal with the solid and hazardous waste we create: *waste management* and *pollution (waste) prevention*. Waste management is a *high-waste approach* that views waste production as an unavoidable product of economic growth. It attempts to manage the resulting wastes in ways that reduce environmental harm, mostly by burying them, burning them, or shipping them off to another state or country. The goal is to move increasing amounts of matter and energy resources through the economy to enhance economic growth (Figure 2-46).

Preventing pollution and waste is a *low-waste approach* that views most solid and hazardous waste either as potential resources (that we should be recycling, composting, or reusing) or as harmful substances that we should not be using in the first place (Figures 11-2 and 11-3). With this approach, taxes and subsidies are used to discourage waste production and encourage waste prevention.

According to the U.S. National Academy of Sciences (Figures 11-2 and 11-3), the low-waste approach should have the following hierarchy of goals: **(1)** *reduce*

Table 11-1 Common Household Toxic and Hazardous Materials

Cleaning Products

Disinfectants

Drain, toilet, and window cleaners

Oven cleaners

Bleach and ammonia

Cleaning solvents and spot removers

Septic tank cleaners

Paint and Building Products

Latex and oil-based paints

Paint thinners, solvents, and strippers

Stains, varnishes, and lacquers

Wood preservatives

Acids for etching and rust removal

Asphalt and roof tar

Gardening and Pest Control Products

Pesticide sprays and dusts

Weed killers

Ant and rodent killers

Flea powder

Automotive Products

Gasoline

Used motor oil

Antifreeze

Battery acid

Solvents

Brake and transmission fluid

Rust inhibitor and rust remover

General Products

Dry cell batteries (mercury and cadmium)

Artist paints and inks

Glues and cements

Source: Data from Science Advisory Board, *Reducing Risks* (Washington, D.C.: Environmental Protection Agency, 1990). Items in each category are not listed in rank order.

---

waste and pollution, **(2)** *reuse* as many things as possible, **(3)** *recycle and compost* as much waste as possible, **(4)** *chemically or biologically treat or incinerate* waste that can't be reduced, reused, recycled, or composted, and **(5)** *bury* what is left in state-of-the-art landfills or aboveground vaults after the first four goals have been met.

Hint: Enter the search terms *waste disposal, ground* using the Subject Guide.

Figure 11-2 Solutions: priorities suggested by prominent scientists for dealing with material use and solid waste. To date, these priorities have not been followed in the United States (and in most countries); most efforts are devoted to waste management (bury it or burn it). (U.S. Environmental Protection Agency and U.S. National Academy of Sciences)

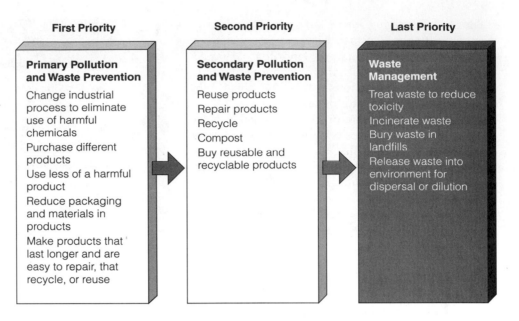

Figure 11-3 Solutions: priorities suggested by prominent scientists for dealing with hazardous waste. To date, these priorities have not been followed in the United States (and in most countries). (U.S. National Academy of Sciences)

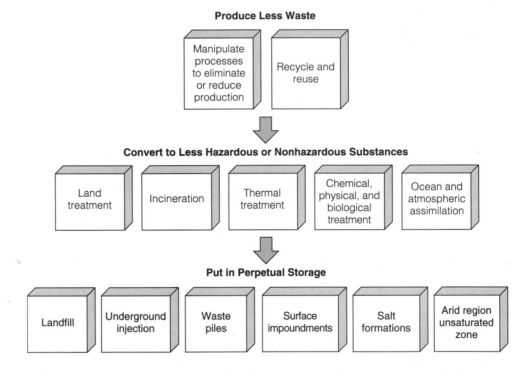

Scientists estimate that in a low-waste society 60–80% of the solid and hazardous waste produced could be eliminated through reduction, reuse, and recycling (including composting). The remaining 20–40% of such wastes would then be treated to reduce their toxicity, and what's left would be burned or buried under carefully regulated conditions. Currently, the order of priorities shown in Figures 11-2 and 11-3 for dealing with solid and hazardous wastes is reversed in the United States (and in most other countries), mostly because the costs of producing and dealing with these wastes are not included in the market prices of products.

Why Is Producing Less Waste and Pollution the Best Choice? A small but increasing number of companies are learning that reducing waste and pollution can be good for corporate profits, worker health and safety, the local community, consumers, and the environment as a whole. Such methods **(1)** save energy and virgin resources by keeping the quality of matter resources

Q: How many people are expected to be infected with HIV by 2000?

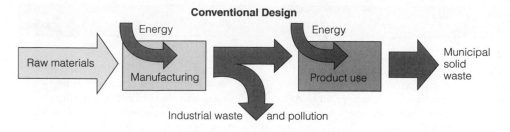

Conventional Design

Raw materials → Manufacturing (Energy) → Product use (Energy) → Municipal solid waste

Industrial waste and pollution

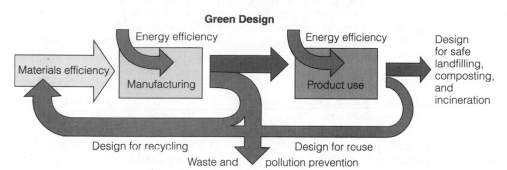

Green Design

Materials efficiency → Manufacturing (Energy efficiency) → Product use (Energy efficiency) → Design for safe landfilling, composting, and incineration

Design for recycling

Design for reuse

Waste and pollution prevention

Figure 11-4 How product design affects throughputs of matter and energy resources and outputs of pollution and solid waste. Most conventional manufacturing processes involve a one-way flow of materials from raw materials to waste and pollution. Green design reduces the overall impact by emphasizing efficient use of matter and energy resources, reduction of outputs of solid and hazardous waste, and reuse and recycling of materials. (Adapted from U.S. Congress Office of Technology Assessment, *Green Products by Design: Choices for a Cleaner U.S. Environment*, Washington, D.C.: Government Printing Office, 1992)

high with a lower input of high-quality energy (Figure 2-4); **(2)** reduce the environmental impacts of extracting, processing, and using resources (Figure 1-11); **(3)** improve worker health and safety by reducing exposure to toxic and hazardous materials; **(4)** decrease pollution control and waste management costs and future liability for toxic and hazardous materials; and **(5)** are usually less costly on a life cycle basis than trying to clean up pollutants and manage wastes once they are produced.

A 1992 study of 181 waste reduction initiatives in 27 U.S. firms found that two-thirds of the initiatives took 6 months or less to implement. One-fourth required no capital investment, two-thirds paid back their capital investments in 6 months or less, and 93% got them back within 3 years.

In 1975 the Minnesota Mining and Manufacturing Company (3M), which makes 60,000 different products in 100 manufacturing plants, began a Pollution Prevention Pays (3P) program. It redesigned equipment and processes, used fewer hazardous raw materials, identified hazardous chemical outputs (and recycled or sold them as raw materials to other companies), and began making more nonpolluting products. By 1995 3M's overall waste production was down by one-third, emissions of air pollutants were reduced by 70%, and the company had saved over $750 million in waste disposal costs.

Solutions: How Can We Reduce Waste and Pollution? There are several ways to reduce waste and pollution or resource throughput (Figure 2-47). One is to *decrease consumption*, which begins by asking whether we really need a particular product (sometimes called *precycling*).

Another way to reduce throughput is to *redesign manufacturing processes and products to use less material* (Figure 11-4). For example, today's cars are lighter because an increasing proportion of their steel parts are being replaced with aluminum (80% from recycled metal) and plastic parts. Another strategy is to *design products that produce less pollution and waste fewer resources when used* (Figure 11-4). Examples include paints that use fewer volatile solvents and using more energy-efficient cars, lights, and appliances (Section 4-2).

Manufacturing processes can be redesigned to produce less waste and pollution (Figure 11-4). Most toxic organic solvents can be recycled within plants or replaced with water-based or citrus-based solvents (Individuals Matter, p.263). Some processes can be redesigned to eliminate washing altogether. For clothes, wet cleaning (with water and steam) and microwave drying can replace dry cleaning with toxic organic solvents. In effect, over the next two decades we need to change the ways we make things (Solutions, p. 316).

People can use less hazardous (and usually cheaper) cleaning products (Table 11-2). Three inexpensive chemicals—baking soda (which can also be used as a deodorant and a toothpaste), vinegar, and borax—can be used for most cleaning and clothes bleaching. People can also use pesticides and other hazardous chemicals only when absolutely necessary and in the smallest amount possible.

Green design and life cycle assessment can help develop products that are easy to repair, reuse, remanufacture, compost, or recycle. Several European auto manufacturers now

Some analysts urge us to bring about an *ecoindustrial revolution* over the next 50 years as a way to help achieve industrial, economic, and environmental sustainability. All industrial products and processes would be redesigned and integrated into an essentially closed system of cyclical material flows.

The goals of this emerging concept of *cleaner production*, or *industrial ecology*, are to reduce resource throughput, waste, and pollution by **(1)** minimizing the input of energy and matter resources and the output of wastes, **(2)** using less material and energy per unit of output, **(3)** substituting less toxic or nontoxic chemicals in manufacturing processes or products, **(4)** recycling or reusing toxic chemicals used or produced in manufacturing processes, and **(5)** developing material exchange systems in which one company's wastes become another company's resources and companies take back packaging and used products from consumers for reuse, recycling, repair, or remanufacturing.

In effect, companies would mimic natural chemical cycles (Section 2-6) and interact in complex *resource exchange webs* similar to food webs in natural ecosystems. With such an exchange and chemical cycling network, producing a large amount of easily reusable waste might sometimes be preferable to designing a more efficient process that produces a small amount of unusable waste.

A prototype of this concept exists in Kalunborg, Denmark, where a coal-fired power plant, an oil refinery, a municipal heating authority, a producer of sulfuric acid, a sheet rock plant, a biotechnology company, local farms, some greenhouses, a fish farm, and several other businesses are working together to exchange and convert their wastes into resources. Ecoindustrial parks are in the planning stages in areas such as Baltimore (Maryland), Rochester (New York), Chattanooga (Tennessee), Halifax (Nova Scotia), and the Brownsville–Matamoros region along the Texas–Mexico border.

Such an ecoindustrial revolution, carried out over the next 50 years, will reduce pollution, improve human health, and help preserve biodiversity and ecological integrity. It will also provide economic benefits to businesses by **(1)** reducing the costs of controlling pollution and complying with pollution regulations, **(2)** improving the health of workers (thus reducing company health-care insurance expenses), **(3)** reducing legal liability for toxic and hazardous wastes, **(4)** stimulating companies to come up with new, environmentally friendly processes and products that can be sold worldwide, and **(5)** giving companies a better image among consumers, based on results rather than on public relations campaigns. Such a revolution would be a win–win situation for everyone—and for the earth.

Critical Thinking

What short- and long-term disadvantages (if any) might there be with an ecoindustrial revolution? Do you believe that it will be possible to phase in such a revolution in the country where you live over the next 2–3 decades? Explain. What are the three most important strategies for doing this?

design their cars for easy disassembly and for reuse and recycling of up to 80% of their parts (75% in the United States), and they are trying to minimize the use of nonrecyclable or hazardous materials.

Products can be designed to last longer. For example, although tires are now being produced with an average life of 97,000 kilometers (60,000 miles), researchers believe this could be extended to at least 160,000 kilometers (100,000 miles).

Eliminating or reducing unnecessary packaging is another important strategy. Here are some key questions that environmentalists believe manufacturers and consumers should ask about packaging: Is it necessary? Can it use fewer materials? Can it be reused? Are the resources that went into it nonrenewable or potentially renewable? Does it contain the highest feasible amount of consumer-discarded (postconsumer) recycled material? Is it designed to be easily recycled? Can it be incinerated without producing harmful air pollutants or a toxic ash? Can it be buried and decomposed in a landfill without producing chemicals that can contaminate groundwater?

Trash taxes can also be used to reduce waste. For example, in 1992 the city of Victoria in British Columbia instituted a trash tax of $1.20–2.10 per bag, along with a strong recycling program. Within a year, household waste generation fell by 18%. In 1986 Denmark imposed a tax on many types of solid waste to promote reuse and recycling and reduce solid waste. By 1996 the country had reduced the amount of waste brought to its municipal landfills and incinerators by 26% and achieved an overall recycling rate of 61%. A related *pay-as-you-throw* system that bases garbage collection charges on the amount of waste a household generates for disposal is being used in an increasing number of communities in the United States for reducing solid waste and encouraging recycling.

Larane, Andre. 1998. "Recycling Is Standard in French Auto Industry." *World Wastes*, vol. 41, no. 1, 8(3).

Table 11-2 Alternatives to Some Common Household Chemicals			
Chemical	**Alternative**	**Chemical**	**Alternative**
Deodorant	Sprinkle baking soda on a damp wash-cloth and wipe skin.	General surface cleaner	Mixture of vinegar, salt, and water.
Oven cleaner	Baking soda and water paste, scouring pad.	Bleach	Baking soda or borax.
Toothpaste	Baking soda.	Mildew remover	Mix ½ cup vinegar, ½ cup borax, and warm water.
Drain cleaner	Pour ½ cup salt down drain, followed by boiling water; or pour 1 handful baking soda and ½ cup white vinegar and cover tightly for one minute.	Disinfectant and general cleaner	Mix ½ cup borax in 1 gallon hot water.
Window cleaner	Add 2 teaspoons white vinegar to 1 quart warm water.	Furniture or floor polish	Mix ½ cup lemon juice and 1 cup vegetable or olive oil.
Toilet bowl, tub, and tile cleaner	Mix a paste of borax and water; rub on and let set one hour before scrubbing. Can also scrub with baking soda and a brush.	Carpet and rug shampoos	Sprinkle on cornstarch, baking soda, or borax and vacuum.
Floor cleaner	Add ½ cup vinegar to a bucket of hot water; sprinkle a sponge with borax for tough spots.	Detergents and detergent boosters	Washing soda or borax and soap powder.
Shoe polish	Polish with inside of a banana peel, then buff.	Spray starch	In a spray bottle, mix 1 tablespoon cornstarch in a pint of water.
Silver polish	Clean with baking soda and warm water.	Fabric softener	Add 1 cup white vinegar or ¼ cup baking soda to final rinse.
Air freshener	Set vinegar out in an open dish. Use an opened box of baking soda in closed areas such as refrigerators and closets. To scent the air, use pine boughs or make sachets of herbs and flowers.	Dishwasher soap	1 part borax and 1 part washing soda.
		Pesticides (indoor and outdoor)	Use natural biological controls.

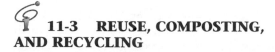

11-3 REUSE, COMPOSTING, AND RECYCLING

What Are the Advantages of Refillable Containers? *Reuse*, a form of waste reduction, extends resource supplies, keeps high-quality matter resources from being reduced to low-quality matter waste, and reduces energy use and pollution even more than recycling. Two examples of reuse are refillable glass beverage bottles and refillable soft-drink bottles made of polyethylene terephthelate (PET) plastic. Studies by Coca-Cola and PepsiCo of Canada show that 0.5-liter (16-ounce) bottles of their soft drinks cost one-third less in refillable bottles.

In 1964, 89% of all soft drinks and 50% of all beer in the United States were sold in refillable glass bottles. By 1997, such bottles made up only about 7% of the beer and soft drink market, and only 10 states even had refillable glass bottles. The disappearance of most local bottling companies has led to a loss of local jobs, income, and tax revenues. Some call for reinstatement of this bottling reuse system in the United States; others say it isn't practical because the system of collections and returns has been dismantled.

Denmark has led the way by banning all beverage containers that can't be reused. To encourage use of refillable glass bottles, Ecuador has a refundable beverage container deposit fee that is 50% of the cost of the drink. In Finland, 95% of the soft drink, beer, wine, and spirits containers are refillable, and in Germany, 73% are refillable.

Another reusable container is the metal or plastic lunchbox that most workers and schoolchildren once used. Sandwiches and refrigerator leftovers can be put in small reusable plastic containers instead of in plastic wrap and aluminum foil (most of which is not recycled). This practice and most forms of reuse save money and reduce resource waste. Another way to reduce resource waste and pollution is by the use of reusable grocery and shopping bags (Solutions, p. 318).

How Can We Recycle Organic Solid Wastes? Community Composting Compost is a sweet-smelling, dark-brown, humuslike material that is rich in organic matter and soil nutrients. It is produced when microorganisms (mostly fungi and aerobic bacteria) in soil break down organic matter such as leaves, food wastes, paper, and wood in the presence of oxygen.

What Kind of Grocery Bags Should We Use?

SOLUTIONS

When you're offered a choice between plastic or paper bags for your groceries, which should you choose? The answer is *neither*. Both are environmentally harmful, and the question of which is the more damaging has no clear-cut answer.

On one hand, plastic bags degrade slowly in landfills and can harm wildlife if swallowed, and producing them pollutes the environment. On the other hand, producing the brown paper bags used in most supermarkets uses trees and pollutes the air and water. Overall, white or clear polyethylene bags require less energy for manufacture and cause less damage to the environment than do paper bags not made from recycled paper.

Instead of having to choose between paper and plastic bags, you can bring your own *reusable* canvas or string containers to the store, and save and reuse any paper or plastic bags you get. To encourage people to bring their own reusable bags, stores in the Netherlands charge for paper or plastic bags.

Critical Thinking

Do you believe that stores should charge for paper and plastic bags as a way to encourage use of reusable bags? Explain.

Biodegradable wastes make up about 35% by weight of municipal solid waste output. Such wastes can be composted by consumers in backyard bins or indoor containers or collected and composted in centralized community facilities (as is done in many western European countries).

The resulting compost can be used as an organic soil fertilizer or conditioner, as topsoil, or as a landfill cover. Compost can also be used to help restore eroded soil on hillsides and along highways, strip-mined land, overgrazed areas, and eroded cropland.

To be successful, a large-scale composting program must overcome siting problems (few people want to live near a giant compost pile or plant), control odors, and exclude toxic materials that can contaminate the compost and make it unsuitable for use as a fertilizer on crops and lawns. Three ways to control or reduce odors for large-scale composting operations are **(1)** enclosing the facilities and filtering the air inside (but residents near large composting plants still complain of unacceptable odors), **(2)** creating municipal compost operations near existing landfills or at other isolated

sites, and **(3)** decomposing biodegradable wastes in a closed metal container in which air is recirculated to give precise control of available oxygen and temperature (a technique that has been used successfully in the Netherlands for 20 years).

What Are the Two Types of Recycling? There are two types of recycling for materials such as glass, metals, paper, and plastics: primary and secondary. The most desirable type is *primary* or *closed-loop recycling*, in which wastes discarded by consumers (*post-consumer wastes*) are recycled to produce new products of the same type (such as newspaper into newspaper and aluminum cans into aluminum cans).

A still useful but less desirable type is *secondary*, or *open-loop, recycling*, in which waste materials are converted into different products. Primary recycling reduces the amount of virgin materials in a product by 20–90%, whereas secondary recycling reduces virgin material by 25% at most.

Environmentalists urge us not to be misled by labels claiming that paper and plastic bags or other items are recyclable. Just about anything is in theory recyclable. What counts is whether an item is actually recycled and whether we complete the recycling loop by buying products using the maximum feasible content of post-consumer recycled materials.

In recent years, recycling of municipal solid waste has increase significantly in the United States and most other developed countries, especially for valuable materials such as aluminum (Case Study, p. 320) and wastepaper. By 1996 the United States had more than 8,800 municipal curbside recycling programs serving 51% of the population. Recent pilot studies in several U.S. communities show that a 60–80% recycling and composting rate is possible.

One way to spur recycling is a *pay-as-you-throw* program that bases garbage collection charges on the amount of waste a household generates for disposal; materials sorted out for recycling are hauled away free. Currently, more than 2,800 communities in the United States have curbside pay-as-you-throw systems. Studies have shown that this is the single best way to boost recycling.

Is Centralized Recycling of Mixed Solid Waste the Answer? Large-scale recycling can be accomplished by collecting mixed urban waste and transporting it to centralized *materials-recovery facilities (MRFs)*. There, machines shred and automatically separate the mixed waste to recover glass (which can be melted and converted to new bottles or to fiberglass insulation), iron, aluminum, and other valuable materials (Figure 11-5); these are then sold to manufacturers as raw materials. The remaining paper, plastics, and other

Q: How many people die of malaria each year?

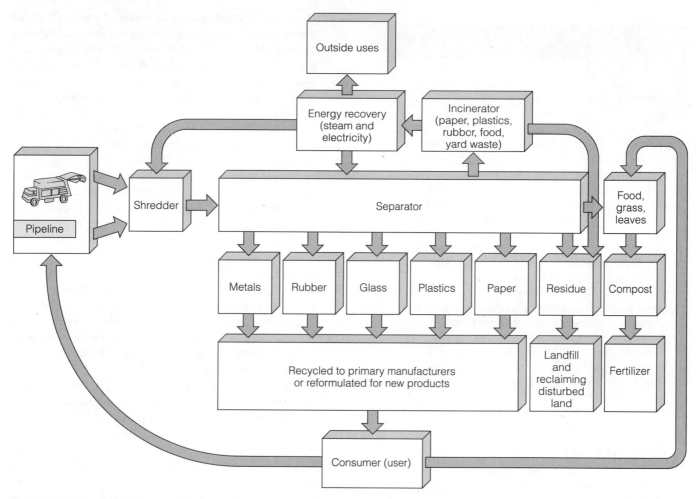

Figure 11-5 Schematic of a generalized materials-recovery facility used to sort mixed wastes for recycling and burning to produce energy. Because such plants require high volumes of trash to be economical, they discourage reuse and waste reduction.

combustible wastes are recycled or burned. The resulting heat produces steam or electricity to run the recovery plant or to sell to nearby industries or homes.

By 1996 the United States had more than 220 materials-recovery facilities, and at least 60 more were in the planning stages. However, such plants are expensive to build and maintain, and once trash is mixed it takes a lot of money and energy to separate it, which is why some MRFs have shut down. MRFs must have a large input of garbage to make them financially successful, so their owners have a vested interest in increased *throughput* of matter and energy resource to produce more trash—the reverse of what scientists believe we should be doing (Figure 11-2).

These facilities also can emit toxic air pollutants and produce a toxic ash that must be disposed of safely. Most communities collecting mixed solid waste can't afford to build or operate high-tech resource recovery facilities. Instead, to meet government-required recy-

cling goals, many communities hire workers to sort the trash by hand. This is a hazardous, low-pay job.

Is Separating Solid Wastes for Recycling the Answer? Many solid-waste experts argue that it makes more sense economically and environmentally for households and businesses to separate trash into recyclable and reusable categories (such as glass, paper, metals, certain types of plastics, and compostable materials) before it is picked up. Compartmentalized city collection trucks, private haulers, or volunteer recycling organizations then pick up the segregated wastes and sell them to scrap dealers, compost plants, and manufacturers. Another alternative (especially in less populated areas) is the establishment of a network of drop-off centers, buyback centers, and deposit-refund programs in which people deliver and either sell or donate their separated recyclable materials.

Recycling Aluminum

Worldwide, the recycling rate for aluminum in 1996 was about 35% (34% in the United States). Recycling aluminum produces 95% less air pollution and 97% less water pollution, and it requires 95% less energy than mining and processing aluminum ore.

In 1994, 62% (compared to 15% in 1973) of aluminum beverage cans in the United States were recycled. Despite this progress, about 38% of the 95 billion aluminum cans produced in 1994 in the United States was still thrown away. Laid end to end, these cans would wrap around the planet more than 120 times.

Recycling aluminum cans is great, but many environmentalists believe that these cans are an example of an unnecessary item that could be replaced by more energy-efficient and less polluting refillable glass or PET plastic bottles—a switch from recycling to reuse that also creates local jobs. One way to encourage this change would be to place a heavy tax on nonrefillable containers and no tax on reusable beverage containers, as is done in at least nine countries.

Critical Thinking

Are you in favor of placing a heavy tax on nonrefillable containers and no tax on reusable containers? Explain.

The source separation approach produces little air and water pollution, reduces litter, and has low start-up costs and moderate operating costs. It also saves more energy and provides more jobs for unskilled workers than centralized MRFs; it also creates three to six times more jobs per unit of material than landfilling or incineration. In addition, separated recyclables are cleaner and can usually be sold for a higher price. Source separation also educates people about the need for waste reduction, reuse, and recycling (Individuals Matter, right).

Aluminum and paper separated out for recycling are worth a lot of money. As a result, in a growing number of cities people steal these materials—from curbside containers set out by residents and from unprotected recycling drop-off centers—and sell them. This undermines municipal recycling programs by lowering the income available from selling these high-value materials.

How Much Wastepaper Is Being Recycled? Paper (especially newspaper and cardboard) is one of the easiest materials to recycle. In 1996 the United States recycled about 40% of its wastepaper (up from 25% in 1989) and within a few years is projected to recycle 50%. At least 10 other countries recycle 50–98% of their wastepaper.

Recycling the Sunday newspapers in the United States alone would save 500,000 trees per week. In addition to saving trees, recycling paper (1) saves energy because it takes 30–64% less energy to produce the same weight of recycled paper as to make the paper from trees, (2) reduces air pollution from pulp mills by 74–95%, (3) lowers water pollution by 35%, (4) helps prevent groundwater contamination by toxic ink left after paper rots in landfills over a 30- to 60-year period, (5) conserves large quantities of water, (6) can save landfill space, (7) creates five times more jobs than harvesting trees for pulp, and (8) can save money.

Buying recycled paper products can save trees and energy and reduce pollution, but it does not necessarily reduce solid waste. Only products made from *postconsumer waste*—waste intercepted on its way from consumer to the landfill—do that.

Most recycled paper is actually made from *preconsumer waste*: scraps and cuttings recovered from paper and printing plants. Because paper manufacturers have always recycled this waste, it has never contributed to landfill problems. Now this paper is labeled "recycled" as a marketing ploy, giving the false impression that people who buy such products (often at higher prices) are helping the solid-waste problem. Most "recycled" paper has no more than 50% recycled fibers, with only 10% from postconsumer waste. Environmentalists propose that the government require companies to report the amount of postconsumer recycled materials in paper and other products and reserve the term *recycled* only for items using postconsumer recycled materials. What do you think?

Is It Feasible to Recycle Plastics? Currently, only about 5% by weight of all plastic wastes and 6% of plastic packaging used in the United States are recycled because there are so many different types of plastic resin. Before they can be recycled, plastics in trash must be separated into different types of resins by consumers or separated from mixed trash (a costly, labor-intensive procedure unless automated separating technologies can be developed). Another problem is that the current price of oil is so low that the price of virgin plastic resins (except for PET, used mostly in plastic drink bottles) is about 40% lower than that of recycled resins. However, in 1998 Chrysler Corporation built an experimental automobile with a body made from plastic derived entirely from recycled PET bottles and expects to sell such vehicles within a decade.

Q: In terms of deaths, what is the world's most dangerous viral disease?

Environmentalists recognize the beneficial qualities of plastics: durability (in products such as car and machine parts, carpeting, toys, furniture, reusable tubs and containers, and refillable bottles), light weight, unbreakability (compared to glass), and in some cases reusability as containers. But many environmentalists believe that some widespread uses of plastics—especially excessive and often unnecessary single-use packaging and throwaway beverage and food containers—should be sharply reduced and replaced with less harmful and wasteful alternatives. What do you think?

Why Don't We Have More Reuse and Recycling? Three factors that hinder recycling (and reuse) are (1) failure to include the environmental and health costs of raw materials in the market prices of consumer items, (2) more tax breaks and subsidies for resource-extracting industries than for recycling industries, and (3) lack of large, steady markets for recycled materials. Recently some critics have claimed that recycling costs more than it's worth (Pro/Con, p. 322).

Analysts have suggested various ways to overcome the obstacles to recycling, including the following:

- Taxing virgin resources and phasing out subsidies for extraction of virgin resources

- Lowering or eliminating taxes on recycled materials based on postconsumer waste content

- Providing subsidies for reuse and postconsumer waste recycling

- Requiring households and businesses to pay directly for garbage collection based on how much they throw away, with lower charges or no charges for materials separated for recycling and composting

- Encouraging or requiring government purchases of recycled products to help increase demand and lower prices

- Viewing landfilling and incineration of solid wastes as last resorts to be used only for wastes that can't be reused, composted, or recycled (Figure 11-2)

- Requiring ecolabels on all products, evaluating life cycle environmental costs and listing preconsumer and postconsumer recycled content

11-4 DETOXIFYING, BURNING, BURYING, AND EXPORTING WASTES

How Can Hazardous Waste Be Detoxified? Denmark has the most comprehensive and effective hazardous-waste detoxification program. Each Danish municipality has at least one facility that accepts paints, solvents, and other hazardous wastes from house-

Source Separation Recycling in Some Georgia Schools

INDIVIDUALS MATTER

In Rome, Georgia, environmental educator Steve Cordle has set up an exciting project that combines the important concept of source separation recycling with environmental education.

He has designed and implemented a source separation recycling program for 17 Floyd Country Schools in Georgia. Before his program, which began in 1992, the schools were a major contributor to the county landfill and hardly any of their waste output was recycled.

During a 7-month period in 1996, approximately 114 metric tons (250,000 pounds) of paper, corrugated cardboard, and aluminum and steel cans were collected in separate containers. The school system stored and sold the items for recycling. During this period the project added $10,139.69 to school funds, although school officials consider the money incidental compared to the educational value of the program.

This important experiment in *environmental education* teaches students (and teachers) from kindergarten through high school about the need for recycling and engages them in hands-on participation in source separation. Mixing and throwing waste materials in garbage cans (which should be called *resource containers*) and hauling them off to a landfill, incinerator, or mixed-resource recycling plant is an out-of-sight, out-of-mind approach that does little to further environmental education.

Cordle is working hard to expand this idea to a larger area and hopefully to Georgia's entire school system. With proper backing and support, this program could become a model for use throughout the United States and other parts of the world.

Critical Thinking

Use the second law of energy (thermodynamics) to explain why a *properly designed* source separation recycling program takes less energy and produces less pollution than a centralized program that collects mixed waste over a large area and hauls it to a centralized facility where workers or machinery separate the wastes for recycling.

holds. Hazardous and toxic waste from industries is delivered to 21 transfer stations throughout the country. All waste is then transferred to a large treatment facility, where about 75% of it is detoxified; the rest is buried in a carefully designed and monitored landfill.

Does Recycling Make Economic Sense?

The answer is yes and no, depending on different ways of looking at the economic and environmental costs and benefits of recycling. Critics contend that recycling **(1)** has become almost a religion that is above criticism regardless of how much it costs communities, **(2)** does not make sense if it costs more to recycle materials than to send them to a landfill or incinerator (as is the case in some areas), **(3)** is often not needed to save landfill space because most areas in the United States are not running out of landfill space, and **(4)** may make economic sense for valuable and easy-to-recycle materials (such as aluminum, mercury, paper, and steel) but not for cheap or plentiful resources (such as glass from silica) and most plastics (which are expensive to recycle).

Many communities established recycling programs with the idea that they should pay for themselves. But recycling proponents argue that *recycling programs should not be judged on whether they pay for themselves any more than are conventional garbage disposal systems based on land burial or incineration.*

Moreover, recycling proponents contend that with full-cost accounting, the net economic, health, and environmental benefits of recycling far outweigh the costs. A study by MIT economist Robert F. Stone revealed that recycling in Massachusetts yields a net social benefit of about $254 per metric ton ($231 per ton) if both the roughly $146 per metric ton ($133 per ton) environmental benefits of recycling and the avoided solid-waste disposal costs are taken into account.

An analysis of various recycling programs revealed that cities and private waste collection firms tend to make money if they have high recycling rates and a single pickup system (for both materials to be recycled and garbage that can't be recycled). Cities cited by critics as losing money on recycling often have expensive dual collection systems and have not designed programs to encourage more cost-effective, high rates of recycling.

Finally, environmentalists point out that the primary benefit from recycling is *not* a reduction in the use of landfills and incinerators. Instead, *the major reasons for recycling are reduced use of virgin resources, reduced throughput of matter and energy resources, and reduced pollution and environmental degradation* (Figure 2-47).

Critical Thinking

Do you believe that recycling programs should be set up and continued only if they make money? Explain.

Some consider biological treatment of hazardous waste, or *bioremediation*, to be the wave of the future for cleaning up some types of toxic and hazardous waste. In this process, microorganisms, usually bacteria, are used to destroy toxic or hazardous substances or convert them to harmless forms. This approach mimics nature by using decomposers to recycle matter.

If toxin-degrading bacteria or fungi can be found or engineered for specific hazardous chemicals, they could clean up contaminated sites at less than half the cost of disposal in landfills, and at only about one-third the cost of on-site incineration. Bioremediation might also clean up contaminated groundwater at an affordable cost by pumping it to the surface, treating it with microorganisms, and then returning it to its aquifer.

Preliminary testing reveals that bioremediation is effective for a number of specific organic wastes, but it does not appear to work very well for toxic metals, highly concentrated chemical wastes, or complete digestion of some complex mixtures of toxic chemicals.

Is Burning Solid and Hazardous Waste the Answer? In 1996 about 15% of the municipal solid waste and 7% of the officially regulated hazardous waste in the United States was burned in incinerators. Most of this mixed municipal solid waste was burned in 150 *mass-burn incinerators* (Figure 11-6), which burn mixed trash without separating out either hazardous materials (such as batteries and polyvinylchloride or PVC plastic materials) or noncombustible materials that can interfere with combustion and pollute the air.

Incinerators are costly to build, operate, and maintain and they create very few long-term jobs. Without continual maintenance and good operator training and supervision, the air pollution control equipment on incinerators often fails, and emission standards are exceeded.

Hazardous-waste expert Peter Montague and some EPA scientists point out that even with advanced air pollution control devices *all incinerators burning hazardous or solid waste release toxic air pollutants.* They also leave a highly toxic ash to be disposed of in landfills that even the EPA concedes will eventually leak.

According to EPA hazardous-waste expert William Sanjour, EPA incinerator regulations don't work because

> *the regulations require no monitoring of the outside air in the vicinity of the incinerator. Because operators maintain the records, they can easily cheat. . . . Government inspectors are poorly trained and have low morale and high turnover. . . . Government inspectors typically work from nine to five Monday through Friday. So if there is anything particularly nasty to burn, it will be done at night or on weekends. When*

Q: In terms of reduced life span, what is the world's greatest risk?

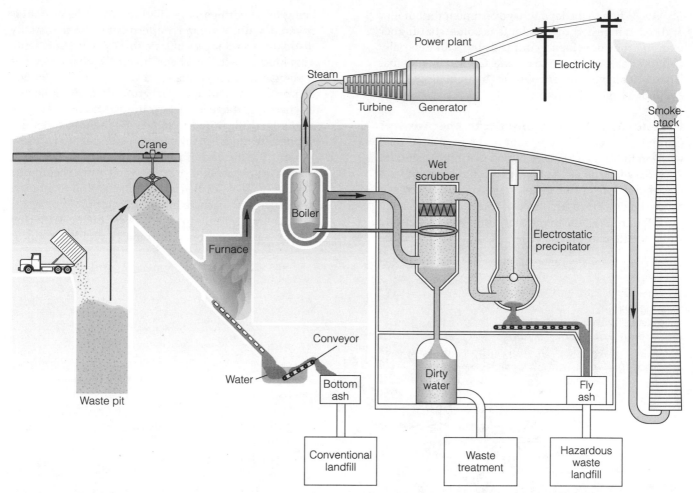

Figure 11-6 Schematic of a waste-to-energy incinerator with pollution controls that burns mixed solid waste and recovers some of the energy to produce steam used for heating or producing electricity. (Adapted from EPA, *Let's Reduce and Recycle*)

complaints come in, . . . the inspector may visit the plant but rarely finds anything. The enforcement officials tend to view the incinerator operator as their client and the public as a nuisance. . . . There is no reward to inspectors for finding serious violations.

In 1993 the *Wall Street Journal* warned that using incinerators to burn municipal trash spells financial disaster for local governments in the United States. Since 1992 Rhode Island and West Virginia have banned solid-waste incineration because of its health threats and high cost. Sweden banned the construction of new incinerators in 1985.

Between 1985 and 1997, several solid-waste incinerators in the United States were shut down because of excessive costs and pollution, and over 280 new incinerator projects were blocked, delayed, or canceled because of intense public opposition and high costs. Most of the plants still in the planning stage may not be

built as communities discover that recycling, reuse, composting, and waste reduction are cheaper, safer, and less environmentally harmful.

Japan depends on incinerators more than any country, burning about 75% of its municipal solid waste in more than 1,850 government-operated incinerators and more than 3,300 privately owned industrial incinerators. Many of these incinerators are built near populous areas.

Recently there has been growing concern in Japan about reported rising infant deaths and health problems in areas downwind of many Japanese incinerators from emissions of toxic dioxins (Section 11-5). The head of a government advisory committee on dioxins found that the concentration of dioxins in the air in Japan is three times that in the United States and some European countries. Dioxins are also found at high levels in the fatty tissues of fish that play an important role in the Japanese diet.

According to the Japanese government the amount of dioxins ingested by residents is somewhat higher than in other developed countries but still generally within levels considered to be safe. Critics contend that Japan is far behind other countries in regulating dioxin emissions from incinerators.

Is Land Disposal of Solid Waste the Answer?
Currently, about 57% by weight of the municipal solid waste in the United States is buried in sanitary landfills. A **sanitary landfill** is a garbage graveyard in which solid wastes are spread out in thin layers, compacted, and covered daily with a fresh layer of clay or plastic foam.

Modern state-of-the-art landfills on geologically suitable sites are lined with clay and plastic before being filled with garbage (Figure 11-7). The bottom is covered with a second impermeable liner, usually made of several layers of clay, thick plastic, and sand. This liner collects *leachate* (rainwater contaminated as it percolates through the solid waste) and is intended to prevent its leakage into groundwater. Collected leachate ("garbage juice") is pumped from the bottom of the landfill, stored in tanks, and sent to a regular sewage treatment plant or an on-site treatment plant. When full, the landfill is covered with clay, sand, gravel, and topsoil to prevent water from seeping in. Several wells are drilled around the landfill to monitor any leakage of leachate into nearby groundwater.

Sanitary landfills for solid wastes offer certain benefits. Air-polluting open burning is avoided. Odor is

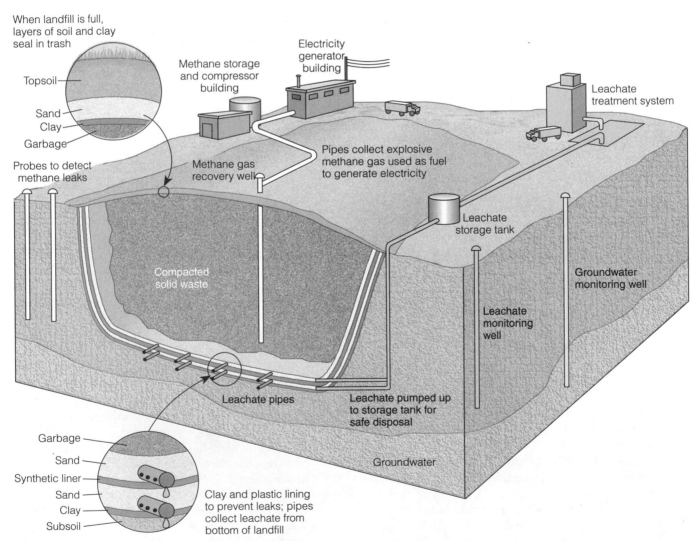

Figure 11-7 State-of-the-art sanitary landfills are designed to eliminate or minimize environmental problems that plague older landfills. Only a few of the 7,500 municipal and industrial landfills in the United States have such a state-of-the-art design, and 85% of U.S. landfills are unlined. Even state-of-the-art landfills will eventually leak, passing both the effects of contamination and cleanup costs on to future generations.

Q: What two gases make up 99% of the volume of air in the atmosphere's innermost layer (troposphere)?

seldom a problem, and rodents and insects cannot thrive. If located properly they can reduce water pollution from leaching, but proper siting is not always achieved. A sanitary landfill can be put into operation fairly quickly, has low operating costs, and can handle a huge amount of solid waste. After a landfill has been filled, the land can be graded, planted, and used as a park, golf course, ski hill, athletic field, or wildlife area, or for some other recreational purpose.

Solid-waste landfills also have drawbacks. While in operation, they cause traffic, noise, and dust; most also emit toxic gases. Paper and other biodegradable wastes break down very slowly in today's compacted and water- and oxygen-deficient landfills. Newspapers dug up from some landfills are still readable after 30 or 40 years; hot dogs, carrots, and chickens that have been dug up after 10 years have not decomposed. Landfills also deprive present and future generations of valuable reusable and recyclable resources and encourage waste production instead of pollution prevention and waste reduction.

The underground anaerobic decomposition of organic wastes at landfills produces explosive methane (a greenhouse gas), toxic hydrogen sulfide gas, and smog-forming volatile organic compounds that escape into the air. Large landfills can be equipped with vent pipes to collect these gases, and the collected methane can be burned in small power plants or fuel cells to produce steam or electricity (Figure 11-7). Methane must be collected in all new landfills in the United States, but thousands of older and abandoned landfills don't have such systems and will emit methane (a potent greenhouse gas; see Figure 9-6c) for decades.

Contamination of groundwater and nearby surface water by leachate from both unlined and lined landfills is another serious problem. Some 86% of the U.S. landfills studied have contaminated groundwater, and a fifth of all Superfund hazardous-waste sites are former municipal landfills that will cost billions of dollars to clean up. Once groundwater is contaminated, it is extremely expensive and difficult—often impossible—to clean up.

Modern double-lined landfills (Figure 11-7), required in the United States since 1997, delay the release of toxic leachate into groundwater below landfills, but they do not prevent it. These landfills are designed to accept waste for 10–40 years, and current EPA regulations require owners to maintain and monitor landfills for at least 30 years after they are closed. However, they could begin to leak after this period, passing the health risks and costs of contamination to future generations.

Since 1979 there has been a sharp drop in the number of solid-waste landfills in the United States as existing landfills reached their capacity or closed; since 1997 only more expensive, state-of-the art landfills (Figure 11-7) are allowed to operate. Although a few U.S. cities, notably Philadelphia and New York, are having trouble finding nearby landfills, there is no national shortage of landfill volume because most of the smaller local landfills are being replaced by larger local and regional landfills.

Is Land Disposal of Hazardous Waste the Answer?
Most hazardous waste in the United States is disposed of by deep-well injections, surface impoundments, and state-of-the-art landfills. In *deep-well disposal*, liquid hazardous wastes are pumped under pressure through a pipe into dry, porous geologic formations or into fracture zones of rock far beneath aquifers tapped for drinking and irrigation water (Figure 10-19). In theory, these liquids soak into the porous rock material and are isolated from overlying groundwater by essentially impermeable layers of rock.

Because this method is simple and cheap, its use is increasing rapidly as other methods are legally restricted or become too expensive. If sites are chosen according to the best geological and seismic data, a number of scientists believe deep wells may be a reasonably safe way to dispose of fairly dilute solutions of organic and inorganic waste; with proper site selection and care, they may be safer than incineration. Furthermore, if some use eventually is found for the waste, it could be pumped back to the surface.

However, many scientists believe that current regulations for geologic evaluation, long-term monitoring, and long-term liability if wells contaminate groundwater are inadequate. Until deep-well disposal is more carefully evaluated and regulated, most environmentalists believe that it should not be allowed to proliferate. What do you think?

Surface impoundments such as ponds, pits, or lagoons (Figure 10-19) used to store hazardous waste are supposed to be sealed with a plastic liner on the bottom. Solid wastes settle to the bottom and accumulate, whereas water and other volatile compounds evaporate into the atmosphere. According to the EPA, however, 70% of these storage basins have no liners, and as many as 90% may threaten groundwater. Eventually all liners will leak, and waste will percolate into groundwater. Moreover, volatile compounds (such as hazardous organic solvents) can evaporate into the atmosphere, promote smog formation, and return to the ground to contaminate surface water and groundwater in other locations.

About 5% by weight of the legally regulated hazardous waste produced in the United States is concentrated, put into drums, and buried, either in one of 21 specially designed and monitored commercial hazardous-waste landfills or in one of 35 such landfills run by companies to handle their own waste. Sweden goes further and buries its concentrated hazardous wastes in underground vaults (Figure 11-8). In the United Kingdom, most hazardous wastes are mixed with household garbage and stored in hundreds of conventional landfills all over the country.

When current and future commercial hazardous-waste landfills in the United States eventually leak and threaten water supplies, many of their operators will declare bankruptcy. Then the EPA will put the landfills on the Superfund list and taxpayers will pick up the tab for cleaning them up. As EPA hazardous-waste expert William Sanjour points out,

The real cost of dumping is not borne by the producer of the waste or the disposer, but by the people whose health and property values are destroyed when the wastes migrate onto their property and by the taxpayers who pay to clean it up. . . . It is better for liners to leak sooner rather than later, because then there will be responsible parties that they can get to clean it up. Liners don't protect communities. They protect the people who put the waste there and the politicians who let them put the waste there, because they are long since gone when the problem comes up.

Some engineers and environmentalists have proposed storing hazardous wastes above ground in large, two-story, reinforced-concrete buildings until better technologies are developed (Figure 11-9). The first floor would contain no wastes but would have

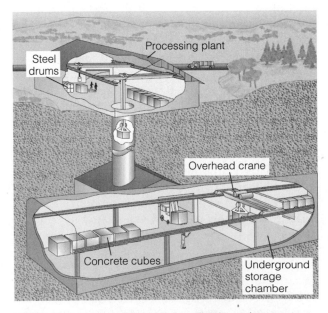

Figure 11-8 Swedish method for handling hazardous waste. Hazardous materials are placed in drums, which are embedded in concrete cubes and stored in an underground vault.

inspection walkways so people could easily check for leaks from the upper story. Any leakage would be collected, treated, solidified, and returned to the storage building. The buildings would have negative air pressure—caused by fans blowing in air and then filtering it—to prevent the release of toxic gases. Proponents believe that such an *in-sight* approach would be cheaper and safer than *out-of-sight* landfills, deep wells, or incinerators for many hazardous wastes. What do you think?

There is growing concern about accidents during some of the more than 500,000 shipments of hazardous wastes (mostly to landfills and incinerators) in the United States each year. Between 1988 and 1992, approximately 34,500 toxic chemical accidents caused 100 deaths, over 11,000 injuries, and evacuation of over 500,000 people. Few communities have the equipment and trained personnel to deal with hazardous-waste spills.

Is Exporting Waste the Answer? Some U.S. cities that lack sufficient landfill space are shipping some of their municipal solid waste to other states or other countries, especially developing countries. However, a few of these states and some developing countries are refusing to accept delivery.

To save money and to escape regulations and local opposition, some U.S. cities and waste disposal companies legally ship large quantities of hazardous waste to other countries. U.S. exports of hazardous wastes can take place without EPA approval because U.S. laws allow exports for recycling; some exported wastes designated for recycling are instead dumped after reaching their destination.

Waste disposal firms can charge high prices for picking up hazardous wastes. If they can then dispose of them (legally or illegally) at low costs, they pocket huge profits. Often officials of developing countries find it difficult to resist the income (often bribes) derived from receiving these wastes.

In 1994, 64 nations met in Basel, Switzerland, and developed a convention banning all exports of hazardous wastes for disposal from developed countries to developing countries (including eastern Europe and the former Soviet Union). By 1998 the convention had been ratified by 118 countries, but not the United States.

An effective worldwide ban on all hazardous-waste exports, including those slated to be recycled (often a sham), would help reduce this transfer of risk and encourage developed countries to reduce their production of such wastes. But this would not end illegal trade in these wastes because the potential profits are much too great. To most environmentalists, the only real solution to the hazardous-waste problem is not to produce it in the first place (pollution prevention).

Q: What is the greenhouse effect?

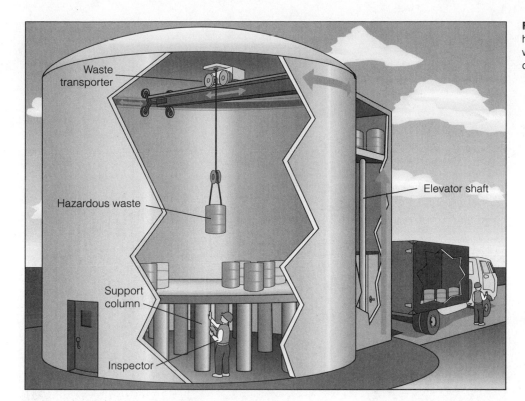

Waste transporter

Hazardous waste

Elevator shaft

Support column

Inspector

11-5 CASE STUDIES: LEAD AND DIOXINS

How Can We Reduce Exposure to Lead? Each year 12,000–16,000 American children under age 9 are treated for acute lead poisoning, and about 200 die. About 30% of the survivors suffer from palsy, partial paralysis, blindness, and mental retardation. Research indicates that children under age 6 and unborn fetuses with fairly low blood levels of lead are especially vulnerable to nervous system impairment, a lowered IQ (by 4–6 points), a shortened attention span, hyperactivity, hearing damage, and behavior disorders.

It is great news that between 1976 and 1992 the percentage of U.S. children ages 1–5 with lead levels above the 10 µg/dL standard dropped from 85% to 6% for white children and from 98% to 21% for black children, preventing at least 9 million childhood lead poisonings. The primary reason is government regulations that phased out leaded gasoline and lead-based solder, a pollution prevention approach instituted primarily because of the pioneering research on lead toxicity in children published in 1979 by Dr. Herbert Needleman.

Even with the encouraging drop in average blood levels of lead, a large number of U.S. children and unborn fetuses still have unsafe levels of lead from exposure to a number of sources (Figure 11-10). Health scientists have proposed a number of ways to help protect children from lead poisoning:

- *Testing all children for lead by age 1.*

- *Banning incineration of solid and hazardous waste or greatly increasing current pollution control standards for old and new incinerators.*

- *Phasing out leaded gasoline worldwide over the next decade.*

- *Testing older housing and buildings for leaded paint and lead dust and removing this hazard.* According to government estimates, 74% of U.S. homes contain lead-based paint.

- *Banning all lead solder in plumbing pipes and fixtures and in food cans.*

- *Removing lead from municipal drinking water systems within 10 years.*

- *Washing fresh fruits and vegetables and hands thoroughly.*

- *Testing ceramicware used to serve food for lead glazing.*

- *Reevaluating the proposed increase in electric cars propelled by lead batteries.* A 1995 study estimated that the manufacture, handling, disposal, and recycling of lead batteries for widespread use in electric vehicles could create up to 60 times the lead pollution caused by vehicles burning leaded gasoline.

A: A natural process in which certain gases trap heat in the lower atmosphere (troposphere)　　　　　CHAPTER 11　　**327**

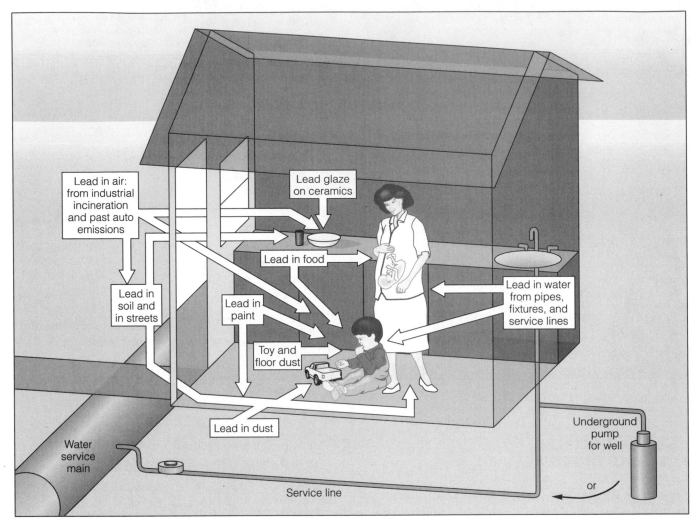

Figure 11-10 Sources of lead exposure for children and fetuses.

Doing most of these things will cost an estimated $50 billion in the United States. But health officials say the alternative is to keep poisoning and mentally handicapping large numbers of children. What do you think?

In the cities of many developing countries where car use is rising, lead levels in the air, water, and blood are rising because leaded gasoline is still used and factories discharge large quantities of lead into waterways. In Chinese cities such as Shanghai, where the roads are clogged with new cars, studies have found that at least 65% of the children have lead levels in their blood higher than the point considered dangerous to mental development.

How Dangerous Are Dioxins? Dioxins are a family of 75 chlorinated hydrocarbon compounds formed as unwanted by-products in chemical reactions (usually at high temperatures, especially in incinerators) involving chlorine and hydrocarbons. One of these compounds, TCDD—sometimes simply called dioxin—is the most harmful and the most widely studied. Dioxins such as TCDD are persistent chemicals that linger in the environment for decades, especially in soil and human fat tissue.

In 1990 representatives of the paper and chlorine industries claimed that the inconsistent results of health studies exonerated TCDD and other dioxins, and they pushed the EPA for a reassessment of the health risks of dioxins. This strategy may have backfired when EPA's 1994 reevaluation indicated that dioxins are an even greater health threat than previously thought.

This new review concluded that dioxin is "likely to present a cancer threat to humans." According to the EPA review, TCDD and several other dioxins in food and air may already be a major cancer hazard; existing levels of dioxins in the U.S. population are now believed

Q: What is a greenhouse gas?

to cause an estimated 2.5–25% of all new cancers, or 26,000–260,000 new cancers each year.

This comprehensive, 3-year, $4-million review by more than 100 scientists around the world also revealed that the most powerful effects of exposure to various dioxins at or near levels already occurring in the U.S. population are seen in the reproductive, endocrine (hormone), and immune systems. These effects may pose an even greater threat to human health (especially for fetuses and newborns) than the cancer-promoting ability of various dioxins. With profits at stake, there is intense industry pressure to soften or discredit this new health reassessment of dioxins.

11-6 HAZARDOUS-WASTE REGULATION IN THE UNITED STATES

What Is the Resource Conservation and Recovery Act? In 1976, the U.S. Congress passed the Resource Conservation and Recovery Act (RCRA, pronounced "RICK-ra"), which was amended in 1984. This law requires the EPA to identify hazardous wastes and set standards for their management, and it provides guidelines and financial aid for states to establish waste management programs. The law also requires all firms that store, treat, or dispose of more than 100 kilograms (220 pounds) of hazardous wastes per month to have a permit stating how such wastes are to be managed.

To reduce illegal dumping, hazardous-waste producers that are granted disposal permits by the EPA must use a cradle-to-grave system to keep track of waste transferred from point of origin to approved off-site disposal facilities. However, the EPA and state regulatory agencies don't have sufficient personnel or money to review the documentation of more than 750,000 hazardous-waste generators and 15,000 haulers each year, let alone detect and prosecute offenders. Environmentalists argue that fines are too small and that culprits don't get jail time for serious violations that can cause severe health problems and even death, sending violators the clear message that pollution pays.

What Is the Superfund Act? In 1980, the U.S. Congress passed the Comprehensive Environmental Response, Compensation, and Liability Act, commonly known as the Superfund program. This law (plus amendments in 1986 and 1990) established a $16.3-billion Superfund financed jointly by federal and state governments and by special taxes on chemical and petrochemical industries (which provide 86% of the funding). The purpose of the Superfund is to identify and clean up abandoned hazardous-waste dump sites

and leaking underground tanks that threaten human health and the environment.

To keep taxpayers from footing most of the bill, cleanups are based on the *polluter-pays principle*. The EPA is charged with locating dangerous dump sites, finding the potentially liable culprits, ordering them to pay for the entire cleanup, and suing them if they don't. When the EPA can find no responsible party, it draws money out of the Superfund for cleanup.

Currently about 1,360 sites are on a National Priority List for cleanup because they pose a real or potential threat to nearby populations; the number may eventually grow to 2,000–3,000. By June 1997 emergency cleanup had been carried out at virtually all sites and more than 400 sites had been cleaned up by containing the wastes to prevent leakage. The EPA has plans to stabilize an additional 500 sites by 2000. Once a site is stabilized, final cleanup (which often involves pumping up and cleaning contaminated groundwater) can take decades. Attempts are being made to find ways to improve the Superfund Act without seriously weakening it (Solutions, p. 330).

The U.S. Office of Technology Assessment and the Waste Management Research Institute estimate that the Superfund list could eventually include at least 10,000 priority sites, with cleanup costs of up to $1 trillion, not counting legal fees. Other studies project only 2,000 sites with estimated cleanup costs of $100–165 billion.

Cleaning up toxic military dumps will cost another $100–200 billion and take at least 30 years. Cleaning up contaminated Department of Energy sites used to make nuclear weapons will cost an additional $200–400 billion and take 30–50 years. In addition, the Department of Interior will need to spend at least $100 billion to clean up 300,000 active and abandoned mine sites, 200,000 oil and gas leases, 3,000 landfills, and 4,200 leaking underground storage tanks on public lands under its jurisdiction.

These estimated costs are only for cleanup to prevent future damage; they don't include the health and ecological costs already associated with such wastes. To environmental scientists and some economists, it is difficult to imagine a more convincing reason for emphasizing waste reduction and pollution prevention (Figures 11-2 and 11-3).

11-7 SOLUTIONS: ACHIEVING A LOW-WASTE SOCIETY

What Is the Role of Grassroots Action? Bottom-Up Change In the United States, local citizens have been stirred into action by proposals to bring incinerators, landfills, or treatment plants for hazardous and

---

A: A gas (such as CO_2 or water vapor) that traps heat in the lower atmosphere (troposphere)

Since the Superfund program began, polluters and their insurance companies have been working hard to do away with the polluter-pays principle at the heart of the program and make it mostly a *public-pays* approach.

This strategy, which is working, has three components: **(1)** Deny responsibility (stonewall) in order to tie up the EPA in expensive legal suits for years, **(2)** sue local governments and small businesses to make them responsible for cleanup, both as a delay tactic and to turn local governments and small businesses into opponents of Superfund's strict liability requirements, and **(3)** mount a public relations campaign declaring that toxic dumps pose little threat, that cleanup is too expensive compared to the risks involved, and that Superfund is unfair to polluters and is wasteful and ineffective.

The strict joint and several liability of the Superfund law may seem unfair to polluters, but the authors of the legislation argued that any other liability scheme wouldn't work.

If the EPA had to identify and bring an enforcement action against every party liable for a dump site, the agency would be overwhelmed with lawsuits, and the pace of cleanup would be still slower. Administrative costs would also rise sharply, causing the program to become more of a *public-pays program* favored by the polluters.

The EPA also points out that the strict polluter-pays principle in the Superfund Act has been effective in making illegal dump sites virtually a thing of the past—an important form of pollution prevention for the future. It has also forced waste producers who are fearful of future liability claims to reduce their production of such waste and to recycle or reuse much more of what they do generate.

A generally accepted proposal for reducing lawsuits and speeding cleanup is to set up an $8-billion Environmental Insurance Resolution Fund funded by insurers for a 10-year period. Companies found liable for cleanup of wastes they disposed of before 1986 would be able to use money from this fund rather than going to their individual insurance companies.

Many analysts argue that sites should be ranked in three general categories: **(1)** sites requiring immediate full cleanup, **(2)** sites considered to pose serious hazard but located sufficiently far from concentrations of people or endangered ecosystems (these sites would receive emergency cleanup and then be isolated by barriers and signs, with more complete cleanup to come later), and **(3)** lower-risk sites requiring only stabilization (capping and containment) and monitoring. Other analysts believe that people and local governments should be involved in cleanup decisions from start to finish.

Critical Thinking

Explain why you agree or disagree with the suggestions listed here for improving the Superfund law. Use the library or the Internet to determine what changes (if any) have been made in the Superfund law and explain why you agree or disagree with these changes.

---

radioactive wastes into their communities. Outrage has grown as numerous studies have shown that such facilities have traditionally been located in communities populated by African-Americans, Asian-Americans, Hispanics, and poor whites.

This loose coalition of grassroots organizations offers the following guidelines for helping to achieve environmental justice for all:

- *Don't compromise our children's futures by cutting deals with polluters and regulators.*

- *Don't be bulldozed by scientific and risk analysis experts.* Risks from incinerators and landfills, when averaged over the entire country, are quite low, but the risks for the people near these facilities are much higher. These people, not the rest of the population,

are the ones whose health, lives, and property values are being threatened.

- *Hold polluters—and elected officials who go along with them—personally accountable because what they are doing is wrong.*

- *Don't fall for the argument that protesters against hazardous-waste landfills, incinerators, and injection wells are holding up progress in dealing with hazardous wastes.* Instead, recognize that the best way to deal with waste and pollution is to produce much less of it. For example, after a charge for each unit of hazardous waste produced in Germany was imposed in 1991, hazardous-waste production fell more than 15% in 3 years.

- *Oppose all hazardous-waste landfills, deep-disposal wells, and incinerators.* This will sharply raise the cost of dealing with hazardous materials, discourage

Q: Is there doubt about the validity of the greenhouse effect?

location of such facilities in poor neighborhoods often populated by minorities, and encourage waste producers and elected officials to get serious about pollution prevention and waste reduction.

- *Recognize that there is no such thing as safe disposal of a toxic or hazardous waste.* For such materials, the goal should be "not in anyone's backyard" (NIABY) or "not on planet earth" (NOPE).

- *Pressure elected officials to pass legislation requiring that unwanted industries and waste facilities be distributed more widely instead of being concentrated in poor and working-class neighborhoods, many populated mostly by minorities.*

- *Ban all hazardous-waste exports from one country to another.*

It is not surprising that representatives of the solid and hazardous waste management industries oppose these tactics and have been able to persuade U.S. federal and state legislators not to put such ideas into practice.

How Can We Make the Transition to a Low-Waste Society? According to physicist Albert Einstein, "A clever person solves a problem, a wise person avoids it." To prevent pollution and reduce waste, many environmental scientists urge us to understand and live by four key principles: **(1)** Everything is connected, **(2)** there is no "away" for the wastes we produce, **(3)** dilution is not the solution to most pollution, and **(4)** the best and cheapest way to deal with waste and pollution is to produce less of it and then reuse and recycle most of the materials we use (Figures 11-2 and 11-3). Some actions you can take to reduce your production of solid waste and hazardous waste are listed in Appendix 4.

The presumption should be that any waste or pollutant is potentially harmful and preventable. . . . With a pollution prevention strategy, human intelligence and creativity as well as science and technology can focus on preventing, eliminating, or reducing the production of all wastes and pollutants.

JOEL HIRSCHORN

CRITICAL THINKING

1. Explain why you support or oppose the following:
   a. Requiring that all beverage containers be reusable
   b. Requiring all households and businesses to sort recyclable materials into separate containers for curbside pickup
   c. Requiring consumers to pay for plastic or paper bags at grocery and other stores to encourage the use of reusable shopping bags

2. For 1 week, keep a list of the solid waste you throw away. What percentage is materials that could be recycled, reused, or burned for energy? What percentage of the items could you have done without in the first place? Tally and compare the results for your entire class.

3. Would you oppose having a hazardous-waste landfill, waste treatment plant, deep-injection well, or incinerator in your community? Explain. If you oppose these disposal facilities, how do you believe that hazardous waste generated in your community and your state should be managed?

4. Give your reasons for agreeing or disagreeing with each of the following proposals for dealing with hazardous waste:
   a. Reducing the production of hazardous waste and encouraging recycling and reuse of hazardous materials by levying on producers a tax or fee for each unit of waste generated
   b. Banning all land disposal and incineration of hazardous waste to encourage recycling, reuse, and treatment and to protect air, water, and soil from contamination
   c. Providing low-interest loans, tax breaks, and other financial incentives to encourage industries producing hazardous waste to recycle, reuse, treat, destroy, and reduce generation of such waste
   d. Banning the shipment of hazardous waste from the United States to any other country
   e. Banning the shipment of hazardous waste from one state to another (probably unconstitutional as long as hazardous waste shipments are classified as interstate commmerce)

5. What hazardous and solid wastes do you create? What specific things could you do to reduce, reuse, or recycle as much of these wastes as possible? Which of these things do you actually plan to do? Do you use any of the alternative chemicals shown in Table 11-2?

A: No; without it, earth would be too cold for life as we know it to exist

12 ENVIRONMENTAL ECONOMICS, POLITICS, AND WORLDVIEWS

When it is asked how much it will cost to protect the environment, one more question should be asked: How much will it cost our civilization if we do not?

GAYLORD NELSON

12-1 ECONOMIC SYSTEMS AND ENVIRONMENTAL PROBLEMS

What Supports and Drives Economies? An **economy** is a system of production, distribution, and consumption of *economic goods:* any material items or services that satisfy people's wants or needs. In an economy, individuals, businesses, and societies make **economic decisions** about what goods and services to produce, how to produce them, how much to produce, how to distribute them, and what to buy and sell.

The kinds of capital that produce material goods and services in an economy are called **economic resources**. They fall into three groups:

- **Earth capital** or **natural resources**: goods and services produced by the earth's natural processes (Figure 1-2). These include the planet's air, water, and land; nutrients and minerals in the soil and deeper in the earth's crust; wild and domesticated plants and animals (biodiversity); and nature's dilution, waste disposal, pest control, and recycling services. There are no other sources for these materials that support all economies and lifestyles.

- **Manufactured capital**: items made from earth capital with the help of human capital. This type of capital includes tools, machinery, equipment, factory buildings, and transportation and distribution facilities.

- **Human capital**: people's physical and mental talents. Workers sell their time and talents for wages. Managers take responsibility for combining earth capital, manufactured capital, and human capital to produce economic goods. In market-based systems, entrepreneurs and investors put up the money needed to produce an economic good with the intention of making a profit on their investment.

What Are the Major Types of Economic Systems? There are two major types of economic systems: centrally planned and market based. In a **pure command economic system**, or **centrally planned economy**, all economic decisions are made by the government. This command-and-control system assumes that government control and ownership of the means of production are the most efficient and equitable way to produce, use, and distribute goods and services.

In a **pure market economic system,** also known as **pure capitalism**, all economic decisions are made in *markets,* in which buyers (demanders) and sellers (suppliers) of economic goods freely interact without government or other interference. All economic resources are owned by private individuals and institutions rather than by the government. All buying and selling is based on *pure competition*, in which no seller or buyer is powerful enough to control the supply, demand, or price of a good. All sellers and buyers have full access to the market and enough information about the beneficial and harmful aspects of economic goods to make informed decisions.

Economists often depict pure capitalism as a circular flow of economic goods and money between households and businesses operating essentially independently of the ecosphere (Figure 12-1). By contrast, environmentalists emphasize the dependence of this or any economic system on the ecosphere (Figure 12-2).

In a pure capitalist system, a business has no legal allegiance to a particular nation, no obligation to supply any particular good or service, and no obligation to provide jobs, safe workplaces, or environmental protection. A company's only obligation is to produce the highest possible short-term economic return (profit) for the owners or stockholders, whose financial capital the company is using to do business.

In reality, all countries have **mixed economic systems** that fall somewhere between the pure market and pure command systems. The economic systems of countries such as China and North Korea fall toward the command-and-control end of the economic spectrum, whereas those of countries such as the United States and Canada fall toward the market-based end of the spectrum. Most other countries fall somewhere in between.

Portney, Paul R. 1998. "Counting the Cost." *Environment*, vol. 40, no. 2, 14(8).

How Do Economic Growth and Sustainability Differ? Virtually all economies seek **economic growth**: an increase in the capacity of the economy to provide goods and services for people's final use. Such growth is usually accomplished by maximizing the flow of matter and energy resources (throughput) by means of population growth (more consumers), more consumption per person, or both (Figure 2-46).

Most economists, investors, and business leaders argue that we must have unlimited economic growth to create jobs, satisfy people's economic needs and desires, clean up the environment, and help eliminate poverty.

Most of these analysts believe that the capacity for economic growth is virtually unlimited because of the earth's vast amount of resources and the ability of human ingenuity to overcome resource shortages and environmental problems through science and technology. These analysts see the earth as an essentially unlimited source of raw materials and the environment as an infinite sink for wastes; technological innovation can overcome any resource or environmental limits. To people with this view, environmentalists put endangered species above endangered people; they threaten jobs and are against the economic growth needed for human survival and gains in the quality of life.

To environmentalists and a small but growing number of economists and business leaders, the notion of *sustainable economic growth* is nonsense because nothing that is based on the consumption of earth capital is sustainable. They believe that if we continue to support economic growth by consuming earth capital instead of living off sustainable *earth income*, then such forms of growth will threaten both business and the planet's life-support systems for humans and other species.

To people with this view, the question is not so much, "To grow or not to grow?" but rather, "How can we grow without plundering the planet?" or "How can we grow as if the earth matters?" or "How can we grow and still leave a flourishing earth to our children and grandchildren?"

Instead of unlimited economic growth, such critics call for **ecologically sustainable development**. This occurs when the total human population size and resource use in the world (or in a region) are limited to a level that does not exceed the carrying capacity of the existing natural capital, and are therefore sustainable. This is accomplished by reducing the throughput of matter and energy resources through economies by *doing the same with less* (Figure 2-47). Proponents of this approach would modify existing economic systems by giving rewards (subsidies) to earth-sustaining businesses and activities and by penalizing (taxing and regulating) earth-degrading

Pure Market Economic System

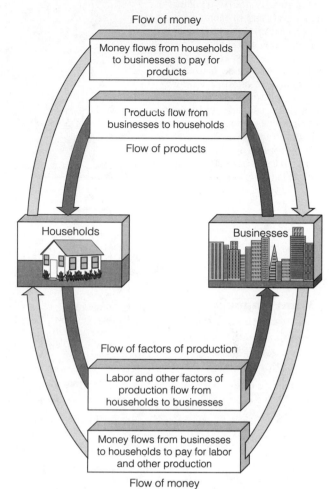

Figure 12-1 In a pure market economic system, economic goods and money would flow between households and businesses in a closed loop. People in households spend money to buy goods that firms produce, and firms spend money to buy factors of production (natural capital, manufactured capital, and human capital). In many economics textbooks, this and other economic systems are depicted, as here, as if they were self-contained and thus independent of the ecosphere—a model that reinforces the idea that unlimited economic growth of any kind is sustainable.

activities, with the overall goal of providing both economic and environmental security.

Lately there has been much talk about so-called *sustainable development*. There is no problem when this term is used as another way to describe *ecologically* sustainable development. However, some environmentalists contend that governments and businesses have jumped on the sustainability bandwagon and used this term as a thinly disguised way to justify or promote unsustainable forms of economic growth and development. To

Figure 12-2 Environmentalists and a small but growing number of economists see all economies as artificial subsystems that depend on resources and services provided by the sun and the ecosphere. A consumer society devoted to unlimited economic growth to satisfy ever-expanding wants assumes that ecosphere resources and services are essentially infinite or that our technological cleverness will allow us to overcome any ecospheric limits. To these analysts such a society is unsustainable.

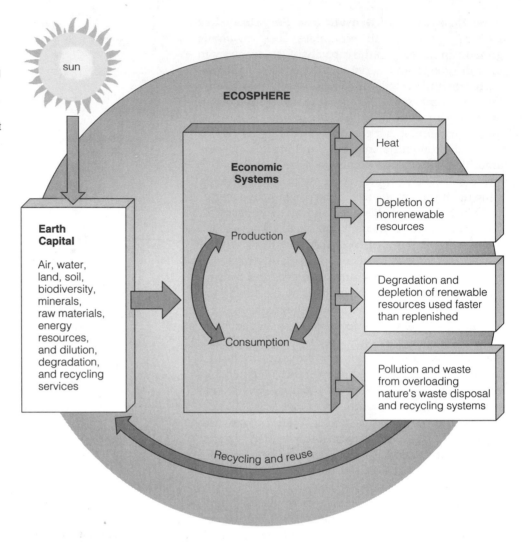

these critics, what we need is the development of ecological sustainability, not sustainable development.

According to systems analysis expert and environmentalist Donella Meadows,

What we need is smart development not dumb growth.... When something grows it gets quantitatively bigger. When something develops it gets qualitatively better. ... Smart development invests in insulation, efficient cars, and ever-renewed sources of energy. It ensures that forests and fields continue to produce wood, paper, and food, recharge wells, harbor wildlife, and attract tourists. Dumb growth crashes around looking for more oil. It clearcuts forests to keep loggers and sawmills going just a few more years until the trees run out. ... It covers the landscape with the same kind of honkytonk ugliness tourists leave home to escape. ... We need to meet dumb growth with smart questions. What really needs to grow? Who will benefit? Who will pay? What will last?

▧ Are GNP and GDP Useful Measures of Quality of Life and Environmental Degradation? We are urged to buy and consume more and more so that the GNP and GDP (p. 5) will rise, making the country where one lives and the world a better place for everyone. The truth is that GNP and GDP indicators are poor measures of human welfare, environmental health, or even economic health, for the following reasons.

GNP and GDP hide the negative effects (on humans and on the rest of the ecosphere) of producing many goods and services. Pollution, crime, sickness, death, and depletion of natural resources are all counted as positive gains in the GDP or GNP. Every time an irreplaceable old-growth forest is felled and a wetland is filled, the GDP and GNP go up. Every time a chemical or radiation causes cancer and the victim is treated, the GNP and GDP go up. They also rise because of the funeral expenses for the 150,000–350,000 people killed prematurely each year in the United States from air pollution. The $2.2 billion that Exxon spent partially cleaning up the oil spill from the *Exxon Valdez* tanker also raised the GDP and GNP, as did the $1 billion spent because of the accident at the Three Mile Island nuclear power plant in Pennsylvania. Pollution is counted as a *triple positive gain.* It is counted as a gain in GDP when it is first produced, counted again when society pays to clean it up partially, and counted as a third gain when people become sick or die from exposure to the pollution.

GNP and GDP don't include the depletion and degradation of natural resources or earth capital on which all economies depend. A country can be headed toward ecological bankruptcy, exhausting its mineral resources, eroding its soils, cutting down its forests, destroying its wetlands and estuaries, and depleting its wildlife and fisheries. At the same time it can have a rapidly rising GNP and GDP, at least for a while, until its environmental debts come due. Any business (or nation) that counted depletion of its capital (assets) as current income would always have a rosy but false picture of its true financial condition.

GNP and GDP hide or underestimate some of the positive effects of responsible behavior on society. More energy-efficient light bulbs, appliances, and cars reduce electric and gasoline bills and pollution, but these beneficial effects register as a decline in GNP and GDP. GDP and GNP indicators also exclude the labor we put into volunteer work, the health care we give loved ones, the food we grow for ourselves, and the cooking, cleaning, and repairs we do for ourselves.

GNP and GDP tell us nothing about economic justice . They don't reveal how resources, income, or the harmful effects of economic growth (pollution, waste dumps, land degradation) are distributed among the people in a country. UNICEF suggests that countries should be ranked not by average per capita GNP or GDP but by average or median income of the poorest 40% of their people.

Solutions: How Can Environmental Accounting Help? Economists have never claimed that GNP and GDP indicators are good measures of environmental health and human welfare, but most governments and business leaders use them that way. Environmentalists and a growing number of economists believe that GNP and GDP indicators should be replaced or supplemented with *environmental indicators* that give a more realistic picture by subtracting from the GDP and GNP things that lead to a lower quality of life and depletion of earth capital.

In 1972, economists William Nordhaus and James Tobin developed an indicator called *net economic welfare (NEW)* to estimate the annual change in a country's quality of life. They calculate a price tag for pollution and other "negative" goods and services included in the GNP and GDP and then subtract them to give the NEW. Economist David Pearce estimates that pollution

and natural resource degradation subtract 1–5% from the GNP or GDP of developed countries (and 5–15% for developing countries), and that in general these percentages are rising.

Economist Robert Repetto and other researchers at the World Resources Institute have developed a *net national product (NNP)* that includes the depletion or destruction of natural resources as a factor in GNP. They have applied this indicator to Indonesia and Costa Rica.

Herman E. Daly, John B. Cobb, Jr., and Clifford W. Cobb have developed an *index of sustainable economic welfare (ISEW)* and applied it to the United States. This comprehensive indicator of human welfare measures per capita GNP adjusted for inequalities in income distribution, depletion of nonrenewable resources, loss of wetlands, loss of farmland from soil erosion and urbanization, the cost of air and water pollution, and estimates of long-term environmental damage from ozone depletion and possible global warming. After rising by 42% between 1950 and 1976, this indicator fell 14% between 1977 and 1990.

A newer indicator is called the *genuine progress indicator (GPI).* When this indicator is applied to the United States, the GPI per person has steadily declined since 1973. Indeed, between 1973 and 1994 the GDP rose from $8,000 to $17,000 per person while the GPI fell from $6,500 to $4,000 per person (Figure 12-3).

Such indicators are far from being perfect and require many crude estimates. However, they are far better than GNP and GDP as measures of life quality and environmental quality. Without such indicators, we tend to disregard harmful things that are happening to people, the environment, and the planet's natural resource base. We fail to see what needs to be done, and we don't have a way to measure what types of policies work. In effect, we are trying to guide national and global economies through treacherous economic and environmental waters at ever-increasing speeds using faulty radar.

The *good news* is that such indicators exist. The *bad news* is that so far they are not widely used.

What Are Internal and External Costs? All economic goods and services have both internal and external costs. For example, the price a consumer pays for a car reflects the costs of the factory, raw materials, labor, marketing, and shipping, as well as a markup to allow the car company and its dealers some profits. After a car is purchased, the buyer must pay for gasoline, maintenance, and repair. All these direct costs, which are paid for by the seller and the buyer of an economic good, are called **internal costs**.

Making, distributing, and using any economic good or service also involve **externalities**: social costs or benefits not included in the market price. For example, if a car dealer builds an aesthetically pleasing

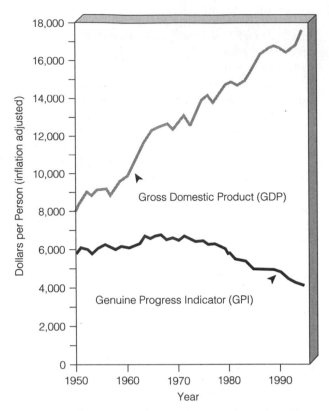

Figure 12-3 Comparison of per capita gross domestic product (GDP) and per capita genuine progress indicator (GPI) in the United States, 1950–94. Units of per capita GDP and GPI are inflation-adjusted using 1982 dollars per capita. (Data from Clifford Cobb, Ted Halstead, and Jonathan Rowe)

showroom and grounds, that is an **external benefit** to people who enjoy the sight at no cost.

On the other hand, extracting and processing raw materials to make and propel cars depletes nonrenewable energy and mineral resources, produces solid and hazardous wastes, disturbs land, pollutes the air and water, contributes to depletion of stratospheric ozone and possible global climate change, and reduces biodiversity and ecological integrity. These harmful effects are **external costs** passed on to workers, the general public, and in some cases future generations. Car owners add to these external costs when they throw trash out of a car, drive a gas-guzzler (which produces more air pollution than a more efficient car), disconnect or don't maintain a car's air-pollution control devices, or don't keep the engine tuned.

Because these harmful costs aren't included in the market price, people don't connect them with car ownership. Still, everyone pays these hidden costs sooner or later, in the form of higher costs for health care and health insurance and higher taxes for pollution control.

To conventional economists, external costs are minor defects in the flow of production and consumption in a self-contained economy not significantly dependent on earth capital (Figure 12-1). They assume that these defects can be rectified using the profits made from additional economic growth in a free-market economy.

To environmentalists and an increasing number of economists and business leaders, harmful externalities are a warning sign that our economic systems are stressing the ecosphere and depleting earth capital (Figures 1-2 and 12-2). They believe that these harmful external costs must be included in the market prices of goods and services—a process economists call *internalizing the external costs*.

12-2 SOLUTIONS: USING ECONOMICS TO IMPROVE ENVIRONMENTAL QUALITY

Should We Shift to Full-Cost Pricing? As long as businesses receive subsidies and tax breaks for extracting and using virgin resources and are not taxed for the pollutants they produce, few will volunteer to reduce short-term profits by becoming more environmentally responsible. Assume you own a company and believe it's wrong to subject your workers to hazardous conditions and pollute the environment beyond what natural processes can handle. Suppose you voluntarily improve safety conditions for your workers and install pollution controls, but your competitors don't. Then your product will cost more than theirs, and you will be at a competitive disadvantage. Your profits will decline; you may eventually go bankrupt and have to lay off your employees.

One way of dealing with the problem of harmful external costs is for the government to levy taxes, pass laws, provide subsidies, or use other strategies that encourage or force producers to include all or most of these costs in the market prices of economic goods and services. Then that price would be the **full cost** of these goods and services: internal costs plus short- and long-term external costs. The two main goals are to **(1)** close the gap between real and false prices by having prices that tell the environmental truth and **(2)** have people and businesses pay the full costs of the harm they do to others and the environment.

Full-cost pricing involves *internalizing the external costs*, which requires government action because few companies will intentionally increase their cost of doing business unless their competitors must do so as well. If the market prices of economic goods reflected all or most of their full estimated cost, economic growth would be redirected. We would increase the beneficial parts of the GNP and GDP (and decrease the harmful parts), increase production of beneficial goods, and raise the net economic welfare. Preventing pollution would become more profitable than cleaning it up; waste reduction,

336

Lerner, Steve. 1998. "The New Environmentalists." *The Futurist*, vol. 32, no. 4, 35(5).

recycling, and reuse would be more profitable than burying or burning most of the waste we produce.

Because external costs would be internalized, the market prices for most goods and services would rise. But the total price we pay would be about the same because the hidden external costs related to each product would already be included in its market price. Using full-cost pricing to internalize external costs provides consumers with information needed to make informed economic decisions about the effects of their lifestyles on the planet's life-support systems.

However, as external costs are internalized, economists and environmentalists warn that governments must reduce income and other taxes and withdraw subsidies formerly used to hide and pay for these external costs. Otherwise, consumers will face higher market prices without tax relief—an unjust policy guaranteed to fail.

Some goods and services would cost less because internalizing external costs encourages producers to find ways to cut costs (by inventing more resource-efficient methods of production); it also encourages producers to offer more earth-sustaining (or *green*) products. Jobs would be lost in earth-degrading businesses, but at least as many (some analysts say more) jobs could be created in earth-sustaining businesses. If this change in the way market prices are established took place over several decades, most current earth-degrading businesses would have time to transform themselves into profitable earth-sustaining businesses.

Full-cost pricing seems to make a lot of sense. So why isn't it more widely done? One reason is that many producers of harmful and wasteful goods would have to charge so much that they couldn't stay in business, or they would have to give up government subsidies and tax breaks that have helped hide the external costs of their goods and services. And why would they want to change a system that's been so good to them? Another problem is that it is hard to put a price tag on many harmful environmental and health effects.

Studies estimate that governments around the globe spend about $600 billion per year to subsidize deforestation, overfishing, overgrazing, unsustainable agriculture, nonrenewable fossil fuels and nuclear energy, groundwater depletion, and other environmentally destructive activities—what environmental expert Norman Myers calls *perverse subsidies*. It is estimated that eliminating these earth-degrading subsidies that distort the global economy would allow a 7% cut in the global tax burden of $7.5 trillion per year and encourage job creation and investment.

Despite the difficulties, proponents believe that full-cost pricing for harmful environmental and health effects deserves a serious try. They argue that doing the best we can to estimate and internalize current external costs is far better than continuing the current pricing system, which gives too little or misleading information about the environmental and health effects of goods and services. The key question is whether the problems with the *tell-the-truth full-cost* pricing system are worse than those with the current *hide-the-true-cost* pricing system.

Should We Rely Mostly on Regulations or Market Forces? Most economists agree that controlling or preventing pollution and reducing resource waste require government intervention in the marketplace. Such government action can take the form of regulation, the use of market forces, or some combination of these approaches.

Regulation is a *command-and-control* approach. It involves enacting and enforcing laws, for example, that set pollution standards, regulate harmful activities, ban the release of toxic chemicals into the environment, and require that certain irreplaceable or slowly replenished resources be protected from unsustainable use (or from any use at all).

Most studies of the effects of environmental regulations in the United States have found that they do businesses very little harm. Indeed, in many cases they have led to improvements in resource-use efficiency, which reduce costs, and to innovative products and industrial processes that increase profits. A recent study by economist Robert Repetto at the World Resources Institute showed that between 1970 and 1990, the U.S. industries that spent the most on pollution control fared significantly better than average in the global marketplace.

However, business leaders and many environmentalists in the United States agree that some pollution control regulations discourage innovation by being too prescriptive and costly, and must be modified. They propose using regulations to set goals but then freeing industries to meet such goals in any way that works.

Market forces can help improve environmental quality and reduce resource waste, mostly by encouraging the internalization of external costs. This is based on a fundamental principle of the marketplace in today's mixed economic systems: *What we reward (mostly by subsidies and tax breaks) we tend to get more of, and what we discourage (mostly by regulations and taxes) we tend to get less of.*

One way to put this principle into practice would be *to phase in government subsidies that encourage earth-sustaining behavior and phase out current perverse subsidies that encourage earth-degrading behavior.* The difficulty with this ecologically and economically appealing *carrot approach* is that removing or adding subsidies involves political decisions; these are easily swayed by powerful economic interests that want to preserve ecologically unsound subsidies to increase their short-term profits.

Another market approach is for the government to *grant tradable pollution and resource-use rights*. For example, a total limit on emissions of a pollutant or use of a resource

Hint: Enter the search terms *environmentalists, innovations* using Key Words.

CHAPTER 12   337

could be set, and the total would be allocated among manufacturers or users by permit. Permit holders not using their entire allocation could use it as a credit against future expansion, use it in another part of their operation, or sell it to other companies. Tradable rights could also be established among countries to preserve biodiversity and reduce emissions of greenhouse gases, ozone-destroying chemicals, lead, and air and water pollutants—chemicals with harmful regional (or global) effects.

Some environmentalists support trading pollution rights as an improvement over the current regulatory approach. Other environmentalists oppose this approach for reasons discussed on p. 276.

Another market-based method is to *enact green taxes or effluent fees* that would help internalize many of the harmful external costs of production and consumption. This method could include taxes on each unit of pollution discharged into the air or water, each unit of hazardous or nuclear waste produced, each unit of virgin resources used, and each unit of fossil fuel used. Experience in the Netherlands and Germany shows that phasing in such taxes can encourage creativity in solving environmental problems and in reducing costs by preventing pollution, using fewer resources, and developing earth-sustaining technologies and products.

There are two major problems with this tax punishment or stick approach. *First,* because the taxes or effluent fees are set politically rather than by markets, elected officials find it easier to aim for popular approval rather than economical and ecological efficiency. *Second,* elected officials are likely to see such taxes as ways of raising revenue instead of improving economic

and ecological efficiency. However, economists point out that *green taxes* on pollution output and resource use would work if they reduced or replaced income taxes or other taxes—and if the poor are given a basic safety net to reduce the regressive nature of consumption taxes on essentials such as food, fuel, and housing.

Charging user fees is another market-based method. For example, users would pay fees to cover all or most costs for grazing livestock, extracting lumber and minerals from public lands, using water provided by government-financed projects, and using public lands for recreation.

Another market approach would *require businesses to post a pollution prevention or assurance bond when they open a mine, plant, incinerator, or landfill.* After a set length of time, the deposit (with interest) would be returned *minus* actual or estimated environmental costs. This approach is similar to the performance bonds contractors are now required to post for major construction projects.

Each of these approaches has advantages and disadvantages (Table 12-1). Currently, in the United States private industry and local, state, and federal government spend about $130 billion a year to comply with federal environmental regulations. Studies estimate that greater reliance on a variety of market-based policies could cut these expenditures by one-third to one-half.

Should We Emphasize Pollution Control or Pollution Prevention? Shouldn't our goal be zero pollution? Ideally, yes; in the real world, not necessarily. First, natural processes can handle some of our wastes

Table 12-1 Economic Solutions to Pollution and Resource Waste

Solution	Internalizes External Costs	Innovation	International Competitiveness	Administrative Costs	Increases Government Revenue
Regulation	Partially	Can encourage	Decreased*	High	No
Subsidies	No	Can encourage	Increased	Low	No
Withdrawing harmful subsidies	Yes	Can encourage	Decreased*	Low	Yes
Tradable rights	Yes	Encourages	Decreased*	Low	Yes
Green taxes	Yes	Encourages	Decreased*	Low	Yes
User fees	Yes	Can encourage	Decreased*	Low	Yes
Pollution prevention bonds	Yes	Encourages	Decreased*	Low	No

*Unless more cost-effective and productive technologies are developed.

Angus, Ian. 1997. "Free Nature." *Alternatives Journal*, vol. 23, no. 3, 18(4).

as long as we don't destroy, degrade, or overload these processes. However, environmentalists argue that harmful chemicals that either cannot be degraded by natural processes or that break down very slowly should not be released into the environment, or they should be released only in small amounts and regulated by special permit.

Second, as long as we continue to rely on pollution control, we can't afford zero pollution. After we've removed a certain proportion of the pollutants in air, water, or soil, the cleanup cost per additional unit of pollutant rises sharply (Figure 12-4). Beyond a certain point, the cleanup costs exceed the harmful costs of pollution. Some businesses could then go bankrupt and some people could lose jobs, homes, and savings. On the other hand, if we don't go far enough, dealing with the harmful external effects of pollution can cost more than pollution reduction.

To find the breakeven point, economists plot two curves: a curve of the estimated economic costs of cleaning up pollution and a curve of the estimated social (external) costs of pollution. Adding the two curves together, we get a third curve showing the total costs. The lowest point on this third curve is the point of the *optimal level of pollution* (Figure 12-5).

On a graph, this looks neat and simple, but environmentalists and business leaders often disagree in their estimates of the harmful costs of pollution. This approach assumes that we know which substances are harmful and how much each part of the environment can handle without serious environmental harm—things we probably will never know precisely.

Most environmentalists and a growing number of economists and business leaders believe that environmental laws should emphasize pollution prevention, which avoids most of the regulatory problems and excessive costs of end-of-pipe pollution control. What do you think?

12-3 SOLUTIONS: REDUCING POVERTY

Does the Trickle-Down Approach to Reducing Poverty Work? Poverty is usually defined as the inability to meet one's basic economic needs. Currently, an estimated 1.3 billion people (70% of them women) in developing countries—one of every five people on the planet—have an annual income of less than $370 per year. This income of roughly $1 per day is the World Bank's definition of poverty.

Poverty causes premature deaths and preventable health problems. It also tends to increase birth rates and often pushes people to use potentially renewable resources unsustainably in order to survive.

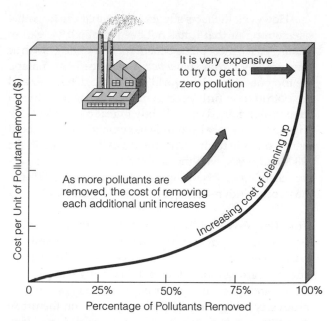

Figure 12-4 The cost of removing each additional unit of pollution rises exponentially, which explains why it is usually cheaper to prevent pollution than to clean it up.

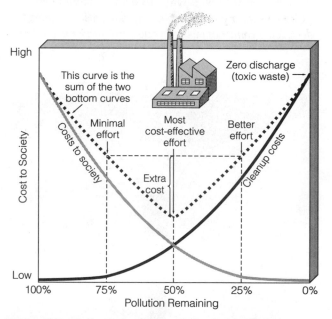

Figure 12-5 Finding the optimal level of pollution. This graph shows the optimal level at 50%, but the actual level can vary depending on the pollutant.

Most economists believe that a growing economy is the best way to help the poor. This is the so-called *trickle-down hypothesis*: Economic growth creates more jobs, enables more of the increased wealth to reach workers, and provides greater tax revenues that can be used to help the poor help themselves.

Hint: Enter the search term *deep ecology* using Key Words.

However, the facts suggest either that the hypothesis is wrong or that it has not been applied. Instead of *trickling down*, most of the benefits of economic growth as measured by income have *flowed up* since 1960, making the top one-fifth of the world's people much richer and the bottom one-fifth poorer; most of those in between have lost or gained only slightly in real per capita GNP (Figure 1-7) and are becoming increasingly fearful about their economic future. This trend has accelerated in the 1980s and 1990s, with the richest fifth of the world's people receiving 82.7% of the world's income in 1991 and the poorest fifth receiving only 1.4% (Figure 12-6).

How Can Poverty Be Reduced? Analysts point out that reducing poverty requires the governments of most developing countries to make policy changes, including shifting more of the national budget to help the rural and urban poor work their way out of poverty and giving villages, villagers, and the urban poor title to common lands and to crops and trees they plant on them.

Analysts also urge developed countries and the wealthy in developing countries to help reduce poverty. Several controversial ways to do this have been suggested. First, *forgive at least 60% of the almost $2 trillion that developing countries owe to developed countries and interna-tional lending agencies.* Some of this debt can be forgiven in exchange for carefully monitored agreements by the governments of developing countries to increase expenditures for family planning, health care, education, land redistribution, protection and restoration of biodiversity, and more sustainable use of renewable resources.

Another proposal is to *increase the nonmilitary aid to developing countries from developed countries.* Proponents say that this aid should go directly to the poor to help them become more self-reliant. Analysts also recommend that we *shift most international aid from large-scale to small-scale projects intended to benefit local communities of the poor.*

Analysts also suggest that *international lending agencies should be required to use a standard environmental and social impact analysis to evaluate any proposed development project.* Proponents believe that no project should be supported unless its net environmental impact using full-cost accounting is favorable, most of its benefits go to the poorest 40% of the people affected, and the local people it affects are involved in planning and executing the project. They call for careful monitoring of all projects and an immediate halt in funding whenever environmental safeguards are not followed. A final proposal is to *establish policies that encourage both developed countries and developing countries to slow population growth and stabilize their populations* (Section 3-3).

12-4 SOLUTIONS: CONVERTING TO EARTH-SUSTAINING ECONOMIES

How Can We Make Working with the Earth Profitable? Greening Business In 1994, entrepreneur and business leader Paul Hawken wrote *Ecology and Commerce*, a widely acclaimed book describing what's wrong with business and how it can be transformed to work with the earth. Hawken and several other business leaders and economists have laid out the following principles for transforming the current earth-degrading economic systems into earth-sustaining, or restorative, economies over the next several decades:

- *Reward (subsidize) earth-sustaining behavior.*

- *Discourage (tax and don't subsidize) earth-degrading behavior.*

- *Use full-cost accounting to include the ecological value of natural resources in their market prices.*

- *Use environmental and social indicators to measure progress toward environmental and economic sustainability and human well-being* (Figure 12-3).

- *Use full-cost pricing to include the external costs of goods and services in their market prices.*

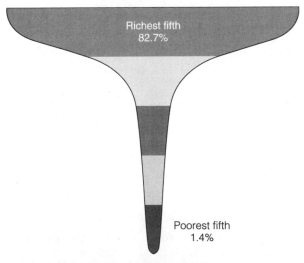

Figure 12-6 Data on the global distribution of income showing that, instead of trickling down, most of the world's income has flowed up, with the richest 20% of the world's population receiving more of the world's income than all of the remaining 80% of the world's population in 1991. Each horizontal band in this diagram represents one-fifth of the world's population. This upward flow of global income has increased since 1960 and has accelerated in the 1980s and 1990s. Do you favor or oppose this trend? Explain. If you oppose this trend, what do you believe are the most important ways to correct it? (Data from UN Development Programme)

Richest fifth 82.7%

Poorest fifth 1.4%

Swift, Byron. 1998. "A Low-Cost Way to Control Climate Change." *Issues in Science and Technology*, vol. 14, no. 3, 75(7).

- *Replace taxes on income and profits with taxes on throughput of matter and energy.*

- *Establish public utilities to manage and protect public lands and fisheries.*

- *Revoke the government-granted charters of environmentally and socially irresponsible businesses.*

- *Make environmental concerns a key part of all trade agreements and of all loans made by international lending agencies.*

- *Reduce waste of energy, water, and mineral resources.*

- *Reduce future ecological damage and repair past ecological damage.*

- *Reduce poverty.*

- *Slow population growth.*

Paul Hawken's simple golden rule for such an economy is, "*Leave the world better than you found it, take no more than you need, try not to harm life or the environment, and make amends if you do.*"

How Can We Make the Transition to an Earth-Sustaining Economy? Even if people believe that an earth-sustaining economy is desirable, is it possible to make such a drastic change in the way we think and act? Some environmentalists, economists, and business leaders say that it's not only possible but imperative and that it can be done over the next 40–50 years. They point out that *the environmental revolution is also an economic revolution that uses a mix of regulations and market-based approaches to reward earth-sustaining behavior by businesses and consumers and discourage earth-degrading behavior.*

According to Paul Hawken, this new approach to economic thinking and actions recognizes that most business leaders are not evil, earth-degrading ogres. Instead, they are trapped in a system that by design rewards them (with the highest profits and salaries, and best chances for promotion) for maximizing short-term profits for owners and investors, regardless of the harmful short- and long-term environmental and social impacts.

Hawken argues that an earth-sustaining economy would free business leaders, workers, and investors from this ethical dilemma and allow them to be financially compensated and respected for doing socially and ecologically responsible work, improving environmental quality, and still making hefty profits for owners and stockholders. Making this shift should also create jobs (Connections, right).

Hawken and others offer the following outline of how the shift from current unsustainable economies to sustainable ones could be made over the next 40–50 years. Over the first 20 years, *all* government subsidies encouraging resource depletion, waste, pollution, and

Jobs and the Environment

CONNECTIONS

Critics of environmental laws and regulations contend that they cause the loss of large numbers of jobs in the United States and other developed countries. In fact, studies show that the opposite is true. U.S. Department of Labor statistics reveal that only 0.1% of the jobs lost in the United States between 1987 to 1990 were the result of environmental regulations.

Instead of causing a net loss of jobs, environmental protection is a major growth industry that creates new jobs. Some 1 million jobs are expected to be added to the U.S. environmental protection industry alone between 1995 and 2005; increasing the aluminum recycling rate in the United States to 75% would create 350,000 more jobs. Collecting and refilling reusable containers creates many more jobs per dollar of investment than using throwaway containers, and most of the new jobs are created in local communities. Reforestation, ecological restoration, sustainable agriculture, and integrated pest management are all labor-intensive activities requiring low to moderate skill levels.

A congressional study concluded that investing $115 billion per year in solar energy and improving energy efficiency in the United States would eliminate about 1 million jobs in oil, gas, coal, and electricity production but would create 2 million other new jobs; investment of the money saved by reducing energy waste could create another 2 million jobs.

Although a host of new jobs would be created in an earth-sustaining economy, jobs would be lost in some industries, regions, and communities. Ways suggested by various analysts to ease the transition include providing tax breaks to make it more profitable for companies to keep or hire more workers instead of replacing them with machines; using incentives to encourage location of new, emerging industries in hard-hit communities, helping such areas diversify their economic base; and providing income and retraining assistance for workers displaced from environmentally destructive businesses (a *Superfund for Workers*).

Critical Thinking

1. What major things (if any) has the federal government in the United States (or the country where you live) done to stimulate the growth of environmental jobs? What major things (if any) has the government done to discourage the growth of environmental jobs?

2. Do you believe that the government should have a significant role in stimulating environmental jobs? Explain.

How Is Germany Investing in the Future and the Earth?

German political and business leaders see sales of environmental protection goods and services—already a roughly $500-billion-per-year business in 1997 (and expected to rise to $572 billion by 2001)—as a major source of new markets and income in the next century because environmental standards and concerns are expected to rise everywhere.

Stricter environmental standards and regulations in Germany have paid off in a cleaner environment and the development of new green technologies that can be sold at home and abroad in a rapidly growing market. Mostly because of stricter air pollution regulations, German companies sell some of the world's cleanest and most efficient gas turbines, and they have developed the world's first steel mill that uses no coal to make steel. Germany sells this improved technology globally.

In 1977 the German government started the Blue Angel product-labeling program to inform consumers about products that cause the least environmental harm. Most international companies now use the German market to test and evaluate green products.

Germany has also revolutionized the recycling business. German car companies are required to pick up and recycle all domestic cars they make. Bar-coded parts and disassembly plants can dismantle an auto for recycling in 20 minutes. Such *take-back* requirements are being extended to almost all products to reduce use of energy and virgin raw materials. Germany plans to sell its newly developed recycling technologies to other countries.

The German government has also supported research and development aimed at making Germany the world's leader in solar-cell technology and hydrogen fuel (Sections 4-3 and 4-4), which it expects will provide a rapidly increasing share of the world's energy. Finally, Germany provides about $1 billion per year in green foreign aid to developing countries. Much of the aid is designed to stimulate demand for German technologies such as solar-powered lights, solar cells, and wind-powered water pumps.

Critical Thinking

Do you agree with Germany's policy of using stricter environmental standards as a way to protect human health and encourage innovation in development of green technologies and products? Explain. Does the country where you live have such a policy?

environmental degradation would be phased out and replaced with taxes on such activities. During that same period, taxes on incomes and profits would be phased out, and new government subsidies would be phased in for businesses built around recycling and reuse, reducing waste, preventing pollution, improving energy efficiency, using renewable energy, protecting biodiversity, and facilitating ecological restoration.

This system for change represents a shift in determining which economic actions are rewarded (profitable) and which ones are discouraged. These plans would be well publicized and would take effect over decades, giving businesses time to adjust. Because businesses go where the profits are, many of today's earth-degrading businesses might well be tomorrow's earth-sustaining businesses—a win–win solution for business, future generations, and the earth. Economic models indicate that after this first phase is completed the entire economy would be transformed within another 30–40 years.

The problem in making this shift is not economics, but politics. It involves the difficult task of convincing business leaders and elected officials to begin changing the current system of rewards and penalties, the profits from which have given them economic and political power. Without the active participation of forward-thinking business and political leaders and political pressure from citizens, there is little hope of making the transition to an earth-sustaining economy. Some countries such as Germany (Case Study, left) and the Netherlands are attempting to develop more sustainable economies.

Forward-looking investors, corporate executives, and the governments of countries such as Japan, Germany, the Netherlands, and the United States recognize that earth-sustaining businesses with good environmental management will prosper as the environmental revolution proceeds. Companies and countries that fail to invest in a green future may find that they do not have a future.

12-5 POLITICS AND U.S. ENVIRONMENTAL POLICY

How Does Social Change Occur in Representative Democracies? **Politics** is the process by which individuals and groups try to influence or control the policies and actions of governments at the local, state, national, or international levels. Politics is concerned with who has power over the distribution of resources and benefits—who gets what, when, and how.

Democracy is literally government "by the people" through elected officials and representatives. In a *constitutional democracy*, a constitution provides the basis of government authority, limits government power by mandating free elections, and guarantees freely expressed public opinion.

Political institutions in constitutional democracies are designed to allow gradual change in order to ensure economic and political stability. Rapid change in the United States, for example, is curbed by the system of checks and balances that distributes power among the three branches of government—executive, legislative, and judicial—and among federal, state, and local governments.

In passing laws, developing budgets, and formulating regulations, government decision makers must deal with pressure from many competing *special-interest groups*. Each group advocates passage of laws favorable to its cause. This includes cheap and better access to resources, subsidies, relief from taxes, and the shaping of regulations to foster the group's goals. Some special-interest groups are *profit-making organizations* such as corporations, and others are *nonprofit, nongovernment organizations (NGOs)*. Examples of NGOs are educational institutions, labor unions, and environmental organizations.

Most political decisions made in democracies result from bargaining, accommodation, and compromise among leaders of competing *elites*, or power brokers. The overarching goal of government by competing elites is maintaining the overall economic and political stability of the system (status quo) by making only gradual change; this goal does *not* involve questioning or changing the rules of the game (the fundamental societal beliefs) that gave the elites their political or economic power.

One disadvantage of this deliberate design for stability is that democratic governments tend to *react* to crises instead of acting to prevent them. This means that there is a built-in bias against policies for protecting the environment because they often call for prevention of crises instead of reaction to them. Also, such policies sometimes call for fundamental changes in societal beliefs that can threaten the power of government and business elites.

Case Study: How Is Environmental Policy Made in the United States? The major function of the federal government in the United States is to develop and implement *policy* for dealing with various issues. This policy is typically composed of various *laws* passed by the legislative branch, *regulations* instituted by the executive branch to put laws into effect, and enough *funding* to implement and enforce the laws and regulations.

The first step in establishing federal environmental policy (or any other policy) is to persuade lawmakers that a problem exists and that the government has a responsibility to find solutions to it. Once over that hurdle, lawmakers try to pass laws to deal with the problem. Most environmental bills are evaluated by as many as 10 committees in both the House of Representatives and the Senate. Effective proposals are often weakened by this fragmentation and by lobbying from groups opposing the law. Nonetheless, since the 1970s a number of environmental laws have been passed in the United States (Appendix 3).

Even if an environmental (or other) law is passed, Congress must appropriate enough funds to implement and enforce it. Indeed, developing and adopting a budget is the most important and controversial thing the executive and legislative branches do. Developing a budget involves answering two key questions: What resource use and distribution problems will be addressed, and how much of the limited tax revenues will be used to address each problem?

Next, regulations for implementing the law are drawn up by the appropriate government department or agency. Groups try to influence how the regulations are written and enforced; some of the affected parties may even challenge the final regulations in court. Finally, the department or agency implements and enforces the approved regulations. Proponents or affected groups may take the agency to court for failing to implement and enforce the regulations or for enforcing them too rigidly.

The net result of this fragmentation, lack of funding, and influence of the regulated on the regulators is *incremental decision making*. As a result, only small changes are usually made in existing policies and programs.

Solutions: How Can Individuals Affect Environmental Policy? A major theme of this book is that individuals matter. History shows that significant change comes from the *bottom up*, not the top down. Without grassroots political action by millions of individual citizens and organized groups, the air you breathe and the water you drink today would be much more polluted, and much more of the earth's biodiversity would have disappeared.

Individuals can influence and change government policies in constitutional democracies in several ways. They can **(1)** vote for candidates and ballot measures; **(2)** contribute money and time to candidates seeking office; **(3)** lobby, write, fax, e-mail, or call elected representatives, asking them to pass or oppose certain laws, establish certain policies, and fund various programs; **(4)** use education and persuasion; **(5)** expose fraud, waste, and illegal activities in government (whistle-blowing); **(6)** file lawsuits; and **(7)** participate in grassroots activities to bring about change or enforce existing laws and regulations.

Solutions: What Are the Three Types of Environmental Leadership? There are three types of environmental leadership. One is *leading by example*, in which people use their lifestyles to show others that change is possible and beneficial.

Environmental Careers

INDIVIDUALS MATTER

Besides committed earth citizens, the environmental movement needs dedicated professionals working to help sustain the earth. In the United States (and in other developed countries), the *green job market* is one of the fastest-growing segments of the economy. Already some 3 million people in the United States are working in the environmental field, and more than 125,000 new positions are being created each year.

Many employers are actively seeking environmentally educated graduates. They are especially interested in people with scientific and engineering backgrounds and in people with double majors (business and ecology, for example) or double minors.

Environmental career opportunities exist in a large number of fields: environmental engineering (currently the fastest growing job market), sustainable forestry and range management, parks and recreation, air and water quality control, solid-waste and hazardous-waste management, recycling, urban and rural land-use planning, computer modeling, ecological restoration, and soil, water, fishery, and wildlife conservation and management.

Environmental careers can also be found in education, environmental planning, environmental management, health, toxicology, geology, ecology, conservation biology, chemistry, climatology, population dynamics and regulation (demography), law, risk analysis, risk management, accounting, journalism, design and architecture, energy conservation and analysis, renewable-energy technologies, hydrology, consulting, public relations, activism and lobbying, economics, diplomacy, development and marketing, publishing (environmental magazines and books), and law enforcement (pollution detection and enforcement teams).

Another involves *working within existing economic and political systems to bring about environmental improvement, often in new, creative ways.* Individuals can influence political elites by campaigning and voting for candidates and by communicating with elected officials. They can also work within the system by choosing an environmental career (Individuals Matter, above).

A third type of leadership involves *challenging the system and basic societal values, as well as proposing and working for better solutions to environmental problems.* Leadership is more than being against something; it also

involves showing people a better way to accomplish various goals. All three types of leadership are needed.

Solutions: How Can We Make Government More Responsive to Ordinary Citizens? According to most political analysts, *the biggest problem that keeps elected officials from being more responsive to the environmental and other needs and problems of ordinary citizens is that the enormous amounts of money needed to run for office or to get reelected come mostly from wealthy and powerful individuals and corporate interests.* Most analysts and about 80% of citizens polled agree that the U.S. political system is based more on money than on citizens' votes—one reason so many Americans don't bother to vote. Even those who do vote feel they're often simply choosing between the lesser of two evils. Many analysts see drastic reform in the way elections to public office are financed as the key to making the government more responsive to ordinary citizens (Solutions, right).

Solutions: What Are the Roles of Mainstream and Grassroots Environmental Groups? Many types of environmental groups are at work at the local, state, national, and international levels (Appendix 1). Environmental organizations range from multimillion-dollar *mainstream* groups, led by chief executive officers and staffed by experts, to *grassroots* neighborhood groups formed to do battle on local environmental issues. More than 8 million U.S. citizens belong to environmental organizations, which together received $894 million in contributions and income in 1994 (up from $630 million in 1990).

Mainstream environmental groups are active primarily at the national level and to a lesser extent at the state level; often they form coalitions to work together on issues. Some mainstream organizations funnel substantial funds to local activists and projects.

Mainstream groups work within the political system; many have been major forces in persuading Congress to pass and strengthen environmental laws and in efforts to fight off attempts to weaken or repeal such laws. However, these groups must continually guard against having their efforts subverted by the political system they work to change and against losing touch with ordinary people and nature in the insulated atmosphere of national and state capitals.

The base of the environmental movement in the United States consists of at least 6,000 grassroots citizens' groups organized to protect themselves against local pollution and environmental damage. According to political analyst Konrad von Moltke, "There isn't a government in the world that would have done anything for the environment if it weren't for the citizen groups." Environmental grassroots groups are also

SOLUTIONS

A growing number of analysts of all political persuasions see *election finance reform* as the single most important way to reduce the influence of so-called money votes. They urge citizens not to get diverted and fragmented over other issues but to focus their efforts on this crucial issue as the key to making government more responsive to ordinary citizens.

One suggestion for reducing undue influence by powerful special interests would be to let the people (taxpayers) *alone* finance all federal, state, and local election campaigns, with low spending limits. Candidates could not accept direct or indirect donations from any other individuals, groups, or parties, with *absolutely no exceptions*.

Once elected, officials could not use free mailing, staff employees, or other privileges to aid their election campaigns, nor could they accept donations or any kind of direct or indirect financial aid from any individual, corporation, political parties, or interest group for their future election campaigns—or for any other reason that could even remotely influence their votes on

legislation. Violators of this new *Public Funding Elections Act* would be barred from the campaign or removed from office. Anyone making illegal donations would be subject to large fines and possible jail sentences.

Having all elections financed entirely by public funds would cost each U.S. taxpayer only about $5–10 per year (as part of their income taxes) for all federal elections (and a much smaller amount for state and local elections)—a small investment for making democracy more responsive to ordinary people.

With such a reform, elected officials could spend their time governing instead of raising money and catering to powerful special interests. Office seekers would not need to be wealthy, and public service could become a way to help the country and the earth, not a lifelong profession or route to wealth and power. Special-interest groups would be heard because of the validity of their ideas, not the size of their pocketbooks.

Proponents of this reform contend that this is an issue that ordinary people of all political persuasions could work together on. From this fundamental politi-

cal reform, other democratic reforms could then flow.

The problem is getting members of Congress (and state legislators) to pass a virtually foolproof and constitutionally acceptable plan that would put them on equal financial footing with their challengers— and that might possibly decrease their chances of getting reelected. Supporters of public financing of all elections argue that the way out of this dilemma is to band together to find and elect candidates who pledge to bring about this fundamental political reform, and then vote them out of office if they don't.

Critical Thinking

1. Do you agree or disagree with this approach to election reform? Explain. What role would you take in bringing about or opposing such a reform?

2. Some of these election reform proposals, especially preventing candidates from using their own money to support their election campaigns, might be unconstitutional. How could the reform process be modified to accommodate such an objection?

---

active on college campuses and public schools across the United States (Individuals Matter, p. 346).

Many grassroots environmental organizations are unwilling to compromise or negotiate (p. 330). Instead of dealing with environmental goals and abstractions, they are fighting perceived threats to their lives, the lives of their children and grandchildren, and the value of their property. They want pollution and environmental degradation, environmental injustices, and violations of their human rights stopped and prevented, not merely controlled.

What Are the Goals and Tactics of the Anti-Environmental Movement? A small but growing number of political and business leaders in the United

States and in other developed countries see improving environmental quality as a way to stimulate innovation, increase profits, and create jobs while working with the earth. However, leaders of some corporations and many people in positions of economic and political power understandably see environmental laws and regulations as threats to their wealth and power, and they vigorously oppose such efforts.

Since 1980 anti-environmentalists in the United States have mounted a massive campaign to weaken or repeal existing environmental laws and to destroy the credibility and effectiveness of the environmental movement.

Whatever your own position, it is useful to understand the following general legal and political tactics being used against the environmental movement. Start

INDIVIDUALS MATTER

Since 1988 there has been a boom in environmental awareness on college campuses and some public schools across the United States. Much of this momentum began in 1989, when the Student Environmental Action Coalition (SEAC) at the University of North Carolina at Chapel Hill held the first national student environmental conference on the UNC campus.

SEAC groups are active on 700 campuses, and the National Wildlife Federation's Campus Ecology Program (launched in 1989) has groups on about 600 campuses.* Most student environmental groups work with members of the faculty and administration to bring about environmental improvements on their own campuses or schools and in their local communities.

*These efforts are described in the book *Ecodemia: Campus Environmental Stewardship at the Turn of the 21st Century* (Washington, D.C.: National Wildlife Federation, 1995) and in the *Campus Environmental Yearbook* published annually by the National Wildlife Federation.

Many of these groups focus on making an environmental audit of their own campuses or schools*; they then use the data gathered to propose changes that will make their campuses or schools more ecologically sustainable, usually saving them money in the process. In addition, students who have learned to do careful research on and develop solutions for environmental problems will be able to use these skills the rest of their lives.

Such audits have resulted in numerous improvements. For example, an energy management plan developed by Morris A. Pierce, a graduate student at the University of Rochester, New York, was adopted by that school's board of trustees. Using this plan, a capital investment of $33 million is projected to save the university $60 million over 20 years. Students have

*Details for conducting campus environmental audits can be found in April Smith and the Student Environmental Action Coalition, *Campus Ecology: A Guide to Assessing Environmental Quality and Creating Strategies for Change* (Los Angeles, Calif.: Living Planet Press, 1993) and Jean Heinze-Fry, *Green Lives, Green Campuses* (Belmont, Calif.: Wadsworth, 1995)—a supplement designed to be used with this textbook.

induced almost 80% of universities and colleges in the United States to develop recycling programs.

Students at Oberlin College in Ohio helped design new sustainable environmental studies buildings. This building **(1)** uses energy efficiently, **(2)** uses solar cells to generate more electricity than they use, **(3)** includes living machines to discharge wastewater at least as pure as the water taken in, **(4)** includes materials grown or manufactured sustainably, **(5)** uses no materials known to cause human health problems, and **(6)** is landscaped in ways that promote biodiversity.

A 1997 report by the National Wildlife Foundation's Campus Ecology Program found that 23 student-researched and -motivated projects had saved the participating universities and colleges $16.3 million. According to this study, implementing similar programs in the nation's 3,700 universities and colleges could not only help improve environmental quality and environmental education but also lead to a savings of more than $2.6 billion.

---

by *establishing an enemy, to create fear and to divert people's attention and energy away from the real issues.* Prey on the fears of ordinary people by labeling environmentalists as the *green menace*—antibusiness, antipeople, antireligious, scaremongering radical extremists who are crippling the economy, hurting small business owners, costing taxpayers billions of dollars in unnecessary regulations, robbing small landowners of the right to do what they want with their land, trying to lock up the natural resources on public lands, and threatening jobs, national security, and traditional values.

When accused of using wild exaggeration and smear tactics against environmentalists to raise money and spread distrust, hate, and fear, Ron Arnold, one of the leaders of the anti-environmental Wise-Use movement (Case Study, right), said, "Facts don't really matter. In politics, perception is reality."

Weaken and intimidate. Infiltrate environmental groups to learn their plans and to create dissension and mistrust. Slap individual activists and small grassroots groups with nuisance lawsuits. Intimidate individual scientists, scientific writers, and mainstream scientific professional groups that stand in the way of corporate anti-environmental goals by lawsuits and threats of lawsuits.

Threaten or use violence. Former Interior Secretary James Watt publicly declared in 1990 that "if the troubles from the environmentalists cannot be solved in the jury box or the ballot box, perhaps the cartridge box should be used." A study by the Center for Investigative Reporting substantiated 124 credible death threats, firebombing, shootings, and assaults on environmental activists in 31 states between 1988 and 1992. These are only the tip of the iceberg because environmentalists have not set up a

Q: Is ozone depletion from CFCs and other human-made chemicals reaching the stratosphere a serious threat?

The Wise-Use Movement

Since 1988, several hundred local and regional grassroots groups in the United States have formed a national anti-environmental coalition called the *Wise-Use movement*. Much of their money comes from real estate developers and from timber, mining, oil, coal, and ranching interests and traditional supporters of right-wing causes.

According to Ron Arnold, who played a key role in setting up this movement, the specific goals of the Wise-Use movement are as follows:

- *Cut all old-growth forests in the national forests and replace them with tree plantations.*

- *Modify the Endangered Species Act so that economic factors override preservation of endangered and threatened species.*

- *Eliminate government restrictions on wetlands development.*

- *Open all national parks, national wildlife refuges, and wilderness areas to oil drilling, mining, off-road vehicles (ORVs), and commercial development.*

- *Do away with the National Park Service and launch a 20-year construction program of new concessions run by private firms in the national parks.*

- *Continue mining on public lands under the provisions of the 1872 Mining Law, which allows mining interests not only to pay no royalties to taxpayers for hard-rock minerals they remove, but also to buy public lands for a pittance.*

- *Recognize private property rights to mining claims, water, grazing permits, and timber contracts on public lands; do not raise fees for these activities. Ideally, sell resource-rich public lands to private enterprise.*

- *Provide civil penalties against anyone who legally challenges economic action or development on federal lands.*

- *Allow pro-industry (Wise-Use) groups or individuals to sue as "harmed parties" on behalf of industries threatened by environmentalists.*

Most anti-environmental groups either affiliated with or generally supportive of the Wise-Use movement and its offshoot the *Alliance for America* (formed in 1991) use environmentally friendly-sounding names.* Here are a few examples: the *U.S. Council for Energy Awareness* (nuclear power industry), *America the Beautiful* (packaging industry), *Partnership for Plastics Progress* (plastics industry), *American Council on Science and Health* (food and pesticide industries), *National Wetlands Coalition* (real estate developers and oil and gas companies), *Global Climate Coalition* (50 corporations and trade associations opposed to reducing fossil-fuel use to slow the rate of global warming), *Blue Ribbon Coalition* (off-road vehicle users and manufacturers who want all public lands opened up to ORVs), *American Forest Resource Alliance* (timber and logging companies), *American Farm Bureau Federation* (agricultural chemicals industries), and *People for the West* (mining industry).

---

*For lists and information about these organizations, see *The Greenpeace Guide to Anti-Environmental Organizations* (1993, Odonian Press, Box 7776, Berkeley, CA 94707); *Masks of Deception: Corporate Front Groups in America* by Mark Megalli and Andy Friedman (1993, Essential Information, P.O. Box 19367, Washington, DC 20036); *Let the People Judge: A Reader on the Wise-Use Movement* (1993, Island Press, Covelo, CA); and *The War Against the Greens* (1994, Sierra Club Books, San Francisco, CA).

The Wise-Use movement also includes the *Sahara Club*, which advocates violence against environmentalists in defense of its members' "right" to dirt-bike in wilderness areas. One of its leaders tells audiences to "Throw environmentalists off the bridge. Water optional." and that "You can't reason with eco-freaks, but you sure can scare them." Another group is *Citizens for the Environment* (an industry-backed education group that counters environmental ideas with such slogans as "Recycling doesn't save forests," "Packaging prevents waste," and "Global warming and ozone depletion are hoaxes").

According to Jay D. Hair, former president of the National Wildlife Federation, the Wise-Use movement is *"merely a wise disguise for a well-financed, industry-backed campaign that preys on the economic woes and fears of U.S. citizens. . . . These organizations with benign-sounding names are not grassroots—they are astro-turf laid down with big corporate money. And they are out to . . . ensure that certain special interests will be allowed to continue to pollute and exploit public resources for private profit."*

According to environmentalist B. J. Bergman, the basic principles of the Wise Use movement are "If it grows, cut it; if it flows, dam it; if it's underground extract it; if it's swampy, fill it; if it moves, kill it."

Critical Thinking

Do you agree or disagree with each of the major goals of the Wise-Use movement? Explain. What positive effects has the Wise-Use movement had on the environmental movement?

---

nationwide network to record, investigate, and publicize the growing number of such incidents.

Influence public opinion. Commission books and articles, establish slick magazines, set up think tanks, and hire PR firms to deny that environmental problems exist or are serious. Use fax and letter-writing campaigns against corporations that provide financial support for TV or radio programs critical of environmentally harmful business practices or anti-environmental concerns. Flood schools with free pamphlets, videos, and other

---

thinly disguised teaching materials that propagandize your position.

Exploit the built-in limitations of science and public ignorance about the nature of science to sow seeds of doubt. Mislead the public by saying that claims against your position have not been absolutely scientifically proven. If that doesn't work, imply that scientific findings that are contrary to your position should not be taken seriously because science can only put forth theories (Section 2-1). These tactics should drive scientists and environmentalists crazy and keep them on the defensive.

Urge the president and Congress to require that all government environmental decisions and regulations be evaluated primarily by cost–benefit and risk–benefit analysis. Then try to load the evaluation panels with experts who support your position. If this doesn't work, challenge the analyses in court. The end result of this version of the *paralysis-by-analysis* strategy will be fewer and weaker environmental, health, and worker safety regulations. All of this would be achieved without trying to repeal existing environmental laws (a politically unpopular cause).

Support unenforceable legislation and regulations. This undermines confidence in government, splits up opposition groups, and helps convince the public that real change is impossible.

When unfavorable environmental laws are passed, urge legislators not to fund the laws. Passing the cost of federally unfunded environmental mandates on to financially strapped states and cities will generate intense political pressure against such laws by citizens, mayors, and governors. Because Congress is unlikely to come up with funding for these laws, they will probably be weakened, unenforced, or repealed.

Require federal and state governments to compensate property owners whenever environmental, health, safety, or zoning laws either limit how the owners can use their property or decrease its financial value. This will make such laws and regulations too expensive to enforce and will keep new ones from being established.

12-6 GLOBAL ENVIRONMENTAL POLICY

Should We Expand the Concept of Security? Countries have legitimate *national security* interests, but without adequate soil, water, clean air, and biodiversity, no nation can be secure. Thus, to a growing number of policy analysts, national security ultimately depends on *global environmental security* based on sustainable use of the ecosphere.

Countries are also legitimately concerned with *economic security.* However, because all economies are supported by the ecosphere (Figure 12-2), economic security also depends on environmental security.

Proponents of putting more emphasis on environmental security propose that national governments have a council of advisers made up of highly qualified experts in environmental, economic, and military security. Any major decision would require integrating all three security concerns. These analysts also call for all countries to make environmental security a major focus of diplomacy and government policy at all levels. What do you think?

What Progress Has Been Made in Developing International Environmental Cooperation and Policy? Since the 1972 United Nations Conference on the Human Environment was held in Stockholm, Sweden, some progress has been made in addressing environmental issues at the global level. Today, 115 nations have environmental protection agencies, and more than 170 international environmental treaties have been signed concerning a variety of issues including endangered species, ozone depletion, ocean pollution, global warming, biodiversity, and export of hazardous waste. The 1972 conference also created the United Nations Environment Programme (UNEP) to negotiate environmental treaties and to help implement them.

In June 1992, the second United Nations Conference on the Human Environment—known as the *Rio Earth Summit*—was held in Rio de Janeiro, Brazil. More than 100 heads of state, thousands of officials, and more than 1,400 accredited NGOs from 178 nations met to develop plans for addressing environmental issues.

The major official results included **(1)** an *Earth Charter*, a nonbinding statement of broad principles for guiding environmental policy that commits countries that sign it to pursue sustainable development and to work toward eradicating poverty; **(2)** *Agenda 21*, a nonbinding detailed action plan to guide countries toward sustainable development and protection of the global environment during the 21st century; **(3)** a *forestry agreement* that is a broad, nonbinding statement of principles of forest management and protection; **(4)** a *convention on climate change* that requires countries to use their "best efforts" to reduce their emissions of greenhouse gases; **(5)** a *convention on protecting biodiversity* that calls for countries to develop strategies for the conservation and sustainable use of biological diversity; and **(6)** establishment of the *UN Commission on Sustainable Development*, composed of high-level government representatives charged with carrying out and overseeing the implementation of these agreements.

Most environmentalists were disappointed because these accomplishments consist of nonbinding agreements without sufficient prods or incentives for their implementation. In addition, countries did not

Bolin, Bert. 1998. "The Kyoto Negotiations on Climate Change: A Science Perspective." *Science*, vol. 279, no. 5349, 330(2).

commit even the minimum amount of money that conference organizers said was needed to begin implementing Agenda 21. Developing countries feel betrayed by the developed world's failure to uphold the promises made at the Rio summit. Their leaders argue that, faced with massive poverty and limited budgets, they cannot reduce pollution and protect biodiversity without outside assistance and transfer of new, more earth-sustaining technologies.

It is discouraging that so little has been done but there is hope for progress in the slowly moving arena of international cooperation. *First*, the conference gave the world a forum for discussing and seeking solutions to environmental problems, and it led to general agreement on some key principles.

Second, paralleling the official meeting was a Global Forum that brought together 18,000 people from more than 1,400 NGOs in 178 countries. These NGOs worked behind the scenes to influence official policy, formulated their own agendas and treaties for helping sustain the earth, learned from one another, and developed a series of new global networks, alliances, and projects. In the long run, these newly formed networks and alliances may play the greatest role in monitoring and supporting the commitments and plans developed by the formal conference.

12-7 ENVIRONMENTAL WORLDVIEWS: CLASHING VALUES AND CULTURES

How Shall We Live? A Clash of Cultures and Values There are conflicting views about how serious our environmental problems are and what we should do about them. These conflicts arise mostly out of differing **environmental worldviews**: how individuals think the world works, what they think their role in the world should be, and what they believe is right and wrong environmental behavior (*environmental ethics*).

People with widely differing environmental worldviews can take the same data, be logically consistent, and arrive at quite different conclusions because they start with different assumptions and are often seeking answers to different questions.

There are many different types of environmental worldviews, as summarized in Figure 12-7. Most can be divided into two groups according to whether they are *individual centered* (atomistic) or *earth centered* (holistic). Atomistic environmental worldviews tend to be *human centered* (anthropocentric) or *life centered* (biocentric, with the primary focus on either individual species or individual organisms). Holistic or ecocentric environmental worldviews are either *ecosystem centered* or *ecosphere* (life-support system) *centered*.

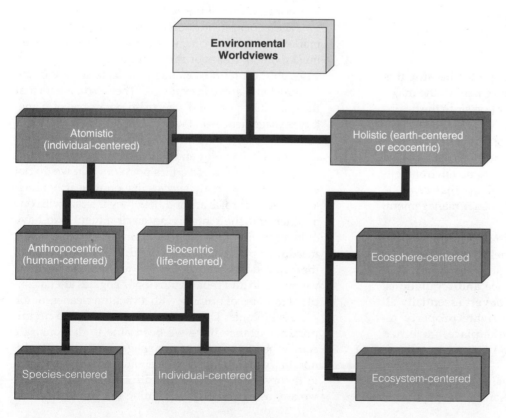

Figure 12-7 General types of environmental worldviews. (Diagram developed by Jane Heinze-Fry)

Hint: Enter the search term *Kyoto protocol* using Key Words.

What Are the Major Human-Centered Environmental Worldviews? Most people in today's industrial–consumer societies have a **planetary management worldview**, which has become increasingly accepted during the past 50 years. According to this human-centered environmental worldview, human beings, as the planet's most important and dominant species, can and should manage the planet mostly for their own benefit. Other species are seen as having only *instrumental value*; that is, their value depends on whether they are useful to us.

The basic environmental beliefs of this worldview include the following:

- *We are the planet's most important species, and we are in charge of the rest of nature.* This idea crops up when people talk about "our" planet or "our" earth and when they talk about "saving the earth."

- *There is always more.* The earth has an essentially unlimited supply of resources, to which we gain access via science and technology. If we deplete a resource, we will find substitutes. To deal with pollutants, we can invent technology to clean them up, dump them into space, or move into space ourselves. If we extinguish other species, we can use genetic engineering to create new and better ones.

- *All economic growth is good, more economic growth is better, and the potential for economic growth is essentially limitless.*

- *Our success depends on how well we can understand, control, and manage the earth's life-support systems for our benefit.*

This worldview is widely supported because it is said to be the primary driving force behind the major improvements in the human condition since the beginning of the industrial revolution.

There are several variations of this environmental worldview. Some people belong to what might be called the *no-problem school*: There are no environmental, population, or resource problems that can't be solved by more economic growth, better management, and better technology.

Another group, *the free-market school*, believes that the best way to manage the planet for human benefit is through a truly free-market global economy with minimal government interference and regulations. Free-market advocates would convert essentially all public property resources to private property resources and let the global marketplace, governed by true free-market competition (pure capitalism), decide essentially everything.

Another variation, held by many people, is that we do have serious environmental, resource, and population problems but that we can deal with them by engaging in *responsible planetary management*. People holding this view follow the pragmatic principle of *enlightened self-interest*: Better earth-care is better self-care. They believe that we can sustain our species with a mixture of market-based competition, better technology, and some government intervention to promote ecologically sustainable forms of economic development, protect environmental quality and private property rights, and protect and manage public and common property resources.

Still another variation of the responsible planetary management environmental worldview is the *spaceship-earth worldview*. In this case, earth is seen as a spaceship—a complex machine that we can understand, dominate, change, and manage in order to prevent environmental overload and provide a good life for everyone.

Yet another group of people advocate that our management of the earth be guided by the principle of *stewardship*: We have an ethical responsibility to be caring and responsible managers or stewards, who tend the earth as if it were a garden. According to this view, we can and should make the world a better place for ourselves and other species through love, care, knowledge, and technology.

Can We Manage the Planet? Some people believe that any human-centered worldview will eventually fail because it wrongly assumes that we now have (or can gain) enough knowledge to become effective managers or stewards of the earth.

These people argue that the unregulated free-market approach won't work because it is based on mushrooming losses of earth capital and because it focuses on short-term economic benefits regardless of the harmful long-term consequences. They also contend that the spaceship-earth and stewardship versions of planetary management won't work because such human constructs are oversimplified and misleading ways to view an incredibly complex and ever-changing planet.

For example, these critics point out that we do not even know how many species live on the earth, much less what their roles are and how they interact with one another and their nonliving environment. We have only an inkling of what goes on in a handful of soil, a meadow, a patch of forest, a pond, or any other part of the earth. Such analysts liken us to technicians who think we can build and repair automobile engines after a couple of minutes of training with a vacuum cleaner motor.

As biologist David Ehrenfeld puts it, "In no important instance have we been able to demonstrate comprehensive successful management of the world, nor do we understand it well enough to manage it even in theory." Environmental educator David Orr says we are losing rather than gaining the knowledge and wisdom needed to adapt creatively to continually changing environmental conditions:

> *On balance, I think, we are becoming more ignorant because we are losing cultural knowledge about how to inhabit our places on the planet sustainably, while impoverishing the genetic knowledge accumulated through millions of years of evolution. . . . Most research is aimed to further domination of the planet. Considerably less of it is directed at understanding the effects of domination. Less still is aimed to develop ecologically sound alternatives that enable us to live within natural limits.*

To critics of the planetary management environmental worldview, our task is not to learn how to pilot spaceship earth, but instead to give up our fantasies of omnipotence and base our actions on earth wisdom—on learning to work with the earth by becoming responsible earth citizens. Even if we had enough knowledge and wisdom to manage spaceship earth, some critics see this approach as requiring us to give up individual freedom in order to survive. Life on spaceship earth under a comprehensive system of planetary management or world government might be very much like the regimented life of astronauts in their capsule. The astronauts have virtually no individual freedom; essentially all of their actions are dictated by a central command (ground control) without which they cannot survive.

Theologian Thomas Berry calls the industrial–consumer society built on the human-centered, planetary management environmental worldview the "supreme pathology of all history":

> *We can break the mountains apart; we can drain the rivers and flood the valleys. We can turn the most luxuriant forests into throwaway paper products. We can tear apart the great grass cover of the western plains, and pour toxic chemicals into the soil and pesticides onto the fields, until the soil is dead and blows away in the wind. We can pollute the air with acids, the rivers with sewage, the seas with oil—all this in a kind of intoxication with our power for devastation. . . . We can invent computers capable of processing ten million calculations per second. And why? To increase the volume and speed with which we move natural resources through the consumer economy to the junk pile or the waste heap. . . . If, in these activities, the topography of the planet is damaged, if the environment is made inhospitable for a multitude of living species, then so be it. We are, supposedly, creating a technological wonderworld. . . . But our supposed progress . . . is bringing us to a wasteworld instead of a wonderworld.*

What Are Some Major Biocentric and Ecocentric Worldviews? Critics of human-centered environmental worldviews believe that such worldviews should be expanded to recognize the *inherent value* of all forms of life (that is, value that exists regardless of these life-forms' potential or actual use to us).

Most people with a life-centered (biocentric) worldview argue that our actions should not lead to the *premature* extinction of species. Some proponents go further and believe that each individual organism, not just each species, has an inherent right to survive.

Trying to decide what types of species or individuals should be protected from premature extinction or death resulting from human activities is an ethical dilemma. It is hard to know where to draw the line and be ethically consistent.

Others believe that we must go beyond this biocentric worldview, which focuses on species and individual organisms. They see our primary role as limiting our actions to those that do not degrade or destroy the earth's life-support systems (and thus threaten the existence of our own species). In other words, they have an *earth-centered*, or *ecocentric*, environmental worldview, devoted to preserving earth's biodiversity and ecological integrity.

Their view is that we are part of, not apart from, the community of life and the ecological processes that sustain all life. Aldo Leopold summed up this idea in 1948: "All ethics rest upon a single premise: that the individual is a member of a community of interdependent parts."

There are many life-centered and earth-centered environmental worldviews, and several of them overlap in some of their beliefs. One ecocentric worldview is the **earth-wisdom worldview**. It is based on the following major beliefs, which are in sharp contrast to those of the planetary management worldview:

- *Nature exists for all of the earth's species, not just for us.* We need the earth, but the earth does not need us.

- *There is not always more.* The earth's limited resources should not be wasted, but instead used sustainably and efficiently for us and all species.

- *Some forms of economic growth are environmentally beneficial and should be encouraged, but some are environmentally harmful and should be discouraged.*

- *Our success depends on learning to cooperate with one another and with the rest of nature by learning how to work with the earth.* Management of resources is essential to human survival. However, such management should involve learning as much as we can about how the earth works, sustains itself, and adapts to changing conditions and then using these lessons from nature to guide our actions.

Why Should We Care About Future Generations?

According to biologist David W. Ehrenfeld, caring about future generations enough not to degrade the earth's life-support systems is important because it gives future generations options for dealing with the problems they will face. He points out that if our ancestors had left for us the ecological devastation we are leaving our descendants, our options for enjoyment—perhaps even for survival—would be quite limited.

And in response to the question, "What can future generations do for us?" Ehrenfeld gives the following answer: "They give us a reason for treating our ecological home respectfully, so that our lives as well as theirs will be enriched."

In thinking about our responsibility toward future generations some analysts believe we should consider the wisdom given to us in the 18th century by the Iroquois Confederation of Native Americans: *In our every deliberation, we must consider the impact of our decisions on the next seven generations.*

Critical Thinking

What obligations, if any, concerning the environment do you have to future generations? Be honest about your feelings. To how many future generations do you have responsibilities? List the most important environmental benefits and harmful conditions passed on to you by the previous two generations.

Are Biocentrists and Ecocentrists Antihuman and Antireligious? Many anti-environmentalists accuse people who promote various biocentric and ecocentric worldviews of being antipeople or against celebrating humanity's special qualities and achievements. Those with life-centered and earth-centered worldviews, to the contrary, consider their environmental worldviews to be profoundly prohuman. To them, recognizing the inherent value of all life and not degrading earth's life-support systems are ways to benefit all people in this and future generations (Connections, above).

Others say we don't need to be biocentrists or ecocentrists to value life or the earth, that human-centered stewardship and responsible planetary-management environmental worldviews also call for us to value individuals, species, and the earth's life-support systems as part of our responsibility as earth's caretakers.

Solutions: What Are Some Ethical Guidelines for Working with the Earth? Ethicists and philosophers have developed a variety of ethical guidelines for living more sustainably on the earth. Such guidelines can be used by anyone, whether they have a human-centered stewardship environmental worldview or a biocentric or ecocentric environmental worldview.

Ecosphere and Ecosystems

- We should try to understand and work with the rest of nature to help sustain the ecological integrity, biodiversity, and adaptability of the earth's life-support systems.

- When we must alter nature to meet our needs or wants, we should carefully evaluate our proposed actions and choose methods that do the least possible short- and long-term environmental harm.

Species and Cultures

- We should work to preserve as much of the earth's genetic variety as possible because it is the raw material for all future evolution and genetic engineering.

- We have the right to defend ourselves against individuals of species that do us harm and to use individuals of species to meet our vital needs, but we should strive not to cause the premature extinction of any wild species.

- The best ways to protect species and individuals of species are to protect the ecosystems in which they live and to help restore those we have degraded.

- No human culture should become extinct because of our actions.

Individual Responsibility

- We should not inflict unnecessary suffering or pain on any animal we raise or hunt for food or use for scientific or other purposes.

- We should leave the earth as good as or better than we found it.

- We should use no more of the earth's resources than we need.

- We should work with the earth to help heal ecological wounds we have inflicted.

Solutions: How Can We Implement Earth Education? Most environmentalists believe that learning how to live sustainably by working with the earth requires a foundation of earth education that relies heavily on an interdisciplinary and holistic approach to learning and a lifelong commitment to such education. Among the most important goals of such an education are the following:

- *Developing respect or reverence for all life.*

- *Understanding as much as we can about how the earth works and sustains itself and using such earth wisdom to guide our lives, communities, and societies.*

- *Understanding as much as we can about connections and interactions*—those within nature, between people and the rest of nature, among people with different cultures and beliefs, among generations, among the problems we face, and among the solutions to these problems.

- *Becoming a wisdom seeker instead of an information vessel.* We need to learn how to sift through mountains of facts and ideas to find the nuggets of knowledge and wisdom that are worth knowing.

- *Understanding and evaluating one's worldview and seeing this as a lifelong process* (Individuals Matter, right).

- *Learning how to evaluate the beneficial and harmful consequences of one's lifestyle and profession on the earth, today and in the future.*

- *Using critical thinking skills to evaluate and resist advertising.* As humorist Will Rogers put it, "Too many people spend money they haven't earned to buy things they don't want, to impress people they don't like."

- *Fostering a desire to make the world a better place.* As David Orr puts it, education should help students "make the leap from 'I know' to 'I care' to 'I'll do something.'"

According to environmental educator Mitchell Thomashow, four basic questions should be at the heart of environmental education:

- Where do the things I consume come from?

- What do I know about the place where I live?

- How am I connected to the earth and other living things?

- What is my purpose and responsibility as a human being?

How we answer these questions determines our *ecological identity*.

Solutions: How Can Direct Experiences Help Us Learn How to Work with the Earth? In addition to top-down formal education, some analysts believe that we need to learn by experience how to walk more lightly on the earth (Connections, p.354). One source of

Questioning and perhaps even changing one's environmental worldview can be difficult and threatening and can set off a cultural *mindquake* that involves examining many of one's most basic beliefs. However, once people change their worldviews, it no longer makes sense for them to do things in the old ways. If enough people do this, then tremendous cultural change, once considered impossible, can take place rapidly.

Most environmentalists urge us to think about what our basic environmental beliefs are and why we have them. They believe that evaluating our beliefs and being open to the possibility of changing them should be one of our most important lifelong activities. What do you think?

wisdom is bottom-up education obtained by *listening to children*. Listen to young children and you will find that many of them believe that much of what we are doing to the earth (and thus to them) is stupid and wrong, and they do not accept the excuses we give for not changing the harmful ways we act toward the earth.

A related goal is to *learn how to live more simply*. Although seeking happiness through the single-minded pursuit of material things is considered folly by virtually every major religion and philosophy, it is preached incessantly by modern advertising. Some affluent people in developed countries, however, are adopting a lifestyle of *voluntary simplicity*, doing and enjoying more with less by learning to live more simply. They agree with Edward Goldsmith that "the more we look at it, the more it is apparent that economic growth is a device for providing us with the superfluous at the cost of the indispensable."

Voluntary simplicity is based on Mahatma Gandhi's *principle of enoughness*: "The earth provides enough to satisfy every person's need but not every person's greed. . . . When we take more than we need, we are simply taking from each other, borrowing from the future, or destroying the environment and other species." It means asking oneself, "How much is enough?" This is not an easy thing to do because affluent people (and nonaffluent people) are conditioned to want more and more, and they often think of such wants as vital needs.

A: Cigarette smoke, radioactive radon gas, and formaldehyde

Learning from the Earth

Formal earth education is important, but many earth thinkers believe that it is not enough. They urge us to take the time to escape the cultural and technological body armor we use to insulate ourselves from nature and to experience nature directly.

They suggest that we reenchant our senses and kindle a sense of awe, wonder, and humility by standing under the stars, sitting in a forest, taking in the majesty and power of an ocean, or experiencing a stream, lake, or other part of untamed nature. We might pick up a handful of soil and try to sense the teeming microscopic life in it that keeps us alive. We might look at a tree, a mountain, a rock, or a bee and try to sense how they are a part of us, and we a part of them as interdependent participants in the earth's life-sustaining recycling processes.

Earth thinker Michael J. Cohen suggests that we recognize who we really are by saying,

I am a desire for water, air, food, love, warmth, beauty, freedom, sensations, life, community, place, and spirit in the natural world. . . . I have two mothers: my human mother and my planet mother, Earth. The planet is my womb of life.

Many psychologists believe that consciously or unconsciously we spend much of our lives in a search for roots—something to anchor us in a bewildering and frightening sea of change. As philosopher Simone Weil observed, "To be rooted is perhaps the most important and least recognized need of the human soul."

Earth philosophers say that to be rooted, each of us needs to find a *sense of place*—a stream, a mountain, a yard, a neighborhood lot, or any piece of the earth we feel truly at one with as a place we know, feel, and love. It can be a place where we live or a place we occasionally visit and experience in our inner being. When we become part of a place, it becomes a part of us. Then we are driven to defend it from harm and to help heal its wounds.

To many earth thinkers, emotionally experiencing our connectedness with the earth leads us to recognize that the healing of the earth and the healing of the human spirit are one and the same. They call for us to discover and tap into what Aldo Leopold calls "the green fire that burns in our hearts" and use this as a force for respecting and working with the earth and with one another.

Critical Thinking

Some analysts believe that learning earth wisdom by experiencing the earth and forming an emotional bond with its life forms and processes is unscientific, mystical poppycock based on a romanticized view of nature. They believe that better scientific understanding of how the earth works and improved technology are the only ways to achieve sustainability. Do you agree or disagree? Explain.

However, voluntary simplicity by those who have more than they need should not be confused with the *forced simplicity* of the poor, who do not have enough to meet their basic needs for food, clothing, shelter, clean water and air, and good health.

Earth care also means *not using guilt and fear to motivate other people to work with the earth and with other people.* We need to nurture, reassure, understand, and love, rather than threaten, one another. Finally, we need to *have fun and take time to enjoy life.* We shouldn't get so intense and serious that we can't laugh every day and enjoy wild nature, beauty, friendship, and love.

Solutions: How Can We Move Beyond Blame, Guilt, and Denial to Responsibility? According to many psychologists, when we first encounter an environmental problem, our initial response is often to

find someone or something to blame—greedy industrialists, uncaring politicians, misguided worldviews. It is the fault of such villains, and we are the victims.

This can lead to despair, denial, and inaction because we feel powerless to stop or influence these forces. There are also so many complex and interconnected environmental problems and conflicting views about their seriousness and possible solutions that we feel overwhelmed and wonder whether there is any way out—another emotion leading to denial and inaction.

Upon closer examination we may realize that we all make some direct or indirect contributions to the environmental problems we face. We don't want to feel guilty or bad about all of the things we are not doing, so we avoid this by not thinking about them—another path leading to denial and inaction.

How do we move beyond immobilizing blame, fear, guilt, and denial to engaging in more responsible

Q: What is the most dangerous indoor air pollutant in developing countries?

environmental actions in our daily lives? Analysts have suggested several ways to do this.

First, we need to recognize and avoid common mental traps that lead to denial, indifference, and inaction. These traps include **(1)** *gloom-and-doom pessimism* (it's hopeless), **(2)** *blind technological optimism* (science and technofixes will always save us), **(3)** *fatalism* (we have no control over our actions and the future), **(4)** *extrapolation to infinity* (if I can't change the entire world quickly, I won't try to change any of it), **(5)** *paralysis by analysis* (searching for the perfect worldview, philosophy, solutions, and scientific information before doing anything), and **(6)** *faith in easy answers*.

Second, we should recognize that no one can even come close to doing all of the things people suggest (or that we know we should be doing) to work with the earth. We should focus our energy on the few things that we feel most strongly about and that we can do something about. Instead of focusing on and feeling guilty about the things we haven't done, rejoice in the good things we have done—and jump in and do more to make the earth a better place.

Third, we should base our actions on a sense of *hope*, which history has shown to be the major energizing force for bringing about change. The secret is to keep our empowering feelings of hope and joy slightly ahead of our immobilizing feelings of despair.

Fourth, it is important to recognize that there is no single correct or best solution to the environmental problems we face. Indeed, one of nature's most important lessons (from evolution) is that preserving diversity, a rainbow of possibilities, is the best way to adapt to earth's largely unpredictable, ever-changing conditions. We are all in this together and need to work together to find a spectrum of flexible and adaptable solutions to the problems we face.

Solutions: What Are the Major Components of an Earth-Wisdom Revolution? Many environmentalists call for us to make a new cultural change in the way we think about and use the earth's endowment of resources. Such an *earth-wisdom* or *environmental revolution* would have several phases. One is an *efficiency revolution* that involves not wasting matter and energy resources, using a combination of technological advances, lifestyle changes, recycling, and reuse.

A second phase is a *pollution prevention or ecoindustrial revolution* (Solutions, p. 316) built on the efficiency revolution. It reduces pollution and environmental degradation by reducing the waste of matter and energy resources (Figure 2-47); it does so by mimicking the earth's chemical cycling processes (Section 2-6) in which each organism's wastes serve as resource inputs for other organisms. Pollution prevention also involves keeping highly toxic substances from being released into the environment by recycling or reusing them within industrial processes, trying to find less harmful or easily biodegradable substitutes, or not producing such substances at all.

A third phase is a *sufficiency revolution*, which means trying to meet the basic needs of all people on the planet and asking how many material things we really need to have a decent and meaningful life.

A fourth phase of this new cultural change is a *demographic revolution* based on bringing the size and growth rate of the human population into balance with the earth's cultural carrying capacity for humans and other life forms.

Opponents of bringing about a new environmental cultural revolution like to paint environmentalists as messengers of gloom and doom and hopelessness. However, *the real message of environmentalism is not gloom and doom, fear, and catastrophe but hope and a positive vision of the future*. This is an exciting message of challenge and adventure as we struggle to find better and more responsible ways to live sustainably on this wondrous planet.

Envision the world as a system of all kinds of matter cycles and energy flows. See these life-sustaining processes as a beautiful and diverse web of interrelationships—a kaleidoscope of patterns, rhythms, and connections whose very complexity and multitude of possibilities remind us that cooperation, sharing, honesty, humility, and love should be the guidelines for our behavior toward one another and the earth.

What has gone wrong, probably, is that we have failed to see ourselves as part of a large and indivisible whole. . . . We have failed to understand that the earth does not belong to us, but we to the earth.

ROLF EDBERG

CRITICAL THINKING

1. The primary goal of all current economic systems is to maximize economic growth by producing and consuming more economic goods. Do you agree with that goal? Explain. What are the alternatives?

2. Do you favor internalizing the external costs of pollution and unnecessary resource waste? Explain. How might it affect your lifestyle? Wildlife? Any children you might have?

3. (a) Do you believe that we should establish optimal levels or zero-discharge levels for most of the chemicals

A: Smoke from unvented or poorly vented stoves for cooking and heating

we release into the environment? Explain. What effects would adopting zero-discharge levels have on your life?

(b) Should we assume that all chemicals we release or propose to release into the environment are potentially harmful until proven otherwise? Explain. What effects would adopting this principle have on your life?

4. Do you agree or disagree with the proposals various analysts have made for sharply reducing poverty discussed on page 340? Explain.

5. Do you agree or disagree with the guidelines for an earth-sustaining economy discussed on pages 340–341? Explain.

6. List all the economic goods you use; then identify those that meet your basic needs and those that satisfy your wants. Identify any economic wants you would be willing to give up, those you believe you should give up but are unwilling to give up, and those you hope to satisfy in the future. List what is likely to make you happy and improve the quality of your life. Relate the results of this analysis to your personal impact on the environment. Compare your results with those of your classmates.

7. What are the greatest strengths and weaknesses of the system of government in your country with respect to protecting the environment, working with the earth, and ensuring environmental justice for all? What substantial changes, if any, would you make in this system?

8. Do you agree or disagree with proposals various analysts have made to improve the U.S. political system by reforming the elective process (Solutions, p 345)? If you agree, what role do you plan to play in bringing about such a change?

9. Do you agree or disagree with each of the major goals of the Wise-Use movement (Case Study, p. 347)? Explain.

10. Which of the tactics, if any, used by the anti-environmental movement (pp. 345–348) do you think are appropriate? Explain. Which, if any, of these same tactics have been used by the environmental movement?

11. What obligations, if any, concerning the environment do you have to future generations? To how many future generations do you have responsibilities? List the most important environmental benefits and harmful conditions passed on to you by the previous two generations.

12. What are the basic beliefs of your environmental worldview? Has taking this course changed your environmental worldview? In what ways?

13. If you knew you were going to die and had an opportunity to address everyone in the world for 5 minutes, what would you say? Write out your 5-minute speech and compare it with those of other members of your class.

PUBLICATIONS, ENVIRONMENTAL ORGANIZATIONS, AND FEDERAL AND INTERNATIONAL AGENCIES*

PUBLICATIONS

The following publications can help you keep well informed and up-to-date on environmental and resource problems. Subscription prices, which tend to change, are not given.

American Biology Teacher Journal of the National Association of Biology Teachers, 11250 Roger Bacon Dr., Rm. 319, Reston, VA 22090; Tel: 703-471-1134

American Forests American Forestry Association, 1516 P St. NW, Washington, DC 20005; Tel: 202-667-3300

Amicus Journal Natural Resources Defense Council, 40 W. 20th St., New York, NY 10011; Tel: 212-727-2700

Annual Review of Energy Department of Energy, Forrestal Building, 1000 Independence Ave. SW, Washington, DC 20585

Audubon National Audubon Society, 950 Third Ave., New York, NY 10022; Tel: 212-832-3200, Web: http://magazine.audubon.org/

BioScience American Institute of Biological Sciences, Central Station, P.O. Box 27417, Washington, DC 20077-0038; Tel: 202-628-1500 or 800-992-2427, Fax: 202-628-1509, E-mail: bioscienc@aibs.org

Conservation Biology Blackwell Scientific Publications, Inc., 52 Beacon St., Boston, MA 02108

Demographic Yearbook Department of International Economic and Social Affairs, Statistical Office, United Nations Publishing Service, United Nations, New York, NY 10017; Tel: 617-253-2889, Fax: 617-258-6779

Discover 114 Fifth Avenue, New York, NY 10011-5690; Tel: 800-829-9132, E-mail: letters@discover.com

Earth Island Journal Earth Island Institute, 300 Broadway, Suite 28, San Francisco, CA 94133; Tel: 415-788-2666

Earth Work Student Conservation Association, P.O. Box 550, Charlestown, NH 03603; Tel: 603-543-1700, Fax: 603-543-1828

The Ecologist MIT Press Journals, 55 Hayward St., Cambridge, MA 02142

Ecology Ecological Society of America, Dr. Duncan T. Patten, Center for Environmental Studies, Arizona State University, Tempe, AZ 85281; Tel: 602-965-3979

EcoNet Institute for Global Communication, 18 De Boom St., San Francisco, CA 94107; Tel: 415-422-0220;

E Magazine 28 Knight St., Westport, CT 06881; Tel: 203-854-5559, Fax 203-866-0602, E-mail: axgm65a @prodigy.com, Web: http://www.emagazine.com

Endangered Species Update School of Natural Resources, University of Michigan, Ann Arbor, MI 48109; Tel: 313-763-3243, Fax: 313-936-2195, E-mail: jfwatson@umich.edu

Environment Heldref Publications, 1319 Eighteenth St. NW, Washington, DC 20036-1802; Tel: 800-365-9753

Environmental Abstracts Annual Bowker A & I Publishing, 121 Chanlon Rd., New Providence, NJ 07974

Environmental Ethics Department of Philosophy, University of North Texas, Denton, TX 76203

Environmental Health Perspectives: Journal of the National Institute of Environmental Health Sciences Government Printing Office, Washington, DC 20402

Everyone's Backyard Citizens' Clearinghouse for Hazardous Waste, P.O. Box 926, Arlington, VA 22216; Tel: 703-237-2249

Family Planning Perspectives Planned Parenthood–World Population, 666 Fifth Ave., New York, NY 10019; Tel: 212-541-7800

Greenpeace Magazine Greenpeace USA, 1436 U St. NW, Washington, DC 20009

Hydrogen Letter 4104 Jefferson St., Hyattsville, MD 20781; Tel: 301-779-1561, Fax: 301-927-6345

International Wildlife National Wildlife Federation, 8925 Leesburg Pike, Vienna, VA 22184; Tel: 703-790-4524

Issues in Science and Technology National Academy of Sciences, 2101 Constitution Ave. NW, Washington, DC 20077-5576; Tel: 213-883-6325, Web: http://utdallas.edu/research/issues

National Geographic National Geographic Society, P.O. Box 2895, Washington, DC 20077-9960; Tel: 202-857-7000

National Library for the Environment National Institute for the Environment Online Library. Web site: http://www.cnie.org

National Parks National Parks and Conservation Association, 1015 31st St. NW, Washington, DC 20007; Tel: 202-223-6722, Fax: 202-659-0650

National Wildlife National Wildlife Federation, 1400 16th St. NW, Washington, DC 20036; Tel: 202-790-4524

Nature 711 National Press Building, Washington, DC 20045

Not Man Apart Friends of the Earth, 530 7th St. SE, Washington, DC 20003; Tel: 202-544-2600, Fax: 202-543-4710

Pesticides and You National Coalition Against the Misuse of Pesticides, 701 E St. SE, Washington, DC 20003; Tel: 703-471-1134

Population and Vital Statistics Report United Nations Environment Programme, New York

North American Office, Publication Sales Section, United Nations, New York, NY 10017

Population Bulletin Population Reference Bureau, 1875 Connecticut Ave. NW, Suite 520, Washington, DC 20009; Tel: 202-483-1100

Rachel's Environment and Health Weekly Environmental Research Foundation, P.O. Box 4878, Annapolis, MD 21403; Tel: 410-263-1584

Rainforest News P.O. Box 140681, Coral Gables, FL 33115

Renewable Energy News Solar Vision, Inc., 7 Church Hill, Harrisville, NH 03450

Renewable Resources 5430 Grosvenor Lane, Bethesda, MD 20814; Tel: 301-493-9101

Rocky Mountain Institute Newsletter 1739 Snowmass Creek Rd., Snowmass, CO 81654

Science American Association for the Advancement of Science, 1333 H St. NW, Washington, DC 20005; Tel: 202-326-6400, Web: http://www.aaas.org

Science News Science Service, Inc., 1719 N St. NW, Washington, DC 20036; Tel: 800-247-2160, Web: http://www.sciencenews.org

Scientific American 415 Madison Ave., New York, NY 10017-1111; Web: http://www.sciam.com

Sierra 730 Polk St., San Francisco, CA 94108

Solar Age Solar Vision, Inc., 7 Church Hill, Harrisville, NH 03450

State of the Environment OECD Publications and Information Center, 1750 Pennsylvania Ave., Suite 1207, Washington, DC 20006 (published annually)

State of the World Worldwatch Institute, 1776 Massachusetts Ave. NW, Washington, DC 20036-1904; Tel: 202-452-1999, Fax: 202-296-7365 (published annually)

Statistical Yearbook Department of International Economic and Social Affairs, Statistical Office, United Nations Publishing Service, United Nations, New York, NY 10017

Technology Review P.O. Box 489, Mount Morris, IL 61054; Tel: 800-877-5320, Fax: 815-734-1127, E-mail: trsubscriptions@mit.edu.ans, Web: http://web.mit.edu/techreview/www/

Vital Signs: The Trends That Are Shaping Our Future Worldwatch Institute, 1776 Massachusetts Ave. NW, Washington, DC 20036-1904; Tel: 202-452-1999, Fax: 202-296-7365 (published annually)

Waste Not Work on Waste USA, 82 Judson, Canton, NY 13617

Wild Earth Cenozoic Society, Inc., P.O. Box 455, Richmond, VT 05477

Wilderness The Wilderness Society, 900 17th St. NW, Washington, DC 20006-2596; Tel: 202-833-2300

Wildlife Conservation New York Zoological Society, 185th St. & Southern Blvd., Bronx, NY 10460; Tel: 718-220-5100, Fax: 718-220-7114

---

*For a more detailed list, consult my longer books, *Environmental Science* and *Living in the Environment* (Pacific Grove, Calif.: Brooks/Cole).

World Rainforest Report Rainforest Action Network, 450 Sansome, Suite 700, San Francisco, CA 94111; Tel: 415-398-4404, Fax: 415-398-2732

World Resources World Resources Institute, 1709 New York Ave. NW, Washington, DC 20006; Tel: 202-638-6300 (published every 2 years)

World Watch Worldwatch Institute, 1776 Massachusetts Ave. NW, Washington, DC 20036-1904; Tel: 202-452-1999, Fax: 202-296-7365, E-mail: wwpub@worldwatch.org

Worldwatch Papers Worldwatch Institute, 1776 Massachusetts Ave. NW, Washington, DC 20036-1904; Tel: 202-452-1999, Fax: 202-296-7365

Yearbook of World Energy Statistics Department of International Economic and Social Affairs, Statistical Office, United Nations Publishing Service, United Nations, New York, NY 10017

ENVIRONMENTAL AND RESOURCE ORGANIZATIONS

For a more detailed list of national, state, and local organizations, see *Conservation Directory* (published annually by the National Wildlife Federation, 1400 16th St. NW, Washington, DC 20036), *Your Resource Guide to Environmental Organizations* (Irvine, CA: Smiling Dolphin Press, 1991), *National Environmental Organizations* (published annually by U.S. Environmental Directories, Inc., P.O. Box 65156, St. Paul, MN 55165), and *World Directory of Environmental Organizations* (published by the California Institute of Public Affairs, P.O. Box 10, Claremont, CA 91711). Also see Environmental Organizations Web site (http://www.econet.apc.org/econet/en.orgs.html/).

American Council for an Energy-Efficient Economy 1001 Connecticut Ave. NW #801, Washington, DC 20036; Tel: 202-429-8873, Web: http://www.aceee.org/

American Forests (formerly American Forestry Association 1516 P St. NW, Washington, DC 20005; Tel: 202-667-3300

American Humane Society 63 Inverness Dr. E, Englewood, CO 80112; Tel: 303-792-9900

American Institute of Biological Sciences, Inc. 730 11th St. NW, Washington, DC 20001-4521; Tel: 202-628-1500, Fax: 202-628-1509, Web: http://www.yahoo.com/Science/Biology/Organizations/Professional/American_Institute_of_Biological_Sciences/

American Solar Energy Society 2400 Central Ave., Suite G, Boulder, CO 80301; Tel: 303-443-3130, Web: http://www.sni.net/solar/

American Water Resources Association 950 Herndon Parkway, Suite 300, Herndon, VA 22070-5528; Tel: 703-904-1255, Fax: 703-904-1228

American Wind Energy Association 777 N. Capitol St. NE, Suite 805, Washington, DC 20002; Tel: 202-408-8988, Web: http://www.igc.apc.org/awea/index.html

Campus Ecology National Wildlife Federation, 8925 Leesburg Pike, Vienna, VA 22184; Tel: 703-790-4318, Web: http://www.nwf.org/nwj/campus

Carrying Capacity Network, Inc. 2000 P St. NW, Suite 240, Washington, DC 20036-5915; Tel: 202-296-4548 or 800-466-4866, Fax: 202-296-4609

Center for Conservation Biology Alice Blandin, *Conservation Biology*, University of Washington, Box 351800, Seattle, WA 98195-1800; Web: http://www.conbio.rice.edu/

Center for Health, Environment, and Justice P.O. Box 6806, Falls Church, VA 22040; Tel: 703-

237-2249, Web: http://essential.org/orgs/cchw/cchw.html

Center for Marine Conservation 1725 DeSales St. NW, Suite 500, Washington, DC 20036; Tel: 202-429-5609, Fax: 202-872-0619, Web: http://www.cmc-ocean.org/

Center for Plant Conservation P.O. Box 299, St. Louis, MO 63166; Tel.: 314-577-9450, Fax: 314-577-9465

Center for Science in the Public Interest 1875 Connecticut Ave. NW, Suite 300, Washington, DC 20009; Tel: 202-332-9110, Fax: 202-265-4594, Web: http://www.cspinet.org/

Clean Water Action Project 1320 18th St. NW, Suite 310, Washington, DC 20036; Tel: 202-457-0336, Fax: 202-457-0287, E-mail: schecter@essential.org

Coastal Society P.O. Box 2081, Gloucester, MA 01930-2081; Tel: 508-281-9209

Conservation International 1015 18th St. NW, Suite 1000, Washington, DC 20036; Tel: 202-429-5660. web: http://www.conservation.org

Council for Economic Priorities 30 Irving Place, New York, NY 10003; Tel: 212-420-1133. Maintains corporate environmental data clearinghouse.

Cultural Survival 11 Divinity Ave., Cambridge, MA 02138; Tel: 617-495-2562

Defenders of Wildlife 1101 14th St. NW, Suite 1400, Washington, DC 20005; Tel: 202-682-9400, Fax: 202-682-1331, E-mail: information@defenders.org

Ducks Unlimited One Waterfowl Way, Memphis, TN 38120-2351; Tel: 901-758-3825, Fax: 901-758-3850

Earth Island Institute 300 Broadway, Suite 28, San Francisco, CA 94133: Tel: 415-788-3666, Fax: 415-788-7324, Web: http://www.earthisland.org/ei

Elmwood Institute P.O. Box 5765, Berkeley, CA 94705; Tel: 510-845-4595

Energy Conservation Coalition 1525 New Hampshire Ave. NW, Washington, DC 20036

Envirolink Network E-mail: admin@envirolink.org, Web: http://www.envirolink.org

Environmental Defense Fund, Inc. 257 Park Ave. South, New York, NY 10010; Tel: 212-505-2100

Environmental Law Institute 1616 P St. NW, Suite 200, Washington, DC 20036; Tel: 202-328-5150, Fax: 202-328-5002

Fish and Wildlife Reference Service 5430 Grosvenor Ln., Bethesda, MD 20814; Tel: 301-492-6403 or 800-582-3421, Fax: 301-564-4059

Friends of the Earth The Global Building, 1025 Vermont Ave., NW, Suite 300, Washington, DC 20005; Tel: 202-783-7400, Fax: 202-783-0444, E-mail: foedc@igc.apc.org, Web: http://www.foe.co.uk/

Global Greenhouse Network 1130 17th St. NW, Suite 530, Washington, DC 20036

Global Tomorrow Coalition 1325 G St. NW, Suite 1010, Washington, DC 20005-3104; Tel: 202-628-4016, Fax: 202-628-4018

Greenpeace, Canada 427 Bloor St., West Toronto, Ontario M5S 1X7

Greenpeace, USA, Inc. 1436 U St. NW, Washington, DC 20009; Tel: 202-462-1177, E-mail: your.name@green2.greenpeace.org and www.greenpeace.org, Web: http://www.greenpeace.org/

Institute for Alternative Agriculture 9200 Edmonston Rd., Suite 117, Greenbelt, MD 20770

Institute for Earth Education Cedar Cove, Box 115, Greenville, WV 24945; Tel: 304-832-6404, Fax: 304-832-6077

Institute for Local Self-Reliance 2425 18th St. NW, Washington, DC 20009; Tel: 202-232-4108, E-mail: ilsr@itp.apc.org

International Institute for Energy Conservation 750 First St. NE, Washington, DC 20002; Tel: 202-842-3388, Fax: 202-842-1565, E-mail: iiec@igc.apc.org

International Planned Parenthood Federation 105 Madison Ave., 7th Floor, New York, NY 10016

International Society for Ecological Economics P.O. Box 1589, Solomons, MD 20688; Tel: 410-326-0794, E-mail: button@cbl.umd.edu

International Union for the Conservation of Nature and Natural Resources (IUCN) 1400 16th St. NW, Washington, DC 20036; Tel: 202-797-5454, Fax: 202-797-5461, E-mail: mail@hg.iucn.ch

Izaak Walton League of America 1401 Wilson Blvd., Level B, Arlington, VA 22209; Tel: 703-528-1818

Land Institute 2440 E. Well Water Road, Salina, KS 67401; Tel: 913-823-5376

League of Conservation Voters 1707 L Street NW, Suite 750, Washington, DC 20036; Tel: 202-785-8683, Fax: 202-835-0491, Web: http://www.lcv.org/

League of Women Voters of the U.S. 1730 M St. NW, Washington, DC 20036; Tel: 202-429-1965, Fax: 202-429-0854, Web: http://acm.cs.umn.edu/~lisi/lwv/main.html

National Audubon Society 700 Broadway, New York, NY 10003-9501; Tel. 212-979-3000, E-mail: mis@audubon.org, web: www.audubon.org/audubon/

National Coalition Against the Misuse of Pesticides 701 E St. SE, Suite 200, Washington, DC 20003; Tel: 202-543-5450, Web: http://www.ncamp.org/

National Environmental Health Association 720 S. Colorado Blvd., South Tower 970, Denver, CO 80222; Tel: 303-756-9090

National Geographic Society 1145 17th St. NW, Washington, DC 20036; Tel: 202-857-7000, Web: http://www.nationalgeographic.com/main.html

National Hydrogen Association 1800 M Street NW, Suite 300, Washington, DC 20036; Tel: 202-223-5547, Fax: 202-223-5537, Web: http://www.ttcorp.com/nha/

National Park Foundation 1101 17th St. NW, Suite 1102, Washington, DC 20036; Tel: 202-785-4500, Web: http://www.nationalparks.org/

National Parks and Conservation Association 1776 Massachusetts Ave. NW, Suite 200, Washington, DC 20036; Tel: 202-223-6722, Fax: 202-659-0650, E-mail: natparks@aol.com, Web: http://www.npca.org/npca/http

National Recycling Coalition 1727 King Street, Suite 105, Alexandria VA 22314; Tel: 703-683-9025, Fax: 703-683-9026, Web: http://www.recycle.net/recycle/index.html

National Toxics Campaign 37 Temple Place, 4th Floor, Boston, MA 02111

National Wildlife Federation 8925 Leesburg Pike, Vienna, VA 22184; Tel 703-790-4000, E-mail: feedback@nws.org, Web: http://www.nwf.org/nwf/home.html

Natural Resources Defense Council 40 W. 20th St., New York, NY 10011; Tel: 212-727-2700; and 1350 New York Ave. NW, Suite 300, Washington,

DC 20005; Tel: 202-783-7800, E-mail: nrdcinfo @igc.apc.org, Web: http://www.nrdc.org

Nature Conservancy 1814 N. Lynn St., Arlington, VA 22209; Tel: 703-841-5300, Fax: 703-841-1283, Web: http://www.tnc.org/

Pesticide Action Network 965 Mission St., No. 514, San Francisco, CA 94103; Web: http://www.panna.org/panna/

Pesticide Information Center P.O. Box 420870, San Francisco, CA 94142-0870; Tel: 415-391-8511

Physicians for Social Responsibility 1101 14th St. NW, 7th Floor, Washington, DC 20005; Tel: 202-898-0150

Planned Parenthood Federation of America 810 Seventh Ave., New York, NY 10019; Tel: 212-541-7800, Web: http://www.plannedparenthood.org

Population Action International (formerly Population Crisis Committee) 1120 19th St. NW, Suite 550, Washington, DC 20036-3605; Tel: 202-659-1833, Web: http://www.populationaction.org/

Population Institute 107 2nd St. NE, Washington, DC 20002; Tel: 202-544-3300

Population Reference Bureau 1875 Connecticut Ave. NW, Suite 520, Washington, DC 20009-5728; Tel: 202-483-1100, Fax: 202-328-3937, E-mail: popref@prb.org, Web: http://www.prb.org/prb/

Public Citizen 215 Pennsylvania Ave. SE, Washington, DC 20003; Web: http://www. essential.org/orgs/public_citizen/

Rainforest Action Network 3450 Sansome St., Suite 700, San Francisco, CA 94111; Tel: 415-398-4404, Fax: 415-389-2732, Web: http://www.ran.org

Rainforest Alliance 65 Bleecker St., New York, NY 10012; Tel: 212-677-1900, Fax: 212-677-2187, E-mail: canopy@cdp.apc.or, Web: http://www. rainforest-alliance.org/

Resources for the Future 1616 P St. NW, Washington, DC 20036; Tel: 202-328-5000, Fax: 202-939-3460, E-mail: info@rrf.org, Web: www.riff.org

Rocky Mountain Institute 1739 Snowmass Creek Rd., Snowmass, CO 81654-9199; Tel: 970-927-3851, Fax: 970-927-3420, Web: http://www.rmi.org/index.html

Rodale Institute 222 Main St., Emmaus, PA 18098; Web: http://www.envirolink.org/seel/rodale/

Scientists' Institute for Public Information 355 Lexington Ave., New York, NY 10017; Tel: 212-661-9110

Sea Shepherd Conservation Society P.O. Box 628, Venice, CA 90294; Tel: 310-301-7325, Fax: 310-574-3161, Web: http://www.seashepherd.org/

Sierra Club 730 Polk St., San Francisco, CA 94109; Tel 415-776-2211, E-mail: information @sierraclub.org, Web: http://www.sierraclub.org

Smithsonian Institution 1000 Jefferson Dr. SW, Washington, DC 20560; Tel: 202-357-2700, Web: http://www.yahoo.com/Government/Agencies /Independent/Smithsonian_Institution/

Soil and Water Conservation Society 7515 NE Ankeny Rd., Ankeny, IA 50021-9764; Tel: 515-289-2331, 800-863-7645, Fax: 515-289-1227

Student Environmental Action Coalition (SEAC) P.O. Box 1168, Chapel Hill, NC 27514; Tel: 919-967-4600, 800-700-SEAC, Fax: 919-967-4648, Web: http://www.seac.org/

Survival International 2121 Decatur Place NW, Washington, DC 20008; Web: http://www. survival.org.uk/

Union of Concerned Scientists Two Brattle Square, Cambridge, MA 02238-9105; Tel: 617-547-5552, Web: http://www.ucsusa.org/

United Nations Population Fund 220 East 42nd St., New York, NY 10017; Tel: 212 297-5020, Fax: 212-557-6416, Web: http://www.unfpa.org/

U.S. Public Interest Research Group 215 Pennsylvania Ave. SE, Washington, DC 20003; Tel: 202-546-9707

Water Pollution Control Federation 601 Wythe St., Alexandria, VA 22314

The Wilderness Society 900 17th St. NW, Washington, DC 20006-2596; Tel: 202-833-2300, Web: http://www.wilderness.org/

Wildlife Conservation Society 185th St. and Southern Blvd., Bronx, NY 10460-1099; Tel 718-220-5100, Fax: 718-220-7114, Web: http://www.wcs.org/

Wildlife Society 5410 Grosvenor Lane, Bethesda, MD 20814; Tel: 301-897-9770

Work on Waste 82 Judson St., Canton, NY 13617

World Resources Institute 1709 New York Ave. NW, 7th Floor, Washington, DC 20006; Tel: 202-638-6300, Web: http://www.wri.org/

Worldwatch Institute 1776 Massachusetts Ave. NW, Washington, DC 20036-1904; Tel: 202-452-1999, Fax: 202-296-7365, E-mail: worldwatch@igc.apc.org and wwpub@igc.apc.org, Web: http://www. worldwatch.org/

World Wildlife Fund 1250 24th St. NW, Suite 500, Washington, DC 20037; Tel: 202-293-4800, Web: http://www.worldwildlife.org/

Zero Population Growth 1400 16th St. NW, Suite 320, Washington, DC 20036; Tel: 202-332-2200, Fax: 202-332-2302, E-mail: zpg@apc.org, web: http://zpg.org

FEDERAL, SCIENTIFIC, AND INTERNATIONAL AGENCIES

Agency for International Development (USAID) State Building, 320 21st St. NW, Washington, DC 20523-0016; Tel: 202-647-1850, Fax: 202-647-8321, Web: http://www.info.usaid.gov/

Bureau of Land Management U.S. Department of Interior, 1620 L St. NW, Rm. 5600, Washington, DC 20240; Tel: 202-208-3801, Web: http://www.blm.gov

Bureau of Reclamation Washington, DC 20240; Web: http://www.usbr.gov/main/

Conservation and Renewable Energy Inquiry and Referral Service P.O. Box 8900, Silver Spring, MD 20907; Tel: 800-523-2929

Department of Agriculture 14th St. and Independence Ave. SW, Washington, DC 20250; Tel: 202-720-8732, Web: http://www.usda.gov

Department of Commerce Herbert C. Hoover Bldg., Rm. 5610, 15th & Constitution Ave. NW, Washington, DC 20230; Tel: 202-219-3605, Web: http://www.doc.gov

Department of Energy Forrestal Building, 1000 Independence Ave. SW, Washington, DC 20585; Web: http://www.doe.gov

Department of Health and Human Services 200 Independence Ave. SW, Washington, DC 20585; Web: http://www.yahoo.com/ Government/Executive_Branch/Departments_ and_Agencies/Department_of_Health_and_ Human_Services/

Department of Housing and Urban Development 451 7th St. SW, Washington, DC 20410; Tel: 202-755-5111, Web: http://www.hud.gov

Department of the Interior 1849 C St. NW, Washington, DC 20240; Tel: 202-208-3100, Web: http://www.usga.gov.doi

Department of Transportation 400 7th St. SW, Washington, DC 20590; Tel: 202-366-4000, Web: http://www.dot.gov

Energy Information Administration Dept. of Energy, National Energy Information Center, Forrestal Bldg., Washington, DC 20585; Tel: 202-586-8800, Web: http://www.yahoo.com/ Government/Executive_Branch/Departments_ and_Agencies/Department_of_Energy/Energy_ Information_Administration/

Environmental Protection Agency 401 M St. SW, Washington, DC 20460; Tel: 202-260-2090, Web: http://www.epa.gov and gopher.epa.gov

Food and Agriculture Organization (FAO) of the United Nations 101 22nd St. NW, Suite 300, Washington, DC 20437; Web: http://www.fao.org/

Food and Drug Administration Department of Health and Human Services, 5600 Fishers Lane, Rockville, MD 20857; Tel: 410-433-1544, Web: http://www.fda.gov

Forest Service P.O. Box 96090, Washington, DC 20090-6090; Tel: 202-205-0957, Web: http://www.fs.fed.us/

Government Printing Office Washington, DC 20402

International Whaling Commission The Red House, 135 Station Rd., Histon, Cambridge CB4 4NP England (0223 233971); Web: http://av. yahoo.com/bin/query?p=International+Whaling +Commission&hc=0&hs=0

Marine Mammal Commission 1825 Connecticut Ave. NW, Rm. 512, Washington, DC 20009; Tel: 202-606-5504, Fax: 202-606-5510, Web: http://www.citation.com/hpages/mmc.html

National Academy of Sciences 2101 Constitution Ave. NW Washington, DC 20418; Web: http://www.nas.edu/

National Center for Atmospheric Research P.O. Box 3000, Boulder, CO 80307; Web: http://www.ucar.edu/

National Lead Information Center Tel: 1-800-LEAD-FYI, Web: http://www.nsc.org/ehc/lead.htm

National Marine Fisheries Service U.S. Dept. of Commerce, Silver Spring Metro Center 1, 1335 East–West Hwy., Silver Spring, MD 20910; Tel: 301-713-2239, Web: http://kingfish.ssp.nmfs.gov/

National Oceanic and Atmospheric Administration Herbert C. Hoover Bldg., Rm. 5128, 14th & Constitution Ave. NW, Washington, DC 20230; Tel: 202-482-3384, Web: http://www.yahoo.com/Government/ Executive_Branch/Departments_and_Agencies/ Department_of_Commerce/National_Oceanic_ and_Atmospheric_Administration/

National Park Service Interior Bldg., P.O. Box 37127, Washington, DC 20013-7127; Tel: 202-208-4747, Web: http://www.nps.gov/

National Renewable Energy Laboratory 1617 Cole Blvd., Golden, CO 80401; Web: http://www. nrel.gov/

National Science Foundation 4201 Wilson Blvd., Arlington, VA 22230; Tel: 703-306-1234, Web: http://www.nsf.gov/

National Solar Heating and Cooling Information Center P.O. Box 1607, Rockville, MD 20850

Nuclear Regulatory Commission
Washington, DC 20555; Tel: 301-415-7000,
Web: http://www.nrc.gov/

Occupational Safety and Health Administration
Department of Labor, 200 Constitution Ave. NW,
Washington, DC 20210; Web: http://www.nrc.gov/

**Office of Ocean and Coastal Resource
Management**   1825 Connecticut Ave. NW, Suite
700, Washington, DC 20235; Tel: 202-208-2553,
Web: http://wave.nos.noaa.gov/ocrm/

**Office of Surface Mining Reclamation and
Enforcement**   Interior South Bldg., 1951
Constitution Ave. NW, Washington, DC 20240;
Web: http://www.osmre.gov/astart3.htm

**Organization for Economic Cooperation and
Development**   U.S. Office: 2001 L St. NW,
Suite 700, Washington, DC 20036; Web: http://
www.oecd.org/

Soil Conservation Service   (now called Natural
Resource Conservation Service) USDA, 14th
and Independence Ave. SW, P.O. Box 2890,
Washington, DC 20013; Tel: 202-720-3210,
Web: http://www.nrcs.usda.gov/

United Nations   1 United Nations Plaza, New
York, NY 10017; Web: http://www.yahoo.com/
Government/International_Organizations/
United_Nations/

United Nations Environment Programme
Regional North American Office, United Nations
Rm. DC 20803, New York, NY 10017; and 1889
F St. NW, Washington, DC 20006; Web: http://
www.yahoo.com/Government/International_
Organizations/United_Nations/Programs/
United_Nations_Environment_Programme_
UNEP_/

U.S. Fish and Wildlife Service   Department
of the Interior, Washington, DC 20240;
Web: http://www.fws.gov/

U.S. Geological Survey   National Center,
Reston, VA 22092; Tel: 703-648-4000

**U.S. Man and the Biosphere (U.S. MAB)
Program**   U.S. MAB Secretariat, OES/EGC/MAB,
Rm. 608, SA-37, Dept. of State, Washington, DC
20522-3706; Web: http://www.mabnetamericas.
org/home2.html

World Bank   1818 H St. NW, Washington, DC
20433; Web: http://www.worldbank.org/

UNITS OF MEASUREMENT

LENGTH

Metric

1 kilometer (km) = 1,000 meters (m)
1 meter (m) = 100 centimeters (cm)
1 meter (m) = 1,000 millimeters (mm)
1 centimeter (cm) = 0.01 meter (m)
1 millimeter (mm) = 0.001 meter (m)

English

1 foot (ft) = 12 inches (in)
1 yard (yd) = 3 feet (ft)
1 mile (mi) = 5,280 feet (ft)
1 nautical mile = 1.15 miles (mi)

Metric-English

1 kilometer (km) = 0.621 mile (mi)
1 meter (m) = 39.4 inches (in)
1 inch (in) = 2.54 centimeters (cm)
1 foot (ft) = 0.305 meter (m)
1 yard (yd) = 0.914 meter (m)
1 nautical mile = 1.85 kilometers (km)

AREA

Metric

1 square kilometer (km^2) = 1,000,000 square meters (m^2)
1 square meter (m^2) = 1,000,000 square millimeters (mm^2)
1 hectare (ha) = 10,000 square meters (m^2)
1 hectare (ha) = 0.01 square kilometer (km^2)

English

1 square foot (ft^2) = 144 square inches (in^2)
1 square yard (yd^2) = 9 square feet (ft^2)
1 square mile (mi^2) = 27,880,000 square feet (ft^2)
1 acre (ac) = 43,560 square feet (ft^2)

Metric-English

1 hectare (ha) = 2.471 acres (ac)
1 square kilometer (km^2) = 0.386 square mile (mi^2)
1 square meter (m^2) = 1.196 square yards (yd^2)
1 square meter (m^2) = 10.76 square feet (ft^2)
1 square centimeter (cm^2) = 0.155 square inch (in^2)

VOLUME

Metric

1 cubic kilometer (km^3) = 1,000,000,000 cubic meters (m^3)
1 cubic meter (m^3) = 1,000,000 cubic centimeters (cm^3)
1 liter (L) = 1,000 milliliters (mL) = 1,000 cubic centimeters (cm^3)
1 milliliter (mL) = 0.001 liter (L)
1 milliliter (mL) = 1 cubic centimeter (cm^3)

English

1 gallon (gal) = 4 quarts (qt)
1 quart (qt) = 2 pints (pt)

Metric-English

1 liter (L) = 0.265 gallon (gal)
1 liter (L) = 1.06 quarts (qt)
1 liter (L) = 0.0353 cubic foot (ft^3)
1 cubic meter (m^3) = 35.3 cubic feet (ft^3)
1 cubic meter (m^3) = 1.31 cubic yard (yd^3)
1 cubic kilometer (km^3) = 0.24 cubic mile (mi^3)
1 barrel (bbl) = 159 liters (L)
1 barrel (bbl) = 42 U.S. gallons (gal)

MASS

Metric

1 kilogram (kg) = 1,000 grams (g)
1 gram (g) = 1,000 milligrams (mg)
1 gram (g) = 1,000,000 micrograms (μg)
1 milligram (mg) = 0.001 gram (g)
1 microgram (μg) = 0.000001 gram (g)
1 metric ton (mt) = 1,000 kilograms (kg)

English

1 ton (t) = 2,000 pounds (lb)
1 pound (lb) = 16 ounces (oz)

Metric-English

1 metric ton (mt) = 2,200 pounds (lb) = 1.1 tons (t)
1 kilogram (kg) = 2.20 pounds (lb)
1 pound (lb) = 454 grams (g)
1 gram (g) = 0.035 ounce (oz)

ENERGY AND POWER

Metric

1 kilojoule (kJ) = 1,000 joules (J)
1 kilocalorie (kcal) = 1,000 calories (cal)
1 calorie (cal) = 4.184 joules (J)

Metric-English

1 kilojoule (kJ) = 0.949 British thermal unit (Btu)
1 kilojoule (kJ) = 0.000278 kilowatt-hour (kW-h)
1 kilocalorie (kcal) = 3.97 British thermal units (Btu)
1 kilocalorie (kcal) = 0.00116 kilowatt-hour (kW-h)
1 kilowatt-hour (kW-h) = 860 kilocalories (kcal)
1 kilowatt-hour (kW-h) = 3,400 British thermal units (Btu)
1 quad (Q) = 1,050,000,000,000,000 kilojoules (kJ)
1 quad (Q) = 2,930,000,000,000 kilowatt-hours (kW-h)

TEMPERATURE CONVERSIONS

Fahrenheit (°F) to Celsius (°C): $°C = \dfrac{(°F - 32.0)}{1.80}$

Celsius (°C) to Fahrenheit (°F): $°F = (°C \times 1.80) + 32.0$

MAJOR U.S. RESOURCE CONSERVATION

GENERAL

National Environmental Policy Act of 1969 (NEPA)

International Environmental Protection Act of 1983

ENERGY

National Energy Act of 1978, 1980

National Appliance Energy Conservation Act of 1987

Energy Policy Act of 1992

WATER QUALITY

Water Quality Act of 1965

Water Resources Planning Act of 1965

Federal Water Pollution Control Acts of 1965, 1972

Ocean Dumping Act of 1972

Safe Drinking Water Act of 1974, 1984

Water Resources Development Act of 1986

Clean Water Act of 1977, 1987

Ocean Dumping Ban Act of 1988

AIR QUALITY

Clean Air Act of 1963, 1965, 1970, 1977, 1990

Pollution Prevention Act of 1990

NOISE CONTROL

Noise Control Act of 1965

Quiet Communities Act of 1978

RESOURCES AND SOLID WASTE MANAGEMENT

Solid Waste Disposal Act of 1965

Resource Recovery Act of 1970

Resource Conservation and Recovery Act of 1976

Marine Plastic Pollution Research and Control Act of 1987

TOXIC SUBSTANCES

Hazardous Materials Transportation Act of 1975

Toxic Substances Control Act of 1976

Resource Conservation and Recovery Act of 1976

Comprehensive Environmental Response, Compensation, and Liability (Superfund) Act of 1980, 1986

Nuclear Waste Policy Act of 1982

PESTICIDES

Federal Insecticide, Fungicide, and Rodenticide Control Act of 1972, 1988

WILDLIFE CONSERVATION

Lacey Act of 1900

Migratory Bird Treaty Act of 1918

Migratory Bird Conservation Act of 1929

Migratory Bird Hunting Stamp Act of 1934

Pittman-Robertson Act of 1937

Anadromous Fish Conservation Act of 1965

Fur Seal Act of 1966

National Wildlife Refuge System Act of 1966, 1976, 1978

Species Conservation Act of 1966, 1969

Marine Mammal Protection Act of 1972

Marine Protection, Research, and Sanctuaries Act of 1972

Endangered Species Act of 1973, 1982, 1985, 1988

Fishery Conservation and Management Act of 1976, 1978, 1982

Whale Conservation and Protection Study Act of 1976

Fish and Wildlife Improvement Act of 1978

Fish and Wildlife Conservation Act of 1980 (Nongame Act)

LAND USE AND CONSERVATION

Taylor Grazing Act of 1934

Wilderness Act of 1964

Multiple Use Sustained Yield Act of 1968

Wild and Scenic Rivers Act of 1968

National Trails System Act of 1968

National Coastal Zone Management Act of 1972, 1980

Forest Reserves Management Act of 1974, 1976

Forest and Rangeland Renewable Resources Act of 1974, 1978

Federal Land Policy and Management Act of 1976

National Forest Management Act of 1976

Soil and Water Conservation Act of 1977

Surface Mining Control and Reclamation Act of 1977

Antarctic Conservation Act of 1978

Endangered American Wilderness Act of 1978

Alaskan National Interests Lands Conservation Act of 1980

Coastal Barrier Resources Act of 1982

Food Security Act of 1985

INDIVIDUALS MATTER: WORKING WITH THE EARTH

This is a list of things individuals can do based on suggestions from a wide variety of environmentalists. It is not meant to be a list of things you must do but a list of actions you might consider. Don't feel guilty about the things you are not doing. Start off by picking the ones that you are willing to do and that you feel will have the most impact. Many of these suggestions are controversial. Carefully evaluate each action to see whether it fits in with your beliefs. Based on practicality and my own beliefs, I do some of these things. Each year I look over this list and try to add several new items.

PRESERVING BIODIVERSITY AND PROTECTING THE SOIL

- *Develop a plan for the sustainable use of any forested area you own.*

- *Plant trees on a regular basis and take care of them.*

- *Reduce the use of wood and paper products, recycle paper products, and buy recycled paper products.*

- *Don't buy furniture, doors, flooring, window frames, paneling, or other products made from tropical hardwoods such as teak or mahogany.* Look for the Good-Wood seal given by Friends of the Earth, or consult the *Wood User's Guide* by Pamela Wellner and Eugene Dickey (San Francisco: Rainforest Action Network, 1991).

- *Don't purchase wood and paper products produced by cutting remaining old-growth forests in the tropics and elsewhere.* Information on such products can be obtained from the Rainforest Action Network, Rainforest Alliance, and Friends of the Earth (Appendix 1).

- *Help rehabilitate or restore a degraded area of forest near your home.*

- *Don't buy furs, ivory products, items made of reptile skin, tortoiseshell jewelry, and materials of endangered or threatened animal species.*

- *When building a home, save all the trees possible.* Require that the contractor disturb as little soil as possible, set up barriers that catch any soil eroded during construction, and save and replace any topsoil removed instead of hauling it off and selling it.

- *Landscape areas not used for gardening with a mix of wildflowers, herbs (for cooking and for repelling insects), low-growing ground cover, small bushes, and other forms of vegetation natural to the area.*

- *Set up a compost bin and use it to produce soil conditioner for yard and garden plants.*

PROMOTING SUSTAINABLE AGRICULTURE AND REDUCING PESTICIDE USE

- *Waste less food.* An estimated 25% of all food produced in the United States is wasted.

- *Eat lower on the food chain* by reducing or eliminating meat consumption to reduce its environmental impact.

- *If you have a dog or a cat, don't feed it canned meat products.* Balanced-grain pet foods are available.

- *Reduce the use of pesticides on agricultural products by asking grocery stores to stock fresh produce and meat produced by organic methods.*

- *Grow some of your own food using organic farming techniques and drip irrigation to water your crops.*

- *Compost your food wastes.*

- *Think globally, eat locally.* Whenever possible, eat food that is locally grown and in season. This supports your local economy, gives you more influence over how the food is grown (by either organic or conventional methods), saves energy required to transport food over long distances, and reduces the fossil-fuel use and pollution. If you deal directly with local farmers, you can also save money.

- *Give up the idea that the only good bug is a dead bug.* Recognize that insect species keep most of the populations of pest insects in check and that full-scale chemical warfare on insect pests wipes out many insects beneficial to us.

- *Don't insist on perfect-looking fruits and vegetables.* These are more likely to contain high levels of pesticide residues.

- *Use pesticides in your home only when absolutely necessary, and use them in the smallest amount possible.*

- *Don't become obsessed with having the perfect lawn.* About 40% of U.S. lawns are treated with pesticides. These chemicals can cause headaches, dizziness, nausea, eye trouble, and even more acute effects in sensitive people (including children who play on treated lawns and in parks).

- *If you hire a lawn-care company, use one that relies only on organic methods, and get its claims in writing.*

SAVING ENERGY AND REDUCING OUTDOOR AIR POLLUTION

- *Reduce use of fossil fuels.* Drive a car that gets at least 15 kilometers per liter (35 miles per gallon), join a car pool, and use mass transit, walking, and bicycling as much as possible. This will reduce emissions of CO_2 and other air pollutants, will save energy and money, and can improve your health.

- *Plant and care for trees to help absorb CO_2.* During its lifetime, the average tree absorbs enough CO_2 to offset the amount produced by driving a car 42,000 kilometers (26,000 miles).

- *Insulate new or existing houses heavily, caulk and weatherstrip to reduce air infiltration and heat loss, and use energy-efficient windows.* Add an air-to-air heat exchanger to minimize indoor air pollution.

- *Obtain as much heat and cooling as possible from natural sources,* especially sun, wind, geothermal energy, and trees.

- *Buy the most energy-efficient homes, lights, cars, and appliances available. Evaluate them only in terms of lifetime cost.*

- *Turn down the thermostat on water heaters to 43–49°C (110–120°F), and insulate hot water pipes.*

- *Lower the cooling load on an air conditioner by increasing the thermostat setting, installing energy-efficient lighting, using floor and ceiling fans, and using whole-house window or attic fans to bring in outside air (especially at night, when temperatures are cooler).*

REDUCING EXPOSURE TO INDOOR AIR POLLUTANTS

- *Test for radon and take corrective measures as needed.*

- *Install air-to-air heat exchangers or ventilate your house regularly by opening windows.*

- *At the beginning of the winter heating season, test your indoor air for formaldehyde when the house is closed up.* To locate a testing laboratory in your area, write to Consumer Product Safety Commission, Washington, DC 20207, or call 301-492-6800.

- *Don't buy furniture and other products containing formaldehyde. Use low-emitting formaldehyde or nonformaldehyde building materials.*

- *Reduce indoor levels of formaldehyde and other toxic gases by growing certain house plants.* Examples are the spider or airplane plant (removes 96% of carbon monoxide), aloe vera (90% of formaldehyde), banana (89% of formaldehyde), elephant ear philodendron (86% of formaldehyde), ficus (weeping fig, 47% of formaldehyde), golden porthos (67% of formaldehyde and benzene and 75% of carbon monoxide), Chinese evergreen (92% of toluene and 81% of benzene), English ivy (90% of benzene), peace lily (80% of benzene and 50% of trichloroethylene), and Janet Craig (corn plant, 79% of benzene). (Toxic removal figures indicate percentage of toxin removed by one plant in a 24-hour period in a 3.4-cubic-meter (12-cubic-foot) space). Plants should be potted with a mixture of soil and granular charcoal (which absorbs organic air pollutants).

- *Consider not using carpeting and using wood or linoleum floors instead. New synthetic carpeting releases vapors from more than 100 volatile organic compounds. New and old carpeting is a haven for microbes (many of them highly allergenic), dust, and traces of lead and pesticides brought in by shoes.*

- *Remove your shoes before entering your house. This reduces inputs of dust, lead, and pesticides.*

- *Test your house or workplace for asbestos fiber levels and for any crumbling asbestos materials if it was built before 1980. Don't buy a pre-1980 house without having its indoor air tested for asbestos and lead. To get a free list of certified asbestos laboratories that charge $25–50 to test a sample, call the EPA's Toxic Substances Control Hotline at 202-554-1404.*

- *Don't store gasoline, solvents, or other volatile hazardous chemicals inside a home or attached garage.*

- *Don't use aerosol spray products and commercial room deodorizers or air fresheners.*

- *If you smoke, do it outside or in a closed room vented to the outside.*

- *Make sure that wood-burning stoves, fireplaces, and kerosene- and gas-burning heaters are properly installed, vented, and maintained. Install carbon monoxide detectors in all sleeping areas.*

SAVING WATER

- *For existing toilets, reduce the amount of water used per flush by putting into each tank a tall plastic container weighted with a few stones or by buying and inserting a toilet dam.*

- *Install water-saving toilets that use no more than 6 liters (1.6 gallons) per flush.*

- *Flush toilets only when necessary. Consider using the advice found on a bathroom wall in a drought-stricken area: "If it's yellow, let it mellow; if it's brown, flush it down."*

- *Install water-saving showerheads and flow restrictors on all faucets. If a 3.8-liter (1-gallon) jug can be filled by your showerhead in less than 15 seconds, you need a more efficient fixture.*

- *Check frequently for water leaks in toilets and pipes, and repair them promptly. A toilet must be leaking more than 940 liters (250 gallons) per day before you can hear the leak. To test for toilet leaks, add a water-soluble vegetable dye to the water in the tank, but don't flush. If you have a leak, some color will show up in the bowl's water within a few minutes.*

- *Turn off sink faucets while brushing teeth, shaving, or washing.*

- *Wash only full loads of clothes; if smaller loads must be washed, use the lowest possible water-level setting.*

- *When buying a new washer, choose one that uses the least amount of water and that fills up to different levels for loads of different sizes. Front-loading models use less water and energy than comparable top-loading models.*

- *Use automatic dishwashers for full loads only. Also, use the short cycle and let dishes air dry to save energy and money.*

- *When washing many dishes by hand, don't let the faucet run. Instead, use one filled dishpan or sink for washing and another for rinsing.*

- *Keep one or more large bottles of water in the refrigerator rather than running water from the tap until it gets cold enough for drinking.*

- *Don't use a garbage disposal system—a large user of water. Instead, consider composting your food wastes.*

- *Wash a car from a bucket of soapy water, and use the hose for rinsing only. Use a commercial car wash that recycles its water.*

- *Sweep walks and driveways instead of hosing them off.*

- *Reduce evaporation losses by watering lawns and gardens in the early morning or evening, rather than in the heat of midday or when it's windy.*

- *Use drip irrigation and mulch for gardens and flower beds. Better yet, landscape with native plants adapted to local average annual precipitation so that watering is unnecessary.*

REDUCING WATER POLLUTION

- *Use manure or compost instead of commercial inorganic fertilizers to fertilize garden and yard plants.*

- *Use biological methods or integrated pest management to control garden, yard, and household pests.*

- *Use low-phosphate, phosphate-free, or biodegradable dishwashing liquid, laundry detergent, and shampoo.*

- *Don't use water fresheners in toilets.*

- *Use less harmful substances instead of commercial chemicals for most household cleaners (Table 11-2).*

- *Don't pour pesticides, paints, solvents, oil, antifreeze, or other products containing harmful chemicals down the drain or onto the ground. Contact your local health department about disposal.*

- *If you get water from a private well or suspect that municipal water is contaminated, have it tested by an EPA-certified laboratory for lead, nitrates, trihalomethanes, radon, volatile organic compounds, and pesticides.*

- *If you have a septic tank, have it cleaned out every 3–5 years by a reputable contractor so that it won't contribute to groundwater pollution.*

REDUCING SOLID WASTE AND HAZARDOUS WASTE

- *Buy less by asking yourself whether you really need a particular item.*

- *Buy things that are reusable, recyclable, or compostable, and be sure to reuse, recycle, and compost them.*

- *Buy beverages in refillable glass containers instead of cans or throwaway bottles. Urge companies and legislators to make refillable plastic (PET) bottles available in the United States.*

- *Use reusable plastic or metal lunch boxes and metal or plastic garbage containers without throwaway plastic liners (unless such liners are required for garbage collection).*

- *Carry sandwiches and store food in the refrigerator in reusable containers instead of wrapping them in aluminum foil or plastic wrap.*

- *Use rechargeable batteries and recycle them when their useful life is over. In 1993 Rayovac began selling mercury-free, rechargeable alkaline batteries (Renewal batteries) that outperform conventional nickel–cadmium rechargeable batteries.*

- *Carry groceries and other items in a reusable basket, a canvas or string bag, or a small cart.*

- *Use sponges and washable cloth napkins, dish towels, and handkerchiefs instead of paper ones.*

- *Stop using throwaway paper and plastic plates, cups, eating utensils, and other disposable items when reusable or refillable versions are available.*

- *Buy recycled goods, especially those made by primary recycling, and then make an effort to recycle them. If you're not buying recycled materials, you're not recycling.*

- *Reduce the amount of junk mail you get. Do this (as several million Americans have done) at no charge by contacting the Mail Preference Service, Direct Marketing Association, 11 West 42nd St., P.O. Box 3681, New York, NY 10163-3861 (212-768-7277) and asking that your name not be sold to large mailing-list companies. Of the junk mail you do receive, recycle as much of the paper as possible.*

- *Buy products in concentrated form whenever possible.*

- *Choose items that have the least packaging—or better yet, no packaging ("nude products").*

- *Don't buy helium-filled balloons that end up as litter. Urge elected officials and school administrators to ban balloon releases except for atmospheric research and monitoring.*

- *Lobby local officials to set up a community composting program.*

- *Use pesticides and other hazardous chemicals (Table 7-1) only when absolutely necessary and in the smallest amount possible.*

- *Use less hazardous (and usually cheaper) cleaning products (Table 11-2).*

- *Don't dispose of hazardous chemicals by flushing them down the toilet, pouring them down the drain, burying them, throwing them into the garbage, or dumping them down storm drains. Consult your local health department or environmental agency for safe disposal methods.*

FURTHER READINGS

INTERNET

For a list of useful Internet sites, hyperlinks to material in this book, and other helpful and interesting information please visit our Web site at http://www.brookscole.com/biology.

GENERAL SOURCES OF ENVIRONMENTAL INFORMATION

Readings

Ashworth, William. 1991. *The Encyclopedia of Environmental Studies*. New York: Facts on File.

Brown, Lester R., et al. Annual. *State of the World*. New York: Norton.

Brown, Lester R., et al. Annual. *Vital Signs*. New York: Norton.

Katz, Michael, and Dorothy Thornton. 1997. *Environmental Management Tools on the Internet: Assessing the World of Environmental Information*. Delray Beach, Calif.: St. Lucie.

Kurland, Daniel J., and Jane Heinze-Fry. 1996. *Introduction to the Internet*. Belmont, Calif.: Wadsworth. Available as a supplement for use with this book.

Lean, Geoffrey, and Don Hinrichsen, eds. 1994. *Atlas of the Environment*. 2d ed. New York: HarperCollins.

Population Reference Bureau. Annual. *World Population Data Sheet*. Washington, D.C.: Population Reference Bureau.

Schupp, Jonathan F. 1995. *Environmental Guide to the Internet*. Rockville, Md.: Government Institutes.

Seager, Joni, et al. 1995. *The New State-of-the-Earth Atlas*. New York: Simon & Schuster.

United Nations. Annual. *Demographic Yearbook*. New York: United Nations.

United Nations Development Programme (UNDP). Annual. *Human Development Program*. New York: UNDP.

United Nations Population Fund. Annual. *The State of World Population*. New York: United Nations Population Fund.

U.S. Bureau of the Census. Annual. *Statistical Abstract of the United States*. Washington, D.C.: U.S. Bureau of the Census.

World Health Organization (WHO). Annual. *World Health Statistics*. Geneva, Switzerland: WHO.

World Resources Institute and International Institute for Environment and Development. Biannual. *World Resources*. New York: Basic Books.

1 / ENVIRONMENTAL PROBLEMS AND THEIR CAUSES

Readings

Bailey, Ronald, ed. 1995. *The True State of the Planet*. New York: Free Press.

Brown, Lester R., and Hal Kane. 1994. *Full House: Reassessing the Earth's Population Carrying Capacity*. New York: Norton.

Commoner, Barry. 1994. *Making Peace with the Planet*, rev. ed. New York: New Press.

Daily, Gretchen, ed. 1997. *Nature's Services: Society's Dependence on Natural Ecosystems*. Covelo, Calif.: Island.

Dunn, Terry L. 1997. *Guide to Global Environmental Issues*. Golden, Colo.: Fulcrum.

Ehrlich, Paul R., and Anne H. Ehrlich. 1990. *The Population Explosion*. New York: Doubleday.

Ehrlich, Paul R., and Anne H. Ehrlich. 1991. *Healing the Planet*. Reading, Mass.: Addison-Wesley.

Ehrlich, Paul R., and Anne H. Ehrlich. 1996. *Betrayal of Science and Reason: How Anti-Environmental Rhetoric Threatens Our Future*. Covelo, Calif.: Island.

Flattau, Edward. 1998. *Tracking the Charlatans: Countering the Eco-Bashers*. New York: Global Horizons.

Gore, Al. 1992. *Earth in the Balance: Ecology and the Human Spirit*. Boston: Houghton Mifflin.

Hardin, Garrett. 1968. "The Tragedy of the Commons." *Science*, vol. 162, 1243–48.

Lubchenco, Jane. 1998. "Entering the Century of the Environment: A New Social Contract for Science." *Science*, vol. 279, 491–97.

Markley, Oliver W., and Walter R. McCuan, eds. 1996. *21st Century Earth: Opposing Viewpoints*. Boston: Greenhaven.

Mazur, Laurie Ann, ed. 1994. *Beyond the Numbers: A Reader on Population, Consumption, and the Environment*. Covelo, Calif.: Island.

Meadows, Donella H., et al. 1992. *Beyond the Limits: Confronting Global Collapse, Envisioning a Sustainable Future*. White River Junction, Vt.: Chelsea Green.

Myers, Norman, ed. 1993. *Gaia: An Atlas of Planet Management*. Garden City, N.Y.: Anchor/Doubleday.

Myers, Norman, and Julian Simon. 1994. *Scarcity or Abundance? A Debate on the Environment*. New York: Norton.

Quinn, Daniel. 1992. *Ishmael*. New York: Bantam/Turner.

Simon, Julian L. 1996. *The Ultimate Resource 2*. Princeton, N.J.: Princeton University Press.

Suzuki, David. 1994. *Time to Change*. Toronto: Stoddard.

Tobias, Michael. 1994. *World War III: Population and the Biosphere at the End of the Millennium*. Santa Fe, N.M.: Bear.

United Nations. 1997. *Global Environmental Outlook*. New York: Oxford University Press.

Wagner, Travis. 1993. *In Our Backyard: A Guide to Understanding Pollution and Its Effects*. New York: Van Nostrand Reinhold.

2 / SCIENCE, MATTER, ENERGY, AND ECOLOGY: CONNECTIONS IN NATURE

Readings

Abramovitz, Janet N. 1996. "Sustaining Freshwater Ecosystems." In Lester R. Brown et al., *State of the World 1996*. Washington, D.C.: Worldwatch Institute, 60–77.

Akin, Wallace E. 1991. *Global Patterns: Climate, Vegetation, and Soils*. Norman: University of Oklahoma Press.

Attenborough, David, et al. 1989. *The Atlas of the Living World*. Boston: Houghton Mifflin.

Baskin, Yvonne. 1997. *The Work of Nature: How the Diversity of Life Sustains Us*. Washington, D.C.: Island.

Begon, Michael, John L. Harper, and Colin R Townsend. 1996. *Ecology: Individuals, Populations, and Communities*, 3d ed. Boston: Blackwell Scientific.

Berenbaum, May R. 1995. *Bugs in the System: Insects and Their Impact on Human Affairs*. Reading, Mass.: Addison-Wesley.

Berner, Elizabeth K., and Robert A. Berner. 1996. *Global Environment: Water, Air, and Geochemical Cycles*. Upper Saddle River, N.J.: Prentice Hall.

Botkin, Daniel. 1990. *Discordant Harmonies: A New Ecology for the Twenty-First Century*. New York: Oxford University Press.

Carey, Stephen S. 1998. *A Beginner's Guide to Scientific Method 2d ed*. Belmont, Calif.: Wadsworth. Available as a supplement for use with this book.

Ehrlich, Anne H., and Paul R. Ehrlich. 1987. *Earth*. New York: Franklin Watts.

Ehrlich, Paul R. 1986. *The Machinery of Life: The Living World Around Us and How It Works*. New York: Simon & Schuster.

Elsom, Derek. 1992. *Earth: The Making, Shaping, and Working of a Planet*. New York: Macmillan.

Goldsmith, Edward, et al. 1990. *Imperiled Planet: Restoring Our Endangered Ecosystems*. Cambridge, Mass.: MIT Press.

Hare, Tony, ed. 1994. *Habitats*. New York: Macmillan.

Heinze-Fry, Jane. 1996. *An Introduction to Critical Thinking*. Belmont, Calif.: Wadsworth. Available as a supplement for use with this book.

Hinrichsen, Don. 1997. *The World's Coastal Seas: Trends, Threats, and Strategies*. Covelo, Calif.: Island.

Kusler, Jon A., et al. 1994. "Wetlands." *Scientific American*, January, 64–70.

Lovelock, James E. 1991. *Healing Gaia: Practical Medicine for the Planet*. New York: Random House.

Lubchenco, Jane. 1998. "Entering the Century of the Environment: A New Social Contract for Science." *Science*, vol. 279, 491–97.

Macmillan Publishing. 1992. *The Way Nature Works*. New York: Macmillan.

McCain, Garvin, and Erwin M. Segal. 1998. *The Game of Science*, 5th ed. Pacific Grove, Calif.: Brooks/Cole. Available as a supplement for use with this book.

Meadows, Donella H., et al. 1992. *Beyond the Limits: Confronting Global Collapse, Envisioning a Sustainable Future*. White River Junction, Vt.: Chelsea Green.

Myers, Norman. 1993. *The Primary Source: Tropical Forests and Our Future*. New York: Norton.

Myers, Norman. 1997. "Mass Extinction and Evolution." *Science*, vol. 278, 597–98.

Nisbet, E. G. 1991. *Living Earth*. New York: HarperCollins.

Pimentel, David, et al. 1997. "Economic and Environmental Benefits of Biodiversity."

BioScience, vol. 47, no. 11, 747–57.

Primack, Richard B. 1995. *A Primer of Conservation Biology.* Sunderland, Mass.: Sinauer.

Rickleffs, Robert E. 1997. *The Economy of Nature: A Textbook in Basic Ecology,* 4th ed. New York: W.H. Freeman.

Rifkin, Jeremy. 1989. *Entropy: Into the Greenhouse World: A New World View.* New York: Bantam.

Smil, Vaclav. 1997. *Cycles of Life: Civilization and the Biosphere.* New York. Scientific American Library.

Smith, Robert L. 1998. *Elements of Ecology,* 4th ed. New York: Harper & Row.

Starr, Cecie, and Ralph Taggart. 1998. *Biology: The Unity and Diversity of Life,* 8th ed. Belmont, Calif.: Wadsworth.

Tudge, Colin. 1991. *Global Ecology.* New York: Oxford University Press.

Vitousek, Peter M., et al. 1997. "Human Domination of Earth's Ecosystems." *Science,* vol. 277, 494–99.

Volk, Tyler. 1998. *Gaia's Body: Toward a Physiology of Earth.* New York: Copernicus.

Weber, Michael L., and Judith A. Gradwohl. 1995. *The Wealth of Oceans: Environment and Development on Our Ocean Planet.* New York: Norton

Williams, Jack. 1992. *The Weather Book.* New York: Vintage Books.

Wilson, E. O. 1992. *The Diversity of Life.* Cambridge, Mass.: Harvard University Press.

Worster, Donald. 1985. *Nature's Economy: A History of Ecological Ideas.* New York: Cambridge University Press.

Zimmerman, Michael. 1995. *Science, Nonscience, and Nonsense: Approaching Environmental Literacy.* Baltimore, Md.: Johns Hopkins University Press.

3 / THE HUMAN POPULATION: SIZE AND DISTRIBUTION

Readings

Aaberley, Doug, ed. 1994. *Futures by Design: The Practice of Ecological Planning.* Philadelphia: New Society.

Abernathy, Virginia D. 1993. *Population Politics: The Choices That Shape Our Future.* New York: Plenum.

Ashford, Lori S. 1995. "New Perspectives on Population: Lessons from Cairo." *Population Bulletin,* vol. 50, no. 1, 1–44.

Badshah, Akhtar A. 1997. *Our Urban Future: Paradigms for Equity and Sustainability.* London: Zed Books.

Beck, Roy. 1996. *The Case Against Immigration: The Moral, Economic, Social, and Environmental Reasons for Reducing U.S. Immigration.* New York: W.W. Norton.

Bongarts, John. 1994. "Population Policy Options in the Developing World." *Science,* vol. 263, 771–76.

Bouvier, Leon F., and Carol J. De Vita. 1991. "The Baby Boom: Entering Midlife." *Population Bulletin,* vol. 6, no. 3, 1–35.

Bouvier, Leon F., and Lindsey Grant. 1994. *How Many Americans? Population, Immigration, and the Environment.* San Francisco: Sierra Club Books.

Brown, Lester R. 1994. *Who Will Feed China?* New York: Norton.

Brown, Lester R., Gary Gardner, and Brian Halweil. 1998. *Beyond Malthus: Sixteen Dimensions of the Population Problem.* Washington, D.C.: Worldwatch Institute.

Brown, Lester R., and Hal Kane. 1994. *Full House: Reassessing the Earth's Population Carrying Capacity.* New York: W.W. Norton.

Calthorpe, Peter. 1993. *The Next American Metropolis: Ecology, Community, and the American Dream.* Princeton, N.J.: Princeton Architectural Press.

Carlson, Daniel, et al. 1995. *At Road's End: Transportation and Land Use Choices for Communities.* Covelo, Calif.: Island.

During, Alan Thein, and Christopher D. Crowther. 1997. *Misplaced Blame, The Real Roots of Population Growth.* Seattle, Wash.: Northwest Environment Watch.

Ehrlich, Paul R., and Anne H. Ehrlich. 1990. *The Population Explosion.* New York: Doubleday.

Ehrlich, Paul R., Anne H. Ehrlich, and Gretchen C. Daily. 1995. *The Stork and the Plow: The Equity Answer to the Human Dilemma.* New York: Putnam.

Engwicht, David. 1993. *Reclaiming Our Cities and Towns: Better Living with Less Traffic.* Philadelphia: New Society.

Erickson, Jon. 1995. *The Human Volcano: Population Growth as a Geologic Force.* New York: Facts on File.

Gordon, D. 1990. *Green Cities: Ecologically Sound Approaches to Urban Space.* New York: Black Rose.

Gordon, Deborah. 1991. *Steering a New Course: Transportation, Energy, and the Environment.* Covelo, Calif.: Island.

Grant, Lindsey. 1992. *Elephants and Volkswagens: Facing the Tough Questions About Our Overcrowded Country.* New York: W.H. Freeman.

Hall, Charles A. H., et al. 1995. "The Environmental Consequences of Having a Baby in the United States." *Wild Earth,* Summer, 78–87.

Hardin, Garrett. 1993. *Living Within Limits: Ecology, Economics, and Population Taboos.* New York: Oxford University Press.

Hardin, Garrett. 1995. *Immigration Reform: Avoiding the Tragedy of the Commons.* Washington, D.C.: Federation for American Immigration Reform.

Harrison, Paul. 1992. *The Third Revolution: Environment, Population, and a Sustainable World.* New York: I.B. Tauris.

Hart, John. 1992. *Saving Cities, Saving Money: Environmental Strategies That Work.* Washington, D.C.: Resource Renewal Institute.

Hartmann, Betsy. 1995. *Reproductive Rights and Wrongs: The Global Politics of Population Control and Contraceptive Choice.* Boston: South End.

Haupt, Arthur, and Thomas T. Kane. 1987. *The Population Handbook,* 4th ed. Washington, D.C.: Population Reference Bureau.

Jiggins, Janice. 1994. *Changing the Boundaries: Women-Centered Perspectives on Population and the Environment.* Covelo, Calif.: Island.

Kay, Jane Holtz. 1997. *Asphalt Nation: How the Automobile Took Over America and How We Can Take It Back.* New York: Crown.

Leisinger, Klaus M., and Karin Schmitt. 1994. *All Our People: Population Policy with a Human Face.* Covelo, Calif.: Island.

Lowe, Marcia D. 1991. *Shaping Cities: The Environmental and Human Dimensions.* Washington, D.C.: Worldwatch Institute.

Lowe, Marcia D. 1994. *Back on Track: The Global Rail Revival.* Washington, D.C.: Worldwatch Institute.

Lutz, Wolfgang. 1994. "The Future of World Population." *Population Bulletin,* vol. 46, no. 2, 1–43.

Lyman, Francesca. 1997. "Twelve Gates to the City." *Sierra,* May/June, 29–35.

MacKenzie, James J., et al. 1992. *The Going Rate: What It Really Costs to Drive.* Washington, D.C.: World Resources Institute.

Martin, Philip, and Elizabeth Midgley. 1994. "Immigration to the United States: Journey to an Uncertain Destination." *Population Bulletin,* vol. 49, no. 2, 1–47.

McFalls, Joseph A., Jr. 1998. "Population: A Lively Introduction." *Population Bulletin,* vol. 53, no. 3, 1–48.

McHarg, Ian L. 1992. *Designing with Nature.* Reprint. New York: Wiley.

McKibben, Bill. 1995. *Hope, Human and Wild: True Stories of Living Lightly on the Earth.* Boston: Little, Brown.

McKibben, Bill. 1998. *Maybe One: A Personal and Environmental Argument for Single-Child Families.* New York: Simon & Schuster.

Nadis, Steve, and James J. MacKenzie. 1993. *Car Trouble.* Boston: Beacon.

Population Reference Bureau. 1990. *World Population: Fundamentals of Growth.* Washington, D.C.: Population Reference Bureau.

Population Reference Bureau. Annual. *World Population Data Sheet.* Washington, D.C.: Population Reference Bureau.

Rabinovitch, Jonas, and Josef Leitman. 1996. "Urban Planning in Curitiba." *Scientific American,* March, 46–53.

Register, Richard. 1992. *Ecocities.* Berkeley, Calif.: North Atlantic Books.

Rocky Mountain Institute. 1995. *A Primer on Sustainable Building.* Snowmass, Colo.: Rocky Mountain Institute.

Russell, Cheryl. 1993. *The Master Trend: How the Baby Boom Generation Is Remaking America.* New York: Plenum.

Ryn, Sin van der, and Stuart Cowan. 1995. *Ecological Design.* Covelo, Calif.: Island.

Simon, Julian L. 1996. *The Ultimate Resource 2.* Princeton, N.J.: Princeton University Press.

Tien, H. Yuan, et al. 1992. "China's Demographic Dilemmas." *Population Bulletin,* vol. 47, no. 1, 1–44.

Todd, John, and Nancy Jack Todd. 1993. *From Ecocities to Living Machines: Precepts for Sustainable Technologies.* Berkeley, Calif.: North Atlantic Books.

Todd, John, and George Tukel. 1990. *Reinhabiting Cities and Towns: Designing for Sustainability.* San Francisco: Planet/Drum Foundation.

Tunali, Odil. 1996. "A Billion Cars: The Road Ahead." *World Watch,* January/February, 24–33.

United Nations. 1991. *Consequences of Rapid Population Growth in Developing Countries.* New York: United Nations.

Visaria, Leela, and Pravin Visaria. 1995. "India's Population in Transition." *Population Bulletin,* vol. 50, no. 3, 1–51.

Walter, Bob, et al., eds. 1992. *Sustainable Cities: Concepts and Strategies for Eco-City Development.* Los Angeles: Eco-Home Media.

Wann, David. 1994. *Biologic: Environmental Protection by Design,* 2d ed. Boulder, Colo.: Johnson Books.

Wann, David. 1995. *Deep Design: Pathways to a Livable Future.* Covelo, Calif.: Island.

4 / ENERGY

Readings

Ahearne, John F. 1993. "The Future of Nuclear Power." *American Scientist,* vol. 81, 24–35.

American Council for an Energy Efficient Economy. 1991. *Energy Efficiency and Environment: Forging the Link.* Washington, D.C.: American Council for an Energy Efficient Economy.

American Council for an Energy Efficient Economy. Annual. *The Most Energy-Efficient Appliances.* Washington, D.C.: American Council for an Energy Efficient Economy.

Beck, Peter. 1994. *Prospects and Strategies for Nuclear Power: Global Boon or Dangerous Diversion?* East Haven, Conn.: Earthscan.

Berger, John. 1997. *Charging Ahead: The Business of Renewable Energy and What It Means to America.* New York: Henry Holt.

Blackburn, John O. 1987. *The Renewable Energy Alternative: How the United States and the World Can Prosper Without Nuclear Energy or Coal.* Durham, N.C.: Duke University Press.

Brower, Michael. 1992. *Cool Energy: Renewable Solution to Environmental Problems,* 2d ed. Cambridge, Mass.: MIT Press.

Chernousenko, Vladimir M. 1991. *Chernobyl: Insight from the Inside.* New York: Springer-Verlag.

Cohen, Bernard L. 1990. *The Nuclear Energy Option: An Alternative for the 90s.* New York: Plenum.

Cole, Nancy, and P. S. Skerrett. 1995. *Renewables Are Ready: People Creating Energy Solutions.* White River Junction, Vt.: Chelsea Green.

Echeverria, John, et al. 1989. *Rivers at Risk: The Concerned Citizen's Guide to Hydropower.* Covelo, Calif.: Island.

Flavin, Christopher. 1996. "Power Shock: The Next Energy Revolution." *World Watch,* January/February, 10–19.

Flavin, Christopher, and Seth Dunn. 1997. *Rising Sun, Gathering Winds: Policies to Stabilize the Climate and Strengthen Economies.* Washington, D.C.: Worldwatch Institute.

Flavin, Christopher, and Nicholas Lenssen. 1994. *Power Surge: Guide to the Coming Energy Revolution.* New York: Norton.

Gever, John, et al. 1991. *Beyond Oil: The Threat to Food and Fuel in Coming Decades.* Boulder: University of Colorado Press.

Gipe, Paul. 1995. *Wind Energy Comes of Age.* New York: Wiley.

Grady, Wayne. 1993. *Greenhome: Planning and Building the Environmentally Advanced House.* Charlotte, Vt.: Camden House Books.

Kachadorian, James. 1997. *The Passive Solar House: Using Solar Design to Heat and Cool Your Home.* White River Junction, Vt.: Chelsea Green.

Lenssen, Nicholas. 1991. *Nuclear Waste: The Problem That Won't Go Away.* Washington, D.C.: Worldwatch Institute.

Lenssen, Nicholas, and Christopher Flavin. 1996. "Meltdown." *World Watch,* May/June, 23–31.

Lovins, Amory B., and L. Hunter Lovins. 1995. "Reinventing the Wheels." *Atlantic Monthly,* January, 75–94.

MacKenzie, James J. 1994. *The Keys to the Car: Electric and Hydrogen Vehicles for the 21st Century.* Washington, D.C.: World Resources Institute.

MacKenzie, James J. 1996. "Heading Off the Permanent Oil Crisis." *Issues in Science and Technology,* Summer, 48–54.

McKeown, Walter. 1991. *Death of the Oil Age and the Birth of Hydrogen America.* San Francisco: Wild Bamboo.

National Academy of Sciences. 1995. *Coal: Energy for the Future.* Washington, D.C.: National Academy Press.

Office of Technology Assessment. 1992. *Building Energy Efficiency.* Washington, D.C.: U.S. Government Printing Office.

Office of Technology Assessment. 1992. *Fueling Development: Energy Technologies for Developing Countries.* Washington, D.C.: U.S. Government Printing Office.

Oppenheimer, Ernest J. 1990. *Natural Gas, the Best Energy Choice.* New York: Pen & Podium.

Potts, Michael. 1993. *The Independent Home: Living Well with Power from the Sun, Wind, and Water.* White River Junction, Vt.: Chelsea Green.

Read, Piers Paul. 1993. *Ablaze: The Story of Chernobyl.* New York: Random House.

Robbins, Elaine, and Stephen Beers. 1998. "Lights Out: The Case for Energy Conservation—It Works, So Why Aren't We Using It? *E Magazine,* 36–41.

Rocky Mountain Institute. 1995. *Homemade Energy: How to Save Energy and Dollars in Your Home.* Snowmass, Colo.: Rocky Mountain Institute.

Savchenko, V. K. 1995. *The Ecology of the Chernobyl Catastrophe.* Pearl River, N.Y.: Parthenon.

Shea, Cynthia Pollock. 1989. "Decommissioning Nuclear Plants: Breaking Up Is Hard to Do." *World Watch,* July/August, 10–16.

Steen, Athena Swentzell, et al. 1994. *The Straw Bale House.* White River Junction, Vt.: Chelsea Green.

Sthcherbak, Yuri M. 1996. "Ten Years of the Chernobyl Era." *Scientific American,* April, 44–49.

Tenenbaum, David. 1995. "Tapping the Fire Down Below." *Technology Review,* January, 38–47.

Toke, David. 1995. *Energy and Environment: The Political and Economic Debate.* Boulder, Colo.: Westview.

Weisman, Alan. 1998. *Gaviotas: A Village to Reinvent the World.* White River Junction, Vt.: Chelsea Green.

Williams, Robert H. 1994. "The Clean Machine." *Technology Review,* April, 21–30.

Winteringham, F. P. W. 1991. *Energy Use and the Environment.* Boca Raton, Fla.: Lewis.

Wouk, Victor. 1997. "Hybrid Electric Vehicles." *Scientific American,* October, 70–74.

Zweibel, Ken. 1990. *Harnessing Solar Power: The Challenge of Photovoltaics.* New York: Plenum.

5 / BIODIVERSITY: SUSTAINING ECOSYSTEMS—FORESTS, RANGE-LANDS, PARKS, AND WILDERNESS

Readings

Abramovitz, Janet N. 1998. *Taking a Stand: Cultivating a New Relationship with the World's Forests.* Washington, D.C.: Worldwatch Institute.

Aplet, Greg, et al., eds. 1993. *Defining Sustainable Forestry.* Covelo, Calif.: Island.

Baskin, Yvonne. 1997. *The Work of Nature: How the Diversity of Life Sustains Us.* Covelo, Calif.: Island.

Callenbach, Ernest. 1995. *Bring Back the Buffalo: A Sustainable Future for America's Great Plains.* Covelo, Calif.: Island.

Collins, N. Mark, et al. 1991. *Conservation Atlas of Tropical Forests: Asia and the Pacific.* New York: Simon & Schuster.

Cronnon, W., ed. 1995. *Uncommon Ground: Toward Reinventing Nature.* New York: Norton.

Daily, Gretchen, ed. 1997. *Nature's Services: Society's Dependence on Natural Ecosystems.* Covelo, Calif.: Island.

Defenders of Wildlife. 1996. *A Status Report on America's Vanishing Habitat and Wildlife.* Washington, D.C.: Defenders of Wildlife.

Devall, Bill, ed. 1993. *Clearcut: The Tragedy of Industrial Forestry.* San Francisco: Sierra Club Books/Earth Island.

Dietrich, William. 1993. *The Final Forest: The Last Great Trees of the Pacific Northwest.* New York: Penguin.

DiSilvestro, Roger L. 1993. *Reclaiming the Last Wild Places: A New Agenda for Biodiversity.* New York: Wiley.

Drengson, Alan, and Duncan Taylor, eds. 1997. *Ecoforestry: The Art & Science of Sustainable Forest Use.* Gabriola Island, Canada: New Society Publishers.

Durning, Alan Thein. 1992. *Saving the Forests: What Will It Take?* Washington, D.C.: Worldwatch Institute.

Ervin, Keith. 1989. *Fragile Majesty: The Battle for North America's Last Great Forest.* Seattle: Mountaineers.

Goodland, Robert, ed. 1990. *Race to Save the Tropics: Ecology and Economics for a Sustainable Future.* Covelo, Calif.: Island.

Head, Suzanne, and Robert Heinzman. 1990. *Lessons of the Rainforest.* San Francisco: Sierra Club Books.

Hirt, Paul W. 1994. *A Conspiracy of Optimism: Management of the National Forests Since World War Two.* Lincoln: University of Nebraska Press.

Jacobs, Lynn. 1992. *Waste of the West: Public Lands Ranching.* Tucson, Ariz.: Lynn Jacobs.

Kellert, Stephen R. 1996. *The Value of Life: Biological Diversity and Human Society.* Covelo, Calif.: Island.

Kelly, David, and Gary Braasch. 1988. *Secrets of the Old Growth Forest.* Salt Lake City: Peregrine Smith.

Kohn, Kathryn, and Jerry M. Franklin. 1997. *Creating a Forestry for the 21st Century.* Covelo, Calif.: Island.

Lansky, Mitch. 1992. *Beyond the Beauty Strip: Saving What's Left of Our Forests.* Gardiner, Maine: Tilbury House.

Leopold, Aldo. 1949. *A Sand County Almanac.* New York: Oxford University Press.

Little, Charles E. 1995. *The Dying of the Trees: The Pandemic in America's Forests.* New York: Viking.

Maser, Chris, et al. 1994. *Sustainable Forestry: Philosophy, Science and Economics.* San Francisco: Rainforest Action Network.

Maybury-Lewis, David. 1992. *Millennium: Tribal Wisdom and the Modern World.* New York: Viking.

Miller, Kenton, and Laura Tangley. 1991. *Trees of Life: Protecting Tropical Forests and Their Biological Wealth.* Boston: Beacon.

Myers, Norman. 1993. *The Primary Source: Tropical Forests and Our Future.* New York: Norton.

Naar, Jon, and Alex J. Naar. 1993. *This Land Is Your Land: A Guide to North America's Endangered Ecosystems.* New York: Harper & Row.

Nash, Roderick. 1982. *Wilderness and the American Mind,* 3d ed. New Haven, Conn.: Yale University Press.

National Academy of Sciences. 1995. *Biodiversity II: Understanding and Protecting Our Biological Resources.* Washington, D.C.: National Academy Press.

National Parks and Conservation Association. 1994. *Our Endangered Parks: What You Can Do to Protect Our National Heritage.* San Francisco: Foghorn.

National Park Service. 1992. *National Parks for the 21st Century.* Washington, D.C.: National Park Service.

Norse, Elliot A. 1990. *Ancient Forests of the Pacific Northwest.* Covelo, Calif.: Island.

Noss, Reed F., and Allen Y. Cooperrider. 1994. *Saving Nature's Legacy: Protecting and Restoring Biodiversity.* Covelo, Calif.: Island.

Oelschlager, Max. 1991. *The Idea of Wilderness from Prehistory to the Age of Ecology.* New Haven, Conn.: Yale University Press.

Office of Technology Assessment. 1992. *Combined Summaries: Technologies to Sustain Tropical Forest Resources and Biological Diversity.* Washington, D.C.: Office of Technology Assessment.

O'Toole, Randal. 1987. *Reforming the Forest Service.* Covelo, Calif.: Island.

Primack, Richard B. 1993. *Essentials of Conservation Biology.* Sunderland, Mass.: Sinauer.

Revkin, Andrew. 1990. *The Burning Season: The Murder of Chico Mendes and the Fight for the Amazon.* Boston: Houghton Mifflin.

Riebsame, William E. 1996. "Ending the Range Wars?" *Environment,* vol. 38, no. 4, 4–29.

Rietbergen, Simon, ed. 1994. *The Earthscan Reader in Tropical Forestry.* San Francisco: Rainforest Action Network.

Rifkin, Jeremy. 1992. *Beyond Beef: The Rise and Fall of the Cattle Culture.* New York: Dutton.

Robinson, Gordon. 1987. *The Forest and the Trees: A Guide to Excellent Forestry.* Covelo, Calif.: Island.

Robbins, Jim. 1994. *Last Refuge: The Environmental Showdown in the American West.* San Francisco: HarperCollins.

Runte, Alfred. 1991. *Public Lands, Public Heritage: The National Forest Idea.* Niwot, Colo.: Roberts Rinehart.

Ryan, John C. 1992. *Life Support: Conserving Biological Diversity.* Washington, D.C.: Worldwatch Institute.

Shiva, Vandana. 1997. *Biopiracy: The Plunder of Nature and Knowledge.* Boston: South End.

Tennebaum, David. 1995. "The Greening of Costa Rica." *Technology Review,* October, 42–52.

Ward, Peter. 1995. *The End of Evolution: Mass Extinctions and the Preservation of Biodiversity*. New York: Bantam.

Wilson, E. O., ed. 1988. *Biodiversity*. Washington, D.C.: National Academy Press.

Wilson, E. O. 1992. *The Diversity of Life*. Cambridge, Mass.: Harvard University Press.

World Resources Institute (WRI), World Conservation Union (IUCN), and United Nations Development Programme. 1992. *Global Biodiversity Strategy*. Washington, D.C.: World Resources Institute.

Worster, Donald. 1994. *An Unsettled Country: Changing Landscapes of the American West*. Albuquerque: University of New Mexico Press.

Zaslowsky, Dyan, and T. H. Watkins. 1994. *These American Lands: Parks, Wilderness, and the Public Lands*, rev. ed. Covelo, Calif.: Island.

6 / BIODIVERSITY: SUSTAINING WILD SPECIES

Readings

Bright, Chris. 1995. "Understanding the Threat of Bioinvasions." In Lester R. Brown et al., *State of the World 1996*. Washington, D.C.: Worldwatch Institute, 95–113.

Cox, George W. 1993. *Conservation Ecology*. Dubuque, Iowa: Wm. C. Brown.

DiSilvestro, Roger L. 1992. *Rebirth of Nature*. New York: Wiley.

Durrell, Lee. 1986. *State of the Ark: An Atlas of Conservation in Action*. New York: Doubleday.

Elderidge, Niles. 1991. *The Miner's Canary*. Englewood Cliffs, N.J.: Prentice Hall.

Falk, Donald, et al. 1996. *Restoring Biodiversity: Strategies for Reintroduction of Endangered Plants*. Covelo, Calif.: Island.

Gibbons, Walt. 1993. *Keeping All the Pieces: Perspectives on Natural History and the Environment*. Washington, D.C.: Smithsonian Institution Press.

Grumbine, R. Edward. 1992. *Ghost Bears: Exploring the Biodiversity Crisis*. Covelo, Calif.: Island.

Hargrove, Eugene C., ed. 1992. *The Animal Rights/ Environmental Ethics Debate: The Environmental Perspective*. Albany: State University of New York Press.

Kellert, Stephen, and E. O. Wilson, eds. 1995. *The Biophilia Hypothesis*. Covelo, Calif.: Island.

Leakey, Richard, and Roger Lewin. 1995. *The Sixth Extinction: Patterns of Life and the Future of Humankind*. New York: Doubleday.

Livingston, John A. 1981. *The Fallacy of Wildlife Conservation*. Toronto: McClelland and Stewart.

Luoma, Jon. 1987. *A Crowded Ark: The Role of Zoos in Wildlife Conservation*. Boston: Houghton Mifflin.

Mann, Charles C., and Mark L. Plummer. 1995. *Noah's Choice: The Future of Endangered Species*. New York: Knopf.

McGinn, Anne Platt. 1998. "Promoting Sustainable Fisheries." In Lester R. Brown et al. *State of the World 1998*. Washington, D.C.: Worldwatch Institute, 59–78.

Myers, Norman. 1994. *A Wealth of Wild Species: Storehouse for Human Welfare*. Boulder, Colo.: Westview.

Nash, Roderick F. 1988. *The Rights of Nature: A History of Environmental Ethics*. Madison: University of Wisconsin Press.

National Academy of Sciences. 1995. *Science and the Endangered Species Act*. Washington, D.C.: National Academy of Sciences.

National Wildlife Federation. 1987. *The Arctic National Wildlife Refuge Coastal Plain: A Perspective for the Future*. Washington, D.C.: National Wildlife Federation.

Passmore, John. 1974. *Man's Responsibility for Nature*. New York: Scribner.

Reisner, Marc. 1991. *Game Wars: The Undercover Pursuit of Wildlife Poachers*. New York: Viking.

Tudge, Colin. 1992. *Last Animals at the Zoo: How Mass Extinction Can Be Stopped*. Covelo, Calif.: Island.

Tuttle, Merlin D. 1988. *America's Neighborhood Bats: Understanding and Learning to Live in Harmony with Them*. Austin: University of Texas Press.

Tuxill, John. 1998. *Losing Strands in the Web of Life: Vertebrate Declines and the Conservation of Biological Diversity*. Washington, D.C.: Worldwatch Institute.

Ward, Peter. 1995. *The End of Evolution: Mass Extinctions and the Preservation of Biodiversity*. New York: Bantam.

Wilson, E. O. 1984. *Biophilia*. Cambridge, Mass.: Harvard University Press.

Wilson, E. O. 1992. *The Diversity of Life*. Cambridge, Mass.: Harvard University Press.

7 / BIODIVERSITY: SUSTAINING SOILS AND PRODUCING FOOD

Readings

Aliteri, Miguel A. 1995. *Agroecology: The Scientific Basis of Alternative Agriculture*, 2d ed. Boulder, Colo.: Westview.

Ausubel, Kenny. 1994. *Seeds of Change*. New York: HarperCollins.

Benbrook, Charles M., et al. 1996. *Pest Management at the Crossroads*. Yonkers, N.Y.: Consumer's Union.

Berrill, Michael. 1997. *The Plundered Seas: Can the World's Fish Be Saved?* San Francisco, Calif.: Sierra Club Books.

Berstein, Henry, et al., eds. 1990. *The Food Question: Profits Versus People*. East Haven, Conn.: Earthscan.

Bongaarts, John. 1994. "Can the Growing Human Population Feed Itself?" *Scientific American*, March, 36–42.

Briggs, Shirley, and the Rachel Carson Council. 1992. *Basic Guide to Pesticides: Their Characteristics and Hazards*. Washington, D.C.: Taylor & Francis.

Brookfield, Harold, and Christine Padoch. 1994. "Appreciating Biodiversity: A Look at the Dynamism and Diversity of Indigenous Farming Practices." *Environment*, June, 6–42.

Brown, Lester R. 1994. *Who Will Feed China?* New York: Norton.

Brown, Lester R. 1996. *Tough Choices: Facing the Challenge of Food Scarcity*. New York: Norton.

Brown, Lester R., and Hal Kane. 1994. *Full House: Reassessing the Earth's Population Carrying Capacity*. New York: Norton.

Care, James R., and Maureen K. Hinkle. 1994. *Integrated Pest Management: The Path of a Paradigm*. Washington, D.C.: National Audubon Society.

Carson, Rachel. 1962. *Silent Spring*. Boston: Houghton Mifflin.

Carter, V. G., and T. Dale. 1974. *Topsoil and Civilization*. Norman: University of Oklahoma Press.

Coleman, Elliot. 1992. *The New Organic Grower's Four-Season Harvest*. White River Junction, Vt.: Chelsea Green.

Denckla, Tanya. 1994. *The Organic Gardener's Home Reference*. Pownal, Vt.: Garden Way Publishing.

Donahue, Roy, et al. 1990. *Soils and Their Management*, 5th ed. Petaluma, Calif.: Inter Print.

Dunning, Alan B., and Holly W. Brough. 1991. *Taking Stock: Animal Farming and the Environment*. Washington, D.C.: Worldwatch Institute.

Edwards, Clive A., et al., eds. 1990. *Sustainable Agricultural Systems*. Ankeny, Iowa: Soil and Water Conservation Society.

Ehrlich, Paul R., et al. 1995. *The Stork and the Plow: The Equity Answer to the Human Dilemma*. New York: Putnam.

Faeth, Paul. 1995. *Growing Green: Enhancing the Economic and Environmental Performance of U.S. Agriculture*. Washington, D.C.: World Resources Institute.

Faeth, Paul, et al. 1993. *Agricultural Policy and Sustainability: Case Studies from India, Chile, the Philippines, and the United States*. Washington, D.C.: World Resources Institute.

Fowler, Cary, and Pat Mooney. 1990. *Shattering: Food, Politics, and the Loss of Genetic Diversity*. Tucson: University of Arizona Press.

Friends of the Earth. 1990. *How to Get Your Lawn and Garden Off Drugs*. Ottawa, Ontario: Friends of the Earth.

Fukuoka, Masanobu. 1985. *The Natural Way of Farming: The Theory and Practice of Green Philosophy*. New York: Japan Publications.

Gardner, Gary. 1996. *Shrinking Fields: Cropland Loss in a World of Eight Billion*. Washington, D.C.: Worldwatch Institute.

Goering, Peter, et al. 1993. *From the Ground Up: Rethinking Industrial Agriculture*. Atlantic Highlands, N.J.: Zed Books.

Gordon, R. Conway, and Edward R. Barbier. 1990. *After the Green Revolution: Sustainable Agriculture for Development*. East Haven, Conn.: Earthscan.

Grainger, Alan. 1990. *The Threatening Desert: Controlling Desertification*. East Haven, Conn.: Earthscan.

Heylin, Michael, ed. 1991. "Pesticides: Costs Versus Benefits." *Chemical and Engineering News*, January 7, 27–56.

Hynes, Patricia. 1989. *The Recurring Silent Spring*. New York: Pergamon.

Jackson, Wes, et al., eds. 1985. *Meeting the Expectations of Land: Essays in Sustainable Agriculture and Stewardship*. Berkeley, Calif.: North Point Press.

Little, Charles E. 1987. *Green Fields Forever: The Conservation Tillage Revolution in America*. Covelo, Calif.: Island.

Mainguet, Monique. 1991. *Desertification: Natural Background and Human Mismanagement*. New York: Springer-Verlag.

Mann, Charles. 1997. "Reseeding the Green Revolution." *Science*, vol. 277, 1038–43.

Marquardt, Sandra. 1989. *Exporting Banned Pesticides: Fueling the Circle of Poison*. Washington, D.C.: Greenpeace.

Mollison, Bill. 1990. *Permaculture*. Covelo, Calif.: Island.

Myers, Norman. 1997. "Population and Food Security." In S. R. Johnson, ed. *Food Security: New Solutions for the 21st Century*. Ames: Iowa State University Press.

National Academy of Sciences. 1992. *Marine Aquaculture: Opportunities for Growth*. Washington, D.C.: National Academy Press.

National Academy of Sciences. 1993. *Pesticides in the Diet of Infants and Children*. Washington, D.C.: National Academy Press.

National Academy of Sciences. 1993. *Soil and Water Quality: An Agenda for Agriculture*. Washington, D.C.: National Academy Press.

National Academy of Sciences. 1993. *Sustainable Agriculture and the Environment in the Humid Tropics*. Washington, D.C.: National Academy Press.

National Academy of Sciences. 1996. *Ecologically Based Pest Management: New Solutions for a New Century*. Washington, D.C.: National Academy Press.

Paddock, Joe, et al. 1987. *Soil and Survival: Land Stewardship and the Future of American Agriculture*. San Francisco: Sierra Club Books.

Pimentel, David, and Marcia Pimentel, eds. 1996. *Food, Energy, and Society*. Niwot: University Press of Colorado.

Pimentel, David, et al. 1995. "Environmental and Economic Costs of Soil Erosion and Conservation Benefits." *Science*, vol. 267, 1117–23.

Platt, Anne. 1996. "IPM and the War on Pests." *World Watch*, March/April, 21–27.

Pretty, Jules N. 1995. *Regenerating Agriculture: Policies and Prospects for Sustainability and Self-Reliance*. London: Earthscan.

Prosterman, Roy L., et al. 1996. "Can China Feed Itself?" *Scientific American*, November, 90–96.

Rifkin, Jeremy. 1992. *Beyond Beef: The Rise and Fall of the Cattle Culture*. New York: Dutton.

Robbins, John. 1992. *May All Be Fed: Diet for a New World*. New York: William Morrow.

Safina, Carl. 1998. *Song for the Blue Ocean: Encounters Along the World's Coast and Beneath the Seas*. New York: Henry Holt.

Shiva, Vandana. 1991. *The Violence of the Green Revolution*. Atlantic Highlands, N.J.: Zed Books.

Soil and Water Conservation Society. 1990. *Sustainable Agricultural Systems*. Ankeny, Iowa: Soil and Water Conservation Society.

Soule, Judith D., and Jon Piper. 1992. *Farming in Nature's Image: An Ecological Approach to Agriculture*. Covelo, Calif.: Island.

Tamsey, Geoff, and Tony Worsley. 1995. *The Food System*. East Haven, Conn.: Earthscan.

Todd, Nancy J., and John Todd. 1984. *Bioshelters, Ocean Arks, City Farming: Ecology as a Basis for Design*. San Francisco: Sierra Club Books.

United Nations Development Fund. 1996. *Urban Agriculture: Food, Jobs, and Sustainability*. New York: United Nations.

Vasey, Daniel E. 1992. *An Ecological History of Agriculture, 10,000 b.c.–a.d. 10,000*. Ames: Iowa State University Press.

Wargo, J. 1996. *Our Toxic Legacy: How Science and Law Fail to Protect Us from Pesticides*. New Haven Conn.: Yale University Press.

Weber, Peter. 1994. *Net Loss: Fish, Jobs, and the Marine Environment*. Washington, D.C.: Worldwatch Institute.

Wild, Alan. 1993. *Soils and the Environment: An Introduction*. New York: Cambridge University Press.

Yepsen, Roger B., Jr. 1987. *The Encyclopedia of Natural Insect and Pest Control*. Emmaus, Pa.: Rodale.

8 / RISK, TOXICOLOGY, AND HUMAN HEALTH

Readings

American Council on Science and Health. 1997. *Cigarettes: What the Warning Label Doesn't Tell You*. Amherst, N.Y.: Prometheus.

Baggs, Sydney, and Joan C. Baggs. 1997. *The Healthy House: Creating a Safe, Healthy, and Environmentally Friendly Home*. New York: Harper-Collins.

Bartecchi, Carl E., et al. 1995. "The Global Tobacco Epidemic." *Scientific American*, May, 44–47.

Chivian, Eric, et al. 1993. *Critical Condition: Human Health and the Environment*. Cambridge, Mass.: MIT Press.

Colborn, Theo, et al. 1996. *Our Stolen Future*. New York: Dutton.

Dadd-Redalla, Debra. 1994. *Sustaining the Earth: Choosing Consumer Products That Are Environmentally Safe for You*. New York: Hearst Books.

Emsley, John. 1994. *The Consumers' Good Chemical Guide: A Jargon-Free Guide to the Chemicals of Everyday Life*. New York: W.H. Freeman.

Environmental Protection Agency. 1990. *Reducing Risk: Setting Priorities and Strategies for Environmental Protection*. Washington, D.C.: Environmental Protection Agency.

Environmental Protection Agency. 1991. March/April issue of the *EPA Journal*. Entire issue devoted to the debate over risk analysis.

Fox, Michael W. 1992. *Superpigs and Wondercorn: The Brave New World of Biotechnology and Where It All May Lead*. New York: Lyons & Burford.

Freudenburg, William R. 1988. "Perceived Risk, Real Risk: Social Science and the Art of Probabilistic Risk Assessment." *Science*, vol. 242, 44–49.

Garrett, Laurie. 1994. *The Coming Plague: Newly Emerging Diseases in a World Out of Balance*. New York: Farrar, Straus & Giroux.

Graham, John D., and Jonathan Baert Weiner, eds. 1995. *Risk vs. Risk: Tradeoffs in Protecting Health and the Environment*. Cambridge, Mass.: Harvard University Press.

Harris, John. 1992. *Wonderwoman and Superman: The Ethics of Human Biotechnology*. New York: Oxford University Press.

Harte, John, et al. 1992. *Toxics A to Z: A Guide to Everyday Pollution Hazards*. Berkeley: University of California Press.

Lappé, Marc. 1995. *The Evolving Threat of Drug-Resistant Disease*. San Francisco: Sierra Club Books.

Levy, Stuart B. 1998. "The Challenge of Antibiotic Resistance." *Scientific American*, March, 46–53.

Lewis, H. W. 1990. *Technological Risk*. New York: Norton.

Moeller, Dade W. 1997. *Environmental Health*, rev. ed. Cambridge, Mass.: Harvard University Press.

Montague, Peter, ed. *Rachel's Environment and Health Weekly* (P.O. Box 5036, Annapolis, MD 21403-7036). (A two-page weekly newsletter explaining health, risk, and environmental issues in an easily understandable manner.)

Nadakavukaren, Anne. 1995. *Our Global Environment: A Health Perspective*, 4th ed. New York: Waveland.

National Academy of Sciences. 1991. *Malaria: Obstacles and Opportunities*. Washington, D.C.: National Academy Press.

National Academy of Sciences. 1992. *Eat for Life: The Food and Nutrition Board's Guide to Reducing Your Risk of Chronic Disease*. Washington, D.C.: National Academy Press.

National Academy of Sciences. 1993. *Science and Judgment in Risk Assessment*. Washington, D.C.: National Academy Press.

National Academy of Sciences. 1996. *Understanding Risk: Informing Decisions in a Democratic Society*. Washington, D.C.: National Academy Press.

Ottoboni, M. Alice. 1991. *The Dose Makes the Poison: A Plain-Language Guide to Toxicology*, 2d ed. New York: Van Nostrand Reinhold.

Piller, Charles. 1991. *The Fail-Safe Society*. New York: Basic Books.

Platt, Anne. 1996. "Water-Borne Killers." *World Watch*, March/April, 28–35.

Platt, Anne. 1996. *Infecting Ourselves: How Environmental and Social Disruptions Trigger Disease*. Washington, D.C.: Worldwatch Institute.

Preston, Richard. 1994. *The Hot Zone*. New York: Random House.

Real, Leslie A. 1996. "Sustainability and the Challenge of Infectious Disease." *BioScience*, vol. 46, no. 2, 88–97.

Regenstein, Lewis G. 1993. *Cleaning Up America the Poisoned*. Washington, D.C.: Acropolis.

Rifkin, Jeremy. 1998. *The Biotech Century: Harnessing the Gene and Remaking the World*. New York: Tarcher/Plenum.

Rodricks, Joseph V. 1992. *Calculated Risks: Understanding the Toxicity and Human Health Risks of Chemicals in the Environment*. New York: Cambridge University Press.

Steingraber, Sandra. 1997. *Living Downstream: An Ecologist Looks at Cancer and the Environment*. Reading, Mass.: Addison-Wesley.

Tenner, Ward. 1996. *Why Things Bite Back: Technology and the Revenge of Unintended Consequences*. New York: Knopf.

U.S. Department of Health and Human Services. Annual. *The Health Consequences of Smoking*. Rockville, Md.: U.S. Department of Health and Human Services.

Whelan, E. M. 1993. *Toxic Terror: The Truth Behind the Cancer Scares*. Buffalo, N.Y.: Prometheus.

Wildavsky, Aaron. 1995. *But Is It True? A Citizen's Guide to Environmental Health and Safety Issues*. Cambridge, Mass.: Harvard University Press.

World Resources Institute. 1994. *Human and Ecosystem Health*. Washington, D.C.: World Resources Institute.

9 / AIR, CLIMATE, AND OZONE

Readings

Alleman, James E., and Brooke T. Mossman. 1997. "Asbestos Revisited." *Scientific American*, July, 70–75.

Balling, Robert C., Jr. 1992. *The Heated Debate: Greenhouse Predictions Versus Climate Reality*. San Francisco: Pacific Research Institute.

Barry, R. C., and J. Chorley. 1987. *Atmosphere, Weather, and Climate*, 5th ed. London: Methuen.

Bates, Albert K. 1990. *Climate in Crisis: The Greenhouse Effect and What We Can Do*. Summertown, Tenn.: Book Publishing Company.

Bridgman, Howard. 1991. *Global Air Pollution: Problems for the 1990s*. New York: Belhaven.

Brookins, Douglas G. 1990. *The Indoor Radon Problem*. Irvington, N.Y.: Columbia University Press.

Bryner, Gary. 1992. *Blue Skies, Green Politics: The Clean Air Act of 1990*. Washington, D.C.: Congressional Quarterly Press.

Burroughs, William. 1997. *Does the Weather Really Matter? The Social Implications of Climate Change*. New York: Cambridge University Press.

Coffel, Steve, and Karyn Feiden. 1991. *Indoor Pollution*. New York: Random House.

Council for Agricultural Science and Technology. 1992. *Preparing U.S. Agriculture for Global Climate Change*. Ames, Iowa: Council for Agricultural Science and Technology.

Edgerton, Lynne T. 1990. *The Rising Tide: Global Warming and World Sea Levels*. Covelo, Calif.: Island.

Flavin, Christopher, and Seth Dunn. 1997. *Rising Sun, Gathering Winds: Policies to Stabilize Climate and Strengthen Economies*. Washington, D.C.: Worldwatch Institute.

Flavin, Christopher, and Odil Tunali. 1996. *Climate of Hope: New Strategies for Stabilizing the World's Atmosphere*. Washington, D.C.: Worldwatch Institute.

French, Hilary F. 1990. *Clearing the Air: A Global Agenda*. Washington, D.C.: Worldwatch Institute.

Gates, David M. 1993. *Climate Change and Its Biological Consequences*. Sunderland, Mass.: Sinauer.

Gelbspan, Ross. 1997. *The Heat Is On: The High Stakes Battle over Earth's Threatened Climate*. Reading, Mass.: Addison Wesley.

Graedel, T. C., and Paul J. Cutzen. 1995. *Atmosphere, Climate, and Change*. New York: W.H. Freeman.

Hedin, Lars O., and Gene E. Likens. 1996. "Atmospheric Dust and Acid Rain." *Scientific American*, December, 88–92.

Houghton, John T. 1997. *Global Warming: The Complete Briefing*. New York: Cambridge University Press.

Intergovernmental Panel on Climate Change (IPCC). 1990. *Climate Change: The IPCC Assessment*. New York: Cambridge University Press.

Intergovernmental Panel on Climate Change (IPCC). 1995. *The Supplementary Report to the IPCC Scientific Assessment*. New York: Cambridge University Press.

Karplus, Walter J. 1992. *The Heavens Are Falling: The Scientific Prediction of Catastrophes in Our Time*. New York: Plenum.

Lee, Henry, ed. 1995. *Shaping Responses to Climate Change*. Covelo, Calif.: Island.

Leffell, David J., and Douglas E. Brash. 1996. "Sunlight and Skin Cancer." *Scientific American*, July, 52–59.

Leggett, Jeremy, ed. 1994. *The Climate Time Bomb: Signs of Climate Change from the Greenpeace*

Database. Amsterdam, Netherlands: Stichting Greenpeace Council.

Leslie, G. B., and F. W. Linau. 1994. *Indoor Air Pollution*. New York: Cambridge University Press.

Lovins, Amory B., et al. 1989. *Least-Cost Energy: Solving the CO_2 Problem*, 2d ed. Andover, Mass.: Brick House.

Lyman, Francesca, et al. 1990. *The Greenhouse Trap: What We're Doing to the Atmosphere and How We Can Slow Global Warming*. Washington, D.C.: World Resources Institute.

MacKenzie, James J., and Mohamed T. El-Ashry. 1990. *Air Pollution's Toll on Forests and Crops*. New Haven, Conn.: Yale University Press.

Makhijani, Arjun, and Kevin Gurney. 1995. *Mending the Ozone Hole: Science, Technology, and Policy*. Cambridge, Mass.: MIT Press.

Manne, Alan S., and Richard G. Richels. 1992. *Buying Greenhouse Insurance: The Economic Costs of CO_2 Emission Limits*. Cambridge, Mass.: MIT Press.

McCormick, John. 1997. *Acid Earth: The Politics of Acid Rain*. London: Earthscan.

McKibben, Bill. 1989. *The End of Nature*. New York: Random House.

Mintzer, Irving, et al. 1990. *Protecting the Ozone Shield: Strategies for Phasing Out CFCs During the 1990s*. Washington, D.C.: World Resources Institute.

Mintzer, Irving, and William R. Moomaw. 1991. *Escaping the Heat Trap: Probing the Prospects for a Stable Environment*. Washington, D.C.: World Resources Institute.

National Academy of Sciences. 1990. *Confronting Climate Change*. Washington, D.C.: National Academy Press.

National Academy of Sciences. 1990. *Sea Level Change*. Washington, D.C.: National Academy Press.

National Academy of Sciences. 1992. *Global Environmental Change*. Washington, D.C.: National Academy Press.

National Academy of Sciences. 1995. *Health Effects of Exposure to Radon: Time for Reassessment?* Washington, D.C.: National Academy Press.

National Audubon Society. 1990. *CO_2 Diet for a Greenhouse Planet: A Citizen's Guide to Slowing Global Warming*. New York: National Audubon Society.

Office of Technology Assessment. 1993. *Preparing for an Uncertain Climate*. Washington, D.C.: U.S. Government Printing Office.

O'Meara, Molly. 1997. "The Risks of Disrupting Climate." *WorldWatch*, November/December, 16–24.

Oppenheimer, Michael, and Robert H. Boyle. 1990. *Dead Heat: The Race Against the Greenhouse Effect*. New York: Basic Books.

Peters, Robert L., and Thomas E. Lovejoy. 1992. *Global Warming and Biological Diversity*. New Haven, Conn.: Yale University Press.

Philander, S. George. 1998. *Is the Temperature Rising? The Uncertain Science of Global Warming*. Princeton, N.J.: Princeton University Press.

Revkin, Andrew. 1992. *Global Warming: Understanding the Forecast*. New York: Abbeville.

Roan, Sharon L. 1989. *Ozone Crisis: The 15-Year Evolution of a Sudden Global Emergency*. New York: Wiley.

Rowland, F. Sherwood, and Mario J. Molina. 1994. "Ozone Depletion: 20 Years After the Alarm." *Chemical and Engineering News*, August 15, 8–13.

Smith, William H. 1991. "Air Pollution and Forest Damage." *Chemical and Engineering News*, November 11, 30–43.

Soroos, Marvin S. 1997. *The Endangered Atmosphere: Preserving a Global Commons*. Columbia: University of South Carolina Press.

Wagner, Travis. 1993. *In Our Backyard: A Guide to Understanding Pollution and Its Effects*. New York: Van Nostrand Reinhold.

Warde, John. 1997. *The Healthy Home Handbook: All You Need to Know to Rid Your Home of Health and Safety Hazards*. New York: Times Books.

Watson, R. B., et al., eds. 1996. *Climate Change 1995: Impacts, Adaptations, and Mitigation of Climate Change: Scientific Technical Analysis*. New York: Cambridge University Press.

10 / WATER

Readings

Abramovitz, Janet N. 1996. "Sustaining Freshwater Ecosystems." In Lester R. Brown et al., *State of the World 1996*. New York: Norton, 60–77.

Ashworth, William. 1986. *The Late, Great Lakes: An Environmental History*. New York: Knopf.

Bolling, David M. 1994. *How to Save a River: A Handbook for Citizen Action*. Covelo, Calif.: Island.

Burger, Joanna. 1997. *Oil Spills*. Rutgers, N.J.: Rutgers University Press.

Clarke, Robin. 1993. *Water: The International Crisis*. Cambridge, Mass.: MIT Press.

Edmonson, W. T. 1991. *The Uses of Ecology: Lake Washington and Beyond*. Seattle: University of Washington Press.

Environmental and Energy Study Institute. 1993. *New Policy Directions to Sustain the Nation's Water Resources*. Washington, D.C.: Environmental and Energy Study Institute.

Environmental Protection Agency. 1990. *Citizen's Guide to Ground-Water Protection*. Washington, D.C.: Environmental Protection Agency.

Falkenmark, Malin, and Carl Widstand. 1992. "Population and Water Resources: A Delicate Balance." *Population Bulletin*, vol. 47, no. 3, 1–36.

Feldman, David Lewis. 1991. *Water Resources Management: In Search of an Environmental Ethic*. Baltimore: Johns Hopkins University Press.

Gardner, Gary. 1995. "From Oasis to Mirage: The Aquifers That Won't Replenish." *World Watch*, May/June, 30–36, 40–41.

Gardner, Gary. 1998. "Recycling Human Waste: Fertile Ground or Toxic Legacy?" *World Watch*, January/February, 28–34.

Gleick, Peter H. 1994. "Water, War and Peace in the Middle East." *Environment*, April, 6–41.

Hansen, Nancy R., et al. 1988. *Controlling Nonpoint-Source Water Pollution*. New York: National Audubon Society and the Conservation Society.

Hillel, Daniel. 1995. *Rivers of Eden: The Struggle for Water and the Quest for Peace in the Middle East*. New York: Oxford University Press.

Hinrichsen, Don. 1998. *Coastal Waters of the World: Trends, Threats, and Strategies*. Covelo, Calif.: Island.

Horton, Tom, and William Eichbaum. 1991. *Turning the Tide: Saving the Chesapeake Bay*. Covelo, Calif.: Island.

Hundley, Norris, Jr. 1992. *The Great Thirst: Californians and Water, 1770s–1990s*. Berkeley: University of California Press.

Ives, J. D., and B. Messeric. 1989. *The Himalayan Dilemma: Reconciling Development and Conservation*. London: Routledge.

Keeble, John. 1991. *Out of the Channel: The Exxon Valdez Oil Spill in Prince William Sound*. San Francisco: HarperCollins.

Knopman, Debra S., and Richard A. Smith. 1993. "20 Years of the Clean Water Act." *Environment*, vol. 35, no. 1, 17–20, 34–41.

Laws, Edward A. 1993. *Aquatic Pollution: An Introductory Text*, 2d ed. New York: Wiley.

Marx, Wesley. 1991. *The Frail Ocean: A Blueprint for Change in the 1990s and Beyond*. San Francisco: Sierra Club Books.

McCully, Patrick. 1996. *Silenced Rivers: The Ecology and Politics of Large Dams*. London: Zed.

McCutcheon, Sean. 1992. *Electric Rivers: The Story of the James River Project*. New York: Paul.

Micklin, Philip P., and William D. Williams. 1996. *The Aral Sea Basin*. New York: Springer-Verlag.

National Academy of Sciences. 1992. *Water Transfers in the West: Efficiency, Equity, and the Environment*. Washington, D.C.: National Academy Press.

Patrick, Ruth, et al. 1992. *Surface Water Quality: Have the Laws Been Successful?* Princeton, N.J.: Princeton University Press.

Pearce, Fred. 1992. *The Dammed: Rivers, Dams, and the Coming World Water Crisis*. London: Bodley Head.

Platt, Anne. 1995. "Dying Seas." *World Watch*, January/February, 10–19.

Platt, Anne. 1996. "Water-Borne Killers." *World Watch*, March/April, 28–35.

Postel, Sandra. 1992. *The Last Oasis: Facing Water Scarcity*. New York: Norton.

Postel, Sandra. 1996. "Forging a Sustainable Water Strategy." In Lester R. Brown et. al., *State of the World 1996*. New York: Norton, 40–59.

Qing, Dai. 1994. *Yangtze! Yangtze! Debate Over the Three Gorges Project*. East Haven, Conn.: Earthscan.

Reisner, Marc, and Sara Bates. 1990. *Overtapped Oasis: Reform or Revolution for Western Water*. Covelo, Calif.: Island.

Sierra Club Defense Fund. 1989. *The Poisoned Well: New Strategies for Groundwater Protection*. Covelo, Calif.: Island.

Starr, Joyce R. 1991. "Water Wars." *Foreign Policy*, Spring, 12–45.

Wagner, Travis. 1993. *In Our Backyard: A Guide to Understanding Pollution and Its Effects*. New York: Van Nostrand Reinhold.

Weber, Michael L., and Judith A. Gradwohl. 1995. *The Wealth of Oceans: Environment and Development on Our Ocean Planet*. New York: Norton.

Weber, Peter. 1993. *Abandoned Seas: Reversing the Decline of the Oceans*. Washington, D.C.: Worldwatch Institute.

Worster, Donald. 1985. *Rivers of Empire: Water, Aridity, and the Growth of the American West*. New York: Pantheon.

11 / SOLID AND HAZARDOUS WASTE

Readings

Ackerman, Frank. 1997. *Why Do We Recycle? Markets, Values, and Public Policy*. Covelo, Calif.: Island.

Alexander, Judd H. 1993. *In Defense of Garbage*. Westport, Conn.: Praeger.

Allen, Robert. 1992. *Waste Not, Want Not*. London: Earthscan.

Bullard, Robert D. 1994. *Dumping in Dixie: Race, Class, and Environmental Quality*, 2d ed. Boulder, Colo.: Westview.

Carless, Jennifer. 1992. *Taking Out the Trash: A No-Nonsense Guide to Recycling*. Covelo, Calif.: Island.

Centers for Disease Control and Prevention. 1991. *Preventing Lead Poisoning in Young Children*. Atlanta: Centers for Disease Control and Prevention.

Christopher, Tom, and Marty Asher. 1994. *The Art of Composting for Your Yard, Your Community, and the Planet*. San Francisco: Sierra Club Books.

Clarke, Marjorie J., et al. 1991. *Burning Garbage in the U.S.: Practice vs. State of the Art*. New York: INFORM.

Cohen, Gary, and John O'Connor. 1990. *Fighting Toxics: A Manual for Protecting Family, Community, and Workplace*. Covelo, Calif.: Island.

Commoner, Barry. 1994. *Making Peace with the Planet*, rev. ed. New York: New Press.

Connett, Paul H. 1992. "The Disposable Society." In F. H. Bormann and Stephen R. Kellert, eds., *Ecology, Economics, Ethics*. New Haven, Conn.: Yale University Press, 99–122.

Denison, Richard A., and John Ruston. 1997. "Recycling Is Not Garbage." *Technology Review*, October, 55–60.

Durning, Alan Thein. 1992. *How Much Is Enough? The Consumer Society and the Future of the Earth.* New York: Norton.

Earth Works Group. 1990. *The Recycler's Handbook: Simple Things You Can Do.* Berkeley, Calif.: Earth Works.

Frosch, Robert A. 1995. "Industrial Ecology." *Environment,* December, 16–37.

Gardner, Gary. 1997. *Recycling Organic Waste: From Urban Pollutant to Farm Resource.* Washington, D.C.: Worldwatch Institute.

Gibbs, Lois. 1995. *Dying from Dioxin.* Boston: South End.

Gibbs, Lois. 1997. *Love Canal: The Story Continues.* Philadelphia: New Society Publishers.

Gordon, Ben, and Peter Montague. 1989. *Zero Discharge: A Citizen's Toxic Waste Manual.* Washington, D.C.: Greenpeace.

Gottlieb, Robert, ed. 1995. *Reducing Toxics: A New Approach to Policy and Industrial Decisionmaking.* Covelo, Calif.: Island.

Harte, John, et al. 1991. *Toxics A to Z: A Guide to Everyday Pollution Hazards.* Berkeley: University of California Press.

Hilz, Christoph. 1993. *The International Toxic Waste Trade.* New York: Van Nostrand Reinhold.

Kane, Hal. 1996. "Shifting to Sustainable Industries." In Lester R. Brown et al., *State of the World 1996.* Washington, D.C.: Worldwatch Institute, 152–67.

Kessel, Irene, and John T. O'Connor. 1997. *Getting the Lead Out: The Complete Resource on How to Prevent and Cope with Lead Poisoning.* New York: Plenum.

Lappé, Marc. 1991. *Chemical Deception: The Toxic Threat to Health and Environment.* San Francisco: Sierra Club Books.

Lipsett, B., and D. Farrell. 1990. *Solid Waste Incineration Status Report.* Arlington, Va.: Citizen's Clearinghouse for Hazardous Waste.

Minnesota Mining and Manufacturing. 1988. *Low- or Non-Pollution Technology Through Pollution Prevention.* St. Paul, Minn.: 3M Company.

Moyers, Bill. 1990. *Global Dumping Ground: The International Traffic in Hazardous Waste.* Cabin John, Md.: Seven Locks.

Needleman, Herbert L., and Philip J. Landrigan. 1995. *Raising Children Toxic Free.* New York: Avon Books.

Nemerow, Nelson L. 1995. *Zero Pollution for Industry: Waste Minimization Through Industrial Complexes.* New York: Wiley.

Office of Technology Assessment. 1992. *Green Products by Design.* Washington, D.C.: Government Printing Office.

Pellerano, Maria B. 1995. *How to Research Chemicals: A Resource Guide.* Annapolis, Md.: Environmental Research Foundation.

Platt, Brenda, et al. 1991. *Beyond 40 Percent: Record-Setting Recycling and Composting Programs.* Covelo, Calif.: Island.

Rathje, William, and Cullen Murphy. 1992. *Rubbish! The Archaeology of Garbage: What Our Garbage Tells Us About Ourselves.* San Francisco: HarperCollins.

Setterberg, Fred, and Lonny Shavelson. 1993. *Toxic Nation: The Fight to Save Our Communities from Chemical Contamination.* New York: Wiley.

Stapleton, Richard M. 1995. *Lead Is a Silent Hazard.* New York: Walker.

Theodore, Louis, and Young C. McGuinn. 1992. *Pollution Prevention.* New York: Van Nostrand Reinhold.

Thomas, William. 1995. *Scorched Earth: The Military's Assault on the Environment.* Philadelphia: New Society.

Water Pollution Control Federation. 1989. *Household Hazardous Waste: What You Should and Shouldn't Do.* Alexandria, Va.: Water Pollution Control Federation.

Young, John E. 1991. *Discarding the Throwaway Society.* Washington, D.C.: Worldwatch Institute.

Young, John E. 1995. "The Sudden New Strength of Recycling." *World Watch,* July/August, 20–25.

12 / ENVIRONMENTAL ECONOMICS, POLITICS, AND WORLDVIEWS

Readings

Aaseng, Nathan. 1994. *Jobs vs. the Environment: Can We Save Both?* Hillside, N.J.: Enslow.

Adler, Jonathan. 1996. *Environmentalism at the Crossroads: Green Activism in America.* Washington, D.C.: Capital Research Center.

Anderson, Terry, and Donald Leal. 1990. *Free Market Environmentalism.* San Francisco: Pacific Research Institute.

Ashworth, William. 1995. *The Economy of Nature: Rethinking the Connections Between Ecology and Economics.* Boston: Houghton Mifflin.

Athansiou, Tony. 1996. *Divided Planet: The Ecology of Rich and Poor.* Boston: Little, Brown.

Ausbel, Kenny. 1997. *Restoring the Earth: Visionary Solutions from the Bioneers.* Tiburon, Calif.: H.J. Kramer.

Barnet, Richard J., and John Cavanagh. 1994. *Global Dreams: Imperial Corporations and the New World Order.* New York: Simon & Schuster.

Beder, Sharon. 1997. *Global Spin: The Corporate Assault on Environmentalism.* White River Junction, Vt.: Chelsea Green.

Berry, Thomas. 1988. *The Dream of the Earth.* San Francisco: Sierra Club Books.

Bossel, Hartmut. 1998. *Paths to a Sustainable Future.* New York: Cambridge University Press.

Bowers, C. A. 1997. *The Culture of Denial: Why the Environmental Movement Needs a Strategy for Reforming Universities and Schools.* Albany: State University of New York Press.

Bramwell, Anna. 1994. *The Fading of the Greens: The Decline of Environmental Politics in the West.* New Haven, Conn.: Yale University Press.

Brick, Phil. 1995. "Determined Opposition: The Wise Use Movement Challenges Environmentalism." *Environment,* vol. 27, no. 8, 17–41.

Brown, Lester R., et al. 1991. *Saving the Planet: How to Shape an Environmentally Sustainable Global Economy.* New York: Norton.

Bryant, Bunyan, ed. 1995. *Environmental Justice: Issues, Policies, and Solutions.* Covelo, Calif.: Island.

Cairncross, Frances. 1992. *Costing the Earth.* Boston: Harvard Business School.

Cairncross, Frances. 1995. *Green, Inc.: A Guide to Business and the Environment.* Covelo, Calif.: Island.

Caldwell, Lynton K. 1997. *International Environmental Policy: From the Twentieth to the Twenty-First Century.* Durham, N.C.: Duke University Press.

Chetlow, Marian R., and Daniel C. Esty. 1997. *Thinking Ecologically: The Next Generation of Environmental Policy.* New Haven, Conn.: Yale University Press.

Clark, Mary E. 1989. *Adriadne's Thread: The Search for New Models of Thinking.* New York: St. Martin's.

Cobb, Clifford, et al. 1995. *The Genuine Progress Indicator.* San Francisco, Calif.: Redefining Progress.

Cohen, Michael J. 1988. *How Nature Works: Regenerating Kinship with Planet Earth.* Walpole, N.H.: Stillpoint.

Cohn, Susan. 1995. *Green at Work: Finding a Business Career That Works for the Environment,* rev. ed. Covelo, Calif.: Island.

Coleman, Daniel A. 1994. *Ecopolitics: Building a Green Society.* New Brunswick, N.J.: Rutgers University Press.

Collett, Jonathan, and Stephen Karakashian, eds. 1995. *Greening the College Curriculum: A Guide to Environmental Teaching in the Liberal Arts Colleges.* Covelo, Calif.: Island.

Daily, Gretchen C., ed. 1997. *Nature's Services: Societal Dependence on Natural Ecosystems.* Covelo, Calif.: Island.

Daly, Herman E. 1996. *Beyond Growth: The Economics of Sustainable Development.* Boston: Beacon.

Daly, Herman E., and John B. Cobb, Jr. 1994. *For the Common Good: Redirecting the Economy Toward Community, the Environment, and a Sustainable Future,* 2d ed. Boston: Beacon.

Desjardins, Joseph R. 1993. *Environmental Ethics.* Belmont, Calif.: Wadsworth.

Devall, Bill, and George Sessions. 1985. *Deep Ecology: Living as if Nature Mattered.* Salt Lake City: Gibbs M. Smith.

Dowie, Mark. 1995. *Losing Ground: American Environmentalism at the Close of the Twentieth Century.* Cambridge, Mass.: MIT Press.

Durning, Alan T. 1989. *Poverty and the Environment: Reversing the Downward Spiral.* Washington, D.C.: Worldwatch Institute.

Durning, Alan T. 1992. *How Much Is Enough? The Consumer Society and the Earth.* New York: Norton.

Eagen, David J., and David W. Orr. 1994. *The Campus and Environmental Responsibility.* San Francisco: Jossey-Bass.

Earley, Jay. 1997. *Transforming Human Culture: Social Evolution and the Planetary Crisis.* Albany, N.Y.: State University of New York Press.

Echeverria, John, and Raymond Booth Eby, eds. 1995. *Let the People Judge: A Reader on the Wise Use Movement.* Covelo, Calif.: Island.

Eden, Sally. 1996. *Environmental Issues and Business: Implications of a Changing Agenda.* New York: Wiley.

Ehrenfeld, David. 1993. *Beginning Again: People and Nature in the New Millennium.* New York: Oxford University Press.

Ehrlich, Paul R., and Anne H. Ehrlich. 1991. *Healing the Planet.* Reading, Mass.: Addison-Wesley.

Ehrlich, Paul R., and Anne H. Ehrlich. 1996. *Betrayal of Science and Reason: How Anti-Environmental Rhetoric Threatens Our Future.* Covelo, Calif.: Island.

Elgin, Duane. 1993. *Voluntary Simplicity: Toward a Way of Life That Is Outwardly Simple, Inwardly Rich,* rev. ed., New York: William Morrow.

Environmental Careers Organization. 1993. *The New Complete Guide to Environmental Careers.* Covelo, Calif.: Island.

Fanning, Odum. 1995. *Opportunities for Environmental Careers.* Lincolnwood, Ill.: VCM Career Horizons.

Ferkiss, Victor. 1993. *Nature, Technology, and Society: Cultural Roots of the Current Environmental Crisis.* New York: New York University Press.

Flattau, Edward. 1998. *Tracking the Charlatans: Countering the Eco-Bashers.* New York: Global Horizons.

Flavin, Christopher, and John E. Young. 1993. "Shaping the Next Industrial Revolution." In Lester R. Brown et al., *State of the World 1993.* New York: Norton, 180–99.

Foreman, Dave. 1990. *Confessions of an Eco-Warrior.* New York: Crown.

Foster, John Bellamy. 1995. *The Vulnerable Planet: A Short Economic History of the Planet.* New York: Monthly Review Press.

French, Hilary E. 1995. *Partnership for the Planet: An Environmental Agenda for the United Nations.* Washington, D.C.: Worldwatch Institute.

George, Susan. 1992. *The Debt Boomerang: How Third World Debt Harms Us All.* Boulder, Colo.: Westview.

Goudy, John, and Sabine O'Hara. 1995. *Economic Theory for Environmentalists.* New York: St. Lucie.

Greenpeace. 1993. *The Greenpeace Guide to Anti-Environmental Organizations.* Berkeley, Calif.: Odonian.

Greider, William. 1992. *Who Will Tell the People?* New York: Simon & Schuster.

Hardin, Garrett. 1986. *Filters Against Folly.* New York: Penguin.

Hawken, Paul. 1993. *The Ecology of Commerce.* New York: HarperCollins.

Hayden, Tom. 1996. *The Lost Gospel of the Earth: A Call for Renewing Nature, Spirit, and Politics.* San Francisco: Sierra Club Books.

Heinze-Fry, Jane. 1994. *Green Lives, Green Campuses.* Belmont, Calif.: Wadsworth. Available as a supplement for use with this book.

Helvarg, David. 1994. *The War Against the Greens.* San Francisco: Sierra Club Books.

Henderson, Hazel. 1996. *Building a Win–Win World: Life Beyond Global Economic Warfare.* San Francisco: Berrett-Koehler.

Hoffman, Andrew J. 1997. *From Heresy to Dogma: An Institutional History of Corporate Environmentalism.* San Francisco, Calif.: New Lexington.

Hunter, J. Robert. 1997. *Simple Things Won't Save the Earth.* Austin: University of Texas Press.

Jansson, Ann Marie, et al., eds. 1994. *Investing in Natural Capital: The Ecological Economics Approach to Sustainability.* Covelo, Calif.: Island.

Kellert, Stephen R. 1996. *The Value of Life: Biological Diversity and Human Society.* Covelo, Calif.: Island.

Keniry, Julian. 1995. *Ecodemia: Campus Environmental Stewardship at the Turn of the Century.* Washington, D.C.: National Wildlife Federation.

Korten, David. 1995. *When Corporations Rule the World.* West Hartford, Conn.: Kumarian.

LaMay, Craig L., and Everette E. Dennis, eds. 1992. *Media and the Environment.* Covelo, Calif.: Island.

Landy, Marc K., et al. 1994. *The Environmental Protection Agency: Asking the Wrong Questions, from Nixon to Clinton.* New York: Oxford University Press.

Learner, Steve. 1997. *Eco-Pioneers: Practical Visionaries Solving Environmental Problems.* Cambridge, Mass.: MIT Press.

Leopold, Aldo. 1949. *A Sand County Almanac.* New York: Oxford University Press.

Lewis, Martin W. 1992. *Green Delusions: An Environmentalist Critique of Radical Environmentalism.* Durham, N.C.: Duke University Press.

Livingston, John. 1973. *One Cosmic Instant: Man's Fleeting Supremacy.* Boston: Houghton Mifflin.

Luke, Timothy W. 1997. *Ecocritique: Contesting the Politics of Nature, Economy, and Culture.* Minneapolis: University of Minnesota Press.

Makower, Joel. 1993. *The E-Factor: The Bottom-Line Approach to Environmentally Responsible Business.* New York: Times Books.

Matre, Steve Van. 1990. *Earth Education.* Greenville, W.V.: Institute for Earth Education.

McKibben, Bill. 1997. *Hope, Human, and Wild: Treading Lightly on the Earth.* St. Paul, Minn.: Hungry Mind.

Meadows, Donella H. 1991. *Global Citizen.* Covelo, Calif.: Island.

Milbrath, Lester W. 1989. *Envisioning a Sustainable Society.* Albany: State University of New York Press.

Milbrath, Lester W. 1995. *Learning to Think Environmentally While There Is Still Time.* Albany: State University of New York Press.

Myers, Norman. 1993. *Ultimate Security: The Environmental Basis of Political Stability.* New York: Norton.

Naess, Arne. 1989. *Ecology, Community, and Lifestyle.* New York: Cambridge University Press.

National Academy of Sciences. 1995. *Assigning Economic Value to Natural Resources.* Washington, D.C.: National Academy Press.

National Wildlife Federation. 1995. *Ecodemia: Campus Environmental Stewardship at the Turn of the 21st Century.* Washington, D.C.: National Wildlife Federation.

National Wildlife Federation. Annual. *Campus Environmental Yearbook.* Washington, D.C.: National Wildlife Federation.

Norton, Bryan G. 1991. *Toward Unity Among Environmentalists.* New York: Oxford University Press.

Ophuls, William, and A. Stephen Boyan, Jr. 1992. *Ecology and the Politics of Scarcity Revisited: The Unraveling of the American Dream.* San Francisco: W.H. Freeman.

Orr, David. 1992. *Ecological Literacy.* Ithaca: State University of New York Press.

Orr, David. 1994. *Earth in Mind: On Education, Environment, and the Human Prospect.* Covelo, Calif.: Island.

Orr, David. 1994. "Love It or Lose It: The Coming Biophilia Revolution." *Orion,* Winter, 8–15.

Palmer, Joy. 1997. *Environmental Education in the 21st Century: Theory, Practice, Progress, and Promise.* New York: Routledge.

Peterson, Tarla Rai. 1997. *Sharing the Earth: The Rhetoric of Sustainable Development.* Columbia: University of South Carolina Press.

Piasecki, Bruce W. 1995. *Corporate Environmental Strategy: The Avalanche of Change Since Bhopal.* New York: Wiley.

Ponting, Clive. 1992. *A Green History of the World. The Environment and the Collapse of Great Civilizations.* New York: St. Martin's.

Porter, Gareth, and Janet Welsh Brown. 1994. *Global Environmental Politics,* 2d ed. Boulder, Colo.: Westview.

Quinn, Daniel. 1992. *Ishmael.* New York: Bantam/Turner.

Raskin, Jamin B., and John Bonifaz. 1994. *The Wealth Primary.* Washington, D.C.: Center for Responsive Politics.

Rees, William, and Mathis Wackernagel. 1995. *Our Ecological Footprint.* Philadelphia: New Society.

Repetto, Robert. 1992. "Accounting for Environmental Assets." *Scientific American,* June, 94–100.

Repetto, Robert. 1995. *Jobs, Competitiveness, and Environmental Regulation: What Are the Real Issues?* Washington, D.C.: World Resources Institute.

Repetto, Robert, et al. 1992. *Green Fees: How a Tax Shift Can Work for the Environment and the Economy.* Washington, D.C.: World Resources Institute.

Rich, Bruce. 1994. *Mortgaging the Earth.* Boston: Beacon.

Rolston, Holmes, III. 1994. *Conserving Natural Values.* New York: Columbia University Press.

Roodman, David M. 1996. *Paying the Piper: Subsidies, Politics, and the Environment.* Washington, D.C.: Worldwatch Institute.

Roodman, David M. 1997. *Getting the Signals Right: Tax Reform to Protect the Environment and the Economy.* Washington, D.C.: Worldwatch Institute.

Roodman, David M. 1998. *The Natural Wealth of Nations: Harnessing the Market for the Environment.* New York: Norton.

Rosenbaum, Walter A. 1998. *Environmental Politics and Solution.* Washington, D.C.: Congressional Quarterly.

Roszak, Theodore. 1992. *The Voice of the Earth.* New York: Simon & Schuster.

Rowell, Andrew. 1997. *Green Backlash: Subversion of the Environmental Movement.* New York: Routledge.

Rubin, Charles T. 1994. *The Green Crusade: Rethinking the Roots of Environmentalism.* New York: Free Press.

Ruckelshaus, William D., and Karl Hausker, eds. 1998. *The Environmental Protection System in Transition: Toward a More Desirable Future.* Washington, D.C.: Center for Strategic and International Studies.

Ryan, John C. and Alan Thein. 1997. *Stuff: The Secret Lives of Everyday Things.* Seattle, Wash.: Northwest Environmental Press.

Sachs, Aaron. 1995. *Eco-Justice: Linking Human Rights and the Environment.* Washington, D.C.: Worldwatch Institute.

Sale, Kirkpatrick. 1993. *The Green Revolution: The American Environmental Movement 1962–1992.* New York: Hill & Wang.

Sanjor, William. 1992. *Why the EPA Is Like It Is and What Can Be Done About It.* Washington, D.C.: Environmental Research Foundation.

Schmidheiny, Stephan. 1992. *Changing Course: A Global Business Perspective on Development and the Environment.* Cambridge, Mass.: MIT Press.

Schumacher, E. F. 1973. *Small Is Beautiful: Economics as if the Earth Mattered.* New York: Harper & Row.

Schwab, Jim. 1994. *Deeper Shades of Green: The Rise of Blue Collar and Minority Environmentalism in America.* San Francisco: Sierra Club Books.

Sessions, George, ed. 1994. *Deep Ecology for the Twenty-First Century.* Boston: Shambhala.

Shabecoff, Philip. 1993. *A Fierce Green Fire: The American Environmental Movement.* New York: Hill & Wang.

Smith, April, and the Student Environmental Action Coalition. 1994. *Campus Ecology: A Guide to Assessing Environmental Quality and Creating Strategies for Change.* Los Angeles: Living Planet.

Smith, Joseph Wayne, et al. 1997. *Healing a Wounded World: Economics, Ecology, and Health for a Sustainable Life.* Westport, Conn.: Praeger.

Stauber, John C., and Sheldon Rampton. 1995. *Toxic Sludge Is Good for You: Lies, Damn Lies, and the Public Relations Industry.* Madison, Wisc.: PR Watch.

Stead, W. Edward, and John Garner Stead. 1992. *Management for a Small Planet.* New York: Sage.

Strong, Susan. 1995. *The GDP Myth: How It Harms Our Quality of Life and What Communities are Doing About It.* Mountain View, Calif.: Center for Economic Conversion.

Suzuki, David. 1998. *The Sacred Balance: Rediscovering Our Place in Nature.* New York: Prometheus.

Swimme, Brian, and Thomas Berry. 1992. *The Universe Story: From the Primordial Flaring Forth to the Ecozoic Era.* New York: HarperCollins.

Taylor, Ann. 1992. *A Practical Politics of the Environment.* New York: Routledge.

Taylor, Paul W. 1986. *Respect for Nature: A Theory of Environmental Ethics.* Princeton, N.J.: Princeton University Press.

Thomas, Lewis. 1992. *The Fragile Species.* New York: Scribner.

Thomashow, Mitchell. 1995. *Ecological Identity: Becoming a Reflective Environmentalist.* Cambridge, Mass.: MIT Press.

Tietenberg, Tom. 1992. *Environmental and Resource Economics,* 3d ed. Glenview, Ill.: Scott, Foresman.

Tucker, Mary Evelyn, and John A. Grim, eds. 1994. *Worldviews and Ecology: Religion, Philosophy, and the Environment.* Maryknoll, N.Y.: Orbis Books.

Turner, R. Kerry, et al. 1994. *Environmental Economics: An Elementary Introduction.* Baltimore: Johns Hopkins University Press.

Van DeVeer, Donald, and Christine Pierce. 1994. *The Environmental Ethics and Policy Book: Philosophy, Ecology, Economics.* Belmont, Calif.: Wadsworth.

Ward, Barbara. 1979. *Progress for a Small Planet.* New York: Norton.

Werbach, Adam. 1997. *Act Now, Apologize Later.* New York: Cliff street Books.

Wilkinson, Todd. 1998. *Science Under Siege: The Politicians' War on Nature and Truth.* New York: Johnson.

World Resources Institute. 1989. *The Crucial Decade: The 1990s and the Global Environmental Challenge.* Washington, D.C.: World Resources Institute.

Worster, Donald. 1993. *The Wealth of Nature: Environmental History and the Ecological Imagination.* New York: Oxford University Press.

GLOSSARY

abiotic   Nonliving. Compare *biotic*.

acclimation   Adjustment to slowly changing new conditions. Compare *threshold effect*.

acid deposition   The falling of acids and acid-forming compounds from the atmosphere to earth's surface. Acid deposition is commonly known as *acid rain*, a term that refers only to wet deposition of droplets of acids and acid-forming compounds.

acid rain   See *acid deposition*.

active solar heating system   System that uses solar collectors to capture energy from the sun and store it as heat for space and water heating. Liquid or air pumped through the collectors transfers the captured heat to a storage system such as an insulated water tank or rock bed. Pumps or fans then distribute the stored heat or hot water throughout a dwelling as needed. Compare *passive solar heating system*.

adaptation   Any genetically controlled trait that helps an organism survive and reproduce under a given set of environmental conditions. It usually results from a beneficial mutation. See *biological evolution, differential reproduction, mutation, natural selection*.

adaptive radiation   Period of time (usually millions of years) during which numerous new species evolve to fill vacant and new ecological niches in changed environments, usually after a mass extinction.

adaptive trait   See *adaptation*.

advanced sewage treatment   Specialized chemical and physical processes that reduce the amount of specific pollutants left in wastewater after primary and secondary sewage treatment. This type of treatment is usually expensive. See also *primary sewage treatment, secondary sewage treatment*.

aerobic respiration   Complex process that occurs in the cells of most living organisms, in which nutrient organic molecules such as glucose ($C_6H_{12}O_6$) combine with oxygen (O_2) and produce carbon dioxide (CO_2), water (H_2O), and energy. Compare *photosynthesis*.

age structure   Percentage of the population (or the number of people of each sex) at each age level in a population.

agroforestry   Planting trees and crops together.

air pollution   One or more chemicals in high enough concentrations in the air to harm humans, other animals, vegetation, or materials. Excess heat or noise can also be considered forms of air pollution. Such chemicals or physical conditions are called air pollutants. See *primary pollutant, secondary pollutant*.

alien species   See *nonnative species*.

alleles   Slightly different molecular forms found in a particular gene.

alley cropping   Planting of crops in strips with rows of trees or shrubs on each side.

alpha particle   Positively charged matter, consisting of two neutrons and two protons, that is emitted as a form of radioactivity from the nuclei of some radioisotopes. See also *beta particle, gamma rays*.

altitude   Height above sea level. Compare *latitude*.

ancient forest   See *old-growth forest*.

animal manure   Dung and urine of animals that can be used as a form of organic fertilizer. Compare *green manure*.

animals   Eukaryotic, multicelled organisms such as sponges, jellyfishes, arthropods (insects, shrimp, lobsters), mollusks (snails, oysters, octopuses), fish, amphibians (frogs, toads, salamanders), reptiles (turtles, alligators, snakes), birds, and mammals (kangaroos, bats, whales, apes, humans). See *carnivores, herbivores, omnivores*.

annual   Plant that grows, sets seed, and dies in one growing season. Compare *perennials*.

aquaculture   Growing and harvesting of fish and shellfish for human use in freshwater ponds, irrigation ditches, and lakes, or in cages or fenced-in areas of coastal lagoons and estuaries. See *fish farming, fish ranching*.

aquatic   Pertaining to water. Compare *terrestrial*.

aquatic life zone   Marine and freshwater portions of the ecosphere. Examples include freshwater life zones (such as lakes and streams) and ocean or marine life zones (such as estuaries, coastlines, coral reefs, and the deep ocean).

aquifer   Underground porous, water-saturated layers of sand, gravel, or bedrock through which groundwater flows.

arable land   Land that can be cultivated to grow crops.

arid   Dry. A desert or other area with an arid climate has little precipitation.

atmosphere   The whole mass of air surrounding the earth. See *stratosphere, troposphere*.

atomic number   Number of protons in the nucleus of an atom. Compare *mass number*.

atoms   Minute units made of subatomic particles that are the basic building blocks of all chemical elements and thus all matter; the smallest unit of an element that can exist and still have the unique characteristics of that element. Compare *ion, molecule*.

autotroph   See *producer*.

background extinction   Normal extinction of various species as a result of changes in local environmental conditions. Compare *mass extinction*.

bacteria   Prokaryotic, one-celled organisms. Some transmit diseases. Most act as decomposers and get the nutrients they need by breaking down complex organic compounds in the tissues of living or dead organisms into simpler inorganic nutrient compounds.

barrier islands   Long, thin, low offshore islands of sediment that generally run parallel to the shore along some coasts.

beta particle   Swiftly moving electron emitted by the nucleus of a radioactive isotope. See also *alpha particle, gamma rays*.

bioaccumulation   An increase in the concentration of a chemical in specific organs or tissues at a level higher than would normally be expected. Compare *biomagnification*.

biodegradable   Capable of being broken down by decomposers.

biodegradable pollutant   Material that can be broken down into simpler substances (elements and compounds) by bacteria or other decomposers. Paper and most organic wastes such as animal manure are biodegradable but can take decades to biodegrade in modern landfills. Compare *degradable pollutant, nondegradable pollutant, slowly degradable pollutant*.

biodiversity   See *biological diversity*.

biogeochemical cycle   Natural processes that recycle nutrients in various chemical forms from the nonliving environment to living organisms, and then back to the nonliving environment. Examples are the carbon, oxygen, nitrogen, phosphorus, sulfur, and hydrologic cycles.

biological community   See *community*.

biological diversity   Variety of different species (*species diversity*), genetic variability among individuals within each species (*genetic diversity*), and variety of ecosystems (*ecological diversity*).

biological evolution   Change in the genetic makeup of a population of a species in successive generations. If continued long enough, it can lead to the formation of a new species. Note that populations, not individuals, evolve. See also *adaptation, differential reproduction, natural selection, theory of evolution*.

biological oxygen demand (BOD)   Amount of dissolved oxygen needed by aerobic decomposers to break down the organic materials in a given volume of water at a certain temperature over a specified time period.

biological pest control   Control of pest populations by natural predators, parasites, or disease-causing bacteria and viruses (pathogens).

biomagnification   Increase in concentration of DDT, PCBs, and other slowly degradable, fat-soluble chemicals in organisms at successively higher trophic levels of a food chain or web. Compare *bioaccumulation*.

biomass   Organic matter produced by plants and other photosynthetic producers; total dry weight of all living organisms that can be supported at each trophic level in a food chain or web; dry weight of all organic matter in plants and animals in an ecosystem; plant materials and animal wastes used as fuel.

biome   Terrestrial region inhabited by certain types of life, especially vegetation. Examples are various types of deserts, grasslands, and forests.

biosphere   Zone of the earth where life is found. It consists of parts of the atmosphere (the troposphere), hydrosphere (mostly surface water and groundwater), and lithosphere (mostly soil and surface rocks and sediments on the bottoms of oceans and other bodies of water) where life is found. It is also called the *ecosphere*.

biotic   Living. Compare *abiotic*.

biotic potential   Maximum rate at which the population of a given species can increase when there are no limits on its rate of growth. See *environmental resistance*.

birth rate   See *crude birth rate*.

bitumen   Gooey, black, high-sulfur, heavy oil extracted from tar sand and then upgraded to synthetic fuel oil. See *tar sand*.

breeder nuclear fission reactor   Nuclear fission reactor that produces more nuclear fuel than it consumes by converting nonfissionable uranium-238 into fissionable plutonium-239.

cancer   Group of more than 120 different diseases, one for each type of cell in the human body. Each type of cancer produces a tumor in which cells multiply uncontrollably and invade surrounding tissue.

carbon cycle Cyclic movement of carbon in different chemical forms from the environment to organisms and then back to the environment.

carcinogen Chemicals, ionizing radiation, and viruses that cause or promote the development of cancer. See *cancer, mutagen, teratogen*.

carnivore Animal that feeds on other animals. Compare *herbivore, omnivore*.

carrying capacity (K) Maximum population of a particular species that a given habitat can support over a given period of time.

cell Smallest living unit of an organism. Each cell is encased in an outer membrane or wall and contains genetic material (DNA) and other parts to perform its life function. Organisms such as bacteria consist of only one cell, but most of the organisms we are familiar with contain many cells. See *eukaryotic cell, prokaryotic cell*.

centrally planned economy See *pure command economic system*.

CFCs See *chlorofluorocarbons*.

chain reaction Multiple nuclear fissions, taking place within a certain mass of a fissionable isotope, that release an enormous amount of energy in a short time.

chemical One of the millions of different elements and compounds found naturally or synthesized by humans. See *compound, element*.

chemical change Interaction between chemicals in which there is a change in the chemical composition of the elements or compounds involved. Compare *physical change*.

chemical evolution Formation of the earth and its early crust and atmosphere, evolution of the biological molecules necessary for life, and evolution of systems of chemical reactions needed to produce the first living cells. These processes are believed to have occurred about 1 billion years before biological evolution. Compare *biological evolution*.

chemical formula Shorthand way to show the number of atoms (or ions) in the basic structural unit of a compound. Examples are H_2O, NaCl, and $C_6H_{12}O_6$.

chemical reaction See *chemical change*.

chemosynthesis Process in which certain organisms (mostly specialized bacteria) extract inorganic compounds from their environment and convert them into organic nutrient compounds without the presence of sunlight. Compare *photosynthesis*.

chlorinated hydrocarbon Organic compound made up of atoms of carbon, hydrogen, and chlorine. Examples are DDT and PCBs.

chlorofluorocarbons (CFCs) Organic compounds made up of atoms of carbon, chlorine, and fluorine. An example is Freon-12 (CCl_2F_2), used as a refrigerant in refrigerators and air conditioners and in making plastics such as Styrofoam. Gaseous CFCs can deplete the ozone layer when they slowly rise into the stratosphere and their chlorine atoms react with ozone molecules. Use of these molecules is being phased out.

chromosome A grouping of various genes and associated proteins in plant and animal cells that carry certain types of genetic information. See *genes*.

clear-cutting Method of timber harvesting in which all trees in a forested area are removed in a single cutting. Compare *seed-tree cutting, selective cutting, shelterwood cutting, strip cutting*.

climate Physical properties of the troposphere of an area based on analysis of its weather records over a long period (at least 30 years). The two main factors determining an area's climate are *temperature*, with its seasonal variations, and the amount and distribution of *precipitation*. Compare *weather*.

coal Solid, combustible mixture of organic compounds with 30–98% carbon by weight, mixed with various amounts of water and small amounts of sulfur and nitrogen compounds. It is formed in several stages as the remains of plants are subjected to heat and pressure over millions of years.

coastal wetland Land along a coastline, extending inland from an estuary that is covered with salt water all or part of the year. Examples are marshes, bays, lagoons, tidal flats, and mangrove swamps. Compare *inland wetland*.

coastal zone Warm, nutrient-rich, shallow part of the ocean that extends from the high-tide mark on land to the edge of a shelflike extension of continental land masses known as the continental shelf. Compare *open sea*.

cogeneration Production of two useful forms of energy, such as high-temperature heat or steam and electricity, from the same fuel source.

commensalism An interaction between organisms of different species in which one type of organism benefits and the other type is neither helped nor harmed to any great degree. Compare *mutualism*.

commercial extinction Depletion of the population of a wild species used as a resource to a level at which it is no longer profitable to harvest the species.

commercial inorganic fertilizer Commercially prepared mixtures of plant nutrients such as nitrates, phosphates, and potassium applied to the soil to restore fertility and increase crop yields. Compare *organic fertilizer*.

common-property resource Resource that people are normally free to use; each user can deplete or degrade the available supply. Most are potentially renewable and are owned by no one. Examples are clean air, fish in parts of the ocean not under the control of a coastal country, migratory birds, gases of the lower atmosphere, and the ozone content of the upper atmosphere (stratosphere). See *tragedy of the commons*.

community Populations of all species living and interacting in an area at a particular time.

community development See *ecological succession*.

competition Two or more individual organisms of a single species (*intraspecific competition*) or two or more individuals of different species (*interspecific competition*) attempting to use the same scarce resources in the same ecosystem.

competitive exclusion principle No two species can occupy exactly the same fundamental niche indefinitely in a habitat where there is not enough of a particular resource to meet the needs of both species. See *ecological niche, fundamental niche, realized niche*.

compost Partially decomposed organic plant and animal matter that can be used as a soil conditioner or fertilizer.

compound Combination of atoms, or oppositely charged ions, of two or more different elements held together by attractive forces called chemical bonds. Compare *element*.

concentration Amount of a chemical in a particular volume or weight of air, water, soil, or other medium.

coniferous evergreen plants Cone-bearing plants (such as spruces, pines, and firs) that keep some of their narrow, pointed leaves (needles) all year. Compare *deciduous plants, evergreen plants*.

consensus science Scientific data, models, theories, and laws that are widely accepted. This aspect of science is very reliable. Compare *frontier science*.

conservation biology Multidisciplinary science created to deal with the crisis of maintaining the genes, species, communities, and ecosystems that make up the earth's biological diversity. Its goals are to investigate human impacts on biodiversity and to develop practical approaches to preserving biodiversity and ecological integrity.

conservation-tillage farming Crop cultivation in which the soil is disturbed little (minimum-tillage farming) or not at all (no-till farming) to reduce soil erosion, lower labor costs, and save energy. Compare *conventional-tillage farming*.

consumer Organism that cannot synthesize the organic nutrients it needs and gets its organic nutrients by feeding on the tissues of producers or other consumers; generally divided into *primary consumers* (herbivores), *secondary consumers* (carnivores), *tertiary (higher-level) consumers*, *omnivores*, and *detritivores* (decomposers and detritus feeders). In economics, one who uses economic goods.

contour farming Plowing and planting across the changing slope of land, rather than in straight lines, to help retain water and reduce soil erosion.

conventional-tillage farming Crop cultivation that involves making a planting surface by plowing land, breaking up the exposed soil, and then smoothing the surface. Compare *conservation-tillage farming*.

coral reef Formation produced by massive colonies containing billions of tiny coral animals, called *polyps*, that secrete a stony substance (calcium carbonate) around themselves for protection. When the corals die, their empty outer skeletons form layers that cause the reef to grow. They are found in the coastal zones of warm tropical and subtropical oceans.

core Inner zone of the earth. It consists of a solid inner core and a liquid outer core. Compare *crust, mantle*.

cost–benefit analysis Estimates and comparison of short-term and long-term costs (losses) and benefits (gains) from an economic decision. If the estimated benefits exceed the estimated costs, the decision to buy an economic good or provide a public good is considered worthwhile.

critical mass Amount of fissionable nuclei needed to sustain a nuclear fission chain reaction.

crop rotation Planting a field, or an area of a field, with different crops from year to year to reduce depletion of soil nutrients. A plant such as corn, tobacco, or cotton, which removes large amounts of nitrogen from the soil, is planted one year. The next year a legume such as soybeans, which adds nitrogen to the soil, is planted.

crude birth rate Annual number of live births per 1,000 people in a geographic area at the midpoint of a given year. Compare *crude death rate*.

crude death rate Annual number of deaths per 1,000 in a geographic area at the midpoint of a given year. Compare *crude birth rate*.

crude oil Gooey liquid consisting mostly of hydrocarbon compounds and small amounts of compounds containing oxygen, sulfur, and nitrogen. Extracted from underground accumulations, it is sent to oil refineries, where it is converted to heating oil, diesel fuel, gasoline, tar, and other materials.

crust Solid outer zone of the earth. It consists of oceanic crust and continental crust. Compare *core, mantle*.

cultural eutrophication Overnourishment of aquatic ecosystems with plant nutrients (mostly nitrates and phosphates) because of human activities such as agriculture, urbanization, and discharges from industrial and sewage treatment plants. See *eutrophication*.

cyanobacteria Single-celled, prokaryotic, microscopic organisms. Before being reclassified as monera, they were called blue-green algae.

DDT Dichlorodiphenyltrichloroethane, a chlorinated hydrocarbon that has been widely used as a pesticide but is now banned in some countries.

death rate See *crude death rate*.

debt-for-nature swap Agreement in which a certain amount of foreign debt is canceled in exchange for local currency investments that will improve natural resource management or protect certain areas in the debtor country from harmful development.

deciduous plants Trees that survive during dry seasons or cold seasons by shedding their leaves. Compare *coniferous evergreen plants*.

decomposer Organism that digests parts of dead organisms and cast-off fragments and wastes of living organisms. A decomposer breaks down the complex organic molecules in those materials into simpler inorganic compounds and then absorbs the soluble nutrients. Most of these chemicals are returned to the soil and water for reuse by producers. Decomposers consist of various bacteria and fungi. Compare *consumer, detritivore, producer*.

deforestation   Removal of trees from a forested area without adequate replanting.

degradable pollutant   Potentially polluting chemical that is broken down completely or reduced to acceptable levels by natural physical, chemical, and biological processes. Compare *biodegradable pollutant, nondegradable pollutant, slowly degradable pollutant*.

degree of urbanization   Percentage of the population in the world, or a country, living in areas with a population of more than 2,500 people (higher in some countries). Compare *urban growth*.

democracy   Government by the people through their elected officials and appointed representatives. In a *constitutional democracy*, a constitution provides the basis of government authority and puts restraints on government power through free elections and freely expressed public opinion.

demographic transition   Hypothesis that countries, as they become industrialized, have declines in death rates followed by declines in birth rates.

depletion time   How long it takes to use a certain fraction—usually 80%—of the known or estimated supply of a nonrenewable resource at an assumed rate of use. Finding and extracting the remaining 20% usually costs more than it is worth.

desalination   Purification of salt water or brackish (slightly salty) water by removing dissolved salts.

desert   Biome in which evaporation exceeds precipitation and the average amount of precipitation is less than 25 centimeters (10 inches) a year. Such areas have little vegetation or have widely spaced, mostly low vegetation. Compare *forest, grassland*.

desertification   Conversion of rangeland, rainfed cropland, or irrigated cropland to desertlike land, with a drop in agricultural productivity of 10% or more. It is usually caused by a combination of overgrazing, soil erosion, prolonged drought, and climate change.

detritivore   Consumer organism that feeds on detritus, parts of dead organisms, and cast-off fragments and wastes of living organisms. The two principal types are *detritus feeders* and *decomposers*.

detritus   Parts of dead organisms and cast-off fragments and wastes of living organisms.

detritus feeder   Organism that extracts nutrients from fragments of dead organisms and the cast-off parts and organic wastes of living organisms. Examples are earthworms, termites, and crabs. Compare *decomposer*.

developed country   Country that is highly industrialized and has a high per capita GNP. Compare *developing country*.

developing country   Country that has low to moderate industrialization and low to moderate per capita GNP. Most are located in Africa, Asia, and Latin America. Compare *developed country*.

development   Change from a society that is largely rural, agricultural, illiterate, and poor, with a rapidly growing population, to one that is mostly urban, industrial, educated, and wealthy, with a slowly growing or stationary population. See *developed country* and *developing country*.

differential reproduction   Phenomenon in which individuals with adaptive genetic traits produce more living offspring than do individuals without such traits. See *natural selection*.

dioxins   Family of 75 different toxic chlorinated hydrocarbon compounds formed as unwanted by-products in chemical reactions involving chlorine and hydrocarbons, usually at high temperatures.

dissolved oxygen (DO) content   Amount of oxygen gas (O_2) dissolved in a given volume of water at a particular temperature and pressure, often expressed as a concentration in parts of oxygen per million parts of water.

DNA (deoxyribonucleic acid)   Large molecules in the cells of organisms that carry genetic information in living organisms.

domesticated species   Wild species tamed or genetically altered by crossbreeding for use by humans for food (cattle, sheep, and food crops), pets (dogs and cats), or enjoyment (animals in zoos and plants in gardens).

dose   The amount of a potentially harmful substance an individual ingests, inhales, or absorbs through the skin. Compare *response*. See *dose–response curve, median lethal dose*.

dose–response curve   Plot of data showing effects of various doses of a toxic agent on a group of test organisms. See *dose, median lethal dose, response*.

doubling time   The time it takes (usually in years) for the quantity of something growing exponentially to double. It can be calculated by dividing the annual percentage growth rate into 70.

drainage basin   See *watershed*.

dredge spoils   Materials scraped from the bottoms of harbors and streams to maintain shipping channels. They are often contaminated with high levels of toxic substances that have settled out of the water. See *dredging*.

dredging   Type of surface mining in which chain buckets and draglines scrape up sand, gravel, and other surface deposits covered with water. It is also used to remove sediment from streams and harbors to maintain shipping channels. See *dredge spoils*.

drought   Condition in which an area does not get enough water because of lower-than-normal precipitation, higher-than-normal temperatures that increase evaporation, or both.

dust dome   Dome of heated air that surrounds an urban area and traps pollutants, especially suspended particulate matter. See *urban heat island*.

dust plume   Elongation of a dust dome by winds that can spread a city's pollutants hundreds of kilometers downwind.

earth capital   The earth's natural resources and processes that sustain us and other species. Compare *human capital, manufactured capital, solar capital*.

earth-wisdom society   Society based on working with nature by recycling and reusing discarded matter; preventing pollution; conserving matter and energy resources by reducing unnecessary waste and use; not degrading renewable resources; building things that are easy to recycle, reuse, and repair; not allowing population size to exceed the carrying capacity of the environment; and preserving biodiversity. See *earth-wisdom worldview*. Compare *matter-recycling society, planetary management worldview*.

earth-wisdom worldview   Beliefs that nature exists for all of the earth's species, not just for us; that we are not in charge of the rest of nature; that there is not always more, and it's not all for us; that some forms of economic growth are beneficial, and some are harmful; that our goals should be to design economic and political systems that encourage earth-sustaining forms of growth and discourage or prohibit earth-degrading forms; and that our success depends on learning to cooperate with one another and with the rest of nature instead of trying to dominate and manage earth's life-support systems primarily for our own use. Compare *planetary management worldview*.

ecological diversity   The variety of forests, deserts, grasslands, oceans, streams, lakes, and other biological communities interacting with one another and with their nonliving environment. See *biological diversity*. Compare *genetic diversity, species diversity*.

ecological efficiency   The percentage of energy transferred from one trophic level to another.

ecological health   The degree to which an area's biodiversity and ecological integrity remain intact. See *biodiversity, ecological integrity*.

ecological integrity   The conditions and natural processes (such as the flow of materials and energy through ecosystems) that generate and main-

tain biodiversity and allow evolutionary change as a key mechanism for adapting to changes in environmental conditions.

ecologically sustainable development   Development in which the total human population size and resource use in the world (or in a region) are limited to a level that does not exceed the carrying capacity of the existing natural capital and is therefore sustainable. Compare *economic growth*.

ecological niche   Total way of life or role of a species in an ecosystem. It includes all physical, chemical, and biological conditions a species needs to live and reproduce in an ecosystem. See *fundamental niche, realized niche*.

ecological resource   Anything required by an organism for normal maintenance, growth, and reproduction. Examples include habitat, food, water, and shelter. Compare *economic resources*.

ecological succession   Process in which communities of plant and animal species in a particular area are replaced over time by a series of different and often more complex communities. See *primary succession, secondary succession*.

ecology   Study of the interactions of living organisms with one another and with their nonliving environment of matter and energy; study of the structure and functions of nature.

economic growth   Increase in the real value of all final goods and services produced by an economy; an increase in real GNP or GDP. Compare *ecologically sustainable development*.

economic resources   Natural resources, capital goods, and labor used in an economy to produce material goods and services. See *earth capital, human capital, manufactured capital*.

economic system   Method that a group of people uses to choose what goods and services to produce, how to produce them, how much to produce, and how to distribute them to people. See *mixed economic system, pure command economic system, pure market economic system*.

economy   System of production, distribution, and consumption of economic goods.

ecosphere   The earth's collection of living organisms interacting with one another and their nonliving environment (energy and matter) throughout the world; all of earth's ecosystems. Also called the *biosphere*.

ecosystem   Community of different species interacting with one another and with the chemical and physical factors making up its nonliving environment.

electromagnetic radiation   Forms of kinetic energy traveling as electromagnetic waves. Examples are radio waves, TV waves, microwaves, infrared radiation, visible light, ultraviolet radiation, X rays, and gamma rays. Compare *ionizing radiation, nonionizing radiation*.

electron (e)   Tiny particle moving around outside the nucleus of an atom. Each electron has one unit of negative charge and almost no mass. Compare *neutron, proton*.

element   Chemical, such as hydrogen (H), iron (Fe), sodium (Na), carbon (C), nitrogen (N), or oxygen (O), whose distinctly different atoms serve as the basic building blocks of all matter. There are 92 naturally occurring elements. Another 18 have been made in laboratories. Two or more elements combine to form compounds, which make up most of the world's matter. Compare *compound*.

endangered species   Wild species with so few individual survivors that the species could soon become extinct in all or most of its natural range. Compare *threatened species*.

energy   Capacity to do work by performing mechanical, physical, chemical, or electrical tasks or to cause a heat transfer between two objects at different temperatures.

energy efficiency   Percentage of the total energy input that does useful work and is not converted into low-quality, usually useless heat in an energy conversion system or process. See *energy quality, net energy.*

energy quality Ability of a form of energy to do useful work. High-temperature heat and the chemical energy in fossil fuels and nuclear fuels are concentrated high-quality energy. Low-quality energy such as low-temperature heat is dispersed or diluted and cannot do much useful work. See *high-quality energy, low-quality energy.*

environment All external conditions and factors, living and nonliving (chemicals and energy), that affect an organism or other specified system during its lifetime; the earth's life-support systems for us and for all other forms of life—another term for describing solar capital and earth capital.

environmental degradation Depletion or destruction of a potentially renewable resource such as soil, grassland, forest, or wildlife by using it at a faster rate than it is naturally replenished. If such use continues, the resource can become nonrenewable (on a human time scale) or nonexistent (extinct). See also *sustainable yield.*

environmental ethics Beliefs about what is right or wrong behavior toward the environment or the earth's life-support systems.

environmental resistance All the limiting factors acting jointly to limit the growth of a population. See *biotic potential, limiting factor.*

environmental science Study of how we and other species interact with one another and with the nonliving environment (matter and energy). It is a physical and social science that integrates knowledge from a wide range of disciplines including physics, chemistry, biology (especially ecology), geology, geography, resource technology and engineering, resource conservation and management, demography (the study of population dynamics), economics, politics, sociology, psychology, and ethics. In other words, it is a study of how the parts of nature and human societies operate and interact—a study of connections and interactions.

environmental worldview How people think the world works, what they think their role in the world should be, and what they believe is right and wrong environmental behavior (environmental ethics). See *earth-wisdom worldview, planetary management worldview.*

EPA Environmental Protection Agency; responsible for managing federal efforts in the United States to control air and water pollution, radiation and pesticide hazards, environmental research, hazardous waste, and solid-waste disposal.

epidemiology Study of the patterns of disease or other harmful effects from toxic exposure within defined groups of people to find out why some people get sick and some do not.

erosion Process or group of processes by which loose or consolidated earth materials are dissolved, loosened, or worn away and removed from one place and deposited in another. See *weathering.*

estuary Partially enclosed coastal area at the mouth of a river where its fresh water, carrying fertile silt and runoff from the land, mixes with salty seawater.

ethics Our beliefs about what is right or wrong behavior.

eukaryotic cell Cell containing a *nucleus*, a region of genetic material surrounded by a membrane. Membranes also enclose several of the other internal parts found in a eukaryotic cell. Compare *prokaryotic cell.*

eutrophication Physical, chemical, and biological changes that take place after a lake, an estuary, or a slow-flowing stream receives inputs of plant nutrients, mostly nitrates and phosphates, from natural erosion and runoff from the surrounding land basin. See *cultural eutrophication.*

eutrophic lake Lake with a large or excessive supply of plant nutrients, mostly nitrates and phosphates. Compare *mesotrophic lake, oligotrophic lake.*

evaporation Conversion of a liquid into a gas.

even-aged management Method of forest management in which trees, sometimes of a single species in a given stand, are maintained at about the same age and size and are harvested all at once. Compare *uneven-aged management.*

evergreen plants Plants that keep some of their leaves or needles throughout the year. Examples are ferns and cone-bearing trees (conifers) such as firs, spruces, pines, redwoods, and sequoias. See *coniferous evergreen plants.* Compare *deciduous plants, succulent plants.*

evolution See *biological evolution.*

exhaustible resource See *nonrenewable resource.*

exotic species See *nonnative species.*

exponential growth Growth in which some quantity, such as population size or economic output, increases by a fixed percentage of the whole in a given time; when the increase in quantity over a long enough time is plotted, this type of growth typically yields a curve shaped like the letter *J*. Compare *linear growth.*

external benefit Beneficial social, health, or environmental effect of producing and using an economic good that is not included in the market price of the good. Compare *external cost, full cost.*

external cost Harmful social, health, or environmental effect of producing and using an economic good that is not included in the market price of the good. Compare *external benefit, full cost.*

externalities Social benefits ("goods") and social costs ("bads") not included in the market price of an economic good. See *external benefit, external cost.* Compare *full cost, internal cost.*

extinction Complete disappearance of a species from the earth. This happens when a species cannot adapt and successfully reproduce under new environmental conditions or when it evolves into one or more new species. Compare *speciation.* See also *endangered species, threatened species.*

family planning Providing information, clinical services, and contraceptives to help people choose the number and spacing of children they want to have.

famine Widespread malnutrition and starvation in a particular area because of a shortage of food, usually caused by drought, war, flood, earthquake, or other catastrophic events that disrupt food production and distribution.

feedlot Confined outdoor or indoor space used to raise hundreds to thousands of domesticated livestock. Compare *rangeland.*

fertilizer Substance that adds inorganic or organic plant nutrients to soil and improves its ability to grow crops, trees, or other vegetation. See *commercial inorganic fertilizer, organic fertilizer.*

first law of energy See *first law of thermodynamics.*

first law of human ecology We can never do merely one thing. Any intrusion into nature has numerous effects, many of which are unpredictable.

first law of thermodynamics In any physical or chemical change, no detectable amount of energy is created or destroyed, but in these processes energy can be changed from one form to another. You can't get more energy out of something than you put in; in terms of energy quantity, you can't get something for nothing (there is no free lunch). This law does not apply to nuclear changes, in which energy can be produced from small amounts of matter. See also *second law of thermodynamics.*

fishery Concentrations of particular aquatic species suitable for commercial harvesting in a given ocean area or inland body of water.

fish farming Form of aquaculture in which fish are cultivated in a controlled pond or other environment and harvested when they reach the desired size. See also *fish ranching.*

fish ranching Form of aquaculture in which members of a fish species such as salmon are held in captivity for the first few years of their lives,

released, and then harvested as adults when they return from the ocean to their freshwater birthplace to spawn. See also *fish farming.*

floodplain Flat valley floor next to a stream channel. For legal purposes the term is often applied to any low area that has the potential for flooding, including certain coastal areas.

flyway Generally fixed route along which waterfowl migrate from one area to another at certain seasons of the year.

food chain Series of organisms in which each eats or decomposes the preceding one. Compare *food web.*

food web Complex network of many interconnected food chains and feeding relationships. Compare *food chain.*

forest Biome with enough average annual precipitation (at least 76 centimeters, or 30 inches) to support growth of various species of trees and smaller forms of vegetation. Compare *desert, grassland.*

fossil fuel Products of partial or complete decomposition of plants and animals that occur as crude oil, coal, natural gas, or heavy oils as a result of exposure to heat and pressure in earth's crust over millions of years. See *coal, crude oil, natural gas.*

Freons See *chlorofluorocarbons.*

freshwater life zones Aquatic systems where water with a dissolved salt concentration of less than 1% by volume accumulates on or flows through the surfaces of terrestrial biomes. Examples are *standing* (lentic) bodies of fresh water such as lakes, ponds, and inland wetlands and *flowing* (lotic) systems such as streams and rivers. Compare *biomes.*

frontier science Preliminary scientific data, hypotheses, and models that have not been widely tested and accepted. Compare *consensus science.*

full cost Cost of a good when its internal costs and its estimated short- and long-term external costs are included in its market price. Compare *external cost, internal cost.*

fundamental niche The full potential range of the physical, chemical, and biological factors a species can use if there is no competition from other species. See *ecological niche.* Compare *realized niche.*

fungi Eukaryotic, mostly multicelled organisms such as mushrooms, molds, and yeasts. As decomposers, they get the nutrients they need by secreting enzymes that speed up the break down the organic matter in the tissue of other living or dead organisms. Then they absorb the resulting nutrients.

fungicide Chemical that kills fungi.

gamma rays A form of ionizing electromagnetic radiation with a high energy content, emitted by the sun and some radioisotopes. They readily penetrate body tissues.

gene mutation See *mutation.*

gene pool The sum total of all genes found in the individuals of the population of a particular species.

generalist species Species with a broad ecological niche. They can live in many different places, eat a variety of foods, and tolerate a wide range of environmental conditions. Examples are flies, cockroaches, mice, rats, and human beings. Compare *specialist species.*

genes Coded units of information about specific traits that are passed on from parents to offspring during reproduction. They consist of segments of DNA molecules found in chromosomes.

genetic diversity Variability in the genetic makeup among individuals within a single species. See *biological diversity.* Compare *ecological diversity, species diversity.*

geographic isolation Separation of populations of a species for fairly long times into areas with different environmental conditions.

geothermal energy Heat transferred from the earth's underground concentrations of dry steam

(steam with no water droplets), wet steam (a mixture of steam and water droplets), or hot water trapped in fractured or porous rock.

GDP See *gross domestic product.*

global warming Warming of the earth's atmosphere as a result of increases in the concentrations of one or more greenhouse gases. See *greenhouse effect, greenhouse gases.*

GNP See *gross national product, per capita GNP.*

grassland Biome found in regions where moderate annual average precipitation (25 to 76 centimeters, or 10 to 30 inches) is enough to support the growth of grass and small plants, but not enough to support large stands of trees. Compare *desert, forest.*

greenhouse effect A natural effect that traps heat in the atmosphere (troposphere) near the earth's surface. Some of the heat flowing back toward space from the earth's surface is absorbed by water vapor, carbon dioxide, ozone, and several other gases in the lower atmosphere (troposphere) and then radiated back toward the earth's surface. If the atmospheric concentrations of these greenhouse gases rise and are not removed by other natural processes, the average temperature of the lower atmosphere will increase gradually.

greenhouse gases Gases in the earth's lower atmosphere (troposphere) that cause the greenhouse effect. Examples are carbon dioxide, chlorofluorocarbons, ozone, methane, water vapor, and nitrous oxide.

green manure Freshly cut or still-growing green vegetation that is plowed into the soil to increase the organic matter and humus available to support crop growth. Compare *animal manure.*

green revolution Popular term for introduction of scientifically bred or selected varieties of grain (rice, wheat, maize) that, with high enough inputs of fertilizer and water, can greatly increase crop yields.

gross domestic product (GDP) Total market value in current dollars of all goods and services produced *within a country* for final use usually during a year. Compare *gross national product.*

gross national product (GNP) Total market value in current dollars of all goods and services produced by an economy for final use usually during a year. Compare *gross domestic product.*

gross primary productivity (GPP) The rate at which an ecosystem's producers capture and store a given amount of chemical energy as biomass in a given length of time. Compare *net primary productivity.*

groundwater Water that sinks into the soil and is stored in slowly flowing and slowly renewed underground reservoirs called aquifers; underground water in the zone of saturation, below the water table. Compare *runoff, surface water.*

gully reclamation Restoring land suffering from gully erosion by seeding gullies with quick-growing plants, building small dams to collect silt and gradually fill in the channels, and building channels to divert water away from the gully.

habitat Place or type of place where an organism or a population of organisms lives. Compare *ecological niche.*

habitat fragmentation Breakup of a habitat into smaller pieces, usually as a result of human activities.

hazardous chemical Chemical that can cause harm because it is flammable or explosive, or that can irritate or damage the skin or lungs (such as strong acidic or alkaline substances) or cause allergic reactions of the immune system (allergens). See *toxic chemical.*

hazardous waste Any solid, liquid, or containerized gas that can catch fire easily, is corrosive to skin tissue or metals, is unstable and can explode or release toxic fumes, or has harmful concentrations of one or more toxic materials that can leach out. See also *toxic waste.*

herbicide Chemical that kills a plant or inhibits its growth.

herbivore Plant-eating organism. Examples are deer, sheep, grasshoppers, and zooplankton. Compare *carnivore, omnivore.*

heterotroph See *consumer.*

high-quality energy Energy that is organized or concentrated and has great ability to perform useful work. Examples are high-temperature heat and the energy in electricity, coal, oil, gasoline, sunlight, and nuclei of uranium-235. Compare *low-quality energy.*

high-quality matter Matter that is organized and concentrated and contains a high concentration of a useful resource. Compare *low-quality matter.*

high-waste society The situation in most advanced industrialized countries, in which ever-increasing economic growth is sustained by maximizing the rate at which matter and energy resources are used, with little emphasis on pollution prevention, recycling, reuse, reduction of unnecessary waste, and other forms of resource conservation. Compare *earth-wisdom society, matter-recycling society.*

human capital Physical and mental talents of people used to produce, distribute, and sell an economic good. Compare *earth capital, manufactured capital, solar capital.*

humus Slightly soluble residue of undigested or partially decomposed organic material in topsoil. This material helps retain water and water-soluble nutrients, which can be taken up by plant roots.

hunter–gatherers People who get their food by gathering edible wild plants and other materials and by hunting wild animals and fish.

hydrologic cycle Biogeochemical cycle that collects, purifies, and distributes the earth's fixed supply of water from the environment to living organisms, and then back to the environment.

hydropower Electrical energy produced by falling or flowing water.

hydrosphere The earth's liquid water (oceans, lakes, and other bodies of surface water and underground water), frozen water (polar ice caps, floating ice caps, and ice in soil known as *permafrost*), and small amounts of water vapor in the atmosphere.

identified resources Deposits of a particular mineral-bearing material of which the location, quantity, and quality are known or have been estimated from direct geological evidence and measurements. Compare *undiscovered resources.*

igneous rock Rock formed when molten rock material (magma) wells up from earth's interior, cools, and solidifies into rock masses. Compare *metamorphic rock, sedimentary rock.* See *rock cycle.*

immigrant species Species that migrate into an ecosystem or are deliberately or accidentally introduced into an ecosystem by humans. Some of these species are beneficial, whereas others can take over and eliminate many native species. Compare *indicator species, keystone species, native species, nonnative species.*

immigration Migration of people into a country or area to take up permanent residence.

indicator species Species that serve as early warnings that a community or ecosystem is being degraded. Compare *immigrant species, keystone species, native species.*

industrialized agriculture Using large inputs of energy from fossil fuels (especially oil and natural gas), water, fertilizer, and pesticides to produce large quantities of crops and livestock for domestic and foreign sale. Compare *subsistence farming.*

industrial revolution Use of new sources of energy from fossil fuels (and later from nuclear fuels) and use of new technologies to grow food and manufacture products.

industrial smog Type of air pollution consisting mostly of a mixture of sulfur dioxide, suspended droplets of sulfuric acid formed from some of the sulfur dioxide, and a variety of suspended solid particles. Compare *photochemical smog.*

infant mortality rate Number of babies out of every 1,000 born each year that die before their first birthday.

infiltration Downward movement of water through soil.

inland wetland Land away from the coast, such as a swamp, marsh, or bog, that is covered all or part of the time with fresh water, excluding streams and lakes. Compare *coastal wetland.*

inorganic fertilizer See *commercial inorganic fertilizer.*

input pollution control See *pollution prevention.*

insecticide Chemical that kills insects.

integrated pest management (IPM) Combined use of biological, chemical, and cultivation methods in proper sequence and timing to keep the size of a pest population below the size that causes economically unacceptable loss of a crop or livestock animal.

intercropping Growing two or more different crops at the same time on a plot. For example, a carbohydrate-rich grain that depletes soil nitrogen and a protein-rich legume that adds nitrogen to the soil may be intercropped. Compare *monoculture, polyculture, polyvarietal cultivation.*

intermediate goods See *manufactured capital.*

internal cost Direct cost paid by the producer and the buyer of an economic good. Compare *external cost.*

interplanting Simultaneously growing a variety of crops on the same plot. See *agroforestry, intercropping, polyculture, polyvarietal cultivation.*

interspecific competition Members of two or more species trying to use the same limited resources in an ecosystem. See *competition, competitive exclusion, intraspecific competition.*

intraspecific competition Two or more organisms of a single species trying to use the same limited resources in an ecosystem. See *competition, interspecific competition.*

intrinsic rate of increase (r) Rate at which a population could grow if it had infinite resources. Compare *environmental resistance.*

inversion See *thermal inversion.*

invertebrates Animals that have no backbones. Compare *vertebrates.*

ion Atom or group of atoms with one or more positive (+) or negative (−) electrical charges. Compare *atom, molecule.*

ionizing radiation Fast-moving alpha or beta particles or high-energy radiation (gamma rays) emitted by radioisotopes. They have enough energy to dislodge one or more electrons from atoms they hit, forming charged ions (in tissue) that can react with and damage living tissue.

isotopes Two or more forms of a chemical element that have the same number of protons but different mass numbers because of different numbers of neutrons in their nuclei.

J-shaped curve Curve with a shape similar to that of the letter *J*; can represent prolonged exponential growth.

kerogen Solid, waxy mixture of hydrocarbons found in oil shale rock. When the rock is heated to high temperatures, the kerogen is vaporized. The vapor is condensed, purified, and then sent to a refinery to produce gasoline, heating oil, and other products. See also *oil shale, shale oil.*

keystone species Species that play roles affecting many other organisms in an ecosystem. Compare *immigrant species, indicator species, native species.*

kinetic energy Energy that matter has because of its mass and speed or velocity. Compare *potential energy.*

labor See *human capital.*

lake Large natural body of standing fresh water formed when water from precipitation, land runoff, or groundwater flow fills a depression in the earth created by glaciation, earth movement, volcanic activity, or a giant meteorite. See *eutrophic lake, mesotrophic lake, oligotrophic lake.*

landfill See *sanitary landfill.*

land-use planning Process for deciding the best present and future use of each parcel of land in an area.

latitude Distance from the equator. Compare *altitude*.

law of conservation of energy See *first law of thermodynamics*.

law of conservation of matter In any physical or chemical change, matter is neither created nor destroyed, but merely changed from one form to another; in physical and chemical changes, existing atoms are rearranged into different spatial patterns (physical changes) or different combinations (chemical changes).

law of tolerance The existence, abundance, and distribution of a species in an ecosystem are determined by whether the levels of one or more physical or chemical factors fall within the range tolerated by the species. See *threshold effect*.

LD$_{50}$ See *median lethal dose*.

leaching Process in which various chemicals in upper layers of soil are dissolved and carried to lower layers and, in some cases, to groundwater.

less developed country (LDC) See *developing country*.

life cycle cost Initial cost plus lifetime operating costs of an economic good.

life expectancy Average number of years a newborn infant can be expected to live.

limiting factor Single factor that limits the growth, abundance, or distribution of the population of a species in an ecosystem. See *limiting factor principle*.

limiting factor principle Too much or too little of any abiotic factor can limit or prevent growth of a population of a species in an ecosystem, even if all other factors are at or near the optimum range of tolerance for the species.

linear growth Growth in which a quantity increases by some fixed amount during each unit of time. Compare *exponential growth*.

liquefied natural gas (LNG) Natural gas converted to liquid form by cooling to a very low temperature.

liquefied petroleum gas (LPG) Mixture of liquefied propane (C_3H_8) and butane (C_4H_{10}) gas removed from natural gas.

lithosphere Outer shell of the earth, composed of the crust and the rigid, outermost part of the mantle outside of the asthenosphere; material found in the earth's plates. See *crust, mantle*.

LNG See *liquefied natural gas*.

loams Soils containing a mixture of clay, sand, silt, and humus. Good for growing most crops.

low-input agriculture See *sustainable agriculture*.

low-quality energy Energy that is disorganized or dispersed and has little ability to do useful work. An example is low-temperature heat. Compare *high-quality energy*.

low-quality matter Matter that is disorganized, dilute, or dispersed or that contains a low concentration of a useful resource. Compare *high-quality matter*.

low-waste society See *earth-wisdom society*.

LPG See *liquefied petroleum gas*.

magma Molten rock below the earth's surface.

malnutrition Faulty nutrition. Caused by a diet that does not supply a person with enough protein, essential fats, vitamins, minerals, and other nutrients needed for good health. Compare *overnutrition, undernutrition*.

mangrove swamps Swamps found on the coastlines in warm tropical climates. They are dominated by mangrove trees, any of about 55 species of trees and shrubs that can live partly submerged in the salty environment of coastal swamps.

mantle Zone of the earth's interior between its core and its crust. Compare *core, crust*. See *lithosphere*.

manufactured capital Manufactured items made from earth capital and used to produce and distribute economic goods and services bought by consumers. These include tools, machinery, equipment, factory buildings, and transportation and distribution facilities. Compare *earth capital, human capital, solar capital*.

manure See *animal manure, green manure*.

mass The amount of material in an object.

mass extinction A catastrophic, widespread, often global event in which major groups of species are wiped out over a short time compared to normal (background) extinctions. Compare *background extinction*. See *adaptive radiation*.

mass number Sum of the number of neutrons and the number of protons in the nucleus of an atom. It gives the approximate mass of that atom. Compare *atomic number*.

matter Anything that has mass (the amount of material in an object) and takes up space. On earth, where gravity is present, we weigh an object to determine its mass.

matter quality Measure of how useful a matter resource is, based on its availability and concentration. See *high-quality matter, low-quality matter*.

matter-recycling society Society that emphasizes recycling the maximum amount of all resources that can be recycled. The goal is to allow economic growth to continue without depleting matter resources and without producing excessive pollution and environmental degradation. Compare *earth-wisdom society, high-waste society*.

maximum sustainable yield See *sustainable yield*.

median lethal dose (LD$_{50}$) Amount of a toxic material per unit of body weight of test animals that kills half the test population in a certain time.

megacities Cities with 10 million or more people.

meltdown The melting of the core of a nuclear reactor.

mesotrophic lake Lake with a moderate supply of plant nutrients. Compare *eutrophic lake, oligotrophic lake*.

metabolism Ability of a living cell or organism to capture and transform matter and energy from its environment to supply its needs for survival, growth, and reproduction.

metamorphic rock Rock produced when a preexisting rock is subjected to high temperatures (which may cause it to melt partially), high pressures, chemically active fluids, or a combination of these agents. Compare *igneous rock, sedimentary rock*. See *rock cycle*.

metastasis Spread of malignant (cancerous) cells from a cancer to other parts of the body.

microorganisms Organisms that are so small that they can be seen only by using a microscope.

mineral Any naturally occurring inorganic substance found in the earth's crust as a crystalline solid. See *mineral resource*.

mineral resource Concentration of naturally occurring solid, liquid, or gaseous material in or on the earth's crust, in a form and amount such that extracting and converting it into useful materials or items is currently or potentially profitable. Mineral resources are classified as *metallic* (such as iron and tin ores) or *nonmetallic* (such as fossil fuels, sand, and salt).

minimum-tillage farming See *conservation-tillage farming*.

mixed economic system Economic system that falls somewhere between pure market and pure command economic systems. Virtually all of the world's economic systems fall into this category, with some closer to a pure market system and some closer to a pure command system. Compare *pure command economic system, pure market economic system*.

mixture Combination of two or more elements and compounds.

model An approximate representation or simulation of a system being studied.

molecule Combination of two or more atoms of the same chemical element (such as O_2) or different chemical elements (such as H_2O) held together by chemical bonds.

monera See *bacteria, cyanobacteria*.

monoculture Cultivation of a single crop, usually on a large area of land. Compare *polyculture, polyvarietal cultivation*.

more developed country (MDC) See *developed country*.

municipal solid waste Solid materials discarded by homes and businesses in or near urban areas. See *solid waste*.

mutagen Chemical or form of ionizing radiation that causes inheritable changes in the DNA molecules in the genes found in chromosomes (mutations). See *carcinogen, mutation, teratogen*.

mutation A random change in DNA molecules making up genes that can yield changes in anatomy, physiology, or behavior in offspring. See *mutagen*.

mutualism Type of species interaction in which both participating species generally benefit. Compare *commensalism*.

native species Species that normally live and thrive in a particular ecosystem. Compare *immigrant species, indicator species, keystone species, nonnative species*.

natural gas Underground deposits of gases consisting of 50–90% by weight methane gas (CH_4) and small amounts of heavier gaseous hydrocarbon compounds such as propane (C_3H_8) and butane (C_4H_{10}).

natural radioactive decay Nuclear change in which unstable nuclei of atoms spontaneously shoot out particles (usually alpha or beta particles), energy (gamma rays), or both at a fixed rate.

natural recharge Natural replenishment of an aquifer by precipitation that percolates downward through soil and rock. See *recharge area*.

natural resource capital See *earth capital*.

natural resources Nutrients and minerals in the soil and deeper layers of the earth's crust, water, wild and domesticated plants and animals, air, and other resources produced by the earth's natural processes. Compare *human capital, manufactured capital, solar capital*. See *earth capital*.

natural selection Process by which a particular beneficial gene (or set of genes) is reproduced more than other genes in succeeding generations. The result of natural selection is a population that contains a greater proportion of organisms better adapted to certain environmental conditions. See *adaptation, biological evolution, differential reproduction, mutation*.

negawatt A watt of electrical power saved by improving energy efficiency.

nematocide Chemical that kills nematodes (roundworms).

net economic welfare (NEW) Measure of annual change in quality of life in a country. It is obtained by subtracting the value of all final products and services that decrease the quality of life from a country's GNP.

net energy Total amount of useful energy available from an energy resource or energy system over its lifetime minus the amount of energy used (the first energy law), automatically wasted (the second energy law), and unnecessarily wasted in finding, processing, concentrating, and transporting it to users.

net primary productivity (NPP) Rate at which all the plants in an ecosystem produce net useful chemical energy; equal to the difference between the rate at which the plants in an ecosystem produce useful chemical energy (primary productivity) and the rate at which they use some of that energy through cellular respiration. Compare *gross primary productivity*.

neutron (n) Elementary particle in the nuclei of all atoms (except hydrogen-1). It has a relative mass of 1 and no electric charge. Compare *electron, proton.*

niche See *ecological niche.*

nitrogen cycle Cyclic movement of nitrogen in different chemical forms from the environment to organisms and then back to the environment.

noise pollution Any unwanted, disturbing, or harmful sound that impairs or interferes with hearing, causes stress, hampers concentration and work efficiency, or causes accidents.

nondegradable pollutant Material that is not broken down by natural processes. Examples are the toxic elements lead and mercury. Compare *biodegradable pollutant, degradable pollutant, slowly degradable pollutant.*

nonionizing radiation Forms of radiant energy such as radio waves, microwaves, infrared light, and ordinary light that do not have enough energy to cause ionization of atoms in living tissue. Compare *ionizing radiation.*

nonnative species Species that migrate into an ecosystem or are deliberately or accidentally introduced into an ecosystem by humans.

nonpersistent pollutant See *degradable pollutant.*

nonpoint source Large or dispersed land areas such as cropfields, streets, and lawns that discharge pollutants into the environment over a large area. Compare *point source.*

nonrenewable resource Resource that exists in a fixed amount (stock) in various places in the earth's crust and has the potential for renewal only by geological, physical, and chemical processes taking place over hundreds of millions to billions of years. Examples are copper, aluminum, coal, and oil. We classify these resources as exhaustible because we are extracting and using them at a much faster rate than they were formed. Compare *potentially renewable resource.*

nontransmissible disease A disease that is not caused by living organisms and does not spread from one person to another. Examples are most cancers, diabetes, cardiovascular disease, and malnutrition. Compare *transmissible disease.*

no-till farming See *conservation-tillage farming.*

nuclear energy Energy released when atomic nuclei undergo a nuclear reaction such as the spontaneous emission of radioactivity, nuclear fission, or nuclear fusion.

nuclear fission Nuclear change in which the nuclei of certain isotopes with large mass numbers (such as uranium-235 and plutonium-239) are split apart into lighter nuclei when struck by a neutron. This process releases more neutrons and a large amount of energy. Compare *nuclear fusion.*

nuclear fusion Nuclear change in which two nuclei of isotopes of elements with a low mass number (such as hydrogen-2 and hydrogen-3) are forced together at extremely high temperatures until they fuse to form a heavier nucleus (such as helium-4). This process releases a large amount of energy. Compare *nuclear fission.*

nucleus Extremely tiny center of an atom, making up most of the atom's mass. It contains one or more positively charged protons and one or more neutrons with no electrical charge (except for a hydrogen-1 atom, which has one proton and no neutrons in its nucleus).

nutrient Any food or element an organism must take in to live, grow, or reproduce.

nutrient cycle See *biogeochemical cycle.*

oil See *crude oil.*

oil shale Fine-grained rock containing various amounts of kerogen, a solid, waxy mixture of hydrocarbon compounds. Heating the rock to high temperatures converts the kerogen into a vapor that can be condensed to form a slow-flowing heavy oil called shale oil. See *kerogen, shale oil.*

old-growth forest Virgin and old, second-growth forests containing trees that are often hundreds, sometimes thousands of years old. Examples include forests of Douglas fir, western hemlock,

giant sequoia, and coastal redwoods in the western United States. Compare *second-growth forest, tree farm.*

oligotrophic lake Lake with a low supply of plant nutrients. Compare *eutrophic lake, mesotrophic lake.*

omnivore Animal that can use both plants and other animals as food sources. Examples are pigs, rats, cockroaches, and people. Compare *carnivore, herbivore.*

open sea The part of an ocean that is beyond the continental shelf. Compare *coastal zone.*

ore Part of a metal-yielding material that can be economically extracted at a given time. An ore typically contains two parts: the ore mineral, which contains the desired metal, and waste mineral material (gangue).

organic farming Producing crops and livestock naturally by using organic fertilizer (manure, legumes, compost) and natural pest control (bugs that eat harmful bugs, plants that repel bugs, and environmental controls such as crop rotation) instead of using commercial inorganic fertilizers and synthetic pesticides and herbicides.

organic fertilizer Organic material such as animal manure, green manure, and compost, applied to cropland as a source of plant nutrients. Compare *commercial inorganic fertilizer.*

organism Any form of life.

other resources Identified and undiscovered resources not classified as reserves. See *identified resources, reserves, undiscovered resources.*

output pollution control See *pollution cleanup.*

overburden Layer of soil and rock overlying a mineral deposit; removed during surface mining.

overconsumption Situation in which some people consume much more than they need at the expense of those who cannot meet their basic needs and at the expense of earth's present and future life-support systems for humans and other forms of life.

overfishing Harvesting so many fish of a species (especially immature fish) that there is not enough breeding stock left to replenish the species, such that it is not profitable to harvest them.

overgrazing Destruction of vegetation when too many grazing animals feed too long and exceed the carrying capacity of a rangeland area.

overnutrition Diet so high in calories, saturated (animal) fats, salt, sugar, and processed foods and so low in vegetables and fruits that the consumer runs high risks of diabetes, hypertension, heart disease, and other health hazards. Compare *malnutrition, undernutrition.*

overpopulation State in which there are more people than can live on earth or in a geographic region in comfort, happiness, and health and still leave the planet or region a fit place for future generations. It is a result of growing numbers of people, growing affluence (resource consumption), or both.

oxygen-demanding wastes Organic materials that are usually biodegraded by aerobic (oxygen-consuming) bacteria if there is enough dissolved oxygen in the water. See also *biological oxygen demand.*

ozone depletion Decrease in concentration of ozone in the stratosphere.

ozone layer Stratospheric layer of gaseous ozone (O_3) that protects life on the earth by filtering out most harmful ultraviolet radiation from the sun.

parasite Consumer organism that lives on or in and feeds on a living plant or animal, known as the *host*, over an extended period of time. The parasite draws nourishment from and gradually weakens its host; it may or may not kill the host. See *parasitism.*

parasitism Interaction between species in which one organism, called the parasite, preys on another organism, called the *host*, by living on or in the host. See *parasite.*

parts per billion (ppb) Number of parts of a chemical found in one billion parts of a particular gas, liquid, or solid.

parts per million (ppm) Number of parts of a chemical found in one million parts of a particular gas, liquid, or solid.

parts per trillion (ppt) Number of parts of a chemical found in one trillion parts of a particular gas, liquid, or solid.

passive solar heating system System that captures sunlight directly within a structure and converts it into low-temperature heat for space heating or for heating water for domestic use, without the use of mechanical devices. Compare *active solar heating system.*

pathogen Organism that produces disease.

per capita GDP Annual gross domestic product (GDP) of a country divided by its total population. See *gross domestic product.*

per capita GNP Annual gross national product (GNP) of a country divided by its total population. See *gross national product.*

percolation Passage of a liquid through the spaces of a porous material such as soil.

perennial Plant that can live for more than 2 years. Compare *annual.*

permafrost Perennially frozen layer of the soil that forms when the water there freezes. It is found in arctic tundra.

perpetual resource See *potentially renewable resource.*

persistent pollutant See *slowly degradable pollutant.*

pest Unwanted organism that directly or indirectly interferes with human activities.

pesticide Any chemical designed to kill or inhibit the growth of an organism that people consider to be undesirable. See *fungicide, herbicide, insecticide.*

pesticide treadmill Situation in which the cost of using pesticides increases while their effectiveness decreases, mostly because the pest species develop genetic resistance to the pesticides.

petrochemicals Chemicals obtained by refining (distilling) crude oil. They are used as raw materials in the manufacture of most industrial chemicals, fertilizers, pesticides, plastics, synthetic fibers, paints, medicines, and many other products.

petroleum See *crude oil.*

pH Numeric value that indicates the acidity or alkalinity of a substance on a scale of 0 to 14, with the neutral point at 7. Acid solutions have pH values lower than 7, and basic or alkaline solutions have pH values greater than 7.

phosphorus cycle Cyclic movement of phosphorus in different chemical forms, from the environment to organisms and then back to the environment.

photochemical smog Complex mixture of air pollutants produced in the lower atmosphere by the reaction of hydrocarbons and nitrogen oxides under the influence of sunlight. Especially harmful components include ozone, peroxyacyl nitrates (PANs), and various aldehydes. Compare *industrial smog.*

photosynthesis Complex process that takes place in cells of green plants. Radiant energy from the sun is used to combine carbon dioxide (CO_2) and water (H_2O) to produce oxygen (O_2) and carbohydrates (such as glucose, $C_6H_{12}O_6$) and other nutrient molecules. Compare *aerobic respiration, chemosynthesis.*

photovoltaic cell (solar cell) Device in which radiant (solar) energy is converted directly into electrical energy.

physical change Process that alters one or more physical properties of an element or compound without altering its chemical composition. Examples are changing the size and shape of a sample of matter (crushing ice and cutting aluminum foil) and changing a sample of matter from one physical state to another (boiling and freezing water). Compare *chemical change, nuclear change*.

phytoplankton Small, drifting plants, mostly algae and bacteria, found in aquatic ecosystems. Compare *plankton, zooplankton*.

pioneer species First hardy species (often microbes, mosses, and lichens) that begin colonizing a site as the first stage of ecological succession. See *ecological succession*.

planetary management worldview Beliefs that we are the planet's most important species; we are in charge of the rest of nature; there is always more, and it's all for us; all economic growth is good, more economic growth is better, and the potential for economic growth is limitless; and our success depends on how well we can understand, control, and manage earth's life-support systems for our own benefit. Compare *earth-wisdom worldview*.

plankton Small plant organisms (phytoplankton) and animal organisms (zooplankton) that float in aquatic ecosystems.

plantation agriculture Growing specialized crops such as bananas, coffee, and cacao in tropical developing countries, primarily for sale to developed countries.

plants (plantae) Eukaryotic, mostly multicelled organisms such as algae (red, blue, and green), mosses, ferns, flowers, cacti, grasses, beans, wheat, rice, and trees. These organisms use photosynthesis to produce organic nutrients for themselves and for other organisms feeding on them. Water and other inorganic nutrients are obtained from the soil for terrestrial plants and from the water for aquatic plants.

point source A single identifiable source that discharges pollutants into the environment. Examples are the smokestack of a power plant or an industrial plant, the drainpipe of a meat-packing plant, the chimney of a house, or the exhaust pipe of an automobile. Compare *nonpoint source*.

poison A chemical that in one dose kills exactly 50% of the animals (usually rats and mice) in a test population (usually 60 to 200 animals) within a 14-day period. See *median lethal dose*.

politics Process through which individuals and groups try to influence or control government policies and actions that affect the local, state, national, and international communities.

pollutant A particular chemical or form of energy that can adversely affect the health, survival, or activities of humans or other living organisms. See *pollution*.

pollution An undesirable change in the physical, chemical, or biological characteristics of air, water, soil, or food that can adversely affect the health, survival, or activities of humans or other living organisms.

pollution cleanup Process that removes or reduces the level of a pollutant after it has been produced or has entered the environment. Examples are automobile emission-control devices and sewage treatment plants. Compare *pollution prevention*.

pollution prevention Process that prevents a potential pollutant from forming or entering the environment or sharply reduces the amounts entering the environment. Compare *pollution cleanup*.

polyculture Complex form of intercropping in which a large number of different plants maturing at different times are planted together. See also *intercropping*. Compare *monoculture, polyvarietal cultivation*.

polyvarietal cultivation Planting a plot of land with several varieties of the same crop. Compare *intercropping, monoculture, polyculture*.

population Group of individual organisms of the same species living in a particular area.

population change An increase or decrease in the size of a population. It is equal to (Births + Immigration) − (Deaths + Emigration).

population crash Large number of deaths over a fairly short time, brought about when the number of individuals in a population is too large to be supported by available environmental resources.

population density Number of organisms in a particular population found in a specified area.

population dispersion General pattern in which the members of a population are arranged throughout its habitat.

population distribution Variation of population density over a particular geographic area. For example, a country has a high population density in its urban areas and a much lower population density in rural areas.

population dynamics Major abiotic and biotic factors that tend to increase or decrease the population size and the age and sex composition of a species.

population size Number of individuals making up a population's gene pool.

potential energy Energy stored in an object because of its position or the position of its parts. Compare *kinetic energy*.

potentially renewable resource Resource that theoretically can last indefinitely without reducing the available supply, either because it is replaced more rapidly through natural processes than are nonrenewable resources or because it is essentially inexhaustible (solar energy). Examples are trees in forests, grasses in grasslands, wild animals, fresh surface water in lakes and streams, most groundwater, fresh air, and fertile soil. If such a resource is used faster than it is replenished, it can be depleted and converted into a nonrenewable resource. Compare *nonrenewable resource*. See also *environmental degradation*.

poverty Inability to meet basic needs for food, clothing, and shelter.

ppb See *parts per billion*.

ppm See *parts per million*.

ppt See *parts per trillion*.

precipitation Water in the form of rain, sleet, hail, and snow that falls from the atmosphere onto the land and bodies of water.

predation Situation in which an organism of one species (the predator) captures and feeds on parts or all of an organism of another species (the prey).

predator Organism that captures and feeds on parts or all of an organism of another species (the prey).

predator–prey relationship Interaction between two organisms of different species in which one organism, called the *predator*, captures and feeds on parts or all of another organism, called the *prey*.

prey Organism that is captured and serves as a source of food for an organism of another species (the predator).

primary consumer Organism that feeds directly on all or parts of plants (herbivore) or on other producers. Compare *detritivore, omnivore, secondary consumer*.

primary pollutant Chemical that has been added directly to the air by natural events or human activities and occurs in a harmful concentration. Compare *secondary pollutant*.

primary sewage treatment Mechanical treatment of sewage in which large solids are filtered out by screens and suspended solids settle out as sludge in a sedimentation tank. Compare *advanced sewage treatment, secondary sewage treatment*.

primary succession Sequential development of communities in a bare area that has never been occupied by a community of organisms. Compare *secondary succession*.

probability A mathematical statement about how likely it is that something will happen.

producer Organism that uses solar energy (green plant) or chemical energy (some bacteria) to manufacture the organic compounds it needs as nutrients from simple inorganic compounds

obtained from its environment. Compare *consumer, decomposer*.

prokaryotic cell Cell that doesn't have a distinct nucleus. Other internal parts are also not enclosed by membranes. Compare *eukaryotic cell*.

protists Eukaryotic, mostly single-celled organisms such as diatoms, amoebas, some algae (golden brown and yellow-green), protozoans, and slime molds. Some protists produce their own organic nutrients through photosynthesis. Others are decomposers and some feed on bacteria, other protists, or cells of multicellular organisms.

proton (p) Positively charged particle in the nuclei of all atoms. Each proton has a relative mass of 1 and a single positive charge. Compare *electron, neutron*.

pure capitalism See *pure market economic system*.

pure command economic system System in which all economic decisions are made by the government or some other central authority. Compare *mixed economic system, pure market economic system*.

pure market economic system System in which all economic decisions are made in the market, where buyers and sellers of economic goods interact freely, with no government or other interference. Compare *mixed economic system, pure command economic system*.

pyramid of energy flow Diagram representing the flow of energy through each trophic level in a food chain or food web. With each energy transfer, only a small part (typically 10%) of the usable energy entering one trophic level is transferred to the organisms at the next trophic level.

radiation Fast-moving particles (particulate radiation) or waves of energy (electromagnetic radiation).

radioactive decay Change of a radioisotope to a different isotope by the emission of radioactivity.

radioactive isotope See *radioisotope*.

radioactivity Nuclear change in which unstable nuclei of atoms spontaneously shoot out "chunks" of mass, energy, or both, at a fixed rate. The three principal types of radioactivity are gamma rays and fast-moving alpha and beta particles.

radioisotope Isotope of an atom that spontaneously emits one or more types of radioactivity (alpha particles, beta particles, gamma rays).

rangeland Land that supplies forage or vegetation (grasses, grasslike plants, and shrubs) for grazing and browsing animals and is not intensively managed. Compare *feedlot*.

range of tolerance Range of chemical and physical conditions that must be maintained for populations of a particular species to stay alive and grow, develop, and function normally. See *law of tolerance*.

real GDP Gross domestic product adjusted for inflation.

real GNP Gross national product adjusted for inflation.

realized niche Parts of the fundamental niche of a species that are actually used by that species. See *ecological niche, fundamental niche*.

real per capita GDP Per capita GDP adjusted for inflation.

real per capita GNP Per capita GNP adjusted for inflation.

recharge area Any area of land allowing water to pass through it and into an aquifer. See *aquifer, natural recharge*.

recycling Collecting and reprocessing a resource so it can be made into new products. An example is collecting aluminum cans, melting them down, and using the aluminum to make new cans or other aluminum products. Compare *reuse*.

reforestation Renewal of trees and other types of vegetation on land where trees have been removed; can be done naturally by seeds from nearby trees or artificially by planting seeds or seedlings.

renewable resource   See *potentially renewable resource.*

replacement-level fertility   Number of children a couple must have to replace themselves. The average for a country or the world is usually slightly higher than 2 children per couple (2.1 in the United States and 2.5 in some developing countries) because some children die before reaching their reproductive years. See also *total fertility rate.*

reproduction   Production of offspring by one or more parents.

reproductive isolation   Long-term geographic separation of members of a particular sexually reproducing species.

reproductive potential   See *biotic potential.*

reserves   Resources that have been identified from which a usable mineral can be extracted profitably at present prices with current mining technology. See *identified resources, other resources, undiscovered resources.*

resource   Anything obtained from the living and nonliving environment to meet human needs and wants. It can also be applied to other species.

resource partitioning   Process of dividing up resources in an ecosystem so that species with similar requirements (overlapping ecological niches) use the same scarce resources at different times, in different ways, or in different places. See *ecological niche, fundamental niche, realized niche.*

respiration   See *aerobic respiration.*

response   The amount of health damage caused by exposure to a certain dose of a harmful substance or form of radiation. See *dose, dose–response curve, median lethal dose.*

reuse   To use a product over and over again in the same form. An example is collecting, washing, and refilling glass beverage bottles. Compare *recycling.*

riparian zones   Thin strips and patches of vegetation that surround streams. They are very important habitats and resources for wildlife.

risk   The probability that something undesirable will happen from deliberate or accidental exposure to a hazard. See *risk analysis, risk assessment, risk–benefit analysis, risk management.*

risk analysis   Identifying hazards, evaluating the nature and severity of risks (*risk assessment*), using this and other information to determine options and make decisions about reducing or eliminating risks (*risk management*), and communicating information about risks to decision makers and the public (*risk communication*).

risk assessment   Process of gathering data and making assumptions to estimate short- and long-term harmful effects on human health or the environment from exposure to hazards associated with the use of a particular product or technology. See *risk, risk–benefit analysis.*

risk–benefit analysis   Estimate of the short- and long-term risks and benefits of using a particular product or technology. See *risk.*

risk communication   Communicating information about risks to decision makers and the public. See *risk. risk analysis, risk–benefit analysis.*

risk management   Using risk assessment and other information to determine options and make decisions about reducing or eliminating risks. See *risk, risk analysis, risk–benefit analysis, risk communication.*

rock   Any material that makes up a large, natural, continuous part of earth's crust. See *mineral.*

rock cycle   Largest and slowest of the earth's cycles, consisting of geologic, physical, and chemical processes that form and modify rocks and soil in the earth's crust over millions of years.

rodenticide   Chemical that kills rodents.

runoff   Fresh water from precipitation and melting ice that flows on the earth's surface into nearby streams, lakes, wetlands, and reservoirs. See *surface runoff, surface water.* Compare *groundwater.*

rural area   Geographic area in the United States with a population of less than 2,500. The number of people used in this definition may vary in different countries. Compare *urban area.*

salinity   Amount of various salts dissolved in a given volume of water.

salinization   Accumulation of salts in soil that can eventually make the soil unable to support plant growth.

saltwater intrusion   Movement of salt water into freshwater aquifers in coastal and inland areas as groundwater is withdrawn faster than it is recharged by precipitation.

sanitary landfill   Waste disposal site on land in which waste is spread in thin layers, compacted, and covered with a fresh layer of clay or plastic foam each day.

scavenger   Organism that feeds on dead organisms that were killed by other organisms or died naturally. Examples are vultures, flies, and crows. Compare *detritivore.*

science   Attempts to discover order in nature and use that knowledge to make predictions about what should happen in nature. See *consensus science, frontier science, model, scientific data, scientific hypothesis, scientific law, scientific methods, scientific theory.*

scientific data   Facts obtained by making observations and measurements. Compare *model, scientific hypothesis, scientific methods, scientific law, scientific theory.*

scientific hypothesis   An educated guess that attempts to explain a scientific law or certain scientific observations. Compare *model, scientific data, scientific law, scientific methods, scientific theory.*

scientific law   Description of what scientists find happening in nature over and over in the same way, without known exception. See *first law of thermodynamics, second law of thermodynamics, law of conservation of matter.* Compare *model, scientific data, scientific hypothesis, scientific methods, scientific theory.*

scientific methods   The ways scientists gather data and formulate and test scientific hypotheses, models, theories, and laws. See *model, scientific data, scientific hypothesis, scientific law, scientific theory.*

scientific model   See *model.*

scientific theory   A well-tested and widely accepted scientific hypothesis. Compare *model, scientific data, scientific hypothesis, scientific methods.*

secondary consumer   Organism that feeds only on primary consumers. Most secondary consumers are animals, but some are plants. Compare *detritivore, omnivore, primary consumer.*

secondary pollutant   Harmful chemical formed in the atmosphere when a primary air pollutant reacts with normal air components or other air pollutants. Compare *primary pollutant.*

secondary sewage treatment   Second step in most waste treatment systems, in which aerobic bacteria break down up to 90% of degradable, oxygen-demanding organic wastes in wastewater. This is usually done by bringing sewage and bacteria together in trickling filters or the activated sludge process. Compare *advanced sewage treatment, primary sewage treatment.*

secondary succession   Sequential development of communities in an area in which natural vegetation has been removed or destroyed but the soil is not destroyed. Compare *primary succession.*

second-growth forest   Stands of trees resulting from secondary ecological succession. Compare *ancient forest, old-growth forest, tree farm.*

second law of energy   See *second law of thermodynamics.*

second law of thermodynamics   In any conversion of heat energy to useful work, some of the initial energy input is always degraded to a lower-quality, more dispersed, less useful energy, usually low-temperature heat that flows into the environment; you can't break even in terms of energy quality; See *first law of thermodynamics.*

sedimentary rock   Rock that forms from the accumulated products of erosion and in some cases from the compacted shells, skeletons, and other remains of dead organisms. Compare *igneous rock, metamorphic rock.* See *rock cycle.*

seed-tree cutting   Removal of nearly all trees on a site in one cutting, with a few seed-producing trees left uniformly distributed to regenerate the forest. Compare *clear-cutting, selective cutting, shelterwood cutting, strip cutting.*

selective cutting   Cutting of intermediate-aged, mature, or diseased trees in an uneven-aged forest stand, either singly or in small groups. This encourages the growth of younger trees and maintains an uneven-aged stand. Compare *clear-cutting, seed-tree cutting, shelterwood cutting, strip cutting.*

septic tank   Underground tank for treatment of wastewater from a home in rural and suburban areas. Bacteria in the tank decompose organic wastes, and the sludge settles to the bottom of the tank. The effluent flows out of the tank into the ground through a field of drain pipes.

sewage sludge   See *sludge.*

shale oil   Slow-flowing, dark brown, heavy oil obtained when kerogen in oil shale is vaporized at high temperatures and then condensed. Shale oil can be refined to yield gasoline, heating oil, and other petroleum products. See *kerogen, oil shale.*

shelterbelt   See *windbreak.*

shelterwood cutting   Removal of mature, marketable trees in an area in a series of partial cuttings to allow regeneration of a new stand under the partial shade of older trees, which are later removed. Typically, this is done by making two or three cuts over a decade. Compare *clear-cutting, seed-tree cutting, selective cutting, strip cutting.*

shifting cultivation   Clearing a plot of ground in a forest, especially in tropical areas, and planting crops on it for a few years (typically 2–5 years) until the soil is depleted of nutrients or the plot has been invaded by a dense growth of vegetation from the surrounding forest. Then a new plot is cleared and the process is repeated. The abandoned plot cannot grow crops successfully for 10–30 years. See also *slash-and-burn cultivation.*

slash-and-burn cultivation   Cutting down trees and other vegetation in a patch of forest, leaving the cut vegetation on the ground to dry, and then burning it. The ashes that are left add nutrients to the nutrient-poor soils found in most tropical forest areas. Crops are planted between tree stumps. Plots must be abandoned after a few years (typically 2–5 years) because of loss of soil fertility or invasion of vegetation from the surrounding forest. See also *shifting cultivation.*

slowly degradable pollutant   Material that is slowly broken down into simpler chemicals or reduced to acceptable levels by natural physical, chemical, and biological processes. Compare *biodegradable pollutant, degradable pollutant, nondegradable pollutant.*

sludge   Gooey mixture of toxic chemicals, infectious agents, and settled solids, removed from wastewater at a sewage treatment plant.

smelting   Process in which a desired metal is separated from the other elements in an ore mineral.

smog   Originally a combination of smoke and fog, but now used to describe other mixtures of pollutants in the atmosphere. See *industrial smog, photochemical smog.*

soil   Complex mixture of inorganic minerals (clay, silt, pebbles, and sand), decaying organic matter, water, air, and living organisms.

soil conservation   Methods used to reduce soil erosion, prevent depletion of soil nutrients, and restore nutrients already lost by erosion, leaching, and excessive crop harvesting.

soil erosion   Movement of soil components, especially topsoil, from one place to another, usually by wind, flowing water, or both. This natural process can be greatly accelerated by human activities that remove vegetation from soil.

soil horizons Horizontal zones that make up a particular mature soil. Each horizon has a distinct texture and composition that varies with different types of soils.

soil permeability Rate at which water and air move from upper to lower soil layers. Compare *soil porosity*.

soil porosity The pores (cracks and spaces) in rocks or soil, or the percentage of the rock's or soil's volume not occupied by the rock or soil itself. Compare *soil permeability*.

soil profile Cross-sectional view of the horizons in a soil.

soil structure How the particles that make up a soil are organized and clumped together. See also *soil permeability, soil texture*.

soil texture Relative amounts of the different types and sizes of mineral particles in a sample of soil.

solar capital Solar energy from the sun reaching the earth. Compare *earth capital*.

solar cell See *photovoltaic cell*.

solar collector Device for collecting radiant energy from the sun and converting it into heat. See *active solar heating system, passive solar heating system*.

solid waste Any unwanted or discarded material that is not a liquid or gas. See *municipal solid waste*.

specialist species Species with a narrow ecological niche. They may be able to live in only one type of habitat, tolerate only a narrow range of climatic and other environmental conditions, or use only one or a few types of food. Compare *generalist species*.

speciation Formation of two species from one species as a result of divergent natural selection in response to changes in environmental conditions; usually takes thousands of years. Compare *extinction*.

species Group of organisms that resemble one another in appearance, behavior, chemical makeup and processes, and genetic structure. Organisms that reproduce sexually are classified as members of the same species only if they can breed with one another and produce fertile offspring.

species diversity Number of different species and their relative abundances in a given area. See *biological diversity*. Compare *ecological diversity, genetic diversity*.

spoils Unwanted rock and other waste materials produced when a material is removed from the earth's surface or subsurface by mining, dredging, quarrying, or excavation.

S-shaped curve Leveling off of an exponential, *J*-shaped curve when a rapidly growing population exceeds the carrying capacity of its environment and ceases to grow.

stratosphere Second layer of the atmosphere, extending from about 17–48 kilometers (11–30 miles) above the earth's surface. It contains small amounts of gaseous ozone (O_3), which filters out about 99% of the incoming harmful ultraviolet (UV) radiation emitted by the sun. Compare *troposphere*.

strip cropping Planting regular crops and close-growing plants, such as hay or nitrogen-fixing legumes, in alternating rows or bands to help reduce depletion of soil nutrients.

strip cutting A variation of clear-cutting in which a strip of trees is clear-cut along the contour of the land, with the corridor narrow enough to allow natural regeneration within a few years. After regeneration, another strip is cut above the first, and so on. Compare *clear-cutting, seed-tree cutting, selective cutting, shelterwood cutting*.

strip mining Form of surface mining in which bulldozers, power shovels, or stripping wheels remove large chunks of the earth's surface in strips. See *surface mining*. Compare *subsurface mining*.

subsidence Slow or rapid sinking down (not slope related) of part of earth's crust.

subsistence farming Supplementing solar energy with energy from human labor and draft animals to produce enough food to feed oneself and family members; in good years there may be enough food left over to sell or put aside for hard times. Compare *industrialized agriculture*.

subsurface mining Extraction of a metal ore or fuel resource such as coal from a deep underground deposit. Compare *surface mining*.

succession See *ecological succession*.

succulent plants Plants, such as desert cacti, that survive in dry climates by having no leaves, thus reducing the loss of scarce water. They store water and use sunlight to produce their food in the thick, fleshy tissue of their green stems and branches. Compare *deciduous plants, evergreen plants*.

surface mining Removing soil, subsoil, and other strata, and then extracting a mineral deposit found fairly close to the earth's surface. Compare *subsurface mining*.

surface runoff Water flowing off the land into bodies of surface water.

surface water Precipitation that does not infiltrate the ground or return to the atmosphere by evaporation or transpiration. See *runoff*. Compare *groundwater*.

sustainability Ability of a system to survive for some specified (finite) time. See *sustainable society*.

sustainable agriculture Method of growing crops and raising livestock based on organic fertilizers, soil conservation, water conservation, biological control of pests, and minimal use of nonrenewable fossil-fuel energy.

sustainable living Taking no more potentially renewable resources from the natural world than can be replenished naturally and not overloading the capacity of the environment to cleanse and renew itself by natural processes.

sustainable society A society that manages its economy and population size without doing irreparable environmental harm by overloading the planet's ability to absorb environmental insults, replenish its resources, and sustain human and other forms of life over a specified period, usually hundreds to thousands of years. During this period it satisfies the needs of its people without depleting earth capital and thereby jeopardizing the prospects of current and future generations of humans and other species.

sustainable yield (sustained yield) Highest rate at which a potentially renewable resource can be used without reducing its available supply throughout the world or in a particular area. See also *environmental degradation*.

synfuels Synthetic gaseous and liquid fuels produced from solid coal or sources other than natural gas or crude oil.

synthetic natural gas (SNG) Gaseous fuel containing mostly methane produced from solid coal.

tailings Rock and other waste materials removed as impurities when waste mineral material is separated from the metal in an ore.

tar sand Deposit of a mixture of clay, sand, water, and varying amounts of a tarlike heavy oil known as bitumen. Bitumen can be extracted from tar sand by heating. It is then purified and upgraded to synthetic crude oil. See *bitumen*.

technology Creation of new products and processes intended to improve our efficiency, chances for survival, comfort level, and quality of life. Compare *science*.

temperature inversion See *thermal inversion*.

teratogen Chemical, ionizing agent, or virus that causes birth defects. See *carcinogen, mutagen*.

terracing Planting crops on a long, steep slope that has been converted into a series of broad, nearly level terraces (with short vertical drops from one to another) that run along the contour of the land to retain water and reduce soil erosion.

terrestrial Pertaining to land. Compare *aquatic*.

tertiary (higher-level) consumers Animals that feed on animal-eating animals. They feed at high trophic levels in food chains and webs. Examples are hawks, lions, bass, and sharks. Compare *detritivore, primary consumer, secondary consumer*.

tertiary sewage treatment See *advanced sewage treatment*.

theory of evolution Widely accepted idea that all life-forms developed from earlier life-forms. Although this theory conflicts with the creation stories of most religions, it is the way biologists explain how life has changed over the past 3.6–3.8 billion years and why it is so diverse today.

thermal inversion Layer of dense, cool air trapped under a layer of less dense, warm air. This prevents upward-flowing air currents from developing. In a prolonged inversion, air pollution in the trapped layer may build up to harmful levels.

threatened species Wild species that is still abundant in its natural range but is likely to become endangered because of a decline in numbers. Compare *endangered species*.

threshold effect The harmful or fatal effect of a small change in environmental conditions that exceeds the limit of tolerance of an organism or population of a species. See *law of tolerance*.

throughput Rate of flow of matter, energy, or information through a system.

throwaway society See *high-waste society*.

tolerance limits Minimum and maximum limits for physical conditions (such as temperature) and concentrations of chemical substances beyond which no members of a particular species can survive. Compare *law of tolerance*.

total fertility rate (TFR) Estimate of the average number of children that will be born alive to a woman during her lifetime if she passes through all her childbearing years (ages 15–44) conforming to age-specific fertility rates of a given year. In simpler terms, it is an estimate of the average number of children a woman will have during her childbearing years.

totally planned economy See *pure command economic system*.

toxic chemical Chemical that is fatal to humans in low doses or fatal to over 50% of test animals at stated concentrations. Most are neurotoxins, which attack nerve cells. See *carcinogen, hazardous chemical, mutagen, teratogen*.

toxicity Measure of how harmful a substance is.

toxicology Study of the adverse effects of chemicals on health.

toxic waste Form of hazardous waste that causes death or serious injury (such as burns, respiratory diseases, cancers, or genetic mutations). See *hazardous waste*.

traditional intensive agriculture Producing enough food for a farm family's survival and perhaps a surplus that can be sold. This type of agriculture requires higher inputs of labor, fertilizer, and water than traditional subsistence agriculture. See *traditional subsistence agriculture*.

traditional subsistence agriculture Production of enough crops or livestock for a farm family's survival and, in good years, a surplus to sell or put aside for hard times. Compare *traditional intensive agriculture*.

tragedy of the commons Depletion or degradation of a resource to which people have free and unmanaged access. An example is the depletion of commercially desirable species of fish in the open ocean beyond areas controlled by coastal countries. See *common-property resource*.

transmissible disease A disease that is caused by living organisms (such as bacteria, viruses, and parasitic worms) and can spread from one person to another by air, water, food, or body fluids (or in some cases by insects or other organisms). Compare *nontransmissible disease*.

transpiration Process in which water is absorbed by the root systems of plants, moves up through the plants, passes through pores (stomata) in their leaves or other parts, and then evaporates into the atmosphere as water vapor.

tree farm Site planted with one or only a few tree species in an even-aged stand. When the stand matures, it is usually harvested by clear-cutting and then replanted. These farms are normally used to grow rapidly growing tree species for fuelwood, timber, or pulpwood. See *even-aged management.* Compare *old-growth forest, second-growth forest, uneven-aged management.*

trophic level All organisms that are the same number of energy transfers away from the original source of energy (for example, sunlight) that enters an ecosystem. For example, all producers belong to the first trophic level and all herbivores belong to the second trophic level in a food chain or web.

troposphere Innermost layer of the atmosphere. It contains about 75% of the mass of earth's air and extends about 17 kilometers (11 miles) above sea level. Compare *stratosphere.*

true cost See *full cost.*

undernutrition Consuming insufficient food to meet one's minimum daily energy requirement for a long enough time to cause harmful effects. Compare *malnutrition, overnutrition.*

undiscovered resources Potential supplies of a particular mineral resource, believed to exist because of geologic knowledge and theory, although specific locations, quality, and amounts are unknown. Compare *identified resources, other resources, reserves.*

uneven-aged management Method of forest management in which trees of different species in a given stand are maintained at many ages and sizes to permit continuous natural regeneration. Compare *even-aged management.*

upwelling Movement of nutrient-rich bottom water to the ocean's surface. This can occur far from shore but usually occurs along certain steep coastal areas, where the surface layer of ocean water is pushed away from shore and replaced by cold, nutrient-rich bottom water.

urban area Geographic area with a population of 2,500 or more people. The number of people used in this definition may vary, with some countries setting the minimum number of people at 10,000–50,000.

urban growth Rate of growth of an urban population. Compare *degree of urbanization.*

urban heat island Buildup of heat in the atmosphere above an urban area. This heat is produced by the large concentration of cars, buildings, factories, and other heat-producing activities. See also *dust dome.*

urbanization See *degree of urbanization.*

vertebrates Animals with backbones. Compare *invertebrates.*

water cycle See *hydrologic cycle.*

waterlogging Saturation of soil with irrigation water or excessive precipitation so that the water table rises close to the surface.

water pollution Any physical or chemical change in surface water or groundwater that can harm living organisms or make water unfit for certain uses.

watershed Land area that delivers water, sediment, and dissolved substances via small streams to a major stream (river).

water table Upper surface of the zone of saturation, in which all available pores in the soil and rock in the earth's crust are filled with water.

weather Short-term changes in the temperature, barometric pressure, humidity, precipitation, sunshine, cloud cover, wind direction and speed, and other conditions in the troposphere at a given place and time. Compare *climate.*

weathering Physical and chemical processes in which solid rock exposed at earth's surface is changed to separate solid particles and dissolved material, which can then be moved to another place as sediment. See *erosion.*

wetland Land that is covered all or part of the time with salt water or fresh water, excluding streams, lakes, and the open ocean. See *coastal wetland, inland wetland.*

wilderness Area where the earth and its community of life have not been seriously disturbed by humans and where humans are only temporary visitors.

wildlife All free, undomesticated species. Sometimes the term is used to describe only free, undomesticated species of animals.

wildlife management Manipulation of populations of wild species (especially game species) and their habitats for human benefit, the welfare of other species, and the preservation of threatened and endangered wildlife species.

wildlife resources Species of wildlife that have actual or potential economic value to people.

wildness Existence of wild gene pools, species, and ecosystems that are completely or mostly undisturbed by human activities. Another term for biodiversity.

windbreak Row of trees or hedges planted to partially block wind flow and reduce soil erosion on cultivated land.

worldview How people think the world works and what they think their role in the world should be. See *earth-wisdom worldview, planetary management worldview.*

zero population growth (ZPG) State in which the birth rate (plus immigration) equals the death rate (plus emigration) so that the population of a geographic area is no longer increasing.

zone of saturation Area where all available pores in soil and rock in the earth's crust are filled by water. See *water table.*

zoning Regulating how various parcels of land can be used.

zooplankton Animal plankton. Small floating herbivores that feed on plant plankton (phytoplankton). Compare *phytoplankton.*

INDEX